Fostering an Ecological Shift Through Effective Environmental Education

Tatjana Kochetkova
O.P. Jindal Global University, India

A volume in the Advances in Environmental Engineering and Green Technologies (AEEGT) Book Series

IGI Global
Publishing Tomorrow's Research Today

Published in the United States of America by
IGI Global
Engineering Science Reference (an imprint of IGI Global)
701 E. Chocolate Avenue
Hershey PA, USA 17033
Tel: 717-533-8845
Fax: 717-533-8661
E-mail: cust@igi-global.com
Web site: http://www.igi-global.com

Library of Congress Cataloging-in-Publication Data

Names: Kochetkova, Tatjana, editor.
Title: Fostering an ecological shift through effective environmental education / edited Tatjana Kochetkova.
Description: Hershey, PA : IGI Global, 2024. | Includes bibliographical references and index. | Summary: "The book's comprehensive approach extends to eco-therapy, forest school programs, and the influence of parents in environmental education. By scrutinizing case studies and global movements, this work illuminates the achievements and challenges of environmental education on both national and global scales"-- Provided by publisher.
Identifiers: LCCN 2024035467 (print) | LCCN 2024035468 (ebook) | ISBN 9798369325773 (hardcover) | ISBN 9798369346303 (paperback) | ISBN 9798369325780 (ebook)
Subjects: LCSH: Environmental education--Case studies. | Environmental education--Moral and ethical aspects--Case studies.
Classification: LCC GF70 .F67 2024 (print) | LCC GF70 (ebook) | DDC 304.2/8071--dc23/eng/20240906
LC record available at https://lccn.loc.gov/2024035467
LC ebook record available at https://lccn.loc.gov/2024035468

This book is published in the IGI Global book series Advances in Environmental Engineering and Green Technologies (AEEGT) (ISSN: 2326-9162; eISSN: 2326-9170)

British Cataloguing in Publication Data
A Cataloguing in Publication record for this book is available from the British Library.

For electronic access to this publication, please contact: eresources@igi-global.com.

Advances in Environmental Engineering and Green Technologies (AEEGT) Book Series

Sang-Bing (Jason) Tsai
Zhongshan Institute, University of Electronic Science and Technology of China, China & Wuyi University, China
Ming-Lang Tseng
Lunghwa University of Science and Technology, Taiwan
Yuchi Wang
University of Electronic Science and Technology of China Zhongshan Institute, China

ISSN:2326-9162
EISSN:2326-9170

Mission

Growing awareness and an increased focus on environmental issues such as climate change, energy use, and loss of non-renewable resources have brought about a greater need for research that provides potential solutions to these problems. Research in environmental science and engineering continues to play a vital role in uncovering new opportunities for a "green" future.

The **Advances in Environmental Engineering and Green Technologies (AEEGT)** book series is a mouthpiece for research in all aspects of environmental science, earth science, and green initiatives. This series supports the ongoing research in this field through publishing books that discuss topics within environmental engineering or that deal with the interdisciplinary field of green technologies.

Coverage

- Contaminated Site Remediation
- Green Technology
- Alternative Power Sources
- Waste Management
- Policies Involving Green Technologies and Environmental Engineering
- Electric Vehicles
- Radioactive Waste Treatment
- Industrial Waste Management and Minimization
- Biofilters and Biofiltration
- Sustainable Communities

IGI Global is currently accepting manuscripts for publication within this series. To submit a proposal for a volume in this series, please contact our Acquisition Editors at Acquisitions@igi-global.com or visit: http://www.igi-global.com/publish/.

The Advances in Environmental Engineering and Green Technologies (AEEGT) Book Series (ISSN 2326-9162) is published by IGI Global, 701 E. Chocolate Avenue, Hershey, PA 17033-1240, USA, www.igi-global.com. This series is composed of titles available for purchase individually; each title is edited to be contextually exclusive from any other title within the series. For pricing and ordering information please visit http://www.igi-global.com/book-series/advances-environmental-engineering-green-technologies/73679. Postmaster: Send all address changes to above address.

Titles in this Series

For a list of additional titles in this series, please visit: www.igi-global.com/book-series

Reshaping Environmental Science Through Machine Learning and IoT
Rajeev Kumar Gupta (Pandit Deendayal Energy University, India) Arti Jain (Jaypee Institute of Information Technology, India) John Wang (Montclair State University, USA) and Rajesh Kumar Pateriya (Maulana Azad National Institute of Technology, India)
Engineering Science Reference • copyright 2024 • 438pp • H/C (ISBN: 9798369323519) • US $295.00 (our price)

Advancements in Climate and Smart Environment Technology
Jamal Mabrouki (Faculty of Science, Mohammed V University in Rabat, Morocco) and Mourade Azrour (Faculty of Sciences and Techniques, Moulay Ismail University of Meknes, Errachidia, Morocco)
Engineering Science Reference • copyright 2024 • 292pp • H/C (ISBN: 9798369338070) • US $295.00 (our price)

Harnessing NanoOmics and Nanozymes for Sustainable Agriculture
Vishnu D. Rajput (Academy of Biology and Biotechnology, Southern Federal University, Rostov-on-Don, Russia) Abhishek Singh (Faculty of Biology, Yerevan State University, Armenia) Karen Ghazaryan (Faculty of Biology, Yerevan State University, Armenia) Athanasios Alexiou (Department of Science and Engineering, Novel Global Community Educational Foundation, Hebersham, Australia) and Abdel Rahman Mohammad Said Al-Tawaha (Department of Biological Sciences, Al Hussein bin Talal University, Jordan)
Engineering Science Reference • copyright 2024 • 531pp • H/C (ISBN: 9798369318904) • US $345.00 (our price)

Machine Learning and Computer Vision for Renewable Energy
Pinaki Pratim Acharjya (Haldia Institute of Technology, India) Santanu Koley (Haldia Institute of Technology, India) and Subhabrata Barman (Haldia Institute of Technology, India)
Engineering Science Reference • copyright 2024 • 333pp • H/C (ISBN: 9798369323557) • US $295.00 (our price)

Convergence Strategies for Green Computing and Sustainable Development
Vishal Jain (Sharda University, India) Murali Raman (Asia Pacific University of Technology, Malaysia) Akshat Agrawal (Amity University, Guragon, India) Meenu Hans (K.R. Mangalam University, India) and Swati Gupta (Center of Excellence, K.R. Mangalam University, India)
Engineering Science Reference • copyright 2024 • 332pp • H/C (ISBN: 9798369303382) • US $270.00 (our price)

Biosorption Processes for Heavy Metal Removal
Pinki Saini (University of Allahabad, India)
Engineering Science Reference • copyright 2024 • 371pp • H/C (ISBN: 9798369316184) • US $275.00 (our price)

IGI Global
Publishing Tomorrow's Research Today

701 East Chocolate Avenue, Hershey, PA 17033, USA
Tel: 717-533-8845 x100 • Fax: 717-533-8661
E-Mail: cust@igi-global.com • www.igi-global.com

Table of Contents

Section 4
Environmental Pedagogy

Detailed Table of Contents

This chapter aims to contribute to building an empirical and conceptual basis for ecopsychology and environmental education by providing a picture of psyche-nature interaction. The chapter contributes to methods of environmental education by elaborating on a series of nature awareness exercises (see Appendix), which happened to be efficient and broadly applicable. It presents reports of these exercises as applied in a workshop with a group of students from the University for Humanistics (Utrecht, The Netherlands) at the Eifel National Park in Germany. The workshop lasted three days. The author intended to clarify the effects of nature awareness on the participants, especially on their subjective well-being and perception of the meaning of life. As a result, a "battery-recharging" effect of nature and its influence on young adults at four levels of awareness (external physical perception, internal (bodily) sensation, emotions, and existential meaning), were established. The chapter focuses on the episode of a night walk.

This chapter focuses on the viability of distance awareness as an indispensable ingredient of form awareness and, subsequently, environmental education. It studies relevant motifs in the work of two post-Romantic German thinkers, Ludwig Klages (1872-1956) and Hermann Friedmann (1873-1957). The main argument comprises two areas of concern: 1) enhancing crisis awareness and reintegration, and 2) promoting poetry education. As a common thread running through these areas, consciousness as distance awareness imposes itself. Distance should not be taken as a quantity but rather as an original quality that justifies itself.

This chapter explores existing theoretical debates on anthropocentrism, intrinsic value theory, and Deep Ecology to establish that human beings need to rethink their relationship with the environment in order to take environmental issues seriously. It argues that the relationship of domination and exploitation with the environment should be replaced with a relationship of care.

The chapter explores the interplay among identity construction, environmentalism, and education within the Tibetan community in exile in India with a focus on the Tibetan Nuns Project. Drawing on Toni Huber's analysis of Tibetan self-representation in exile, the chapter examines the emergence of a modern, liberal Shangri-la image, informed by complex translational politics influenced by Western political traditions. The Tibetan exiled elite, through prolonged contact with the globalized cultural environment, adapted Western representational styles to reshape their self-image, challenging traditional perceptions of Tibet. Statements on the so-called 'environmentalist' Tibetan identity emphasises a worldview and modes of behaviour that are described in terms of 'completely modern' concepts and language with terms like 'ecological balance,' interdependence, and 'preservation.'

Trees play a profound role in the spiritual rituals of the Shakta Tantric tradition, acting as a catalyst for the practitioner's spiritual transformation. Yet, the practice of tree worship often seems obscure to the modern mind, necessitating innovative approaches to integrate its essence into contemporary educational frameworks. This chapter endeavors to redefine worship as an artistic process that could potentially cultivate empathy, highlighting its capacity to deepen human connections with the environment by exploring the implications of adopting a tree-centric or arboromorphic perspective within educational frameworks.

Hindu philosophy renders a distinctive perspective concerning the way humans interact with the environment, as it is firmly anchored in old scriptures and customs. A belief that all life is interconnected and the understanding that the environment is an essential component of the cosmic order is paramount to this viewpoint. This chapter penetrates the fundamental ideas of Hindu philosophy that form the basis of environmental consciousness. It encompasses concepts like Dharma, an acronym for righteous duty, Karma, which means action and consequences, and Ahimsa, which means non-violence. The chapter inquires about the ethical underpinnings of environmental stewardship in Hindu philosophy while illustrating people's duty to preserve the universe's delicate balance with animals and plants. It gazes at how Vedic philosophies can promote thankfulness, environmental consciousness, and a sense of oneness with the natural world. The chapter further explores the symbolic meaning of customs such as Yajna (sacrificial offerings) and how they can encourage reciprocity with the environment.

This chapter deals with the interplay of practice of yoga and environmental education. It analyzes in detail the intricacies of both the disciplines and puts forward the connection of yoga with the environment. It dives deeper into the different schools of yoga and how they are related in promoting environmental awareness. This chapter also discusses the philosophical foundations of yoga and explains how the core principles and beliefs of yoga relate to one's understanding of the natural world. It further explores the integration of yoga with environment education, discussing in detail the various educational approaches to it along with numerous initiatives running worldwide to bridge the gap between yoga and environmental education. This chapter emphasizes the need to utilize the combined potential of yoga and environmental education towards a sustainable future.

This chapter delves into the pivotal role of Indigenous Knowledge Systems (IKS) under the framework of ethnoscience in transforming global environmental education. Scholars document the ongoing sixth mass extinction of biodiversity, necessitating urgent collaborative action. Ethnoscience, emerging from joint efforts of pedagogy and UN-led organizations, offers promise in addressing environmental crises. Since the 1990s, ethnoscience studies have highlighted sustainable indigenous management systems. Traditional Environmental Knowledge (TEK), a subset of IKS, imparts sustainable practices and holistic perspectives across generations. TEK fosters global conservation awareness and ecosystem restoration. Building on indigenous cosmovisions like "use the fruits, don't use the wood," a Model of Integrated Eco-Education (IEE) is proposed. This transdisciplinary model bridges the "knowledge gap" between science and ethnoscience aiming to cultivate eco-conscious global citizens. Efforts target multicultural environmental education in both developing and developed nations, emphasizing students' potential to drive ecological transformation.

The interaction dynamics upheld by Colombian indigenous communities with nature have been grounded in environmental principles derived from their ancestral knowledge and worldview. In the present day, these principles, which fostered the preservation and attentive guardianship of their natural surroundings, also permeate indigenous schools and their educational methods by teaching their myths, rituals, and agricultural heritage. This chapter examines the environmental principles and the pedagogical techniques upheld by the Arhuaco indigenous schools in the Sierra Nevada de Santa Marta in Colombia. It will delve into the Arhuaco worldview and the potential parallels between Arhuaco's environmental principles and the philosophy "Ecosophy," which promotes harmonious and wise ecological behavior. Along the same line, integrating experiential activities in nature into classrooms will also be highlighted as a reasonable means for nurturing a sense of stewardship and empathy towards the natural world in children.

This chapter delves into the importance of environmental education in Honduras, a country with vast biodiversity facing many environmental challenges. It explains the historical background of environmental education and recognizes the significance of indigenous knowledge and its relationship with the environment. Additionally, it explores existing initiatives towards environmental education. The chapter highlights some difficulties when implementing environmental education, such as limited funding, inadequate infrastructure, and language or cultural barriers. Despite these challenges, it emphasizes the importance of a holistic approach to the adaptation and reformation of the educational system. Finally, the conclusion emphasizes the need for collaborative efforts between government agencies, educational institutions, and community organizations to collectively take part in effectively implementing environmental education initiatives to build a greener and more resilient future for Honduras.

The chapter argues that human life is fundamentally attached to nature and that the latter's language plays a crucial role in defining our identity. Indigenous communities, in particular, have deep knowledge of their environment rooted in their traditions and cultural identity. They have a reciprocal relationship with nature, operating in tandem with it and never otherwise. This knowledge reflects a delicate balance required to maintain life on this planet. Thus, the chapter questions the current education system, which primarily focuses on scientific principles, solutions, and figures, and highlights how indigenous

Rarely are English language arts (ELA) educators actively encouraging outdoor exploration in the learning process. Yet, ELA skills (writing, reading, researching, and communication) can be used to communicate elements of environmental literacy for the 21st century learner in Virginia (VA) and beyond. This can be done in alignment with VA's standards of learning (SOL) through nature-based learning (NBL), which encourages thinking beyond science, technology, engineering, and math (STEM). Including ELA skills and strategies, historically not the focal point of environmental education (EE) initiatives, encourages outdoor exploration in 6-12 secondary ELA curriculum development. Educators, nature centers, parents, and community leaders can all be stakeholders in communicating elements of environmental literacy through NBL-ELA instruction. Therefore, this chapter promotes NBL as a sound option for improved teaching and learning when creating and implementing interdisciplinary curriculum. It provides research-based tools, strategies, tips, and resources useful for various ELA stakeholders.

This chapter explores the pivotal role of special education in promoting environmental education, fostering heightened nature awareness, and integrating sustainable practices. It underscores the importance of inclusive environmental education, tailored to meet the diverse needs of students, particularly those with disabilities. Through sensory experiences in nature and hands-on learning, special education enables a deeper connection between students and the environment. The integration of sustainable practices, such as waste reduction, energy conservation, and sustainable gardening, forms the core of this inclusive education approach. By empowering students to actively engage with and apply sustainable principles in their daily lives, special education not only imparts knowledge but also instills a sense of responsibility towards environmental preservation. Ultimately, special education serves as a catalyst for nurturing environmentally conscious individuals, ensuring no one is excluded from contributing to a greener and more sustainable future.

Foreword

I must compliment Prof. Tatjana Kochetkova, who has been very passionate about the cause of environment protection and has been exploring eco-centric philosophies, nature-conducive cultures, and their practices, and what concerns us the most here, the methodologies involved in nature education for the benefit of educators, students, and the general public. This anthology *Fostering an Ecological Shift Through Effective Environmental Education* is a testimony and tribute to her unswerving commitment to this cause.

The environment ought to be of foremost concern to mankind in this Anthropocene age. As a corollary, environmental education is, by all accounts, the need of the hour. However, it cannot just be imparted in educational institutions. Society, on the whole, has to be made aware of the importance of eco-friendly lifestyles and practices. In this, social media can play an important role even as the laws governing the concerned stakeholders need to be attuned to this cause.

Now, while social media is free for all and can be availed of by interested environmentalists, the action through law requires political willpower which is yet to manifest fully, what with the dependence of politicians on MNCs playing havoc with the environment for their profits, or with the autocrats eyeing domination of the world through violence, with no concern for the environment. Self-centered politics plagues international relations and so, when delegates at conferences like COPs mull over clean technologies, the conversation turns into business negotiations and concerns for profits of individual nations score over ideals about conservation-linked collective welfare of humankind. This self-interest needs to be curbed to some extent if we mean to save mankind from the looming disaster. Those who do not have a voice in decision-making continue to suffer silently or squeak sometimes *a la* little Greta Thunberg!

Come to think of it: Haven't we moved away from nature in the race to be civilized? Don't we need to learn from ancient societies whose remnants are still extant? Likely, the practices of the tribals or the cultures that are still possessed by past practices provide an effective antidote to the depletion and degeneration of the environment. Such a perception might offend the sensibilities of the modern man whose mind is submerged in scientific reason and who boasts of knowledge to reach the stars, yet to write off the cultures of ancient societies is to be unscientifically arrogant, for a true scientist will have his mind open toward what lies unexplored or unsolved in any culture as also in nature.

We have seen the catastrophic effects of unchecked rape of nature through modern technology. The answer lies in adopting a holistic approach in which there is scope to amalgamate current scientific knowledge with a sense of humility toward nature. Unless we inculcate -- what our ancient cultures have practiced -- a sense of reverence for nature, we probably cannot thwart the cataclysm staring us in the face. If one believes that the call to emotions is irrelevant in our times, then one just has to remind oneself that romance, friendship, parenthood, and family are still valued and enjoyed. God save us from the day when humans are churned out of a factory like heartless products as Aldous Huxley portrays in his novel *Brave New World*!

Call it *Dharti Maa*, as the Indians call it, or the Gaia as the ancient Greeks called it, or the dust unto which we finally return, as the Christians know, our lives are a part of this universe with its *panchbhootas* – earth, air, water, fire, and space. And to tend to these elements is to care for our own lives. It is as simple as that.

The present book covers the ancient green practices of various communities like Buddhism, Yoga, the Hindu concepts of Dharma, Tantra, Karma, etc., and the non-mainstream native approaches conceptualized and tried in other societies. It also deals with ideas like distance awareness and models like TEK, PET, IEE, 'ecosophy' etc. Various aspects of the issue of environmental protection – historical, theoretical, political, social, technological, etc. have been discussed in the book by experts and intellectuals, all of which make it a delightful reading with lots of takeaways for concerned readers.

The burden of this wonderful anthology, planned and prepared diligently, containing enlightening essays is to turn the focus from anthropocentrism to ecocentrism in which environment educators play a key role. The book, therefore, rightly dwells on topics ranging from teaching philosophy, methodologies, and curricula that aim to involve teachers even as these are sought to be made student-centric. For the uninitiated ones, there is a powerful appeal that can't be missed. It is a valuable compilation that can serve as a resource book as much for the curious reader as for the seasoned pedagogue.

Jagdish Batra
O.P. Jindal Global University, Sonipat, India

Preface

Nature education emerged in response to the paramount challenge of our era: the imperative to halt the degradation of nature and reclaim equilibrium within our life-sustaining foundation—the biosphere. Over decades of scientific inquiry and environmental advocacy, a gradual awakening unfolded among the public, illuminating the indispensable role of the planetary biosphere as the essential life support system for humankind.

However, the public has not yet been awoken enough to ensure the factual implementation of sustainability in the policy. While sustainability is a fashionable concept, widely advertised by governments and corporations, still the ongoing discourse on sustainability bears an undertone of insincerity. While entrepreneurs and policymakers claim allegiance to sustainability, reality begs the question: are they truly committed? What began as a policy approach has evolved into a global advertisement, permeating social policies and corporate agendas. Amidst the urgency of environmental challenges such as climate change, resource scarcity, and biodiversity loss, the essence of sustainability seems lost.

One glaring issue lies in policymakers' adherence to a human-centric perspective of sustainable development, which overlooks the interests of non-human species and ecosystems. The human-centric perspective on sustainability views it through a lens of trade-offs and rational choices and sees technology as the only way to achieve it. This perspective until now has failed to restore the balance in the biosphere – i.e. to stop biodiversity loss or anthropogenic greenhouse emissions. To ensure the stability of planet's ecosystems, we need to shift towards an eco-centric perspective, which renders intrinsic value in the biodiversity and biosphere's flourishing for their own sake, and not as merely human resource.

The same Enlightenment legacy that enabled us to develop technology and gain partial autonomy from nature, having environmental devastation as its side effect, can also help to overcome this crisis by employing our critical thinking capacity. Instead of viewing nature and humanity as separate entities, the way out could be to see them as one indivisible complex whole, which they truly are. The challenge lies in forging consciously developed connections with nature, which makes it necessary to give a central role to ecocentric environmental education.

To reverse the environmental crisis and promote a sustainable shift in our economy and technology, a fundamental cultural transformation is imperative. This transformation involves a shift in personal identity, moving from the self-centered "Ego" to the ecologically conscious "Eco." Achieving this shift is possible through environmental education, spanning primary, secondary, and adult learning. Indeed, scholars widely accept environmental education as an effective method for fostering personal and social growth while encouraging environmental conservation.

Regrettably, ecocentric environmental philosophical perspectives have not yet sufficiently permeated contemporary scientific paradigms. Visionaries like Arne Naess and Ken Wilber have long advocated for adopting an ecocentric or synergistic worldview, emphasizing the interconnectedness of all life. The ecological crisis cannot be resolved through technological means alone; it represents a deeper crisis of worldview, necessitating a shift towards recognizing humanity's integral role within the broader community of life.

This paradigm shift, from reductionism to an interconnected worldview, is already underway socially, exemplified by initiatives like organic farms, transition towns, and integrating traditional knowledge into modern environmental policies, a subject that we discuss in detail here. By embracing a new ethic of sustainable thought and action centered on interconnectedness, we can go beyond fashion and achieve genuine sustainability.

Nature education's pioneers – writers Aldo Leopold and Rachel Carson – were driven by the realization that environmental education is a crucial factor in shifting to sustainability. This is why early nature writers, who resisted environmental destruction, were so keen on promoting nature learning and psychological reconnection to nature. Thus, Rachel Carson, who raised public awareness of the dangers of DDT use, eventually leading to the ban on DDT, also dedicated some of her books to children's nature education.

Indeed, nature education needs to address the root cause of today's environmental degradation, namely: the structure of the modern self, currently disconnected from nature. One of the purposes of nature education should be to initiate a shift from Ego to Eco, i.e. to reconnect humans to nature psychologically. This requires more than only nature science education. While the science of nature education is a necessary part of the sustainable shift, it is by itself not enough. Indeed, a sustainable shift means not only an increase in knowledge about ecosystems but also a change in perspective. We should shift away from seeing nature – the biosphere – as an object, external to us, and towards embracing the biosphere as part of the self, i.e., away from a human-centric and towards a nature-centric approach.

The cultural shift to the Ecocentric self, while supported by policy, economic, and technological transformations, can produce a sustainable reorientation in society. This cannot be achieved by technological advances alone because the latter is driven by societal demand. Hence, a shift to clean technologies will happen only in response to societal reorientation toward ecological stability. This is already going on, to an extent.

This book addresses the major problem of the shift from Ego to Eco by addressing the inner side of nature education. It describes the methods for psychological and spiritual reconnection with nature through education. It considers its spiritual and philosophical foundation, as well as its roots in the traditional cultures of Tibet and India, the ancient cultures of Maya, and other indigenous cultures of South America: Columbia, Peru, and Honduras. It also looks at the variety of audiences and aspects of environmental education – from teaching young children and including nature in language education to specialized education to university campus gardens. This makes it relevant to nature educators of all age groups.

The contents of this book will help you, as an environmental educator, to broaden your teaching methodology, as well as to enlarge your insights into the theoretical basis of environmental education. This book will benefit those who may be struggling with nature deficit by showing some practical ways of reconnecting with nature: nature awareness techniques. It can also benefit eco-therapists, and everybody who wishes to reconnect to nature as well as parents who want to bring up their children connected to nature. The teachers will benefit from this book by finding numerous practical activities for environmental education as well as an explanation of their theoretical background.

We hope that the topics and materials discussed in the various chapters of this book will provide you with the necessary basis for improving your teaching of nature-related topics.

In the first section, we discuss the philosophy of environmental education. The chapter "Effects of Nature on Psychological Well-being and Perception of the Meaning of Life" by Tatjana Kochetkova contributes to ecopsychology and environmental education by providing a picture of psyche-nature interaction and introducing a series of nature-awareness practices that promote *gestalt* awareness of nature along with personal growth and self-realization. Adult nature learners can use them in a wide variety of contexts – from wild surroundings to parks and semi-natural areas.

In his chapter on distance awareness, Rico Sneller presents it as an indispensable ingredient of ecological awareness, looking at the work of two German thinkers, Ludwig Klages (1872-1956) and Hermann Friedmann (1873-1957), who highlight a tendency in the Greco-European tradition to privilege tact and tangibility over vision, or contemplation; in other words, distance awareness. Claiming that distance awareness is crucial to environmental education, Rico Sneller proposes that distance awareness should be actively promoted in education through enhancing environmental awareness and reintegration and promoting poetry education.

In "Re-imagining the Relationship between Self and Environment," Somosri Hare describes key debates concerning the relationship between humans and the environment, such as anthropocentrism and ecocentrism. She explores human-nature relations first theoretically and concludes with a survey section reviewing the ramifications of environmental ethics for public perception and suggestions for environmental education.

The second section of the book is dedicated to spiritual approaches to environmental education, viewing Tibetan Buddhist, Hinduist, and Yogic perspectives. It starts with the chapter on Transnational Buddhism and Ecological Awareness, by Swati Chawla. She discusses how Tibetan Buddhists living in India, especially through the Tibetan Nuns Project, are shaping their identity, caring for the environment, and educating themselves. The chapter challenges the old views of Tibetan culture, considering how Tibetans in India are well-acknowledged with global culture as well as their native traditions, while they're using Western ways of presenting themselves to change how they see themselves and how others see them. The chapter is also unique and relevant in its description of Green Buddhist practices, which are applicable in the broad context of reconnection to nature.

In her artistic chapter, "Re-defining Tree Worship as Artistic Becoming for Environmental Education," Shruti Jain reveals to the reader the profound role that trees play in the Shakta Tantric traditions, serving as catalysts for the spiritual transformation of the practitioner. The chapter offers an innovative approach to integrating the essence of tree worship into contemporary educational frameworks, exploring the implications of a tree-centric perspective within education.

The chapter on "Karma and Conservation" by Siddharth Kanojia describes how ancient Indian wisdom enables environmental consciousness by looking into traditional concepts such as Dharma, Karma, and Ahimsa, which can promote eco-friendly and nature-reconnected ways of life, facilitating a shift towards sustainable culture.

Finally, in the chapter on "Yoga and Environmental Education," Pranjal Khare and Anny Gabriela Molina Ochoa dive into the different schools of yoga and how they promote environmental awareness. They also discuss the philosophical foundations of yoga and explains how the core principles and beliefs of yoga relate to one's understanding of the natural world. It further explores the integration of yoga with environmental education, discussing in detail the various educational approaches to it along with numerous global initiatives to bridge the gap between yoga and environmental education. It looks into how yoga practice promotes awareness of our bodies as being integral parts of nature.

The third section of the book is dedicated to the contributions of indigenous and traditional environmentalism to contemporary sustainability. It demonstrates the relevance of indigenous and traditional knowledge for environmental education. The section starts with a grounded examination of how Indigenous Knowledge Systems (IKS), can play a crucial role in reshaping global environmental education, conducted by ethnoscientists Jan J. Slikkerveer and Sander Gellaerts. They highlight the potential of IKS to drive a shift towards more rational and sustainable use of natural resources on a worldwide scale. They introduce a TEK-focused extension of the indigenous principle "use the fruits, not the wood, for the sake of future generations," aiming to advance global education for sustainable development. The chapter seeks to initiate an ecological shift in human interactions with ecosystems through an Integrated Eco-Education Model (IEE) – the transdisciplinary model that is tailored for implementation in multicultural settings worldwide, cultivating students' ability to become environmentally aware global citizens.

Sara Acevedo Garzon in her chapter "The "Ecosophy" Concept Incubated in the Indigenous Schools of Colombia" examines the environmental principles and the pedagogical techniques upheld by the Arhuaco indigenous schools in the Sierra Nevada de Santa Marta in Colombia. It delves into the Arhuaco worldview and the parallels between Arhuaco's environmental principles and Ecosophy. The author argues that integrating experiential activities in nature and classrooms is also a reasonable means for nurturing a sense of stewardship and empathy towards the natural world in children.

In "Forgotten Realms: Mainstreaming Indigenous Knowledge into Environmental Education" Sunidhi Setia and Darshna Gupta question our current education system, which primarily focuses on scientific principles, solutions, and figures, and highlight ways in which Indian indigenous knowledge can contribute to our environmental education.

Further, Pranjal Khare argues for the need to integrate Traditional Ecological Knowledge (TEK) with Environmental Education in India. His chapter throws light on the concept, meaning, and significance of TEK, discussing the prevalent TEK systems in India in detail and emphasizing the need to preserve and protect it.

The chapter "Expanding Environmental Education in Honduras" delves into the importance of environmental education in Honduras, where, not unlike in India, there is great biodiversity but also many environmental challenges. It explores initiatives towards environmental education as well as difficulties in their implementation and emphasizes the need for collaborative efforts between government agencies, educational institutions, and community organizations to build a greener and more resilient future for Honduras.

The fourth section of the book is dedicated to environmental pedagogy in the context of various target groups: young learners, university and college students, community members, and special learners.

Sibylle Gruber in her chapter "Narratives of Hope: Teachers' and Students' Roles in Climate Justice Activism" addresses the crucial need to shift from narratives of climate despair and disaster toward the narratives of hope, action, and activism. She encourages educators to become teacher-researchers who can use inquiry-based and critical self-reflection to promote hope and constructive climate activism in students and teachers.

In the chapter "Growing Pains: Moving Campus Gardens Beyond Administrative Barriers" Amara Szrom with her ample pedagogical experience provides recommendations of how university campus gardens can strengthen their programs and establish themselves as vital student resource spaces. To explore the variables contributing to thriving and well-integrated university garden programs, 19 in-depth interviews were conducted with university program coordinators, administrators, and key stakeholders in the United States, including the University of New Mexico, where the author managed the campus garden and outdoor classroom.

The chapter on "Educating the local population about the need for environmental protection through Participatory Educational Theatre (PET)," elaborates on the practice of participatory educational theatre (PET) as an effective tool in implementing socio-environmental projects across the globe. The genesis and evolution of PET from traditional theatrical forms are also showcased in this chapter of the book. The benefit of PET is turning the passive audiences into active participants, which helps implement social projects. PET also caters to more acceptability of any project's critical take-home message, as it was tested by authors in the disaster-vulnerable Indian Sundarbans.

The chapter by Katrina Brugman, "The Power of Nature-Based Learning in Language Arts Curriculum," addresses the gap in the current English language arts curriculum: it rarely encourages active outdoor exploration in the learning process. To fill this gap, her chapter promotes nature-based learning as a sound option for improved teaching and learning and describes research-based tools, strategies, tips, and resources useful for various English language arts stakeholders.

The chapter Histories, Theories, and Contemporary Considerations of Nature-Based Early Childhood Education: Roots to Branches by Stacey Alfonso explores how history, philosophy, and theories behind forest schools have shaped the growth of nature-based early childhood education (NBECE), particularly in the US and UK. The author describes how exposing kids to nature helps their development and makes them feel connected to nature and care about the environment, providing special attention to indigenous and local communities.

Finally, in "Empowering Special Education," Finita Roy delves into how special education by empowering students to actively engage with and apply sustainable principles in their daily lives, not only imparts knowledge but also instills a sense of responsibility towards environmental preservation.

In conclusion, *Fostering an Ecological Shift Through Effective Environmental Education* presents a range of new approaches to environmental education. It develops a philosophical framework for ecocentric environmental education, which connects self-realization, deep ecology, and personal development with awareness of nature.

The book argues that a basis of recovery from the environmental crisis is a shift towards nature-centered culture and values, which can be achieved through ecocentric environmental education. Even though current technological developments allow sustainable economic reorientation towards a carbon-free and closed-cycle economy, its practical achievement necessitates a cultural reorientation towards an ecocentric self, including the recognition of ourselves as part of the biosphere. This book presents an overview of educational strategies toward such a sustainable shift, based on the recovery of indigenous knowledge and wisdom traditions from the most ancient world civilizations, such as Tibetan, ancient Indian, and South American Arhuaco cultures.

It will be of interest to scholars and practitioners of nature education, as well as parents willing to bring their children into connection with nature, to those who want to avoid nature deficit disorder, and to reconnect with nature on the path towards self-realization.

Tatjana Kochetkova
O. P. Jindal Global University

Acknowledgment

The authors would like to acknowledge the great input of all individuals who made this book possible, particularly the indigenous peoples living and striving in rural communities around the world.

Dr. Tatjana Kochetkova would like to acknowledge Prof. Dr. Jagdish Batra for his consistent support and great inspiration. Also, gratitude goes to Dr. Sergio Meira, and Mrs. Victoria Meira for their presence in her life and love.

Dr. Shruti Jain extends her heartfelt gratitude to Dr Tatjana Kochetkova and Dr Anway Mukhopadhyay for their kind support and valuable input. A special thanks to Shantanu Saha for sharing his first-hand experiences of sacred groves and sparking the author's interest in the subject.

Prof. Dr. Jan Slikkerveer and dr. Sander Gellaerts wish to express their gratitude for the great inspiration and encouragement of H.R.H. Prof. Dr. Anak Agung Gde Agung, Mrs. Dr. Martha Tilaar, Mrs. Prof. Dr. Vina Adriany, and Dr. Kurniawan Saefullah in Indonesia. Special thanks go to Mrs. Drs. Mady K.L. Slikkerveer MA for her unflagging support.

Section 1
Philosophy of Environmental Education

Chapter 1
Effects of Nature on Psychological Well-Being and Perception of the Meaning of Life:
The Case of a Night Walk

Tatjana Kochetkova
O.P. Jindal Global University, India

ABSTRACT

This chapter aims to contribute to building an empirical and conceptual basis for ecopsychology and environmental education by providing a picture of psyche-nature interaction. The chapter contributes to methods of environmental education by elaborating on a series of nature awareness exercises (see Appendix), which happened to be efficient and broadly applicable. It presents reports of these exercises as applied in a workshop with a group of students from the University for Humanistics (Utrecht, The Netherlands) at the Eifel National Park in Germany. The workshop lasted three days. The author intended to clarify the effects of nature awareness on the participants, especially on their subjective well-being and perception of the meaning of life. As a result, a "battery-recharging" effect of nature and its influence on young adults at four levels of awareness (external physical perception, internal (bodily) sensation, emotions, and existential meaning), were established. The chapter focuses on the episode of a night walk.

INTRODUCTION

A Survey of the Relevant Research

The research on the psychological bond of humans with the natural environment, the subject of ecopsychology, is still in its infancy. The earliest attempts to study nature as a psychological factor came from the related area of deep ecology, founded by Naess and further advanced by Devall and Sessions (1985,

DOI: 10.4018/979-8-3693-2577-3.ch001

2021) and Mattews (2001). More than other types of eco-philosophy, deep ecology pays attention to the experience of nature and its spiritual value. The idea of a broad synthesis of ecology and psychology was proposed by Theodore Roszak, who introduced the new approach in *The Voice of the Earth* (2001). Since then, several books and articles on the subject have been published, with significant contributions such as *Contesting Earth's Future*: (Zimmerman, 1994). A series of relevant articles on ecopsychology appeared in the *Journal of Humanistic Psychology* in 2001.

Despite these courageous and fruitful attempts, the fundamental question 'In which way does the natural environment influence humans and what is the role of nature for the human psyche?' is still open. The urge to address this question was the primary reason for the Eifel case study. The methodology of this study was taken from the study of Patterson, et al. (1998) and White, D., & Hendee, J. (2000). The methodology included a series of assignments on the perception of nature (the exercises are described in the attachment). Experiences were noted by participants in personal diaries which we used later for analysis.

There are many trustworthy claims that contact with nature has positive effects on human subjective well-being and that it is of essential psychological and spiritual value. For instance, the book *The Last Child in the Woods* (2005) argues that the experience of nature is vital for healthy child development, while the deficit of nature is linked to issues such as obesity, attention disorders, and depression. However, there is a need for more support and theoretical explanation for this phenomenon. As a contribution to fill these gaps, I conducted this qualitative study.

Questions and Goals of This Study

The aim of this study is firstly, to extend and elaborate methods of environmental education, specifically the *exercises on nature-connection* —techniques for focusing awareness. Some of them, i.e. gestalt practices, were introduced by the author, and others suggested by Michel Cohen (1997). One can see the entire list of practices described in the Appendix in a replicable form.

Secondly, the study intended to explore the psychological effects of short-term awareness of wilderness on one's subjective well-being (Diener & Suh, 2000) and one's sense of the meaning of life. It is based on the more general hypothesis that there are positive relationships between the experience of wilderness and the development of the self, feelings of community, and spiritual development (White Hendee, 1999). I restricted the study of the psychological effects of nature to two variables (subjective well-being and the sense of meaning in life

The study aimed at:

(i) testing the hypothesis that there are positive effects of the awareness of wilderness on the subjective sense of well-being and meaning in life.
(ii) understanding the nature of the psychological need for wilderness.

Although there is a substantial amount of data on the psychological benefits of nature, (Fisher, A. (2002), Riley, M. F., & Hendee, J.C, (1999), R, White, D., & Hendee, J. (2000), Young, R. & Crandall, R. (1984) Wicks C, Barton J, Orbell S, Andrews L. (2022), we still lack theory that explains the findings. This chapter aims to contribute to such a theory.

Further, nature reconnection boosts personal development, making it a part and parcel of holistic education. Hence, eco-psychology is directly related to educational experiences, as discussed by Robbins, J. (2020), Toth, N., Little, L., Read, J. C., Fitton, D., & Horton, M. (2013)., Miao, R. E. & Cagle, N. L. (2020), Johnstone, C. (2020).

This case study aims specifically at the effects of nature on the sense of well-being and meaning in life as applicable to environmental education.

PSYCHOLOGICAL WHOLENESS

The Notion of Wholeness

Psychological wholeness is a complex notion, studied by different intellectual traditions. For instance, feminist, humanistic, and transpersonal psychologies cherish the idea of growth of the individual towards wholeness. The idea itself has a much longer tradition, going back to Plato, Indian Upanishads, Sufism, and Kabala philosophy. As one might expect, these approaches diverge in their account of the nature of wholeness. One major issue concerns the controversy between emphasis on harmony and paradoxical tension.

The more traditional approach perceives wholeness as a state of the subject in which each psychological element is harmoniously related to all others, and the subject's consciousness and the unconscious as a fully integrated entity (Kochetkova, 2007). It implies self-acceptance and a sense of direction as its components. Human beings are seen in this tradition as having an inborn tendency towards wholeness, which is essential for optimal psychological functioning.

This approach was in the 20th century represented by Carl Gustav Jung and more so by Roberto Assagioli. Assagioli sees modern man as fragmented into sub-personalities without an inner core or direction. Consequently, human potential is diminished, resulting in lower subjective well-being (Assagioli, 1976). Assagioli developed a psychosynthesis approach to counter this fragmentation (Ibid). His view, however, is criticized as overoptimistic, especially by Schneider (2004), who thought that tragic contradictions are insurmountable. Namely, Shneider argues that his existential approach presents "a fuller, more dynamic understanding of wholeness" than the traditional understanding of wholeness in terms of harmony alone, as reflected in the writings of Rollo May, Ernest Becker, and Paul Tillich (2004). Schneider's later version would speak of wholeness as a dialectic between self and non-self, separation and tension, ever-evolving end yet energizing, inspiring.' (Ibid)

Harmony or Tragedy

The critique of the traditional approach may be that harmony is not a real state of any human subject, but rather an ideal model, unattainable in actual life. This model tends to neglect the insurmountable tragedy and 'godlike-wormlike' duality of the human condition (Kochetkova; 2007). We tend to go along with this critique, even though the notion of harmony has its credits for real human subjects: it is a powerful attractor that can guide psychological growth.

Existential psychology, introduced by Rollo May, offers an alternative to this approach, stating that inner growth comes from confronting the pain of existence rather than escaping into pleasures or positive thinking (May, Angel, and Ellenberger; 2004). According to them, genuine joy can emerge from an appreciation of life's agonies (May, Angel, and Ellenberger; 2004). One of the most recent examples of thinking along these lines is proposed by Kirk J. Schneider in *Rediscovery of Awe*. His book grounds paradoxical wholeness in the inherently tragic nature of the human condition, the paradoxical coexistence of bliss and dread of life in the face of death and non-being (Scheider, 2004). Schneider sees wholeness not as the negation of tension between the opposites, but as retaining this tension and embracing what will never be reconciled (Ibid, 2004). As an example of the impossibility of reconciling the opposites, Schneider quotes one of his clients:

"My experience of tragic optimism, it's the paradox. It is the awful and very freeing parts of my life — my grief [at dad's] death and in this sense that I'm free. They don't seem to go together although I experience them simultaneously" (2004).

This expansion of the psyche to encompass the irreconcilable opposites of suffering and fragility together with freedom, happiness, and the gift of life, reminds us of the negative dialectic of Adorno (2003), for whom the opposites are never really transcended or synthesized, but achieve a flexible unity. In contrast, the traditional view of wholeness is the Hegelian synthesis of opposites in which they lose their separate character in becoming the poles of an indivisible whole.

However, in my view, the existential paradoxical fluid center is a realistic response to those actual life situations in which it is impossible to achieve Hegelian synthesis (1966). I find it a real-life approach, working in actual therapy, and attainable by real people. However, reaching for Hegelian synthesis is also a factor of psychological life. Though unattainable, it is an ideal image, capable of enhancing psychological growth (Kochetkova, 2007).

For Schneider's fluid center approach (2004), tragedy is insurmountable, but it can be awe-provoking and a source for transformation. From this perspective, an important dimension of human experience and the meaning of life is lost, if the tragic character of life is completely missing. One can agree with this, but the dream of life without tragedy is also rooted in the human psyche and never dies. Even the existentialist Lev Shestov (2011), whose reflections originated precisely from the paradoxical nature of the human condition, once confessed about the never dying longing to simply destroy the tragic condition of life. The awareness of the impossibility of this does not prevent people from dreaming of it and from acting to minimize tragedy in their lives.

A CASE STUDY

Participants

I called for students to participate in a connection-to-nature workshop at the Eifel Mountain in Germany for three days via email within the University. Thirteen students responded from the University for Humanist Studies in Utrecht, The Netherlands. The participants were in their early twenties. Only some were experienced hikers. The group consisted of three men and ten women. This distribution was coincidental but reflected the population of the University for Humanist Studies. The camping place in the study area was found at random via the internet. The choice was for an area with only a few inhabitants and, during that time of the year, few visitors.

The group went on a 6-hour journey by train from Utrecht (the Netherlands) to Bad Bentheim (Germany) and then by bus to the village of Geroldstein. After two hours of walking, we reached a camping place, which served as a base camp.

Figure 1. Maar Lake

The Eifel Area

Eifel is the name of a plateau in Germany that is bordered on the southeast by the Mosel River, on the west by the Ardennes and Belgium, and on the northeast by the Rhine. It consists mainly of barren uplands covered with moors. The Volcanic Park area includes approximately 240 volcanic crates. [1] The heart of the Eifel region is called the Vulkan or Volcanic Eifel. It is full of volcanic lakes, in German called "Maar". When a volcano erupts under a lake, there is almost an explosion of water that comes to boiling within seconds. As a result, a perfect round hole is formed, which fills up with water again: a Maar. [2] In time, most of those lakes are filled up with soil again, but there are still almost 80 of them. They are surrounded by forests. In this semi-wild area, the lakes, forests, and meadows closely border farmlands. Due to the time of year (early autumn), when this study took place, the area was almost deserted. During our group walks, we crossed fields of crops as well as grassland and wooden areas. We regularly encountered herds of cattle.

The Workshop

The core of the workshop consisted of a series of exercises directed at levels of connection to nature, which I describe fully in the Appendix. First, the participants filled in a questionnaire on subjective well-being (Tov & Diener, 2013). At the end of their stay in the Eiffel area, they did so again. This was done to obtain background data about the general psychological condition of participants. Additionally, this was to have an instrument for measuring possible changes resulting from their stay in nature.

Each day started with bodily awareness exercises. After that, the changing focus during the day was on emotional response, aesthetic perception, and existential response. In these exercises, participants were asked to be aware of their environment in a way that was as free as possible from cultural clichés, stereotypes, and preconceived categories of experience. The participants kept written records of all their

Figure 2. The walk in the Eifel Area

experiences and impressions. Sometimes they were invited to exchange their ideas and feelings with each other.

One of the emotional response exercises, for instance, was as follows. The participants got instructions to look into their memory for earlier pleasant and unpleasant experiences with nature and to write them down. They got to think about what sensations gave the emotional quality to their experience. They were also asked to remember if they were ever taught to have worthwhile sensations and experiences with nature.

The exercise concerning aesthetic perception focusing on the "awareness continuum" is one example done in the natural environment. It could be done either by individual participants silently or verbally expressed in a group. If the exercise is done in a group, then participants are only allowed to communicate the contents of their awareness. (Fagan & Shephard, 1970: 85). The purpose of thisexercise is to stop the 'autopilot' and to stop labeling without consciousnesses. Experiences of nature are also frequently automatic — we give labels to sensory perceptions like "birds singing", "wind blowing" or "green grass" and then stop paying attention to them. In reality, birds may sing in thousands of different ways, the colors of grass may be various types, and every particular sensation is unique. Yet it can be perceived only if we give our full attention to it.

"Practice of the awareness continuum is a kind of training in applied phenomenology; an attempt to concentrate without judgment or labeling on every new figure which becomes of interest without preconception or expectation. Frequently, human beings engage habitually in behaviors, attitudes, and feelings stripped of conscious awareness. The awareness continuum exercise is particularly designed to heal this lack of mindfulness and to restore richness and vitality to living.' (Clarkson, 1989: 79).

The existential dimension was brought to the foreground through an instruction to reflect on the meaning of the 'sunrise experience.' The group rose early before sunrise and walked uphill to a quiet spot where they could oversee the surroundings and wait for the sunrise. They were asked to be silent and let the experience sink in. Afterward, they were asked to reflect on the experience. The reflections varied from describing the direct sensations (emotional response) to contemplating an attitude to nature and existing as a human being (new cosmic beginning, eternal return, source of life). To add, the experience of the night walk included a 4-hour walk with attention instructions, a relaxation practice, and a guided fantasy of going back in time, described in the next section.

After the workshop, written records of connection-to-nature experiences and some additional interview data were analyzed according to the methods used by Young and Crandall (1984) and White and Hendee (1999). All participants gave their consent to analyze and share their reports in anonymized form.

Four categories of analysis were determined in advance: body-in-nature (awareness of physical sensations in reaction to the natural surroundings), emotional response/feelings (feelings and emotions related to being in nature), aesthetic perception (evaluation of nature from an aesthetic viewpoint: appreciation or aversive reactions), and existential response (reflections on the experience of being). The notebooks provided further data, which led to the identification of two additional categories: symbolical interpretation of the night and memories of other nights.

Importantly, all these practices are broadly applicable to diverse adult audiences in nature education. They are safe, have positive effects, and, as far as the author can see, are not culture-specific. For example, in my more recent experience at another university in India, students had also similar positive responses to these exercises, when I replicated some of them at the campus park.

Night Walk

While the reader can find all the details of practices in the Appendix, the complete analysis of students' reports about them would have taken more space than we have here. Therefore, instead of describing all the exercises, I will focus on one of them: the night walk. The quotes from Whitman and Bulgakov (see epigraph) remind us of what Plato told us long ago: night is a time of visions and revelations. Throughout history, night is often seen as the time of true insights: the time when deceptions and illusions fall apart, and the truth reveals itself. Can nature at night help to reveal psychological truth? I engaged in the night walk intending to illuminate answers to this question.

The group that was camping on the Eiffel did a night walk of about six kilometers through hills and forests towards a Maar (volcano lake). The trip started when darkness fell at approximately 10 pm. The walk to the 'Maar' took about one hour and a half. We walked through fields and forests, climbing and descending. The group stopped for a while near a beautiful stone area. They stayed at the Maar for about two hours. They participated in a fantasy, guided by the researcher, imagining this crater millions of years ago when the volcano was still active. Then they meditated for some time sitting in silence.

We did not expect to meet any wild animals because we walked as a group, so when suddenly in the hatches an unknown animal made some noise near one of the participants, she screamed from fright. The encounter was later reflected in the participant's diaries and made an impression on the group. The return road was easier because part of it followed a trail. Around two o'clock in the morning, the group reached the camping place and participants wrote keywords about their experience, followed by a more detailed description early the next morning.

After relaxation instructions and attention-stimulating instructions, the guided fantasy went more or less along the following lines:

Imagine yourself going back in time thousands of years to the era of the dinosaurs. The volcanoes you have seen around you are still active, the climate is different, and you live from gathering herbs and roots and the occasional game. How is it to live in this environment? Are you on the edge, relaxed, or otherwise? Feel the ground, smell the air, then touch the trees and stones. Imagine how you live your life in these primitive conditions. Try to blend in with the surroundings. How is this experienced?

These instructions were given with pauses in between so that participants could visualize the picture in their minds. In the end, participants were asked to make contact with the sounds, the wind, and the smell of the environment in the here and now to return to the contemporary situation from their imagination. The contact with the immediate environment helps to come back to external perception. Immediately afterward, they were asked to keep silent and note the experiences in the diaries.

ANALYSIS OF THE NIGHT WALK

Body-in-Nature

The night walk did not leave any of the participants physically untouched. They described the night in their notebooks in terms of polarities. One of these was the polarity between tiredness and alertness. It turned out that tiredness and alertness are compatible. This is because three participants reported feeling both feelings at the same time.

Four participants reported feeling tense and tired. They found it too dark to walk, expressing that it was physically tiresome under those conditions to climb the hills covered with forest. Three participants complained about the temperature being so low. One stated that "the cold that I felt, I could not ignore" (Participant's notebooks)

For some of the participants, the strangeness of the situation of walking at night in a new natural area with classmates brought 'extra alertness' and joy of walking (participants' notebooks). Of course, they knew in advance what they were going to be doing yet walking in unknown hills at night steel felt strange. The special nature of night walking was associated by six participants with some tension which had a stirring effect on consciousness; it brought about an alertness that was enjoyed by them, as they reported.

The Emotional Response/Feelings

The three categories of emotional response, aesthetic experience, and existential response are so intertwined that it is sometimes hard to set them apart. Yet, we begin with emotional experience as an introduction to the other two. The emotional response turned out to have three major components: feelings of safety, fear, and thrill.

The darkness aroused fear in the majority of participants (eight people). This was accentuated by an incident in the forest when one participant screamed because she sensed an animal coming very close to her. Everybody heard the scream, and this added to the tension caused by the dark. Five of them, however, had feelings of both safety and fear. Darkness sometimes acts as a refuge, because it makes one invisible. Some participants said that they were not afraid of the dark as such, but of people who might misuse it, for instance.

A general response to the darkness seemed to be a heightened state of alertness. Out of twelve participants, only one reported being too sleepy to pay attention to the surroundings. Seven pointed out that they were more energetic and alert than usual. Some participants reported being both alert and tired. The presence of the group, according to the participant's reports, diminished the fear of darkness and fostered feelings of security.

A third component was reported as well. Three participants mentioned a sense of adventure, novelty, and mystery. This caused a thrill. This sense of adventure appears to be linked to an existential response (i.e. reflections on the experience of being, perception of meaningfulness of life), as we shall see in the section on it. For two participants, the night walk stimulated their intuition, energy, and feelings of excitement. An explanation for this might be that night is not an ordinary time for activity, but when a well-motivated activity does happen at night, it can have psychological advantages: better concentration (three participants) and vivid imagination.

For some participants, the night is a time when hidden resources and creative capacities can be drawn upon. One participant specifically mentioned intuitive perception at night: "I was looking with something above my eyes" (Participants notebooks). A general response, despite fear and tiredness, was typically a sense of joy, enhanced by the beauty of the night.

Aesthetic Perception

The night walk appealed strongly to aesthetical perception. This undoubtedly had to do with the beauty of that particular night. When the group left the highway for the track on the slopes, the darkness was still mixed with some light from civilization. The slopes looked silvery under the huge, clear, starry sky, and were covered with aromatic herbs. The view was open and broad. After some time, the track rose and led into a thick forest where one could see only trees and some pieces of sky. When the group reached the side of the lake, a chaotic mass of volcanic rocks and a huge human-made cross were seen. In the moonlight, the stones had a deep red color. The track led to a second lake, surrounded by stones and this time a grass-covered beach.

The Beautiful and the Sublime Experience of Nature

For the description of the aesthetical response, I use two categories in this qualitative study— the beautiful and the sublime. The concept of the sublime has two meanings: the classical and the Gnostic (Doran, 2015). The classical concept was introduced by the ancient Roman author Longinus (2000). According to him, "The human soul rises spontaneously by means of that which is sublime. It flies high and is filled with joy, joy, and pride as if what it perceives were its own doing" (Longinus, 2000). Sublime art originates, according to Longinus, in the greatness of thoughts and genuine and inspiring emotion, along with the right composition and the proper style (Longinus spoke about literature) (Longinus, 2000).

Originally, the sublime was understood as a quality of the arts. The Romantic poets, like Percey Shelley, William Wordsworth, and Fedor Tytchev, discovered that nature could have the same effect of elevating and filling the psyche with joy and wonder (Dumler-Winckler, E., 2015), Thompson, C., 2005). They conceptualized the sublime as a quality of natural phenomena. The Romantics applied the category of the sublime to mountains, waterfalls, storms, volcanoes, stars, and other impressive natural phenomena, while still preserving the core of the sublime as a wonderful and elevating quality.

However, there is also another, Gnostic idea of the sublime as 'the overcoming of the difference between good and evil and between beautiful and ugly, the transcendence of the opposites' (Elders, 2001). Here the sublime is no more wonderful, but it is transcendent to the opposites of beautiful and ugly so that it can be also destructive and scary, as manifested by the Gnostic symbol of Abraxas, God, and Satan at once (Hesse, 2017).

These students were all familiar with the concept of the sublime. Nevertheless, the sublime response of seven participants went mostly without using the word. Only one person wrote about 'sublime', and in a non-traditional sense. However, five participants mentioned the beauty of the night walk, and six participants had a sublime perception of it: they felt awe, reverence, and emotional understanding that goes beyond rational thought or language. The beautiful aspects included the stars and the entire landscape ('I find it all very beautiful') said one participant (Participants reports). Starry sky, aromatic herbs, dark forest, and hills – all this together created a harmony of a specific, mysterious, and awe-provoking beauty.

Most participants felt the sublime in a classical sense. They perceived it as an uplifting and liberating feeling. One participant mentioned sublime in the second, Gnostic sense of the word: 'There the nature is so much bigger than I am and not only beautiful but also a bit scary. Then is nature not only beautiful but also sublime' (Participants reports). One might conclude that the participant refers to the Gnostic meaning of the sublime, the concept he is acquainted with (Participant reports).

The Sublime and Self-Realization

In three cases, the feeling of the sublime is related to self-realization (participants 8, 9, 12). In two cases, it is a positive relation, while in the other case, it is a negative one. 'Self-realization' is used here as a clear vision of oneself, as self-reflection rather than self-actualization. Self-realization refers to having a clear sense of identity, and an insight into who and what one is: in three cases, participants mentioned this insight along with the sublime response to the night.

The sublime response of one participant was that "the starry sky gives me an open, big feeling where I feel humble". It is connected with self-realization because "important things in my life became clearer, I had more insight, it all became much simpler. I can see much better what I want. I am closer to myself, to my life path."

The sublime was felt as being overwhelmed by the light of the stars and by darkness. "Overwhelming, this is a good word for my experience of the past night. The starlight was overwhelming; the darkness was intensive, and the overgrowth everywhere around us was unavoidable, full of power. Suddenly nature gave little away to her daily beauty; all colors, depth, and forms remained hidden".

At the same time, this participant felt existential strangeness and even inadequacy: "I felt myself at the same time a being of nature, belonging there ... and a sense of wonderment of what maybe should be home but what is not that at all. I am an onlooker and at a loss about the language of nature, the road to the right spots is unknown to me... I felt a piercing cold and felt as if my eyes had become useless. The night is the state of reality, the situation where humans don't belong. The night is strange to our being".

One participant described the stars as sublime: "The most important, most impressive elements were stars. I loved them when I was younger... They are still important for me; I feel myself still involved." For this participant the night was also connected with being oneself more clearly: "At night I am stronger and more secure... Night is my freedom – dangerous, but alive. It is my true self. It makes me deeper in myself."

Yet, the sublime response is not always or automatically linked with self-realization: out of six participants with a sublime response, only three also had self-realization, while three others didn't. This suggests a flexible link between the sublime and self-realization, which is possible because the sublime makes it easier to differentiate between essential and non-essential, deep and superficial.

The Existential Response

Here I interpret as existential responses ones that involve the relation between the deeper, truer self and the world as a whole. All but one of the participants responded existentially to the night. Even though all participants were students of the University for Humanist Studies, trained to think existentially, this cannot be the only reason, because then in the sunrise study, we would expect the same frequency of response.

I suppose that the night itself seems to invite an existential response; it is "a manifestation of reality", as several participants mentioned. One can see three broad types of relation between the self and the world, as reported by the participants:

- A distance from the world (7 participants).
- A unity with the world (4 participants).
- A state in which the external world seems to be non-existent (2 participants).

One may define the distance as a state of opposition between self and world, irreconcilable states of self and world. Seven participants out of thirteen related to the world as *distanced* from the self; three of these participants linked this with self-awareness. People mostly experiece distance as fragility of self and the opposition between one's mood and the emotional mood of the night. For instance: "My attention was also drawn by the surroundings which radiated for me an unattainable condition of calmness and quietness; so little was happening there, in contrast with the fight which is going in me."

"Alone, left to my own devices I feel myself smaller, more fragile when it is so dark". 'The darkness felt like a blanket which took away my sight and security; the darkness is governed by the right of the strongest and the right to eat and to be eaten". "In this night experience, I was intensely aware of my mortality and the dark sides of nature and humans"; "The lack of daylight!!!!!!!!"

"The lack of daylight makes it visible the power of night and the relentless nature'; 'at night there is so much unknown to me... I notice that I am not adapted for it. I have a feeling that there is something I have lost, I could have done this some time ago, I am no more trained'; 'If I look deeply, the world looks very threatening, comparable with the night forest. Then I feel small and powerless, and everything is dangerous" (participant's reports).

Sometimes the experience of distance does not exclude the feeling of unity with the world as well, as in the following case: 'The darkness directed me to myself and brought me to nature. At this moment, I was connected with nature and in this way with the assumed 'meaning' that is in nature. From there I can give a clearer meaning to my life. On the other hand, the same participant felt the night as the loss of control over reality and the following from this confrontation with 'the smallness and humbleness of myself as human. I am only a small creature that moves in the all-encompassing darkness.' In this case, the desire to have control over reality and the failure of control lead to the consequent feeling of smallness, which is a manifestation of an existential distance from the world. The point is that if one feels deep unity with the world, there can be hardly a feeling of smallness and fragility, because then the grandeur of the starry sky, for instance, makes the subject feel freedom and space, not the fragility of self. This happens when the subject concentrates not on the boundaries of the ego, but on the world around. However, if the ego boundaries are contrasted with the world, the feeling of fragility is almost unavoidable.

This dilemma might be solved if one recognizes oneself in the world, and then the infinity of the cosmos becomes somehow intrinsic to the 'thinking rid', which stops being 'rid' and becomes everything else. This existential unity is a strong factor of psychological growth and a signifier of psychological health, as it was understood in humanistic psychology.

"During the trip, I was rather emotional, and tears went down my cheeks. I felt enormously at home, back to where we all came from, to the source of wonder, the source of life. I experienced that I am a part of this planet and play there a certain role and function. It is very healing to feel embraced by this enormous whole, to be back home (made a lot of impression on me, a lot of tears and joy)" according to reports of the one of participants.

The unity between the self and the world was mentioned by four participants. It was associated with the feelings of connectedness, of being embraced by the greater whole. 'I feel myself protected under all these stars. It is so huge. It is so big that you cannot feel anything else but surrender to being overwhelmed. The night for this participant is the time to distance oneself from the pressures of the day, and to have a quiet and concentrated look at life. 'I feel connected with my surroundings': 'I was very much in contact with nature, at least the separation between me and nature was gone; now I feel neither safe nor unsafe, I don't need both; the world is quiet, and it looks as if it renews itself during the night. In the past, to stay eye-to-eye with the night meant safety. I can do everything without being seen (To dance in the forest, to make contact with elements, trees, plants, animals, and nature); the night is still a warm quiet blanket covering me; from under this blanket I can use my alertness and calmness to clean up inside and to connect with nature.'

The third relation between the self and the world was the impression of *the non-existence of the external world.* This happened in the two cases and it was linked with creativity at night. External sensations are ignored and it is the best condition for the concentration at own ideas. Thus, '...the night is the time when I am at my best. I feel inspired, I am fully awake. For me the night is the time to work things out, to produce. All other sensations cease to exist.

The same link between creativity, and the ability to ignore the external world at night is specific for another participant:

"Borders looked less important; as if they could suddenly disappear. By looking at my surroundings I could see all kinds of things: like Rorschach's surroundings, a paradise for the imagination. I felt the presence of others but less intense than during the day; I felt bigger and more powerful, as if I created this night, only for myself." (from participants reports)

This relation is clear. It is typical for states of high creativity to ignore the external world and sensations and focus on the task at hand. For some people, the night allows them to escape social time and pressure and helps to unleash imagination and creativity.

Most participants reported their perception of the night as deep, true, and authentic: 12 out of 13 participants. The night was perceived as a deep layer of reality. It was seen as the entity that nibbles away the masks and exposes the true self. It appeared as an eye-opener: it was mentioned that night raises the awareness of the dark sides of our psyche. In the existential response, the night was described as exposing illusions and revealing the depth of reality. It was experienced with awe, as the primordial source, as being encompassed by the cosmic whole. It reminds one of the similar night images by Mikhail Bulgakov, quoted in the epigraph to this text.

The Night as a Symbol

Three participants interpreted the night walk explicitly in a symbolic way. 'If I look at what night symbolizes for me, I can say that it is intuition, perception with my other senses and with my intuition, and with eyes that are different from the eyes that I use during the day.' 'For me, night symbolizes the unknown and dark side of human nature and existence. It also symbolizes peace, turning off the light, and quietness. Ambiguous: frightening, unknown, and peaceful, and stillness (…). The blackness of the night symbolizes death and danger. But it is a fairy-like world as well (…). It was also associated with mystery, and seen as 'wild, free, irrational.'

The night as a symbol has several meanings, and all of them have something in common. In one way or another, they refer to the unconscious in the sense used by Jung and Assagioli: a dynamic power that is unknown, elusive, and uncontrollable. Thus, intuition and imagination are cognitive functions primarily related to the unconscious. Perhaps due to this spiritual dimension, the night is associated with several deities, from Indian Bhairavi and Kali to Greek Nyx and Hekate.

The symbol of the night is inseparable from its counterpart, the day. This is frequently the case, as there are similar opposite pairs such as life and death, wakefulness and sleep, and yin and yang. In this polarity, consciousness is on the side of the day, and the unconscious belongs to the night. The association between night and the unconscious represents the night's connection to creativity, intuition, and imagination. What is invisible but perceived, is mysterious. The participants experienced the night as a deeper layer of reality, as a state of being in which the truth reveals itself. One can see here the connection between night and dreams on the one hand and the unconscious and inner truth/insight on the other. This polarity appears in Nietzsche's writings as opposition between the Dionysian and the Apollonian forces in arts and culture (Nietzsche, 1993). At the same time, the romantic poet Fyodor Tuytchev discovered in his observations of Nature the same Apollonian and Dionysian forces, reflected also in our inner world: "while day illumines and exposes the consciousness and encourages social interaction, night conceals and allows for freedom of the unconscious in individualized dreams and visions." (Wade, 2012)

One of Tuytchev' s masterpieces, *A Vision*, stands particularity in tune with the participants' reports, linking the Night to silence, wonderment and the unexplainable:

A Vision

There is an hour at night when all the world is silent.

Sights are seen. Miracles are done.
The living horse and chariot of creation
stampede the heavens in unbridled run.

....
Night draws in, thick Chaos heavy on the seas.
Oblivion presses on the earth, like Atlas.
Alone on the Muses virgin soul
in seer-dreams the gods inflict unease.

Tyutchev, English Translation by F. Jude (2000)

The night is often associated with sadness and melancholy as well. However, there was no such connection in the notebooks of the participants. Rather, they portrayed the night as a complement to the daily side of life.

CONCLUSION

The sublime response to nocturnal nature can have practical implications for all kinds of psychological work. The night experience can sometimes promote a change in perspective and foster self-realization, which is reminiscent of Mikhail Bulgakov's image of the night as exposing deceptions and illusions, deepening our vision of the world and of ourselves. The night seems favorable to some alternative states of mind, including awakened intuition, imagination, extrasensory perception (participant 1), feeling like a bodiless spirit or a ghost (participant 12), and experiencing the oneness with the cosmos by several participants. Night walks in beautiful natural surroundings can become an impressive meaning-giving factor. The experience of the sublime, according to Maslow and others, has therapeutic and developmental effects and enhances subjective well-being.

Without its unconscious part, the human psyche would be flat, and meaningless. As night symbolically represents the unconscious part of the human psyche and broadly appeals to the unconscious, one might apply the experience of nocturnal nature to work with problems rooted in the unconscious. The experience of existential unity, analyzed here, is healing. At the same time, it is also necessary to recognize the opposite experience of distance and separateness, of 'thinking rid' (Pascal, *Pensees* 347). It provokes self-confrontation, which is of great psychological value.

Both the entire workshop and specifically the night walk have direct applications for environmental education. Teachers can incorporate them as part of environmental education programs in their original or modified form. First of all, the night walk experience, as described here, may require considerable effort to reproduce. However, conducting workshops with shorter, easier nature walks with a similar focus on awareness, during diurnal hours, may be more suited to being a productive part of an environmental education program. They can be practiced even at nearby parks with minimal transportation time. There is a similar practice in Japan, called "forest bathing" or shinrin-yoku, with established therapeutic and wellness effects (Miyazaki, 2021).

Secondly, the night walk was only a part of a three-day workshop with a series of exercises (see Appendix). Judging from the participants' written and spoken reports, one can conclude that these nature-connection activities had a positive impact on their energy levels – a "battery-recharging effect," while also stimulating their self-awareness, and having a moderately positive effect on their subjective well-being. There were no negative effects: all participants visibly enjoyed the workshop. These activities can be applied in a broad range of nature education programs.

ACKNOWLEDGMENT

The author would like to acknowledge Prof. Dr. Jagdish Batra for his support. Also, gratitude goes to Dr. Sergio Meira, and Mrs. Victoria Meira for their love and inspiration.

REFERENCES

Adorno, Th. (2003). *Negative Dialectics*. Continuum. doi:10.4324/9780203479605

Assagioli, R. (1976). *Psychosynthesis. A Collection of Basic Writings*. Penguin books.

Baumeister, R. (1991). *Meanings of Life*. The Guilford Press.

Burton, L. M. (1981). *A critical analysis and review of the research on Outward Bound and related programs.* [Unpublished doctoral dissertation, Rutgers University, New Brunswick, NJ].

Clarkson, P. (1989). *Gestalt Counselling in Action*. SAGE Publications.

Cohen, M. (1997). *Reconnecting with Nature. Finding wellness through restoring your bond with the Earth*. Ecopress.

Devall, B., & Sessions, G. (1985). *Deep Ecology: Living As if Nature Mattered*. Gibbs M. Smith, Inc.

Dewall, B. (2021). *Living Deep Ecology: A Bioregional Journey* (S. C. Chew, Ed.). Lexington Books.

Diener, E., & Suh, E. (2000). *Culture and Subjective Well-being*. The MIT Press. doi:10.7551/mitpress/2242.001.0001

Doran, R. (2015). *The Theory of the Sublime from Longinus to Kant*. Cambridge University Press. doi:10.1017/CBO9781316182017

Driver, B., Nash, R., & Haas, G. (1987). Wilderness benefits: a state of knowledge review. In *Proceedings: national wilderness research conference: issues, state-of-knowledge, future directions*. R. C. Lucas compiler. Gen. Tech. Rep. INT-220. Ogden, UT: U.S. Dept. of Agriculture, Forest Service, Intermountain Research Station.

Dumler-Winckler, E. (2015). "Romanticism as Modern Re-Enchantment: Burke, Kant, and Emerson on Religious Taste" Journal for the History of Modern Theology / Zeitschrift für Neuere Theologiegeschichte, 22(1). doi:10.1515/znth-2015-1001

Duncan, G. *The Psychological Benefits of Wilderness*. ePsychology. http://ecopsychology.athabascau.ca/Final/duncan.htm

Elders, F. (2001). *The Sublime and the Beautiful: On Ontology and Creative Imagination*. VUB University Press.

Fagan & Shephard. (1970). *Gestalt therapy now*. Harper & Row.

Fisher, A. (2002). *Radical Ecopsychology. Psychology in the service of Life*. State University of New York Press. doi:10.1353/book4515

Frankl, V. (2000). *Man's Search For Meaning, An Introduction to Logotherapy*. Touchstone / Simon & Schuster.

Hegel, F. (1966). *Science of Logic*. George Allen and Unwin.

Hesse, H. (2017). *Demian*. Penguin Classics.

Johnstone, C. (2020). *Nature Rx*. Chesapeake Bay Program. https://www.chesapeakebay.net/news/blog/nature_rx

Kaplan, R., & Kaplan, S. (1989). *The experience of nature: a psychological perspective*. Cambridge University Press.

Kochetkova, T. (2007). *Search for Authentic Spirituality in Modern Russian Philosophy*. Edwin Mellen Press.

Kuhn, J. L. (2001). Toward an Ecological Humanistic Psychology. *Journal of Humanistic Psychology, 41*(2), 9–24. doi:10.1177/0022167801412003

Longinus. (2000). *Het Sublime (The sublime)*. Historische Uitgeverij.

Louv, R. (2005). *Last Child in the Woods. Saving Our Children from Nature-Deficit Disorder*. Algonquin Books.

Maslow, A. (1987). *Motivation and Personality*. HarperCollins Publishers.

Mathews, F. (2001). Deep Ecology. In D. Jamieson (Ed.), *A Companion to Environmental Philosophy* (pp. 218–232). Blackwell. doi:10.1002/9780470751664.ch15

May, A. (1958/2004). *Existence*. Rowman & Littelfield Publishers: Lanham.

McDonald, B., Guldin, R., & Wetherhill, R. (1985). The spirit in the wilderness: The use and opportunity of wilderness experience for spiritual development. In R. Lucas, (Ed.) *Proceedings of the National Wilderness Research Conference: Issues, state of knowledge, future directions*, (pp. 294-319). IEEE.

Mehrabian, A., & Russel, J. (1974). *An Approach to Environmental Psychology*. The MIT Press.

Miao, R. E., & Cagle, N. L. (2020). The role of gender, race, and ethnicity in environmental identity development in undergraduate student narratives. *Environmental Education Research, 26*(2), 171–188. doi:10.1080/13504622.2020.1717449

Miyazaki, Y. (2021). Walking in the Woods: Go back to nature with the Japanese way of shinrin-yoku, Hachette UK.

Naess, A. (1989). *Ecosophy, Community, Lifestyle. Outline of an Ecosophy*. Cambridge University Press. doi:10.1017/CBO9780511525599

Nietzsche, F. (1993) The Birth of Tragedy Out of the Spirit of Music, ed. Michael Tanner, transl. Shaun Whiteside, Penguin Books Ltd.

Pascal, B. (1995). *Pensees* (A. J. Krailsheimer, Trans.). Penguin Classics.

Patterson, M. E., Watson, A. E., Williams, D. R., & Roggenbuck, J. R. (1998). A hermeneutic approach to studying the nature of wilderness experiences. *Journal of Leisure Research*, *30*(4), 423–452. doi:10.1080/00222216.1998.11949842

Perls, F. S. (1972). *Gestalt Theory Verbatium*. Bantam Books.

Plato. (1966). Plato in Twelve Volumes, (trans. H. Fowler). Harvard University Press; London, William Heinemann Ltd.

Richards, R. (2001). A New Aesthetic for Environmental Awareness Chaos Theory, The Beauty of Nature, and Our Broader Humanistic Identity. *Journal of Humanistic Psychology*, *41*(2), 59–95. doi:10.1177/0022167801412006

Riebel, L. (2001). Consuming the Earth: Eating Disorders and Ecopsychology. *Journal of Humanistic Psychology*, *41*(2), 38–58. doi:10.1177/0022167801412005

Riley, M. F., & Hendee, J. C. (1999). Wilderness vision quest. Clients, motivations, and reported benefits from an urban-based program 1988-1997. In A. E. Watson, G.H Aplet & J.C., Hendee, *Personal, societal and ecological values of wilderness: Sixth Word Wilderness Congress Proceedings on Research Management and Allocation*, Bangalore, India.

Robbins, J. (2020). *How Emersion in Nature Benefits Your Health*. Yale School of Environmental Studies.

Rossman, B., & Ulehla, Z. J. (1977). ' Psychological reward values associated with wilderness use: A bifunctional-reinforcement approach.'. *Environment and Behavior*, *9*(1), 41–65. doi:10.1177/001391657791003

Russel, K. C., & Hendee, J. C. (1997). Testing Wilderness Discovery: A wilderness experience program for youth at risk in the Federal Job Corps. *Technical Report 24*, Moscow.

Ryff, C. D. (1989). Happiness is everything, or is it? Explorations on the meaning of psychological well-being. *Journal of Personality and Social Psychology*, *57*(6), 1069–1081. doi:10.1037/0022-3514.57.6.1069

Ryff, C. D. (2014). Psychological well-being revisited: Advances in the science and practice of eudaimonia. *Psychotherapy and Psychosomatics*, *83*(1), 10–28. doi:10.1159/000353263 PMID:24281296

Ryff, C. D., & Keyes, C. L. M. (1995). The structure of psychological well-being revisited. *Journal of Personality and Social Psychology*, *69*(4), 719–727. doi:10.1037/0022-3514.69.4.719 PMID:7473027

Schneider, Kirk J., Rediscovery of Awe: Splendour, Mystery and the Fluid Centre of Life. City, ST. (2004). Paragon House.

Sheldon, K. M., & Kasser, T. (2001). Goals, Congruence, and Positive Well-Being: New Empirical Support for Humanistic Theories. *Journal of Humanistic Psychology*, *41*(1), 30–50. doi:10.1177/0022167801411004

Shestov, L. (2011). *All Things Are Possible*. Girvin Press.

Thompson, C. (2005). Travel writing. In N. Roe (Ed.), *Romanticism: an Oxford guide*. Oxford University Press. doi:10.1093/oso/9780199258406.003.0035

Toth, N., Little, L., Read, J. C., Fitton, D., & Horton, M. (2013). Understanding teen attitudes towards energy consumption. *Journal of Environmental Psychology*, *34*, 36–44. https://www.sciencedirect.com/science/article/pii/S0272494412000722. doi:10.1016/j.jenvp.2012.12.001

Tov & Diener. (2013), *Subjective Well-Being*. Research Collection School of Social Sciences. https://ink.library.smu.edu.sg/soss_research/1395/ Archived 2020-12-02 at the Wayback Machine

Tyutchev, F. (2000). *The Complete Poems of Tyutchev in An English Translation by F. Jude*. The University of Durham.

Veenhoven, R. (2000). The Four Qualities of Life. *Journal of Happiness Studies*, *1*(1), 1–39. doi:10.1023/A:1010072010360

Vogler, J. (2012). *Self-Actualization and Peak Experiences in Outdoor Recreation*. TigerPrints. https://tigerprints.clemson.edu/all_theses/1485

Wade, C. (2012). *The Harmony of the Abyss: Translating Tiutchev and Mandelstam*. Wesleyan University.

White, D., & Hendee, J. (2000) Primal hypotheses: The relationship between naturalness, solitude and the wilderness experience benefits of the development of self (DOS), development of community (DOC), and spiritual development (SD). In McCool, S., Cole, D., O'Loughlin, J. *Wilderness science in a time of change conference—Volume 3: Wilderness as a place for scientific inquiry*. Proceedings RMRS-P-15-VOL-3. Ogden, UT: U.S. Department of Agriculture, Forest Service, Rocky Mountain Research Station.

Wicks, C., Barton, J., Orbell, S., & Andrews, L. (2022, August). Psychological benefits of outdoor physical activity in natural versus urban environments: A systematic review and meta-analysis of experimental studies. *Applied Psychology. Health and Well-Being*, *14*(3), 1037–1061. doi:10.1111/aphw.12353 PMID:35259287

Young, R., & Crandall, R. (1984). Wilderness use and self-actualization. *Journal of Leisure Research*, *16*(2), 149–160. doi:10.1080/00222216.1984.11969582

Yunt, J. D. (2001). Jung's Contribution to Ecological Psychology. *Journal of Humanistic Psychology*, *41*(2), 96–121. doi:10.1177/0022167801412007

Zimmerman, M. E. (1994). *Contesting Earth's future: Radical ecology and postmodernity*. University of California Press. doi:10.1525/9780520919228

ENDNOTES

1. See https://www.eifel.info/en/nature/unesco-volcanic-eifel-geopark
2. Ibid, https://www.eifel.info/en/nature/unesco-volcanic-eifel-geopark

APPENDIX: THE WORKSHOP'S ACTIVITIES

This appendix contains a description of the activities carried out during the workshop. Activities designed by Cohen (1997) are marked with an asterisk; the ones left unmarked were designed by the author.
First day: Arrival, Activity 1 (author's original practice of perceptual experiences of nature), and then three of Cohen's activities.
Activity 1. Experiences of the body with nature: developing sensory awareness
We stop mental dialogue by concentrating totally on the body itself and the way the body contacts the environment — the border between the body and the environment. For instance, the feeling of bare feet on the grass (if the weather is sufficiently warm), of the wind on the face and hands, of sunrays on the face, of the smell of the air, and the difference between how we breathe the city air and the air in the forest. How does breathing in the forest differ from breathing in the city?

"Most of us have lost large areas of sensory awareness and proprioceptive sensitivity in our bodies. These losses to our awareness often represent solutions to conflicts, traumas, or deprivations. In the counseling some of these difficulties can be brought into awareness, (...), and in this way, the missing parts of the person as a body-mind can begin to be reclaimed". (Clarkson, 1989: 82-83)

"Concentrate on your 'body' sensation as a whole. Let your attention wander through every part of your body. How much of yourself can you feel? To what degree and with what accuracy and clarity does your body — and thus you — exist? Notice pains, aches, and twinges ordinarily ignored. What muscular tensions can you feel?

Attending to them, permit them to continue, and do not attempt prematurely to relax them. Try to shape their precise limits. Notice your skip sensations. Can you feel your body as a whole? Can you feel your head-to-torso position? Where are your genitals? Where is your chest? Your limbs?" (Pearls et al., 1951/1969: 86), (quoted after Clarkson, 1989: 83)

The second step is concentrating on the borders between the body and the environment — on skin sensations of the ground, the grass, the surfaces of the trees, the temperature, the wind, the sun on one's face, or drops of rain. How does your nose feel when touching flowers, how do your hands feel when touching trees or stones?

The third step is to feel the body as a whole in the immediate surroundings: the simultaneous awareness of the body, the place where it is, and how it interacts. For instance, what is a body-in-the-forest experience? What is a body-as-a-whole-in-the-mountain feeling?

Afterwards, the participants were invited to do three following activities from Cohen (1997):

External Sensations

Experiencing the present or Awareness Continuum

This exercise (designed by the author) should be done in the natural environment. It could be done either by individual participants silently or verbally expressed in a group. If it is done in a group, the

Table 1. Following activities from Cohen (1997)

#	Time	In group or alone	Brief description
1	30 min.	Alone	The subjects will try to remember if they ever had a pleasant experience with nature and write it down. They will think about what sensations made the experience attractive. They will try to decide if they were ever taught to have worthwhile sensations in experiences with nature. (Cohen, 1997)
2	1/2 to 1 hour	Alone	The subjects go to a natural area and enjoy its beauty in silence. During these ten minutes, they will try to feel the natural area and themselves without using words or concepts. (Cohen, 1997) Afterward, they write down all important feelings and perceptions.
3	1 hour	In pairs	Two participants will go together to a natural area. Without talking (i.e. no language), one of them will close his/her eyes (no sight). The other person will become a nature guide, holding the other's hand and supporting him or her in any necessary way. The guide will request the other participant to imagine that s/he has just walked out of a spaceship into the unknown. For five minutes the second participant is led to sense, touch, feel, taste, smell, roll in, crawl on the ground, hug surrounding objects and occasionally see glimpses of the different natural parts of this area. After doing this activity for five minutes the participants switch their roles. (Cohen, 1997) The goal of the exercise is to activate sensory awareness of nature. *

only action allowed by the exercise is that of communication of the contents of awareness. (Fagan & Shephard, 1970: 85.)

The purpose of the exercise is to stop the 'autopilot' and to connect to nature with full awareness. This is because experiences of nature are also frequently automatic — we give labels to sensory perceptions like "birds singing", "wind blowing" or "green grass" and then stop paying attention to them. But in reality, birds may sing in thousands of different ways, the color of grass may be various types, and every particular sensation is unique. Yet it can be perceived only if we give our full attention to it.

"Practice of the awareness continuum is a kind of training in applied phenomenology, an attempt to concentrate without judgment or labeling on every new figure which becomes of interest without preconception or expectation. Frequently human beings engage habitually in behaviors, attitudes, and feelings stripped of conscious awareness. People eat, make love, go to work, etc. However, in all important ways, they are unconscious or unaware of their here-and-now experiences, their recall is barren and their attention is everywhere else but on the vividness of their life's every passing moment. The awareness continuum is particularly designed to heal this lack of mindfulness and to restore richness and vitality to living. Practicing the awareness cycle means excluding nothing, — wishes, thoughts, bodily feelings, sensations from the environment, temperature changes, and voluntary and involuntary actions all before judging labeling, or categorizing. It is very similar to the Zen practice of 'mindfulness' and the way to get into this technique is to start verbalizing without exception to every experience 'Now I am aware....'. (Clarkson, 1989: 79)

Auditory/visual Registration

"In this practice, we are sitting in the garden, relaxed and listening to the sound of birds. Then, after they have gone, we close our eyes and try to again hear their song (...). To people living in the midst of

Table 2. Second day: **:** *7 of Cohen's exercises and then the external sensation of nature*

#	Time	Alone / group	Brief description of the exercise
1	30 min.	In pairs	This activity is directed at exercising the sense of touch. The participant selects six sticks or rocks, shuts his eyes, and then picks one of them, marks it with a pencil, and then tries to differentiate it from other sticks, using his sense of touch *. Afterward, the participants were asked to write down the most important things they learned from this activity. (Cohen, 1997). *
2	30 min.	alone	This activity concentrates on the sense of breathing. Participants are invited to select a green plant, and stop breathing consciously, to notice the feeling of suffocation, then to start breathing again, while embracing the part of the plant that produced oxygen. They will then release the plant, stop breathing, and then repeat the whole process. They will try the same thing holding a deodorant and discuss whether it feels the same as the plant. (Cohen, 1997) *
3	20 min.	alone	"Unity with nature". Participants think of people they like and find out which qualities they like in them, and then look at the natural area to see if it has similar qualities. *(Cohen, 1997)
4	20 min.	alone or in a group.	"Separation from nature". Participants wrap a towel or a sweater around one of their hands, shut their eyes, and touch gently natural things with both bare and wrapped hands. They repeat this procedure with open eyes and compare the sensations from both hands. * (Cohen, 1997)
5	1 hour	1st part alone and 2nd part with a group	Part one: Participants go to an attractive natural object (plant, animal, mineral) and think and write down what they like about this natural attraction and why. * Part two: Participants substitute their names for the names of the natural objects in Part 1. For example: "I like the tree because it is beautiful and strong", becomes "I like myself because I am beautiful and strong". They then try to evaluate whether these judgments feel right. * (Cohen, 1997) The same activity can be done with the inner nature of other people instead of natural objects. The participants were asked what effect this activity had on their sense of self-worth.
6	1 hour	Alone	The activity is directed at activating the old mammalian brain for sensing nature. It comprises feeling one's senses without using language labels for them and going to a natural area to feel it in a non-verbal way, without pronouncing the names of the sensations in the mind. * (Cohen, 1997)
7	1 hour	Alone	The goal of this activity is to express the previous non-verbal perception in words, and by doing this to break with the standards and clichés, but to learn to be attentive to one's sensations first of all. The experience shall be written down afterward. * (Cohen, 1997)

the city, this auditory recall might be particularly helpful, bringing back the sense of, and the contact with, living nature". (Assagioli: 155)

"The purpose, the procedure, and the effects are parallel to those of mental photography. It is possible in this way to re-integrate and re-evoke some of the attendant emotions which one experiences while actually listening to the birds" (Assagioli: 155).

Table 3. Third day: Five of Cohen's exercises and emotional perception of nature

#	Time	Alone/In group	Brief description of the exercise
1	1 hour	Alone	"Natural Attractions". This exercise requires remembering one's problems or disturbing experiences and keeping them in mind, giving full attention to the natural area and sensing it as brightly as possible. Finally, participants have to verbalize the experience. One of the possible effects is that the sensory experience of nature with a problem in mind will stimulate participants to take a new perspective on the problem or even to find out how to solve it. * (Cohen, 1997)
2	1 hour	Alone	This exercise consists of repeating the previous three activities one after another and noticing that they have comforting and joyful effects. * (Cohen, 1997)
3	15 min.	Alone	"Respecting natural attractions". "Go to an attractive natural area, go to a few objects or aspects of nature one at a time, and hold each object for one minute, but make sure that you leave it as you found it. Note what thoughts and feelings come to mind during the minute you are in this balanced relationship with each of the natural objects." * (Cohen 1997, p.180)
4	30 min.	Alone	"Nature as a Guide." (Cohen) Participants go to a natural area and follow the natural attractions spontaneously. For 15 minutes or more, they walk through this natural area while continuously following the immediate natural attractions. *
5	30 min.	In pairs	"Sensing like nature." (Cohen, 1997). This exercise requires going to an attractive natural area, switching off verbal thinking for a short while, walking through this attractive natural area for 15 minutes, and setting free one's imagination concerning all surrounding natural objects. Participants then try to feel which inner thoughts and feelings are evoked by the sensations of nature. The exercise is directed at activating the participants' unconscious through sensing nature. *

Emotional Perception of Nature: The Final Exercise (Author's Original Design)

Gestalt Identification and Dialogue

It is a common event in gestalt training that a person identifies with either a dream character or with another person or object. "This involves the client experientially identifying with an image, an aspect of herself, a dream component, a feeling or even another person, to evoke increased awareness of the topic at hand" (Whitmore, 1991: 56).

This technique was applied to connect with selected natural objects which one finds attractive. They can be trees, dunes, sea, flowers, birds, squirrels, or anything else. The participants will choose a natural object that attracts them and imagine themselves being this object. They will verbalize how it feels to be this object.

The justification of this exercise is that, by identifying with natural objects, participants can find out the feelings, thoughts, or any other mental events that were previously hidden from their actual awareness. It is an exercise in self-knowledge.

Chapter 2
Enhancing Ecological Awareness as Distance Awareness

Rico Sneller
Academy Depth Psychology, Leiden University, The Netherlands

ABSTRACT

This chapter focuses on the viability of distance awareness as an indispensable ingredient of form awareness and, subsequently, environmental education. It studies relevant motifs in the work of two post-Romantic German thinkers, Ludwig Klages (1872-1956) and Hermann Friedmann (1873-1957). The main argument comprises two areas of concern: 1) enhancing crisis awareness and reintegration, and 2) promoting poetry education. As a common thread running through these areas, consciousness as distance awareness imposes itself. Distance should not be taken as a quantity but rather as an original quality that justifies itself.

INTRODUCTION

More than ever before, human beings tend to prefer proximity over distance. Is it not so that what is close at hand can be reached, grasped, if not understood and, therefore, *possessed*? 'Concepts' are objects possessed, since conceived, by the mind, and their conception is made possible by our previous intellectual grasp. When we think we understand something, we first bring it within our reach.

I think the current ecological crisis is partly due to the self-centred essence of what we believe to be 'understanding'. Is not nearness an implicit assumption of 'understanding,' and does not such nearness secretly presuppose an all-comprehensive self (a 'subject') as its control centre? Might it not be requisite to consider forms of reality awareness which rest on distance just as much as on proximity?

In this article, I will focus on the viability of distance awareness as an indispensable ingredient of ecological awareness. I will first study some motifs in the work of two post-Romantic German thinkers, Ludwig Klages (1872-1956) and Hermann Friedmann (1873-1957). Being in many respects predecessors to the 20th century deep ecologist Arne Naess, they both highlight a tendency in the Greco-European tradition to privilege tact and tangibility over seeing, vision, or contemplation; in other words, distance awareness. Whereas Klages aimed at resituating physical touch on the broader field of contemplation,

DOI: 10.4018/979-8-3693-2577-3.ch002

Friedmann argued that scientific 'hapticism' should ultimately be transformed into an optic approach. What these thinkers have in common is a recurrent insistence on distance as a legitimate constituent of human consciousness. While science often reduces the latter's content to mere 'evidence' or 'sense data,' Klages and Friedmann claim that consciousness is permeated by a distance awareness which excludes any alleged 'givenness' (i.e., availability) of data. What seems to be 'given' to the senses is immediately withdrawn. The horizons of vision and contemplation absorb any attempt to retain and only provide a view of *forms*. I will try to make this more concrete toward the end of this chapter.

My thesis will be that distance awareness is crucial to environmental education. Other than the word 'environment' suggests, our environment is more than just what surrounds us – it equally decentres us and stimulates anything *centrifugal* in our minds. While always already there, this centrifugality, or distance awareness, should be actively promoted in education. This can be done in a variety of ways. My proposal comprises two areas of concern: 1) enhancing crisis awareness and reintegration (or what the Polish psychiatrist Kazimierz Dabrowski calls 'personality shaping through positive disintegration'), and 2) promoting poetry education. As a common thread running through these areas, consciousness as distance awareness imposes itself. Distance should not be taken as a quantity but rather as an original quality that justifies itself.

Distance as a Way of Seeing Things

Struggling with scientific positivism, early 20th-century philosophy tried to reconsider our standard approach to 'reality'. In various ways, it insisted that our senses differ when giving a picture of being: seeing, for example, might penetrate deeper than touching. Moreover, some philosophers even claimed that human sensibility does not exhaust consciousness; contemplation, for example, allows for a deeper grasp of reality than mere sight or hearing. Phenomenology, with its principle of intentionality, was one (albeit the most influential) of the many philosophical reassessments of the constitution of what is called 'reality'. In his *Time and Free Will: An Essay on the Immediate Data of Consciousness*, the French philosopher Henri Bergson (1859-1941) argued for various states of consciousness and the distinction between outward and inward perception of things. The German psychologist-philosopher Theodor Lipps (1851-1914) insisted on inner experience and an empathic approach, without which reality remains inaccessible. (Lipps, 1883)[12] To mention a third example, the German psychiatrist-philosopher Karl Jaspers (1883-1969) studied the variety of mindsets which can be distinguished *behind* religious or philosophical systems (objectifying, self-reflective, and enthusiastic). (Jaspers, 1919)

In the context of this chapter, I will resort to two influential thinkers who reconsider how we build reality upon perception: Ludwig Klages and Hermann Friedmann. Both were contemporaneous, and both have been largely forgotten today. The difficulty of their writings is probably one of the reasons for this oblivion. Yet, I believe that motifs in their work can help us reflect on the nature of the environment and the urgency of creating an 'understanding' of it through education.

Contemplation

Let us start with Ludwig Klages, an anti-rationalist philosopher of life. The main distinction which goes through Ludwig Klages' work is one between *spirit* and *soul*. In what follows, I will draw on Klages' Der Geist als Widersacher der Seele ('The Spirit as Adversary of the Soul'), published between 1929 and 1932.

'Spirit' designates an antagonistic, life-hostile force, usually considered under the more respectable names of 'reason' or 'will'. Klages laments that this antagonistic force has left nothing unscathed by its destructivity throughout the history of Western philosophy and religion. It mainly interrupts vital continuities (life, experience, feeling, instinct, etc.). Our rationalised world, with its languishing flora and fauna, polluted environment, etc., is the product of spirit.

'Soul' is the term Klages reserves for life itself, its vital continuities and connections *in* or *by* which we live. 'Soul' expresses the unbreakable bond between living organisms and life; the latter is experienced by the former simply by *being* undergone or 'lived'. Instincts are the links through which life affects living beings. This affection is pre-reflective, pre-rational, and pre-volitional. Reflection or will always come too late.

What is more, they retard. Their retardation can only harm life and transform it into artificial constructs: concepts, systems, units, goals, etc. Rationally conceived concepts, for example, interrupt the lived experience from which they have been taken; willed, goal-oriented actions, for another example, interrupt smooth, continuous agency (the essence of which is passivity, 'drivenness,' rather than activity or wilfulness).

How does Klages interpret the human being as part and parcel of life itself? Before its molestation by spirit, he contends, the human being consists of an original polarity. This polarity is represented by what we term 'body' and 'soul'. 'Body,' Klages argues, is just the name for the capacity to have close experience or to sense (*empfinden*). The body is our organ of proximity, and its primary focus is on touching what is near. The alleged 'presence,' or the here-and-now as the primary unity of space and time, thrives on it: without a body, we would lack immediate spatiotemporal orientation. True, the corporeal preference for tangibility cannot fail to allow immediately for the subsequent disruptive intervention of the spirit, which isolates and objectifies and creates singular 'things'. However, tangibility preference in itself need not pose a problem as long as it is counterbalanced by the other polar orientation: soul and contemplation. Whereas the body's original 'myopia' is entailed by its grasping the proximate, the soul, in turn, allows for distance awareness; the 'soul,' that is, life as the original unity of (1) the living environment and (2) the living human's instinctive reaching out for it. The human being has an inbuilt susceptibility to distance, and this susceptibility is actualised by 'contemplation' (*schauen*). Without contemplation, a living being would be groping in the dark like an animal. Animals do not wholly lack contemplation, Klages notes, but they are poor in it and primarily turned towards close, physical experience (*Empfindung*).[3]

Contemplation deepens immediate sense perception, Klages continues, and even precedes it. It allows the senses to differentiate into feeling, tasting, smelling, hearing, and seeing. Not just this; it equally deepens pleasure and pain, allowing the latter to transform into (what we are used to call) 'happiness,' the former into (what is commonly termed as) 'suffering'. Contemplation marks progress in the global development of life. Whereas the animal kingdom is governed by close experience (*Herrschaft des Empfindens*), rising humanity is awakened by distance awareness and contemplation.

Although an in-depth study of Klages' conception of contemplation (*schauen, Schauung*) would prove highly fertile for ecological responsibility awareness, it runs the risk of leading us too far beyond the purpose of this chapter. Yet, some details are interesting enough to be mentioned here. For example, Klages has an enigmatic view of the *content* of contemplation. Actual being, Klages claims – i.e., being which is contemplated rather than conceived – consists of *images*. Close experience and touching only lead to a tangible world of things and objects (eventually, *conceived*, objecti-*fied* objects, i.e., thoughts or ideas). It is the world with which we are all too familiar, in the light of both 'science' and 'common sense' (these remarkably hardly differ in this respect). Actual being or reality, on the contrary, transcends the object world and becomes manifest in the form of unfathomable *images*. What does Klages mean by this? These images, he explains, rather than still being objects at another, more ethereal level, consist of concatenated *similarities* (*Ähnlichkeiten*). I will try to explain this with an example. Looking at a landscape, one may distinguish objects: sun, mountains, sea, trees, etc. Instead, one may equally be struck by connecting 'threads' running through it, e.g., the moist (in sea, clouds, flood, mud), the rough (rocks, wind), the steep (mountain tops, cliffs), the embarrassing or astounding (solar or lunar eclipses, animal cruelty, sunset), the sublime (vastness of ocean or desert), the liquid, the solid, etc. What our grammar merely qualifies as 'adjectives,' subordinated to the alleged grammatical primacy of 'nouns' or 'substantives,' may have 'ontological' priority throughout. Steepness, for instance, is *before* the mountain top, roughness or flatness precedes the sea, and gloominess is the original dimension in which a forest becomes first manifest. In the light of the topic of this chapter, it is interesting to note that original images not only precede the object world (which relies on objecti-*fication*); they contain an inbuilt distance. Image contemplation comes down to distance awareness. Any environmental concern which is not contemplative is wrong-headed.

Still, close experience or tangible sensing (*empfinden*) matters as well. While the Greco-European world suffers from having overestimated the weight of tangibility and is lost in its objectification, the Chinese tradition (Klages explicitly mentions Daoism), on the contrary, lacks the dimension of tangible proximity and favours pure contemplation. As a result, Klages states, it is stuck in a world of atrophied bodies. Its neglect of tangibility led to a reification of chimerical appearances, in other words, a sheer image world.[4] True, Klages argues (1981),

"form [Gestalt] is no original property of an object, but it is," he continues, "the sole secondary property without whose co-presence the original property would sink in the abyss of sheer possibilities."

In other words, without tangible form, without any delineation, things would remain purely virtual.

"In so far one could agree with Aristotle," he goes on, "had he but acknowledged the independence of matter, rather than pretending to exhaust its essence by determining it as a sheer condition of possibility for form actualisation". (p. 963)

By (1) interpreting entities as a combination of form and matter and (2) taking both 'form' *and* 'matter' as purely intelligible notions, Aristotle failed to do justice to the fundamental alterity matter expresses. Matter's *tangibility*, while enabling the spirit (thinking) to dissolve it into separate entities, nevertheless represents the inexhaustible resistance reality offers to our apprehending or comprehending grasp. Our touching does not, in the least, alter nature's final impenetrability.

In sum, Klages' thinking addresses two issues worth considering: 1) the idea of natural phenomena's inbuilt distance, as appears from contemplation, and 2) the alterity or otherness of matter, as becomes manifest in the resistance it inexorably opposes to our grasp.

Optics

We have seen that Klages rejects the spirit (i.e., reason or will) as it dissolves original experience and destroys life. The implication of this rejection is far-reaching: science is essentially a product of the spirit and is likely to be rejected. Klages' contemporary Hermann Friedmann shares many views with Klages, except for rejecting science.

"Contrary to Klages," Friedmann writes in his autobiography (1950), "I do not consider the 'spirit as the adversary of life'. The spirit – a part of life – is the vigilant, protective, restorative friend of life. When life errs or goes wrong, the spirit will arise and rectify: it confronts the 'time' with its 'concepts'. However, where life – which equally comprehends the everyday spiritual life – follows the right way, even if it is dark, the conscious, scientific spirit submits itself voluntarily to the general course." (p. 220, my trans.).

Despite the complete oblivion into which Friedmann's work has fallen – and unjustly so – its shared concern with Klages's opens up new perspectives on ecological awareness as distance awareness. It is striking that already in 1930 when an extended reprint of his main work *Die Welt der Formen* ('The World of Forms') appeared, Friedmann offered a thorough reading of Klages' notions of 'image' and 'similarity;' Friedmann must have studied Klages' *Der Geist als Widersacher der Seele* immediately after the publishing of its volumes, roughly at the same time.

Even before Husserl wrote his critique of the European sciences – their failure to provide meaning and orientation to modern people (Husserl, 1962) – Friedmann expressed himself similarly. His remedy, however, without necessarily opposing Husserl's, seems much more concrete. In line with Klages' notion of contemplation, Friedmann argues for an *optic* transformation of the prevailing *haptic* approach. The haptic approach, he claims, is governed by touch, whereas the optical approach is characterised by seeing. Let us take a closer look.

Friedmann is certainly not unique in globally assessing European philosophy's history in negative terms. His diagnosis, however, could prove useful for ecological awareness enhancement, especially since this diagnosis, rather than focusing on isolated flaws, more generally addresses the way consciousness is resorted to. European history, Friedmann claims, demonstrates a rise and fall of *form awareness*. Whereas Greek culture subsumed human affect life under the banner of an alleged universal symmetry, and medieval Christianity attempted to reconcile Indo-Germanic and Christian beliefs in higher form constructions (cf. Thomas Aquinas), Modernity presents a collapse of higher form awareness. It is telling, Friedmann argues (1930), that the birth hour of modern science coincides with baroque ("will to subjectivity") and rococo ("capricious decoration of the interior"). (p. 15)[5] Reality splits into two parts, studied by science and art, respectively. "'Science and art' become cultural correlatives [*das geistige Begriffenpaar*]" and "an epoch of arts succeeds the epoch of form". (ibid., p. 14; my trans) As a result, Friedmann continues, moral sensitivity diminishes, and inner tensions aggravate. "Consequently, the ethos of the majority was not stimulated, while its pathos was accumulated in the meantime." (ibid., p. 16; my trans.) Other than Klages, Friedmann believes that this gradual development was not fated but voluntary; it rested on free choice. Therefore, he adds, it entails guilt.

"An essentially goal-oriented sensuous-cultural worldview is guilty when it does not simultaneously intend to provide intellectual, moral and aesthetic compensation for its purely goal-oriented mindset; it is guilty and contradictory, since without such compensation, it will ultimately destroy itself." (ibid., p. 25; my trans.)

What is it that entails the gradual splitting of reality? Friedmann's diagnosis resembles Klages's to the extent that it points to an increasing emphasis on tangibility at the cost of vision.[6]

"The mechanically conceived world is brought about when we interpose our inner experiences reliant on tangibility as objective relations between elements, planets, atoms, and electrons. When summoned to give a mechanical explanation of everything, we are equally summoned to assess all global relations as tangible and exclusively so." (ibid., p. 17; my trans.)

Friedmann calls these touch-based relations *haptic*, as opposed to vision-based, optic relations. Whereas haptic relations are predominant in the sciences and only lead to fragmentation, optic relations compensate for such fragmentation. They tend to integrate insofar as they rely on perception of form (*Formwahrnemung*), the experience of coexistence (*Koexistenzerlebnis*), and the particular faculty to integrate available single perceptions and to interpolate fragmented, incomplete perceptions (*das besondere Vermögen der Integration vorhandener Einzelwahrnehmungen und der Interpolation lückenhafter Wahrnehmungsinhalte*). (ibid., p. 27; my trans.)

The notion of 'form' is crucial here, as a form is the optic approach's counterpart. It is vital for my thesis in this chapter (ecological awareness entails distance awareness). True, philosophical *key* notions are often also notoriously obscure. But still, the obscurity of 'form' need not create an impasse. For it might be that, to understand and do justice to form, we would have to reconsider the *mindset* from which we try to make sense of it. What is 'form,' how does it warrant distance, and ultimately, how do we resort to form awareness to assure ecological awareness? These are questions that I will now address.

"Life," Friedmann contends, "searches form everywhere." "Where there is life, there is form." "Reversely, form is always struggling towards life." (ibid., p. 29; my trans.) A form is a remote completeness accessible to vision yet out of our reach and, therefore, our control. In Friedmann, a major paradigm for forms are living organisms and their study in morphology. Living organisms never fail to be susceptible to (haptic) physiological analysis and yet visibly recede from reduction to its tangible constituents. In a way, touching bodies alienates us from their form, whereas seeing their form at a distance enables us to access it.

Another major paradigm for form awareness in Friedmann is art. Its dynamic resembles biological morphology. Whereas a (haptic) analysis of paint brush techniques and canvas structure does not in the least reveal anything as to the painting's true causality, Friedmann argues, the "rectitude of the perspective," "the harmony of the colours," and "the (con)sequence of its group relationships" are far more important. These "obey another, optic causality, which is in another domain than the logic of tangibility". (ibid., p. 31; my trans.)

Generally speaking, whereas a haptic approach is governed by the will to conquer, an optic approach allows the form to act on the seer and be interiorised. While a haptic approach is, paradoxically, ultimately concerned with *self*-assurance (i.e. the assurance of a self *facing*, or *faced* with, the object), an optic approach is *only* concerned with the form seen.

Friedmann's proposal to extend biological and aesthetic morphology to being at large and to approach the latter optically has exciting implications. It makes the world plastic, i.e., changeable on higher levels. The world's plasticity would create a rupture with the world's alleged 'determination' by 'laws' of nature, to which we are used (as appears from the 'unacceptable' difficulties quantum mechanics confronts us with). It entails overcoming strict subject-object opposition, favouring an ontological mutuality of form and vision. The former (sic) forms (moulds, harmonises) the latter, yet not causally ('causality' is haptic).

Figure 1. Painting by German painter Caspar David Friedrich

Another kind of causality seems to be at stake, transcending ordinary causality. It is paradoxical, once more, that this optic causality makes a far *more profound* impact than the ordinary, albeit at a distance.

Distance Awareness in Education

Let us now return to my central thesis in this chapter, i.e., that distance awareness is crucial to environmental education. Each in their own way, Klages and Friedmann made a case for such distance or form awareness, and I consider their account valuable for my purpose here. How exactly? Whereas the word 'environment' seems to imply the circular motion of 'surrounding,' this surrounding *need not be centripetal*, as one might hastily conclude from a predominantly ego-centred perspective. Once we conceive of the environment as primarily *centrifugal* – and there is no reason not to do this – we might become sensitive to our environment's inbuilt distance. The more all-comprehensive our environment is, the less it can be comprehended by our mind.

Several paintings of the 19th century German painter Caspar David Friedrich illustrate the receding sense of humanity, in favour of dawning horizons and always broader perspectives witnessed by the spectators. Instead of focusing on, e.g., the woman as the alleged centre of Friedrich's painting *Woman In The Setting Sun*, one could equally emphasise the openness and vastness of the space surrounding her. One might even argue that the 'sun' of the focalising (centripetal) human gaze itself is setting.

Could it be that our schooling system, insofar as it aims to teach students to understand, equally prepares them for environmental unconcern or indifference? If this is true, the question will be: How do we re-engage education in distance awareness? Obviously, many answers can be given to this question. For the sake of my argument, I cannot pretend to be all-comprehensive in my account. As said at the beginning of this chapter, my proposal consists of two main areas of attention: 1) the enhancement of crisis awareness and reintegration and 2) the promotion of poetry education. Consciousness as distance awareness imposes itself in these two fields, albeit in different ways. Distance should not be taken as a quantity but as an original, self-explanatory quality. Klages and Friedmann have argued that distance awareness amounts to *form* awareness. 'Form' equals *visible* distance.

Crisis Awareness and Reintegration

As a rule, psychological crises are assessed negatively in many Western societies. People suffering from mental problems should recover as soon as possible, if need be, by psychological or psychiatric treatment. Without denying the despair that mental issues can bring to people, it should not be neglected that any form of psychopathology reflects a particular societal structure and value system. We do not even have to resort to ethno-psychiatry to show this (Kleinman, 1987; also see Sass, 2017); in *spatial* terms, cultural anthropology only aggravates the predicament of diagnostics today. Already ongoing developments in academic psychology and psychiatry (such as the DSM sequence) make clear that approaches to pathology types also vary over time.

Several contemporary philosophers and psychologists have suggested that mental pathologies may have *existential* dimensions. Karl Jaspers, psychiatrist and philosopher, coined the prolific term *Existenzerhellung* ('clarification of existence'). Analysts like Alfred Adler, Erich Fromm, Victor Frankl, Kazimierz Dabrowski, Roberto Assagioli, Carl Gustav Jung, and Ludwig Binswanger, to mention a few, drew on this notion. In his world-famous *Man's Search for Meaning*, Frankl writes (2006):

"We should not, then, be hesitant about challenging man with a potential meaning for him to fulfil. It is only thus that we evoke his will to meaning from its state of latency. I consider it a dangerous misconception of mental hygiene to assume that what man needs in the first place is equilibrium or, as it is called in biology, 'homeostasis,' i.e., a tensionless state. What man actually needs is not a tensionless state but rather the striving and struggling for a worthwhile goal, a freely chosen task. What he needs is not the discharge of a tension at any cost but the call of a potential meaning waiting to be fulfilled by him." (p. 105)

Dabrowski (2015) even highlights the positive value of what he terms 'disintegration'.

"In order to educate himself," Dabrowski states, "a man should, as it were, split himself into a subject and an object – that is, he should disintegrate. He must be the one who educates and the one who is educated and he must isolate in himself the active entity and the one which is subordinated to it." (p. 37)

Dabrowski is referring explicitly to crisis experiences or inner conflicts, the outcome of which might transform humans on several levels: instincts, feelings, will, intellect, and even spiritual life. To take the *intellect* as an example: crisis and disintegration could stimulate our thinking towards becoming more creative and original, and less rigid. It might become more mature and susceptible to integration in the personality as a whole. Disintegration, then, may turn out to be incipient integration on a higher level.

Again, the potentially existential meaning of psychic suffering does not necessarily minimise the latter, and medical treatment may be inevitable. Yet, these new perspectives on mental crises could well transform over-medicalised views of psychic suffering. Quoting one of his schizophrenia patients, the American philosopher-psychiatrist Louis Sass (2017) writes that "[o]nly those who have lost reality and lived for years in the Land of cruel, inhuman Enlightenment, can truly taste the joy in living and prize the transcendent significance of being a part of humanity." (p. 294)

Let me add one hypothesis to those offered by the thinkers mentioned above about the existential dimension of this suffering. This hypothesis could elucidate the relationship between psychopathology, distance awareness, and ecological concern. I would argue that insofar as its existential dimensions are fathomed, psychic suffering could be assuaged as this 'fathoming' itself amounts to enhanced distance awareness.[7]

"Everywhere where the human being is confronted with 'gestalt' [*Gestalt*: form, shape], 'mind' [*Gemüt*] arises in the human being's inwardness," writes Hermann Friedmann (1956). (p. 17) However, such a confrontation might be too intense. For, Friedmann continues,

"[i]t could also happen that the earthly human being [der irdische Mensch], upon approaching the supernatural's image [dem Bilde des Überirdischen] too closely, perishes from the pathos of his own indignity." (ibid., p. 28)

How should we interpret this? From what follows, we may infer that Friedmann thinks of those who confuse the mirroring of their mind (*Gemüt*) with the "prolific gestalt" (*spendende Gestalt*) from which this mind draws. Those profoundly introspective authors (Augustin, Rousseau, Goethe) or artists (Dürer, Rubens) run this risk. Friedmann states that it would be sheer madness to uncover the so-called 'repressions'. "'Repression'," he argues, "is perhaps the most vital achievement of natural healing power; whereas 'bringing to consciousness' is perhaps the most life-threatening danger." (ibid., p. 45)

Friedmann may seem to exaggerate in his critique of uncovering repression. However, I would agree with him if his critique regards the *identification* of form/gestalt, rather than its contemplation and our ensuing inner transformation. Could it be that psycho*pathology*, i.e., the suffering in it, consists of the perilous attempt to *pervade* distance and *appropriate* it? Could it be that psychic healing, rather than re-covering the uncovered, consists of maintaining and preserving distance and *keeping* contemplated forms at a distance?

If this is true, one may identify at least two practical implications. Firstly, education systems (schools, universities) should not primarily regard their pupils and students suffering from mental 'issues' as obstacles to be removed. First and foremost, psychic crises are perhaps indicative of rising distance awareness, the concomitant suffering being the result both of preceding repression and renewed confrontation. Secondly, I believe that *capable* students who are liable to "positive disintegration" (Dabrowski) and susceptible to forms should not only be encouraged to function as ecological cheerleaders. It would be wise to equally save them from over-enthusiast endeavours *into* the distant realm of forms.[8] Educational institutions should monitor psychic suffering and psychopathology amongst their students as warning signals of their own increasing myopia.

Poetry Education

The second means of ecological awareness enhancement in education could be found in poetry. The 19th-century German philosopher Carl du Prel (1833-1899) not only insists on poetry's virtue to strengthen memory and fantasy, but he goes as far as qualifying poetry in terms of "paleontological worldview". The poem renders primordial (ecological) ways of viewing reality. Its poetic character, however, relies on the poet's self-surrender to their object. Poetry thrives on the transformation of subject by object. Let us briefly follow Du Prel's argument and see how poetry can strengthen ecological concern, particularly how poetry can enhance susceptibility to distance; not necessarily poetry of nature but any form of poetry.

Interestingly, the primordiality of poetry reflects a primordiality of vision. It reminds us of Klages and Friedmann, who connected viewing contemplation to distance awareness. Drawing on the linguistic researcher Lazarus Geiger, Du Prel argues that human evolution was promoted by "the enhanced development of vision [*des Gesichtssinnes*: 'the sense of viewing'], the strengthened capability to discern external dimensions of things". This enhanced viewing capability should not be confused with the mere act of seeing. Following Du Prel's account (Du Prel, 1880), one gets the impression that a form of contemplation is meant, which aligns with Klages and Friedmann. The essence of vision is discernment. Once humans can discern, two interrelated faculties arise, memory and fantasy. These are the two roots of poetry. (pp. 96f)[9]

"On [these faculties]", Du Prel (1880) concludes, "as opposed to the faculty of abstraction, rests the proper talent of the poet and the artist as such. And when, already in general, we find forms and remnants in our languages informing us about the primitive people's way of thinking – which is hardly disentangled from contemplation [Anschauung] –, it should not surprise that in the language of the poet, whose thoughts arbitrarily changes into images, such remnants will be more numerous; they are paleontological elements, which more or less clearly betray us a long lost worldview. In the true poet, the ocular animal will betray itself more clearly than the causal animal." (p. 97; my trans.)

We can already infer from Du Prel's account of poetry, should it make sense, that poetry approximates reality better than abstraction and thinking in terms of concepts. But there is more. Approximating reality in poetry animates reality and *transforms* the subject (poet or reader). Poetry reconnects subject and object by animating both simultaneously.

"[T]he highest poetry is not attained in lyrics," Du Prel concludes (1880), "as long as the poet is exposed to the phenomenon only as form. He should animate nature and lose himself to the object. The lyricist's subjectivity should be reflected in the fact that nature becomes the symbol of his inwardness, not just the folly of his vain ego – which is so characteristic of modern lyricism." (p. 154; my trans.)

As an illustration, Du Prel quotes (in German) the following passage from Shelley's *Alastor; or, The Spirit of Solitude*:

[…] The oak,
Expanding its immense and knotty arms,
Embraces the light beech. The pyramids
Of the tall cedar overarching, frame
Most solemn domes within, and far below,
Like clouds suspended in an emerald sky,
The ash and the acacia floating hang
Tremulous and pale. Like restless serpents, clothed
In rainbow and in fire, the parasites,
Starred with ten thousand blossoms, flow around
The grey trunks, and, as gamesome infants' eyes,

With gentle meanings, and most innocent wiles,
Fold their beams round the hearts of those that love,
These twine their tendrils with the wedded boughs
Uniting their close union; […]

One may reject as being 'too speculative' the premises upon which Du Prel's account of poetry rests: "the same is recognised by the same," "we are akin to everything which surrounds us, and the slumbering inside of nature only becomes more extraverted in us, and more abundantly unfolded" (ibid., p. 156; my trans.). However, I would ask, does not such a critique itself tacitly assume – and unjustly so – the ultimate reliability of a *denaturalised* subject, the very subject whose alienation from nature has only increased our current ecological predicament?[10]

How is distance awareness involved in poetry? I think that distance is innate to the relation between the poet and their environment. Re-reading Shelley's poem quoted by Du Prel, we may notice telling imageries like "The oak, / *Expanding* its *immense* and knotty arms," "The *pyramids* / Of the *tall* cedar *overarching*," "*far below*, / Like *clouds suspended* in an emerald *sky*," "*rainbow*," "*Starred*" (my italics). These are all spatial references implying distance. Interestingly, properties of space and extension are equally reflected in the poet's (or the seer's) inwardness: "frame / Most solemn *domes within*," "Fold their beams *round the hearts of those that love*," "Uniting their close *union*" (my italics). Once the environment becomes animated by the poet's vision, an original polarity manifests itself, which can only be experienced as distance, or, as Maria Zambrano puts it, as "a wide open space, deserted [*un ancho espacio abierto, desierta*]".[11]

Regarding education, is it a coincidence that the poem refers to the observant eyes of children ("*infants'* eyes")? These eyes are both "gamesome" and wily but also "innocent" and "gentle". They are susceptible to centrifugal endeavours into the open. "Like restless serpents, clothed / In rainbow and in fire, the parasites, / Starred with ten thousand blossoms, flow around / The grey trunks, and, as gamesome infants' eyes, / With gentle meanings, and most innocent wiles, / Fold their beams round the hearts of those that love". The movement described here is circular ("flow around / The grey trunks"). The apparent sense of touch is superseded by feeling or vision ("Fold their beams round the hearts"). This enables "gamesome eyes," if need be, to mould or remould what they see.

Poetry deserves a central place in our teaching systems. Learning poems by heart should be seen as meritorious and not only as a waste of time. I am convinced that minds sensitised for poetry cannot fall short of intuiting nature's treasures.

CONCLUSION

In this chapter, I have argued that ecological awareness entails distance awareness. Distance thrives on vision rather than on touching. Both Ludwig Klages and Hermann Friedmann outlined an ontology of forms. For such an ontology to make sense, the problematic nature of tangibility as constitutive of reality should be emphasised. Tangibility, being innocent in itself, is susceptible to pretensions of comprehensibility, if not all-comprehensiveness. Such pretensions, explicit or implicit, have gradually ravaged our environment.

I have also argued that the notion of 'environment' itself, with its connotations of circularity, need not imply centripetality. If *environment* comes down to what surrounds *me* or *us*, it will be subservient to my or our point of view. Instead, one might imagine an environment whose essence is to enlarge my perspective and dissolve my boundaries.

As two possibly relevant educational concerns, I have proposed 1) reconsidering psychic crises and 2) promoting poetry. Rather than damaging life, psychic crises may testify to incipient reintegration on a higher level and include an increased ecological sensitivity. Poetry, on the other hand, insofar as it entails distance awareness, could reframe the reading mind and deepen it.

These two concerns are far from exhausting the means of developing education toward enlarged ecological sensitivity. Yet, I believe that they are indispensable. "Everywhere where the human being is confronted with 'gestalt', 'mind' arises in the human being's inwardness." (Friedmann)

REFERENCES

Capra, Fritjof (2013). Deep Ecology: Educational Possibilities for the Twenty-First Century. *The NAMTA Journal, 38*(1).

Dabrowski, K. (2015). *Personality Shaping Through Positive Disintegration*. Red Pill Press. (Original work published 1967)

Derrida, J. (1992). Che cos'è la poesia? Points de suspension. Entretiens. Galilée.

Diehm, C. (2006). Arne Naess and the Task of Gestalt Ontology. *Environmental Ethics*, *28*(1), 21–35. doi:10.5840/enviroethics200628137

du Prel, C. (1880). *Psychologie der Lyrik*. Günter Verlag.

Frankl, V. (2006). *Man's Search for Meaning*. Beacon Press.

Friedmann, H. (1930). *Die Welt der Formen. System eines morphologischen Idealismus*. C.H. Beck'sche Verlagsbuchhandlung.

Husserl, E. (1962/1936). Die Krisis der europäischen Wissenschaften und die transzendentale Phänomenologie. Husserliana Band VI2. Martinus Nijhoff. Trans. The Crisis of European Sciences and Transcendental Phenomenology: An Introduction to Phenomenological Philosophy. Northwestern University Press, 1989.

Id. (1950). *Sinnvolle Odyssee. Geschichte eines Lebens und einer Zeit 1873-1950*. Beck'sche Verlagsbuchhandlung.

Id. (1956). *Das Gemüt. Gedanken zu einer Thymologie*. C.H. Beck.

Id. (1981. 1929-1932). *Der Geist als Widersacher der Seele*. Bouvier Verlag.

Jaspers, K. (1919). *Psychologie der Weltanschauungen*. Springer. doi:10.1007/978-3-662-05505-2

Kleinman, A. (1987). Anthropology and psychiatry. The role of culture in cross-cultural research on illness. *The British Journal of Psychiatry*, *151*(4), 447–454. doi:10.1192/bjp.151.4.447 PMID:3447661

Lipps, T. (1883). *Grundtatsachen des Seelenlebens*. Cohen.

Naess, A. (1989). Ecosophy and Gestalt Ontology. *The Trumpeter 6*(4).

Sass, L. (2017, 1992). Madness and Modernism. Insanity in the Light of Modern Art, Literature, and Thought. Oxford University Press.

Steinbeck, A. (1836). *Der Dichter ein Seher, oder über die innige Verbindung der Poesie und der Sprache mit dem Hellsehen*. Verlag Georg Joachim Göschen.

Zambrano, María (2006/1939). *Filosofía y poesiá*, Fondo de cultura Económica Mexico. Trans. D. Ohmans (2008). *Philosophy and Poetry*.

ENDNOTES

1 Even though neither Klages nor Friedmann adopts Naess' term *gestalt* (Friedmann even criticises it as still being too objectivist), their thinking strongly resembles gestaltism. (Naess, 1989; Diehm, 2006; Capra, 2013)

2 A contemporary Indian scholar influenced by Lipps is Vilaynur S. Ramachandra.

3 Cf. "the animal is characterised by a predominantly corporeal, the human being by a predominantly psychical [*seelisches*] understanding of meaning". (Klages, 1981, p. 376; my trans.)

4 "Meanwhile, the Indian systems are condensations of the mixture of Aryan thinking with Asian thinking; therefore, they are no pure evidence of Asiatism. The latter is represented only by the Chinese Dao. For here, we find a reverence of the universe's *phenomena* – a more beautiful reverence seems hardly thinkable – connected with a degree of *dis*belief in the [material] world of resistances; a western mind could only re-*think* yet not re-*experience* such disbelief. The Indian sage *wants* 'salvation,' possibly *wants* 'nothing,' the real Daoist, on the contrary, does not want at all!" (Klages, 1981, p. 340; my trans.).

5 "The birth hour of our free science is in this common dissociation of occidental life." (Ibid.; my trans.)

6 Cf Naess on our increasing lack of gestalt awareness (Naess, 1989).

7 'The deeper you go, the more you know.' (Hamilton Bohannon)

8 Cf. John Keats, *On First Looking into Chapman's Homer*: "Much have I traveled in the realms of gold […] // Yet never did I breathe its pure serene // Till I heard Chapman speak out loud and bold". Or Arthur Rimbaud, *Une saison en enfer*: "Un soir, j'ai assis la Beauté sur mes genoux. - Et je l'ai trouvée amère. - Et je l'ai injuriée."

9 Seemingly drawing on different 'ontological' premises, Jacques Derrida equally insists on memory as the heart of poetry. Cf. Derrida (1992, p. 304).

10 The 19th century German physician-philosopher Albrecht Steinbeck criticises the irony with which we frequently react to the sensitive poet who pretends to hear nature's profound language. We disregard "that the soul, in its poetic liberation, feels and thinks more intensely and clearly the truth which the visionary seer clearly contemplates and expresses." As an example, Steinbeck refers to "an age-old Indian poem" (i.e., the Charta Bhade) which clairvoyantly reveals pure botanic knowledge. (Steinbeck, 1836, p. 249; my trans.)

11 "And for that, the poet remains empty, in readiness, always. His soul comes of resemble a wide open space, deserted. For there are presences that cannot descend into that which is populated by

others… Deserted, empty; because only when that presence arrives, all the others will arrive as well; only with its plenitude and light will things take on shape and meaning [*cuerpo y sentido*]." (Zambrano, 2006, p. 108; trans. D. Ohmans, 2008).

Chapter 3
Re-Imagining the Relationship Between Self and Environment:
Perspectives on Ethical Education

Somosri Hore
O.P. Jindal Global University, India

Anasuya Agarwala
Gargi College, University of Delhi, India

Idris Hassan Bhat
SRM University, India

ABSTRACT

This chapter explores existing theoretical debates on anthropocentrism, intrinsic value theory, and Deep Ecology to establish that human beings need to rethink their relationship with the environment in order to take environmental issues seriously. It argues that the relationship of domination and exploitation with the environment should be replaced with a relationship of care.

INTRODUCTION

The human quest for the dominance of nature has put human survival at risk. We have reached a stage where we are debating the climate crisis and equating its potential to a nuclear crisis. A recent study finds that Greenhouse Gases (GHG) has led to a 1.1°C increase in the average temperature since the pre-industrial age (Luomi, 2020, pp. 1). Signatories to the Paris Agreement 2015 have pledged to maintain the global average temperature rise below 2°C; but even immediately halting emission production will increase temperatures by 3°C. (Luomi, 2020 pp. 1) Whether humans will be wiped out by the excessive heat caused by climate change or not, its destructive signs are clear to us. We are witnessing climate change in the form of melting ice caps and glaciers, river and lake water pollution, unbreathable air quality, dying forests and ocean-life, submerging of cities, and much more. (see Larr and Neidell (2016)

DOI: 10.4018/979-8-3693-2577-3.ch003

for the effect of climate change on pollution; see Staudt et al (2013) and for the effect of climate change on biodiversity; see Thornes (2016) on the effect of climate change on animals). These horrendous environmental changes have generated global debates that necessitate a rethink of our relationship with the environment. Such rethinking is already taking place at many levels: the international community is negotiating limited use of fossil fuels and pushing towards using alternative green sources of energy, and academic disciplines are debating what went wrong in our interaction with the environment and what can be done to improve such relationships.

The discipline of environmental ethics arose from such discussions in the 1970s (Luomi, 2020, pp.2). Environmental Ethics studies the conceptual foundation of environmental values as well as concrete issues like policies, attitudes, and actions required to protect the environment.

This chapter offers anthropocentric, intrinsic value theory, and the perspective of Deep Ecology to rethink the relationship between self and environment in order to change people's detrimental attitudes towards the environment and inspire care for it. Section one explains the anthropocentric theory to argue that we should care for the environment to ensure our survival, to overcome our speciesist behaviour, and to exercise rights along with obligations. Section two explains the intrinsic value theory that argues that the environment ought to be protected and respected because it is sentient, has goods of its own, is beautiful, and exists for its own sake. Section three explains Deep Ecology and its concept of the ecological self that sees itself as a part of a whole and encourages a life of harmony with the world. The fourth presents findings of a survey conducted to gauge the knowledge, intention, and commitment of ordinary citizens towards environmental care. The concluding section combines the theoretical part of this chapter with the findings of the survey to make recommendations for environmental education.

ANTHROPOCENTRISM AND ETHICS

The anthropocentric view in environmental ethics argues that humans are superior to the non-human world. This view considers all members of the human species as deserving of equal moral worth but does not extend this morality to other species or the environment at large. However, it offers a pragmatic and utilitarian argument for the care and protection of nature. The argument states that we should protect the environment because our survival depends upon it. That is, even if we can't grant moral consideration to non-human living and non-living entities, we must still protect them because they have instrumental value for us. For example, we need water, plants, animals, and other entities for food; minerals, oil, and fossil fuels for energy; sand, metals, wood for building infrastructure; and so on.

The anthropocentric view has been justified by various philosophers through time. Thomas Aquinas, writing in the thirteenth century for example, says "that which acts only when moved by another, is like an instrument; whereas that which acts by itself, is like a principal agent." (Aquinas, 2010, pp. 63) The implication from Aquinas is that human beings, by virtue of possessing the capability to act freely, possess the right to direct the actions of other beings who are devoid of such capabilities. Since the environment is devoid of any such capabilities, it is merely an instrument and subject to slavery (Aquinas, 2010, pp. 63). Rene Descartes (1596 – 1650 CE) similarly justifies the view that humans carry an existence that is special and superior to non-human animals on the ground that they possess the capacity for rationality and independent decision-making whereas non-human animals lack such capacities. (Descartes, 2010, pp. 69)

Whereas Aquinas and Descartes give primacy to humans over non-human animals by virtue of the fact that humans are free and rational agents, John Locke (1632 – 1704 CE) makes a theological argument for the same. Nature, according to Locke, is a resource bestowed by God to humans for fulfilling their needs and comforts (Locke, 2010, pp. 77). Other Western philosophers present similar instrumentalist views of nature. Immanuel Kant (1724 – 1804 CE), for instance, writes that, "Animals are not self-conscious and are there merely as a means to an end" (Kant, 2010, pp. 82).

The anthropocentric view of nature has been criticised for being speciesist. Scholars have compared this speciesism with racism and sexism. Just as racism and sexism denies equal value to people of colour and women, speciesism denies equal value to non-human entities.

Anthropocentrism is Not Speciesism

An alternative reading of the anthropocentric view of nature comes from scholars who deny that anthropocentrism is necessarily or implicitly speciesist and chauvinistic. One such scholar is Onora O' Neill (1997). She agrees that anthropocentrism is human centric but argues that it is not specieciest. She writes:

The term speciesism, which was coined by analogue with terms such as racism or chauvinism, is usually used as a label for unjustified preference for human species. (O'Neill, 1997, pp. 128).

O'Neill uses the utilitarians as an example of an anthropocentrism that is devoid of speciesism. She argues that Utilitarians are anthropocentric in maintaining that only human beings are properly agential in their capacity for ethical reasoning, moral responsibility, and moral rule-following. (O'Neill, 1997, pp. 129). But also argues that they cannot be called 'speciesist' because they accord moral recognition or standing to all sentient beings including non-human animals based on their ability to suffer or feel pain.

Her own position is that it is difficult to locate moral value in the environment. She argues against the intrinsic value theory of nature that assigns rights to nature (to be discussed in the next section) but maintains, alongside anthropocentrism, that moral values can be located only in humans which makes humans morally obliged towards nature. (O'Neill, 1997, pp. 128). She thus shifts the discourse from the rights of nature to human moral obligation. Her argument is that nature needs protection, not because it has or does not have intrinsic value, but that moral agents have moral obligations.

Humans Have Moral Obligations: Rights and Obligations

Onara O'Neill (1997) argues that rights have a necessary connection with obligation in the sense that rights bereft of obligations make them meaningless and futile. A person or an entity may have rights but they would become meaningful only if they are respected by an agent. So rights necessarily imply obligations. (O'Neill, 1997, pp. 133)

As humans are the only agents in the world, they are the only ones who bear obligations. She believes that shifting the moral focus from rights to obligation enhances the scope of the moral accountability of humans.(O'Neill, 1997, pp. 133).

O'Neill's recognition of our moral obligation to the environment includes two categories of obligations: fundamental and institutional. (O'Neill, 1997, pp. 135). Fundamental obligations are essentially universal obligations that would apply to all environmental phenomena. Institutional obligations, in contrast, are obligations that are context-sensitive. But Onara O'Neill is insightful enough to accommodate exceptions to universal obligations, claiming instead that they are not absolute. The ethical principle 'to not harm others' may, for example, face exceptional circumstances which may render it non-universalizable: it may be morally acceptable to breach the principle of no harm in special circumstances such as self-defence (O'Neill, 1997, pp. 135). The implication of this for environmental ethics is that "an obligation-based ethics will not prescribe unlimited care for the environment" and will allow for special circumstances such as animal slaughter for food, etc. (O'Neill, 1997, pp. 136).

Like O'Neill, Hayward (1997) argues in favour of anthropocentrism but distinguishes it from speciesism and human chauvinism. He considers the latter offensive for it accords a superior position to human beings and considers human interest supreme but he retains a commitment to anthropocentrism for pragmatic reasons like human survival. He says that anthropocentric motivation – the motivation to save the environment for the sake of sustaining our own life – is the biggest motivation to act in favour of the environment. His commitment to anthropocentrism is also present in his belief that self-love is the first step to loving others and thus that there is no harm in loving one's own species and then caring for others (Kopnina et.al, 2021, pp. 8).

Both Hayward and O'Neill support the anthropocentric view in environmental ethics. O'Neill talks about human obligation being the precondition for environmental concern and Hayward considers anthropocentric motivation the highest motivation to protect it. But even if we separate anthropocentrism from chauvinism or speciesism, the fundamental implication of the term anthropocentrism remains intact in the sense that humans remain at the centre of all environmental protection.

Apart from being human centric, anthropocentrism is also criticised on the ground that it merely motivates to protect those aspects of the environment that promote human welfare, leaving the rest unprotected. (Kopnina et.al, 2021, pp. 8)

Positions like Onara O'Neill's encourage non-speciesism but one may ask why human speciesism is morally wrong? On the other side of the value debate in environmental ethics are ethicists that argue that speciesism is wrong because nature has intrinsic value. These tell us that nature, like the human species, has intrinsic worth and ought to be respected for its own sake.

INTRINSIC VALUE OF NATURE AND ETHICS

Environmental ethicists have formulated theories to justify the intrinsic value of nature and to convince us that we should care for the environment for its own sake. They consider anthropocentric views, even those that demonstrate a respect for the environment, insufficient as they prioritise humans over non-human entities.

Arguments that show that nature has intrinsic value span from anthropocentric interpretations, subjective interpretations, non anthropocentric interpretations, and objective interpretations of intrinsic values. (see Hargrove 1992 for a detailed account of these interpretations) This chapter focuses on the objective interpretation of intrinsic value as it is the most robust formulation of the view that nature has intrinsic value. That is to say that it is the only position that precludes anthropocentrism and human subjectivism thereby excluding humans altogether in evaluations of nature. It makes a human-independent claim that nature must be respected for its own sake and for no other end than itself.

The Intrinsic Value of Nature

John O'Neill (1992) offers a useful taxonomy for understanding the different definitions of intrinsic value. This is not a question of demonstrating that nature has intrinsic value but clarifying what the term intrinsic value might mean. He enumerates, broadly, three senses of the term.

The first, *Intrinsic Value 1*, understands intrinsic value simply as non-instrumental value. An object or entity can have intrinsic value or instrumental value. An object has intrinsic value if it is an end in itself. For example, human happiness, or pleasure, are valuable as ends in themselves. They are not pursued as means for some other end. An object has instrumental value if it is useful for some other object. Slaves, for example, had instrumental value for their masters. Their care and protection, if any, was derived by virtue of the use the slaves had for their masters. Ethicists who use this definition of intrinsic value attempt to show that nature has intrinsic value that is non-instrumental. John O'Neill (1992) quotes the author of Deep Ecology, Arne Naess, as an exemplar of this understanding of intrinsic value:

The well-being of non-human life on earth has value in itself. This value is independent of any instrumental usefulness for limited human purposes. (J. O'Neill 1992, pp. 119)

Naess continues:

'Well-being and flourishing' of human and non-human life have intrinsic value; the richness and diversity of life forms contribute to the realisation of these values and are therefore also intrinsic values. (Naess 1995, pp. 68)

The intrinsic values Naess specifies are 'well-being", "flourishing", "richness" diversity." Like human happiness, or pleasure, these are pursued and valued for themselves by different natural entities. The problem with this definition is that it is too narrow in specifying only the characteristic of being non-instrumental. Richness and Diversity, and for that matter, even the property of Wilderness, may be non-instrumental but they may also be relational properties in the sense that they are valuable because they are valuable to human valuers. Wilderness for example, is a relational property in the sense that "untouched by humans" is non-instrumental but has value in virtue of the relation it has to humans and their activities. (J. O'Neill 1992 pp. 125) Similarly, richness and diversity may be non-instrumental properties of nature but valuable only because they are valued by human valuers. It might then become a human dependent evaluation of nature. It will be shown later that richness, diversity, etc are actually objective properties of nature that need no human valuers to be considered valuable.

The second, *Intrinsic value 2* might be a better definition because it excludes even relational properties thereby excluding human subjectivism. It is used to denote the non-relational properties of an object or the value an object has solely in virtue of its intrinsic properties. (J. O'Neill 1992, pp. 120) He quotes the philosopher G. E. Moore:

To say a kind of value is intrinsic means merely that the question whether a thing possesses it, and in what degree it possesses it, depends solely on the intrinsic nature of the thing in question. (J. O'Neill 1992, pp. 120)

To hold that non-human beings have intrinsic value given this use of intrinsic value is to hold that the value they have depends solely on their non-relational properties. For example, polar bears are an *endangered* species. Endangered – the threat of extinction – is a fact that belongs to the natural world. It has no relation to human beings. In this view, nature is intrinsically valuable in its own right, by virtue of its own properties. Animal rights activists for example would use this definition of intrinsic value to argue for rights such as life, liberty, and the right to territory for animals independently of humans. (James Rachels, 1976, pp. 12)

While this definition precludes human beings in the sense that it emphasizes non-relational properties, it can invite subjectivism because humans are the one's ascribing value to the properties (of life, liberty, and territory). Even though these properties might be facts of nature, and thus have independent value outside of human evaluation, this sense of intrinsic value forgets to consider the independence of these values by merely rendering them non-relational without also calling them objective.

The third sense of intrinsic value, *Intrinsic value 3*, uses it as a synonym for 'objective value'. (J. O'Neill 1992, pp 120) These ethical objectivists hold that the evaluative properties of objects are real properties of objects. That is, they are properties that objects possess independently of the valuations of human valuers. Objectivists deny the subjectivist view that the source of all values lies in valuers. John O'Neill offers two formulations for objective intrinsic values:

1. A real property is one that can be characterised without reference to the experiences of a being who might experience the object

and

2. The evaluative properties of objects can be characterised without reference to evaluating agents. (J. O'Neill 1992, pp. 126)

Rarity, diversity, flourishing, endangered, etc are objective properties of nature. They are valuable in the biological, evolutionary, and natural context, independent of human valuers. Definitions of intrinsic value that mention only the conditions of being non-instrumental and/or non-relational limit a fuller sense of the term intrinsic value.

There are at least 4 arguments that use intrinsic value in this objective non-human sense.

Sentience: The logical end of the argument from sentience is animal rights. Numerous philosophers and social commentators have made the attempt to demonstrate that if we are able to ascribe basic rights to life, liberty, and property to human beings, we can do so for many higher animals. Peter Singer and Tom Ragan (1989) exemplify utilitarian principles to argue for animal rights. They maintain that differences in kind between other animals and human beings, for certain moral and legal purposes, are differences that do not matter. What is important is the similarities between human beings and animals. Singer makes a strong argument by drawing an analogy between speciesism – which highlights differences between other animals and human beings – and racism and sexism. Admitting differences between human and animal, man and woman, and different races is irrelevant because what matters for moral consideration is the capacity to suffer, and in this, all sentient beings are similar. (Singer 1989) Rights activists like Singer and Ragan ground animal rights on the capacity to suffer. Since animals have the capacity to suffer, they are entitled to rights.

Even if we give up the controversial rights claim that relates animal rights with human rights, sentience is an independent non-relational property belonging to non-human animals. It is grounded in the animal physiology and biology and is thus an objective fact.

The criteria for sentience complies with the first definition of intrinsic value as it is non-instrumental, it complies with the second, in being a non-relational property, and it complies with the objectivist standard because it is a fact of biology to say that sentient beings avoid pain and pursue pleasure. The problem with this view is that it is too narrow. It excludes non sentient entities like lower animals, plants and abiotic natural phenomena.

Inherent worth: Intrinsic value theorists like Rolston (1988) and Taylor (1984) argue that inherent worth implies the value something has simply in virtue of the fact that it has a good of its own. Goods may be of two kinds, metaphorical goods such as happiness and misery, and biological goods such as being well or ill, flourishing, etc. (John O'Neill 1992, pp. 129) John O'neil follows Von Wright in discarding metaphorical goods for non-human living beings and limits himself to biological goods when he says that living things have goods of their own (John O'Neill 1992, pp. 129). An otter needs clean rivers, filled with food, and overgrown banks where it can raise its cubs. It needs healthy fur to protect them from the cold. These conditions are needed for otters to thrive and flourish. Trees need soil filled with nutrients, water, and sunlight to flourish. Deprived of these conditions, they die or deteriorate, they are defective, stunted, or abnormal. Thriving and flourishing are biological goods. To say that otters, trees, and all other nonhuman living things have goods of their own is to say that they have a biological interest to be well and flourish. We can say of such entities that having a good means having biological life. (John O'Neill 1992 pp. 129) Specifying conditions necessary for a life to flourish makes no reference to the experience or evaluations of human valuers. These goods are rather given by a species' specific characteristic features (John O'Neill 1992, pp. 130). Statements about what is good for a living thing or what is conducive to its flourishing, that is, placing it in a biological context, does not depend on a reference to human observers. (John O'Neill, 1992, pp. 131).

John O'Neill and Wright's biological grounding of intrinsic value is not limited to a non-instrumental, non-relational intrinsic value paradigm, and thus invites no subjectivism. It is objective in the biological context of *life*. However, though it is more expansive than the criteria of sentience and includes lower organisms and plants, it is limited to biologically living entities and thus excludes abiotic natural phenomena.

Beauty: There is no doubt that human observers seek aesthetic pleasure from nature. We experience the sublime in expansive landscapes like mountains, oceans, and deserts; we find rainforests, flowers, and sunsets beautiful; we wonder and marvel at the complexity and geometrical symmetry of snowflakes and beehives, etc. Nature has inspired poetry and painting, intrigued mathematicians, borne romanticism as a genre in literature and philosophy, motivated architecture, excited naturalists, etc. All this proves the subjective, non-instrumental value of beauty in nature. But the beauty debate in the philosophy of art, its aesthetic value as subjective or objective, can be applied to the beauty of nature. While aesthetic relativists argue that beauty is subjective, dependent on individuals or culture, aesthetic objectivists argue that beauty is a property of the object, waiting to be discovered. How else does one explain the variety of beautiful art, Omar Khayam's Rubaiyat, Rumi's poetry, Michelangelo's David, being appreciated universally by intelligent people from across cultures? How do we debate and argue and try to convince others that Bach's Bouree is beautiful? Does the beauty of nature require a human consciousness or is it an objective property that needs no evaluators? Some argue that truth, symmetry, and harmony, are characteristics that constitute beauty. These may be found in the artwork or in nature. The debate in the philosophy of art remains unresolved. But it opens the possibility that beauty is an objective property of natural objects and needs no human observer to be appreciated.

The intrinsic value of nature as possessing the property of beauty, if true, is non instrumental, non-relational, and objective. It is wider than the previous two because it is not limited to living things but includes abiotic entities such as oceans, mountains, snowflakes, caves, etc.

Existence: The existential interpretation of the intrinsic value of nature states that entities in nature have intrinsic value merely by virtue of natural existence. Simply for the fact that they exist. Not because they have inherent properties or goods of their own but simply because they exist. This might seem like a religious argument that God made 'all things bright and beautiful' and so we have no right to destroy what we didn't make. But it can be given a secular interpretation. Natural entities, the living and nonliving, have intrinsic worth and are deserving of moral consideration, protection, and preservation, simply because they are complex systems that have natural existence.

These four arguments establish that nature has objective intrinsic value, independent of human evaluation. But does the existence of the intrinsic value of objects oblige us towards a moral consideration of those objects?

Onara O'Neill offers a Kantian argument for moral obligation, irrespective of whether nature has instrumental or intrinsic value. Plato had asked in the *Republic*, why one should be good when it is so beneficial to be bad. Even though we possess the capacity for morality, it is difficult to convince ordinary people to be good. The objective existence of intrinsic value is a strategy for the moral consideration of nature. Most people protect what they deem to have value, especially intrinsic value. We pursue pleasure, happiness, freedom, dignity, and other goods that have intrinsic value. It is not a stretch then to expect that recognizing the intrinsic value of nature will not motivate people to respect it. It provides a stronger motivation than the Kantian good will to be a good person. It is just an inherent property of our moral capacities that the recognition of an intrinsic value demands moral consideration. If we recognize that nature has intrinsic value, we are automatically drawn to respect it.

These rationalistic approaches of the anthropocentric and intrinsic value theory may be complemented by a more holistic and phenomenological understanding of the relationship between self and the environment. Deep Ecology offers just such a holistic understanding.

Deep Ecology and the Ecological Self

The term 'Deep ecology' was conceptualised by the Norwegian philosopher of environment, Arne Naess, who coined this term specifically to distinguish it from the commonly prevalent 'shallow ecology' of his time. In contrast to 'shallow ecology' which emphasised techniques and strategies useful for the preservation of the environment, 'deep ecology' was conceived with the purpose of investigating deeper questions regarding our relationship with the environment and questions regarding our socio-economic growth in order to penetrate the root cause of ecological crisis. Shallow ecology expresses, in a sense, the concerns of the anthropocentric use-theory, while deep ecology has a more holistic concern with the environment. Naess writes, "Shallow ecology tends to talk about resources for humans, whereas in deep ecology we talk about resources for each species." (Bodian, 1982. pp 3).

Ecological Self

Deep ecology challenges the conventional liberal concept of the self as a finite, individual entity bound by time and space and self-interest. It reimagines the self as part of a larger environment that extends not just to other people and communities, but to all living beings, inanimate natural phenomena, and even systems that go up to the biosphere. (In its later articulation, it includes and is identical to a universal consciousness.(Callicott, pp. 2017. pp. 15).

Ecosophists in the 1970s such as Naess and Callicott believed that a reconstruction of the concept and experience of self was the most viable and efficacious basis for an environment philosophy (Callicott 2017 pp. 236). Naess in particular offered a critical explanation of a deep ecological self that stood in opposition to the West's assumption of the rational monadic self. In the "Shallow and the Deep" he articulated a version of 'ecological/relational self' that was naturalistic. He wrote:

Rejection of the man-in-environment image in favour of the relational, total-field image. Organisms as knots in the biospherical net or total field of intrinsic relations. An intrinsic relation between two things a and b is such that the relation belongs to the definitions or basic constitutions of a and b, so that without the relation, a and b are no longer the same things. The total-field model dissolves not only the man-in-environment concept, but every compact thing-in-milieu concept—except when talking at a superficial or preliminary level of communication (Naess, 1973, pp. 95).

The relationship between persons and environment *qua* ecological self is understood as one of *constituting*. The notion of constituting describes an internal relationship between two things such that *a* and *b* cease to be *a* and *b* without each other. In contrast, the monadic notion of the self describes an external relationship between two entities. Callicott offers useful examples to differentiate internal and external relationships. Balls on a pool table, he says, share external relations with each other. No matter how much they hit each other, each ball remains itself without any change in its internal constitution. Planets and their gravitational fields, on the other hand, exemplify internal relations. The earth, for example, took a spherical shape because of its gravitational force (Callicott 2017, pp. 238). Deep ecology's concept of the ecological self describes the relationship between the self and environment as an internal relationship where the self and environment constitute each other.

The ecological self involves a self-realisation of the concept in terms of understanding and experience. One must understand the relation between self and world and one must experience oneself in these relational terms. "One is and one experiences one self to be a 'knot in the biospherical net of intrinsic relations.' Or better: as a node or nexus in a skein of internal socio-environmental relations. One experiences 'the beauty and complexity of nature[as]continuous with[one]sel[f].' (Callicott 2017, pp. 238-239). Seeing oneself in continuity with one's environment in Delhi, for example, includes the impact and influence of society – one's family, friends, community, etc but also the city's natural environment – ruins, gardens, pollution, weather, etc. Undo this knot that is me in the socio-biospherical total-field, and there is nothing left of my self.

It is only in his later writings that Naess is influenced by Gandhi's reading of ancient Indian philosophy. The ancient Indian philosophical school of Advaita Vedanta understands the self as Atman – the individual soul or self – that is considered identical to Brahman, the ultimate reality or cosmic consciousness. According to this philosophy, the apparent diversity in the world is neither real nor unreal (maya), and the true nature of reality is non-dual (advaita). Atman, in its essence, is understood as pure consciousness, infinite, and beyond all attributes (nitya-shuddha-buddha). It is eternal, unchanging, and unaffected by the fluctuations of the material world (jagat).

Such an infinite and universal self, however, can be realised only through following a path to Moksha or liberation from the cycle of birth and death (Samsara). The process of self-realization involves transcending the identification with the body and mind and recognizing the unity of Atman with Brahman. This realisation leads to a state of bliss (ananda) and freedom from the limitations of the material world.

Gandhi adopts this notion of this universal Self to argue for true liberation beyond the political liberation from British Colonialism. Naess writes that Gandhi does not refer to the

'narrow' self (jiva) of egocentric interest 'but to 'the supreme or universal Self—the atman...Through the wider Self every living being is connected intimately, and from this intimacy follows the capacity of identification... No moralizing is needed, just as we need not morals to make us breathe.(Naess 1986, pp. 39).

The metaphysical reconceptualization of the relation between man and environment through the concept of Atman involves an *identification* that replaces Naess' earlier description of a *constitutional* internal relationship between self and environment (Callicott 2017, pp. 240). This identification is not to be understood merely theoretically, but as the Indian tradition insists, must also be experienced through dedicated meditation and self-reflection so that identification becomes a natural inclination. The experience of identification naturalises ethics so that there is no requirement for moral judgement in our dealings with the environment. (Talukder 2016, pp. 222).

Naess however allows for the difficulty involved in achieving this level of metaphysical union through intense meditation and self reflection. He champions the idea that identification with others may manifest through empathy to circumvent this difficulty, (Callicott 2017, pp. 245; Talukder 2016, pp. 222). Identification, Naess says, is a situation which "elicits intense empathy" (Naess 1986, 227). He cites an example from personal experience to illustrate what he means by empathy. He once saw a dying flea that had accidently jumped into a vial of acid. Naess says, "Naturally, what I felt was a painful sense of compassion and empathy" (Naess 1986, 227). Even though he was not able to save the flea, he psychologically felt its pain and suffering and felt deep compassion and empathy by identifying with the flea. Identification then can be realised through empathy: "identification means not to *alienate* from others." (Talukder 2016, pp. 222).

Even if we are sceptical about the metaphysical claims of deep ecology, we might still agree with other ecosophists and adopt a socio-biospheric total-field theory of self along with its prescription for. self-realization. This non-metaphysical version of the ecological self wins its argument for the need for environmental ethics by calling it self-defence.

The defence of the incredibly complex global natural ecosystem, the biosphere, and the incredibly complex global socio-economic society, which is embedded in the biosphere, is self-defence (Callicott 2017, pp. 247).

SURVEY

A public survey was conducted in January 2024 in New Delhi, India to gauge the knowledge, intent, and commitment of ordinary people towards the environment. A questionnaire was distributed that included questions related to the awareness, concern, and accountability of people residing in the city towards environmental ethics. We received a total of 150 responses from educated adults between the age-group of 18- 65 years.

They survey revealed that 40% of the respondents had received some level of formal education in environmental ethics and about 50% include discussions about the current state of the environment with their peers. On the question of the value of nature: 29% of the responses selected the option of aesthetic beauty; 54% selected instrumental value; 27% selected spiritual value; 49% selected preserving nature for its own sake; and 84% selected the option that nature and humans are part of a single ecosystem. On intent towards environmental care: 22% were prepared to donate money for environmental causes and 64% pledged collective action and organization towards environmental concerns. On commitment towards environmental care: 74% had consciously altered their lifestyles towards a healthy environment.

The survey suggests one broad trend: generalizing the result, most people have a basic level of awareness and knowledge about the current environmental crisis and are concerned about it, but they lack serious individual and collective commitment towards its protection. Although 50% of the total number of respondents present an intention towards the well-being of the environment for its instrumental value and 50% for its intrinsic value, there isn't a 100% commitment towards environmental protection in the form of a conscious lifestyle modification or collective action: only about 40% of respondents consciously claim to segregate wet waste from dry waste; 49% claim to use public transport; and 77% claim to use glass containers instead of plastic containers. Further, hardly any of them have participated or collectively organized (protests, demonstrations, campaigns, signature campaigns, etc.) towards environmental protection.

The mismatch between knowledge, intent, and commitment can be accounted for by positing that inadequate knowledge leaves ordinary people ignorant of the real causes of environmental degradation and thus the necessary steps required for environmental care. The mismatch between intent and serious commitment may be explained by either an ignorance of accurate knowledge leading to a withholding of necessary action, or to a mass psychology wherein ordinary individuals with good intentions feel morally satisfied in merely possessing an intention towards the environment. It is not sufficient to possess merely intellectual virtue or knowledge. Knowledge is truly valuable only when it is actualized in practical morality.

CONCLUSION AND RECOMMENDATIONS

The survey reveals gaps in the knowledge, intent, and commitment of ordinary citizens towards the environment and environmental ethics. Knowledge includes an understanding of ecology and the causes and consequences of environmental degradation; good intentions assume a moral education; and serious commitment requires the cultivation of strong wills, passion, and action. An educational curriculum founded on the theoretical perspectives offered in this chapter will bring knowledge, good intentions, and serious commitment together to constitute the kind of care required for engaging with environmental issues.

Knowledge is necessarily scientific in the sense that it requires a scientific understanding of ecology but it remains incomplete without a further understanding of the value of nature. The anthropocentric and intrinsic value theories presented in this chapter offer good insights for the value of nature. Anthropocentric theories teach us the value of nature for our sake; intrinsic value theories teach us the value of nature for its sake. These are important perspectives that appeal to a rational mind. By the standards of a reasonable mind, we tend to protect and pursue what we value. Rationality is an effective rhetorical device for persuading students and ordinary people that we ought to take environmental concerns seriously.

While the anthropocentric and intrinsic value perspectives aid knowledge of the value of nature that may then motivate intent and action towards its care, Deep Ecology and its message of the interconnectedness between humans and the environment is a truly radical approach to environmental care. The anthropocentric and intrinsic value theories assume a distance between the rational knowing subject and a passive environment so that we remain, despite our best efforts, alienated from nature. Deep Ecology's socio-biospheric total-field theory of self posits a relationship of constitution between humans and the environment so that they cease to be what they are without each other. It reimagines the relationship between self and environment in terms of identity and directs attention to the holistic development of an ecological self. The focus here is the development of an enlarged self.

The primary purpose behind education today is preparing students to enter the workforce so that they can be economically productive. Jiddu Krishnamurthi believed that this capitalist education made students mechanical by only acting on their productive skills. (Krishnamurti, 1953, pp. 17). We agree with Krishnamurthi that education should be more holistic and encompass the social and existential aims of human life such as freedom, love, care, responsibility, creativity, self-awareness, critical thinking, and peace. (Krishnamurti, 1953, pp. 18) We thus recommend curricula that promotes the idea of a holistic development of the self in terms of our rational, moral, and existential being but further recommend the inclusion of an ecological self so that we may identify, rather than alienate, ourselves with the larger environment. A key tool for the development of an ecological self is the use of empathy. Empathy may be cultivated through an exposure to plant and animal life: gardening and nature walks promote exposure to insect and plant life so that their interdependence may be learnt; visits to reserves and conservation parks and zoos produce an affinity to wild and endangered animals so that we care about habitat loss; and feeding of birds and stray animals fosters a caring and friendly relation so that we may not fear animals. Furthermore, it is imperative to facilitate students' participation in outdoor excursions such as camping, hiking, and birdwatching. Emphasizing meditation in natural environments is also crucial. Introducing therapies like ecotherapy, wilderness therapy, and forest bathing into students' lives from an early age is paramount. Originating in Japan during the 1980s, forest bathing emerged as a response to growing awareness about the adverse effects of urban living, including depression, distraction, and physical discomfort. Forest bathing, as a medicinal practice, enables individuals to immerse themselves in the tranquility of nature, positively influencing their mental state and alleviating the urban hustle. This sensory journey encourages individuals to appreciate their surroundings, from the melodic chirping of birds to the gentle rustling of foliage and babbling streams. Inhaling fresh, fragrant air while absorbing the intricate details of the forest floor and the canopy above creates a profound sensory experience. Research from Japan indicates that forest bathing enhances sleep quality, mood, concentration, and stress management. Prolonged stress can precipitate various health issues, including anxiety, depression, and insomnia, as well as physiological ailments like hypertension, muscle tension, and compromised immune function. Disconnecting from modern technology and urban sprawl to spend time in nature can effectively mitigate the physiological and psychological impacts of stress. Consequently, integrating these practices can foster students' deeper connection with nature, nurturing a sense of responsibility and stewardship towards the environment.

A proper understanding of ecology as well as a holistic cultivation of the self will impart the knowledge, morality, passion and persistence of will to seriously commit and act towards environmental care.

The theoretical perspectives offered in this chapter as well as the survey and recommendations speak to the issue of environmental care by individuals. Environmental care is important to the extent that it is a starting point to engage with the environment on terms other than exploitation or indifference. But the issue of environmental protection, conservation, and climate change is more serious and requires more accountability than good intentions and changed lifestyles. Robin Stolnit, writing for the *Guardian* in 2021 quotes an article on "The Carbon Footprint Sham"[1]

British Petroleum, the second largest non-state owned oil company in the world, with 18,700 gas and service stations worldwide, hired the public relations professionals Ogilvy & Mather to promote the slant that climate change is not the fault of an oil giant, but that of individuals. It's here that British Petroleum, or BP, first promoted and soon successfully popularized the term "carbon footprint" in the early aughts. The company unveiled its "carbon footprint calculator" in 2004 so one could assess how their normal daily life – going to work, buying food, and (gasp) traveling – is largely responsible for heating the globe. (as cited in Stolnit, 2022, para 3)

While this chapter inspires care for the environment, and while climate-conscious decisions feel good, it is not entirely up to individual virtue to save the planet. The real culprits of climate change are capitalist corporations motivated by profit to overproduce, overmine, and overmilitarize the planet. (Guha, 1989, pp. 2) We thus recommend disseminating environmental critiques of capitalism.[2] We further recommend exposure to the history of people's movements taken against governments and corporations to inspire and inform collective action and organization against those responsible for climate change.

REFERENCES

Aquinas, T. (2010). Humans as Moral End. In D. R. Keller (Ed.), *Environmental Ethics: the big questions* (pp. 63–65). Wiley Blackwell. doi:10.5422/fordham/9781531502041.001.0001

Bacon, F. (2010). The Mastery of Nature. In D. R. Keller (Ed.), *Environmental Ethics: the big questions* (pp. 65–69). Wiley-Blackwell.

Baxter, B. (1999). Environmental Ethics-Values or Obligation? A Reply to O'Neill. *Environmental Values*, *8*(1), 107–112. doi:10.3197/096327199129341743

Beckerman, W., & Pasek, J. (2010). In Defence of Anthropocentrism. In David R. Keller (eds.) Environmental Ethics: the big questions.(pp. 83-89). Wiley-Blackwell.

Berry, T. (1993). The Viable Human. Michael E. Zimmerman, J Baird Callicott, George Sessions G., Warren K. J. et al., (eds.) (1997). Environmental Philosophy From Animal Rights to Radical Ecology (pp.171-182) Prentice Hall

Bodian, S. (1982). Simple in Means, Rich in Ends: A Conversation with Arne Naess. *OpenAirPhilosophy.org, 3*

Callicott, J. B. (1989). *In Defense of the Land Ethic*. SUNY Press.

Callicott, J. B. (2017). Notes on Self-Realization. *Worldviews, 21*(3), 235-250.

Descartes, R. (2010). Nonhumans as Machines. In D. R. Keller (Ed.), *Environmental Ethics: the big questions* (pp. 69–72). Wiley-Blackwell.

Guha, R. (1989). Radical Environmentalism and Wilderness Preservation: A Third World Critique. *Environmental Ethics, 11*(1), 71–83. doi:10.5840/enviroethics198911123

Hargrove, E. C. (1992). Weak Anthropocentric Intrinsic Value. *The Monist, 75*(2), 183–207. doi:10.5840/monist19927529

Kant, I. (2010). Indirect Duties to Nonhumans. In D. R. Keller (Ed.), *Environmental Ethics: the big questions* (pp. 82–83). Wiley-Blackwell.

Kopnina, H., Washington, H., Taylor, B., & Piccolo, J. (2021). Anthropocentrism: More than Just a Misunderstood Problem. *The International Journal of Ecopsychology, 3*, 3–23. https://digitalcommons.humboldt.edu/ije/vol3/iss1/4

Krishnamurti, J. (1953). *Education and the Significance of Life* (M. Lutyens, Ed.). Harper & Row.

Larr and Neidell (2016). Pollution and Climate Change. *Children and Climate Change, 26*. doi:10.1353/foc.2016.0005

Locke, J. (2010). Nature as Economic Resource. In D. R. Keller (Ed.), *Environmental Ethics: the big questions* (pp. 77–82). Wiley-Blackwell.

Luomi, M. (2020). *Global Climate Change Governance: The search for effectiveness and universality.* International Institute for Sustainable Development (IISD). https://www.jstor.org/stable/resrep29269

Machan, T. R. Do Animals have Rights. In J. Rachels & S. Rachels (Eds.), *The Right Thing to Do: Basic Readings in Moral Philosophy* (pp. 45–59). Mcgraw Hill.

Mill, J. S. (2010). The Amoral Status of Nature. In D. R. Keller (Ed.), *Environmental Ethics: the big questions* (pp. 73–77). Wiley-Blackwell.

Naess, A. (1973). The shallow and the deep, long-range ecology movement. A summary. *Inquiry, 16*(1), 95 — 100. doi:10.1080/00201747308601682

Naess, A. (1986). *The Basic Principles of Deep Ecology.*

Naess, A. (1993). Simpleis Means, Rich in Ends. Michael E. Zimmerman, J Baird Callicott, George Sessions, Karen J. Warren et al., (eds.) *Environmental Philosophy From Animal Rights to Radical Ecology*, (pp.182-193)

Naess, A. (1995). *Deep Ecology for the 21st Century, Readings on the Philosophy and Practice of the New Environmentalism.* George Sessions (Ed.) Shambhala.

Newton, I. (2010). Mechanistic Metaphysics. In D. R. Keller (Ed.), *Environmental Ethics: the big questions* (pp. 72–73). Wiley-Blackwell.

O'Neill, J. (1992). The Varieties of Intrinsic Value. *The Monist*, *75*(2), 119–137. doi:10.5840/monist19927527

O'Neill, O. (1997). Environmental Values, Anthropocentrism and Speciesism. *Environmental Values*, *6*(2), 127–142. doi:10.3197/096327197776679121

Rolston, H. III. (1988). *Environmental Ethics*. Temple University Press.

Sessions, G. (1993). Introduction. Michael E. Zimmerman, J Baird Callicott, George Sessions, Karen J. Warren et al., (eds.) *Environmental Philosophy From Animal Rights to Radical Ecology*, (pp.161-170).

Shafer-Landau, R. (2018). *Lliving Ethics: An Introduction with Readings*. Oxford University Press.

Singer, P. (1989). Are all Animals Equal. In T. Regan & P. Singer (Eds.), *Animal Rights and Human Obligations* (pp. 215–226). Oxford University Press.

Staudt, L., & Howard, B. (2013). The added complications of climate change: Understanding and managing biodiversity and ecosystems. *Frontiers in Ecology and the Environment*, *11*(9), 494–501. doi:10.1890/120275

Stolnit, R. (2021, August 23). Big oil coined 'carbon footprints' to blame us for their greed. Keep them on the hook. *Guardian*. https://www.theguardian.com/commentisfree/2021/aug/23/big-oil-coined-carbon-footprints-to-blame-us-for-their-greed-keep-them-on-the-hook

Taylor, P. W. (1984). Are Humans Superior to Animals and Plants. *Environmental Ethics 6*, 2. 149-160.

Thornes (2016). Animals and Climate Change Author. *Journal of Animal Ethics*, *6*, 81-88. doi:10.5406/janimalethics.6.1.0081

Thornes (2016). Animals and Climate ChangeAuthor. *Journal of Animal Ethics*, *6*, 81-88. doi:10.5406/janimalethics.6.1.0081

Union of Concerned Scientists. (2018) *Heat Waves and Climate Change The Effects of Worsening Heat on People, Communities, and Infrastructure*. Union of Concerned Scientists. https://www.jstor.org/stable/resrep24150

Weber, T. (1999). Gandhi and Deep Ecology. *Journal of Peace Research*, *36*(3), 3. doi:10.1177/0022343399036003007

ENDNOTES

1 https://www.theguardian.com/commentisfree/2021/aug/23/big-oil-coined-carbon-footprints-to-blame-us-for-their-greed-keep-them-on-the-hook

2 https://www.theguardian.com/environment/2021/oct/30/capitalism-is-killing-the-planet-its-time-to-stop-buying-into-our-own-destruction

Section 2
Spirituality and Environmental Education

Chapter 4
Transnational Buddhism and Ecological Awareness: The Case of Tibetan Nuns in Exile

Swati Chawla
https://orcid.org/0000-0002-7480-6563
O.P. Jindal Global University, India

ABSTRACT

The chapter explores the interplay among identity construction, environmentalism, and education within the Tibetan community in exile in India with a focus on the Tibetan Nuns Project. Drawing on Toni Huber's analysis of Tibetan self-representation in exile, the chapter examines the emergence of a modern, liberal Shangri-la image, informed by complex translational politics influenced by Western political traditions. The Tibetan exiled elite, through prolonged contact with the globalized cultural environment, adapted Western representational styles to reshape their self-image, challenging traditional perceptions of Tibet. Statements on the so-called 'environmentalist' Tibetan identity emphasises a worldview and modes of behaviour that are described in terms of 'completely modern' concepts and language with terms like 'ecological balance,' interdependence, and 'preservation.'

INTRODUCTION

The chapter explores the interplay among identity construction, environmentalism, and education within the Tibetan community in exile in India with a focus on the Tibetan Nuns Project. Drawing on Toni Huber's analysis of Tibetan self-representation in exile, the chapter examines the emergence of a modern, liberal Shangri-la image, informed by complex translational politics influenced by Western political traditions (Huber, 2001). A feature of this image is the reinvention of a 'modern, liberal Shangri-la image' which has its precedents in the discourse of early modernity (characterised by what Huber calls the 'three powerful -isms' of colonialism, orientalism and nationalism) and the discourse of liberal and social protest movements of environmentalism, pacifism, human rights and feminism in the industrialised West. The Tibetan exiled elite, through prolonged contact with the globalised cultural environment,

DOI: 10.4018/979-8-3693-2577-3.ch004

adapted Western representational styles to reshape their self-image, challenging traditional perceptions of Tibet. Statements on the so-called 'environmentalist' Tibetan identity emphasise a worldview and modes of behaviour that are described in terms of 'completely modern' concepts and language with terms like 'ecological balance,' interdependence, and 'preservation' (Huber, 2001). The chapter delves into the historical roots of these identity claims, tracing them back to the mid-1980s when the Dalai Lama actively sought international support for the Tibetan struggle against Chinese rule; he was awarded the Nobel Peace Prize in 1989. It is focused on the Tibetan Nuns Project, an institution that highlights the importance of traditional Buddhist sangha for global peace and harmony. Founded in the 1980s, the project aims to combine old monastic traditions with modern educational practices. The chapter examines documentaries, pamphlets, newsletters, interviews, and other literature produced by the Tibetan Nuns Project, concentrating on the nuns' ecologically responsible practices, such as recycling projects, concern for animals, and vegetarianism. It demonstrates how a reinterpretation of Buddhist monasticism, concurrent with Tibetan membership in international Buddhist groups advocating nonviolence, contributes to projecting an essential nonviolent and ecologically conscious Tibetan monastic identity.

Tibetan Exile and Freedom Movement

Tibet, often referred to as the world's 'Third Pole'— a term used by glaciologists for the Hindu Kush Himalayas and the Tibetan Plateau— holds the key to crucial river systems sustaining over a quarter of the world's population, particularly in South Asia (Dechen Palmo, 2019; Gordon, 2010). These include the Brahmaputra, Indus, Karnali, and Kosi rivers, which sustain parts of northern and eastern India. As the leader of the Tibetan Government in Exile after his escape to India in March 1959 (he relinquished the position in 2011), the Fourteenth Dalai Lama (b. 1935) has often highlighted Tibet's importance to these ecosystems in his political statements. In the Dalai Lama's Strasbourg Proposal (1988) and Five Point Peace Plan (1987), he transcended narrow regional interests and framed the high plateau as a potential ecological sanctuary for all humankind. The Tibetan exile administration has also sounded the alarm on the global repercussions of Chinese actions on the fragile ecosystem of the Tibetan plateau, encompassing issues such as the dumping of nuclear waste, militarisation, river damming, mineral exploitation, water pollution, large-scale population transfer, and the forced sedentarisation of nomadic (indigenous) populations.

Tibetan refugees in South Asia are settled in 46 settlements, comprising of 24 agricultural, 16 agro-industrial and 10 handicraft units; 29 of these are in India, 10 in Nepal and 7 in Bhutan (Bernstorff & von Welck, 2004, p. 125). The Tibetan community in exile is often identified as one of the most resilient and successful refugee groups in the world. Today, individual Tibetans and their families have reached a high degree of economic self-sufficiency and independence. As "a testament to political will," a largely illiterate society has been transformed into a fully literate one within two generations. Furthermore, the Dalai Lama has established a democratic government directly elected by the Tibetan diaspora by reducing his own power(Bernstorff & von Welck, 2004; Norbu, 2001). Scholars have contrasted this community with others in exile and hint at a possible difference crucial for their success: "It was surely a wise decision to create settlements, (located on land leased to the community by the governments of the various states), where the refugees could rebuild their lives. This contrasts with other refugee groups who have lived in provisional camps for decades entirely dependent on assistance" (Bernstorff & von Welck, 2004, pp. 1–2).

The role of Tibetan nuns and laywomen in the struggle for political and religious autonomy is well documented. However, as monastics, they had occupied a marginal position to monks and curricula in nunneries did not prepare them for teaching or higher ecclesiastical and leadership positions (Chawla 2018; 2019; Schneider 2022). Nunneries in Tibet were fewer, poorer, and less prestigious than monasteries. Most nuns were taught exclusively by male teachers, and many received no formal education. To this day, all ten ecclesiastical representatives in the current Tibetan Parliament-in-Exile are monks.

The exegetical debate around full ordination within the *vinaya* (monastic law) has been addressed in the vast scholarship on Buddhist nuns across traditions, on female religious practitioners more generally, on Tibetan Buddhist nuns, and Tibetan religious institutions within Tibet and the diaspora (Gyatso, 2010; Hannah, 2012; Härkönen, 2023; Havnevik, 1994; Melnick Dyer, 2022; Padma'tsho (Baimacuo) & Jacoby, 2020; Roloff, Carola, 2020). Nevertheless, little scholarly work has focused on Tibetan Buddhist nuns who live in exile in India, and fewer still take a longer historical view of how the condition of exile led to a reconfiguration of the role of the female monastic, which in turn spurred pedagogical and curricular innovations in nunneries. The most prominent among these has been the institution of the Geshe-ma (the equivalent of a PhD) degree in 2013, which ensured that female monastics acquired the same level of education as their male counterparts for the first time in the history of Tibetan Buddhist monastic institutions within the Gelug sect (Chawla, 2018, 2019; Napper, 2020; Schneider, 2022; Salgado, 2013); the Khen-mo degree in the Nyigma, Kagyu, and Shakya traditions has a comparable contemporary social history (Liang, 2021; Liang & Taylor, 2020). The prism of forced dislocation and loss is insufficient to understand the social dynamics of Tibetans in India, and exile has instead served as an opportunity for Tibetan Buddhist nuns to reconfigure their position in society agentively (Chawla, 2010, 2015).

Environmental Protection as a Cornerstone of Tibetan Struggle

The Dalai Lama's Strasbourg Proposal, presented to the European Parliament in 1988, outlines a path for Tibetan autonomy within the framework of the Chinese constitution, emphasising the preservation of Tibet's unique culture, religion, and environment, and a symbiotic connection between the natural environment of Tibet and its culture and religion (The Fourteenth Dalai Lama of Tibet, 1988). A year earlier, in his Five Point Peace Plan, the Dalai Lama had advocated turning the entire Tibetan plateau into a zone of peace and sustainability (The Fourteenth Dalai Lama of Tibet, 1987). These ideas about the management of Tibet's natural resources are informed by Buddhist principles of compassion, non-violence, and inter-dependence; the Dalai Lama has often credited Indian philosophical and political traditions for introducing them, and for maintaining their relevance in the contemporary world, particularly in the anti-colonial struggle under the stewardship of Mahatma Gandhi (Chawla & Balasubramaniam, 2021; Gandhi, 2014).

The Dalai Lama has long contended that "Tibetans have adhered to spiritual and environmental values in order to maintain the delicate balance of life... inspired by Buddha's message of non-violence and compassion." They have respected all life and "abandon[ed] war as an instrument of national policy." Speaking "as the leader of the Tibetan people and as a Buddhist monk devoted to the principles of a religion based on love and compassion," the Dalai Lama envisions Tibet as a "zone of peace... in which humanity and nature can live together in harmony," and as a "self-governing democratic political entity founded on law by agreement of the people for the common good and the protection of themselves and their environment, in association with the People's Republic of China" (The Fourteenth Dalai Lama of Tibet, 1988).

LITERATURE REVIEW

This chapter is informed by the cross-pollination among scholarly work on environmental sustainability, religion (notably Buddhism), and human rights. Through their mythologies and ways of life, religious traditions have always engaged with questions about the relationship of human beings to the natural world. Indeed, religious ideas have influenced the contemporary environmental movement, which gathered momentum in the 1960s and 1970s, soon after the International Bill of Human Rights was established. "Green Buddhism" and "engaged Buddhism" are now institutionalised as fields of academic enquiry in addition to rights-based advocacy, and the contributions of different Buddhist traditions to the Earth Charter are also well documented in scholarship (Badiner, 1990; Batchelor & Brown, 1992; Chapple, 1993; Clugston & Holt, 2012; Kaza & Gottlieb, 2006; Morgante, 1997; Nhât Hanh, 2009; Rome, 2009; Sponberg, 1998; Toh, 2012; Tucker & Williams, 1998).

Human rights are impossible without a healthy and nurturing environment; thus, their fulfilment requires protecting the environment. Religious rights, a subsection of human rights, may be at stake if, for example, the sacred water is polluted or the trees used for religious ceremonies are felled (Jenkins, 2012). Furthermore, scholars, activists, and, increasingly, the legal system have considered whether nature itself might be a bearer of legal rights (Jenkins, 2012; O'Donnell, 2018; Waldau, 2003). In this context, it is instructive to read the Dalai Lama's call for environmental justice—encompassing the protection of river systems, plant- and animal life—as a call for the preservation of human rights. Environmental protections are also needed to safeguard the rights of communities, especially indigenous communities, who often have a close relationship with the natural environment. Political and scientific leaders in the modern world have recognised this, often when spurred by environmental activism from religious communities (Dwivedi, 2004; Jenkins, 2012). For example, Jain activists in Western India have publicly advocated against the mistreatment of animals (Evans 2014).

This chapter is informed by recent work about Tibetan exile and Tibetan Buddhism, most notably Toni Huber's analysis of Tibetan self-representation in exile, which examines the emergence of a modern, liberal Shangri-la image informed by complex translational politics and Western political traditions (Huber, 2001). The latter has its precedents in the discourse of liberal and social protest movements of environmentalism, pacifism, human rights, and feminism in the industrialised West. Tibetologist Tsering Shakya has identified four aspects of the western approach to the study of Tibet: missionary, diplomatic, journalistic, and social scientific, and added a fifth: travelogue (Shakya, 1994). The early accounts by missionaries, ethnographers, travellers, journalists, and diplomats laid the ground for the first histories of Tibet. The authors' employment in the colonial administration explains the works' emphasis on regional rivalries in Asia-- between China, Britain, and Russia-- and the task of delimiting the boundaries of Tibet. Subsequent generations of historians of Tibet in the West consulted British archives, and were likewise focused on questions of sovereignty, treaty relationships with British India, China, and other Himalayan powers, and, most importantly, on the boundary. Shakya pointed out that "Western studies on Tibet are rooted in the historical encounter between the West and Tibet," and "remain defined by their origins" in the encounter with Tibetan Buddhism and the expansion of colonial interests across the Himalayas (Shakya, 1994). The reinvention of Tibetan Buddhism, informed by the broad transnational movement often described as 'Buddhist Modernism,' plays a pivotal role in these identity constructions. Characterised by a rational and undogmatic reinterpretation of Buddhism, this movement rejects traditional superstitions and ritual elements perceived as incompatible with modern liberal values (Barstow, 2019; Geary, 2008; Lopez, Jr., 1998/2018; McMahan, 2008). In part through prolonged contact with the globalised cultural environment, the Tibetan exile elite adapted Western representational styles to reshape their self-image. Thus, the sincere urgency about Tibet's importance as 'the roof of the world' is nevertheless also an appeal to garner patronage and international support for Tibet's political struggle against China.

Ecological Education Within the Tibetan Nuns Project

The preservation of Tibetan culture and its fragile ecosystem are important anchoring points for the Central Tibetan Administration. It has, under the aegis of the Dalai Lama, also founded several institutions to preserve different facets of Tibetan culture: Tibetan Institute for Performing Arts (1959), Tibet House (1965), the Central Institute for Higher Tibetan Studies (1967), and the Library of Tibetan Works and Archives (1971; with the museum opening to public in 1974), Norbulingka Institute (1995).

The Tibetan Nuns Project (TNP) is a relatively recent initiative of the exile government, founded in 1987 by Rinchen Khando Choegyal, formerly the Minister of Education in the Central Tibetan Administration, and the present Dalai Lama's sister-in-law. The TNP currently supports over 800 nuns and seven nunneries in India, as well as nuns some living on their own and in retreat, and its headquarters are in Dolma Ling Nunnery near Dharamsala. In addition to providing financial assistance and educational opportunities for Tibetan Buddhist nuns, the Project is invested in initiatives stressing a greater participation of nuns in the Buddhist sangha, and operates under the joint auspices of the Tibetan Women's Association and the Religious Department. TNP nunneries have also opened their doors to nuns from India (particularly Kinnaur, Ladakh and Spiti Valley) and countries in Europe and North America, many of whom came there as a result of the international visibility of Tibetan Buddhism. The TNP in India is both the catalyst for and the result of this worldwide interest.

An analysis of TNP's own literature about the educational initiatives in nunneries reveals a heavy influence of the scholarship on the historical encounter between the West and Tibet, which has "remain[ed] defined by their origins" in West's encounter with Tibetan Buddhism (Huber, 2001; Shakya, 1994). The ecologically informed monastic education and daily practice within the TNP is rooted in an understanding of Tibetan identity and the character of traditional Tibetan society and culture: "Tibetans are an essentially peaceful people, who never developed an army of their own," "Environmentalism is an innate aspect of Tibetan culture," and "Women in traditional Tibet enjoyed a higher degree of equality than in other Asian societies" (Huber, 2001, p. 357). Huber has argued that these notions of Tibet are "unprecedented and distinctly modern" and should be "understood as the products of complex translational politics of identity within which populations such as the Tibetan exiles . . . are increasingly representing themselves and being represented by others" (p. 357).

As Eric Hobsbawm pointed out in *The Invention of Tradition* (1983), these self-representations are traceable to a dateable past: the mid-1980s in this case when the Dalai Lama actively sought support for the Tibetans' "peaceful" struggle against the Chinese and was awarded the Nobel Peace Prize in 1989. Commenting on the representational style and agenda of this self-image, Huber outlines four characteristics that underscore an influence of Western political traditions facilitated by the experience of exile and subsequent contact with the world (Huber, 2001, pp. 358–359). The first feature of this image is the "reinvention" of a "modern, liberal Shangri-la image" which has its precedents in the discourse of early modernity (characterised by what Huber calls the "three powerful -isms" of colonialism, orientalism and nationalism") and the discourse of liberal and social protest movements of environmentalism, pacifism, human rights and feminism in the industrialised West. The second characteristic of this image is that it was "not spontaneously born out of a groundswell of changes in the general consciousness of the majority of the Tibetan exiled population," but shared the "same general type of parentage as most identity images in emergent nationalisms throughout the developing world." They are the creation of the political and intellectual elite in exile composed of "educated and cosmopolitan Tibetans who have learned and skilfully adapted a repertoire of modern representational styles and strategies during the course of their *enforced and prolonged contact with the modern world*" (Huber, 2001, p. 358; added emphasis). This contact was necessitated by the condition of exile and the experience of the diaspora, but the exiled elite could adapt these new self-images "only after they gained access to and began to draw heavily upon the globalised production and flow of cultural resources and institutions offered by the contemporary transnational cultural environment." Finally, Huber asserts that the new Tibetan exile identity claims represent an appropriation of the hitherto Western "myth of Tibet" that objectified the Tibetan as "Other." The Tibetan's "creative reflection back to the West" makes these identity claims appealing and acceptable to Westerners (Huber, 2001, pp. 358–359).

Another consequence of this influence is the TNP's celebration of the exiled community's interaction with the outside world and the need to make this contact available to the nuns. In the film *Dolma Ling Nunnery: Educating for the Future* produced by Nunnery's media center, the narrator points out that "a phone booth and internet café [in Dolma Ling] provide the nuns with access to the world." Immediately after images of the internet café with nuns working on computers, we hear the loud mooing of a cow, accompanied by the visual of a nun bathing and feeding the cows. The narrator points out that "an important responsibility at Dolma Ling is taking care of our thirteen happy healthy cows so they produce the wholesome milk that is used in the kitchen." Therefore, Dolma Ling is not only a space equipped with modern amenities, but is simultaneously attentive to the environment, often construed as the putative other of a modern lifestyle.

In a similar film made by the Dolma Ling Nuns' Film Department called *Dolma Ling Institute and Nunnery: Educating for the Future*, the narrator alludes to the waste disposal system of Dolma Ling that segregates and recycles the waste, the cows that produce all the milk required for the two hundred nuns and the home-grown manure for the fields; thereby underlining the ecologically responsible behaviour of the nuns. In the film Tibetan Nuns Project: Educating for the Future, produced by Dolma Ling Films Department and Faux Reel Films (2009), Phillipa Russell, then the Architectural Supervisor at TNP, points out that the project is serving the needs of the whole world through nurturing Tibetan nuns: "A balanced sangha will generate peace and harmony in the world." Choegyal underlines the importance of preserving the monastic tradition through TNP: "The Tibetans in exile have so many windows to see into the cause of Tibet. And I think the TNP is one of them. When you say with the hope of going in to free Tibet one day, with what do you go? I don't expect any of us to be taking back gold and silver. But at least let's take back our culture and identity" (*Tibetan Nuns Project*). The film also stresses the importance of a modern curriculum for the nunneries: "[T]hey [nunneries] do not have these very strong traditions [like they have] in particular monasteries, in a sense you have more freedom to add more subjects, do different things, work with something and find whether it works or suits the nuns or not, so they have more in terms of English and modern curricula subjects" (Russell).

After a long shot of nuns chanting for "all sentient beings," Elizabeth Napper, then the co-director of TNP points out against the visual backdrop of skyscrapers and a bejewelled American flag that the systems in the West work so efficiently that people take them for granted and get tired of them: "Getting out of Western life and investing time and effort in something like this puts things in perspective . . . half the population [of the world] is women and it is important to educate them for everyone" (*Tibetan Nuns*

Project). Statements like these on the "environmentalist" Tibetan identity stress an innate worldview and modes of behaviour that are always described in terms of "completely modern" concepts and language with terms like "ecological balance" and "preservation" (Huber 360). Typical of the invented tradition as theorised by Hobsbawm, the contemporary validity and expediency of this approach to the environment is stressed as much as its antiquity. Tibetans claim a tradition of this approach to the environment:

[W]e Tibetans have always been aware of the interdependent nature of this world. We know that . . . [f]or most of Asia, Tibet's environment has always been of crucial importance. And so for centuries Tibet's ecosystem was kept in balance and alive out of a common concern for humanity. (qtd. in Huber 360; added emphases)

As my emphases in the above quotation point out, the antiquity (or eternity) of an innate concern for the environment is expressed in Buddhist terms: "all humanity" and "interdependence." The subtext of this stress on Tibet's importance as "the roof of the world" and an "ecosystem under threat" is an appeal to garner patronage for international support for Tibet's struggle against China.

CONCLUSION: TOWARDS AN ECOLOGICALLY CONSCIOUS 'BUDDHIST MODERNISM'

Far from being confined to the nuns' philosophical and religious education in the classroom, ecological responsibility is a cornerstone of the daily practice of life in TNP nunneries, which in turn offers useful lessons for other educational spaces. Examples abound in their annual newsletters about how small initiatives contribute to a harmonious relationship between human beings and other sentient beings, among neighbours, and (Indian) hosts and (Tibetan) guests. When the monsoon season in 2021 wreaked havoc in Dharamsala, the nuns worked with surrounding communities to repair water lines and channels (TNP Newsletter, 2021, p. 3), while the nuns in Tilokpur Nunnery routinely work to clean the river and contribute to the local environment around the nunnery by recycling and planting trees (TNP Newsletter, 2017, p. 3). In the same year, they installed solar lights at Shugsep Nunnery themselves, to provide green energy for the premises (TNP Newsletter, 2021, p. 54). In 2018, nuns in Sherab Choeling Nunnery and Sakya College for Nuns began growing spinach and other leafy vegetables in their greenhouses to combat the iron deficiency in their diets (TNP Newsletter, 2018, p. 2). Many nunneries such as Dolma Ling and Sherab Choeling also take care of cows, who in turn provide for their dairy requirements (TNP Newsletter, 2017, p. 3).

The traditions outlined above have been made possible by a reinvention of the category of Tibetan Buddhism itself. Described by Heinz Becheret as "Buddhist Modernism," the reinvention is characterised by a reinterpretation of Buddhism as an essentially rational and undogmatic religion. It has also been revitalized by an encounter between the Tibetan and other South Asian Buddhist traditions. The project involved rejecting as "corruptions or distortions" almost all "superstitions" and traditional ritual elements that were unpalatable to the West. Finally, as Huber carefully documents, the Tibetans' membership in new international Buddhist groups like the World Buddhist Sangha Council (founded in 1950) led to the "projection of an essential nonviolent Tibetan national identity". The Council's professed aim was "to oppose war and to contribute towards achieving world peace 'by spreading the Buddha's message of compassion and wisdom against violence and materialist thinking devoid of moral values'" (Huber,

2001, pp. 361–363). Characterized by a rational and undogmatic reinterpretation of Buddhism and the idea that Buddhism is amenable to various kinds of social reform, this movement rejects traditional superstitions and ritual elements perceived as incompatible with modern liberal values.

REFERENCES

Badiner, A. H. (Ed.). (1990). *Dharma Gaia: A Harvest of Essays in Buddhism and Ecology*. Parallax Press.

Barnett, R. (Ed.). (1994). *Resistance and Reform in Tibet*. University of Indiana Press.

Barstow, G. (Ed.). (2019). *The Faults of Meat: Tibetan Buddhist Writings on Vegetarianism*. Wisdom Publications.

Batchelor, M., & Brown, K. (Eds.). (1992). *Buddhism and Ecology*. Cassell.

Bernstorff, D., & von Welck, H. (Eds.). (2004). *Exile as Challenge: The Tibetan Diaspora*. Orient Longman. https://books.google.co.in/books?id=_lLiIvbf9S4C&printsec=frontcover&redir_esc=y#v=onepage&q&f=false

Chapple, C. K. (1993). *Nonviolence to Animals, Earth, and Self in Asian Traditions*. State University of New York Press.

Chawla, S. (2010). *"Like a Sudden Recovery of a Forgotten Memory": The Invention of Tradition Among Exiled Tibetans*. University of Delhi.

ChawlaS. (2015). The Female Renunciant in Exile: (Re)-Invention, Translation, Empowerment [University of Virginia]. doi:10.18130/V31H49

Chawla, S. (2018, September 25). From Illiteracy to PhD: How Exiled Tibetan Buddhist Nuns Are Ensuring Gender Parity. *The Wire*. https://thewire.in/religion/from-illiteracy-to-phd-how-exiled-tibetan-buddhist-nuns-are-ensuring-gender-parity

Chawla, S. (2019, March 15). The exiled daughters of Tibet seek equality. *Firstpost*. https://www.firstpost.com/world/the-exiled-daughters-of-tibet-seek-equality-6268131.html

Chawla, S., & Balasubramaniam, M. (2021, February 24). Writing more of India into Tibetology. *The Hindu*. https://www.thehindu.com/opinion/op-ed/writing-more-of-india-into-tibetology/article33917089.ece

Clugston, R., & Holt, S. (Eds.). (2012). *Exploring Synergies between Faith Values and Education for Sustainable Development*. Earth Charter International and Earth Charter Center for Education for Sustainable Development at University of Peace. https://www.iau-hesd.net/sites/default/files/documents/febr_2_2012.pdf

Palmo, D. (2019, November 1). Tibet's Rivers Will Determine Asia's Future. *The Diplomat*. https://thediplomat.com/2019/11/tibets-rivers-will-determine-asias-future/

Dwivedi, O. P. (2004). Satyagraha for Conservation: Awakening the Spirit of Hinduism. In R. S. Gottlieb (Ed.), *This Sacred Earth: Religion, Nature, Environment* (2nd ed.). Routledge.

Evans, B. (2014). Engaged Jain Traditions and Social Non-Violence: Ethnographic Case Studies of Lay Animal Activists and Service-Oriented Nuns. *Cross Currents*, *64*(2), 202–218.

Gandhi, G. (2014, October 14). Honoured Guest: The gem that India will not call by its name. *The Telegraph Online*. https://www.telegraphindia.com/opinion/honoured-guest-the-gem-that-india-will-not-call-by-its-name/cid/1312859

Geary, D. (2008). Destination Enlightenment: Branding Buddhism and Spiritual Tourism in Bodhgaya, Bihar. *Anthropology Today*, *24*(3), 11–14. doi:10.1111/j.1467-8322.2008.00584.x

Gordon, S. (2010). Major Asian Rivers of the Plateau of Tibet: The Basics. *Education About Asia*, *15*(3), 15–18.

Gyatso, J. (2010). Female Ordination in Buddhism: Looking into a Crystal Ball, Making a Future. In T. Mohr & J. Tsedroen (Eds.), *Dignity and Discipline: Reviving Full Ordination for Buddhist Nuns* (pp. 1–21). Wisdom Publications.

Hannah, M. (2012). Colliding Gender Imaginaries: Transnational Debates about Full Ordination for Tibetan Buddhist Nuns. *Asian Journal of Women's Studies*, *18*(4), 7–44. doi:10.1080/12259276.2012.11666134

Härkönen, M. (2023). *Power and Agency in the Lives of Contemporary Tibetan Nuns: An Intersectional Study*. Equinox. doi:10.1558/isbn.9781800503021

Havnevik. (1994). The Role of Nuns in Contemporary Tibet. In R. Barnett (Ed.), *Resistance and Reform in Tibet* (pp. 259–266). University of Indiana Press.

Huber, T. (2001). Shangri-la in Exile: Representations of Tibetan Identity and Transnational Culture. In T. Dodin & H. Rather (Eds.), *Imagining Tibet—Perceptions, Projections, and Fantasies*. Wisdom Publications.

Jenkins, W. (2012). Religion and Environmental Rights. In J. Witte & M. C. Green (Eds.), *Religion and Human Rights: An Introduction* (pp. 330–345). Oxford University Press.

Kaza, S., & Gottlieb, R. S. (2006). The Greening of Buddhism. In *The Oxford Handbook of Religion and Ecology* (pp. 184–206). Oxford University Press., doi:10.1093/oxfordhb/9780195178722.003.0008

Liang, J. (2021, December 9). Buddhist nuns and female scholars are gaining new leadership roles, in a tradition that began with the ordination of Buddha's foster mother. *The Conversation*. http://theconversation.com/buddhist-nuns-and-female-scholars-are-gaining-new-leadership-roles-in-a-tradition-that-began-with-the-ordination-of-buddhas-foster-mother-172477

Liang, J., & Taylor, A. S. (2020). Tilling the Fields of Merit: The Institutionalization of Feminine Enlightenment in Tibet's First Khenmo Program. *Journal of Buddhist Ethics*, *27*, 231–262.

Lopez, D. S., Jr. (2018). *Prisoners of Shangri-La: Tibetan Buddhism and the West*. University of Chicago Press. https://press.uchicago.edu/ucp/books/book/chicago/P/bo26798165.html

McMahan, D. L. (2008). *The Making of Buddhist Modernism*. Oxford University Press.

Melnick Dyer, A. (2022). *The Tibetan Nun Mingyur Peldrön: A Woman of Power and Privilege*. University of Washington Press.

Morgante, A. (1997). *Buddhist perspectives on the Earth Charter*. Boston Research Center for the 21st Century. https://search.worldcat.org/title/38283659

Napper, E. (2020, November 11). *The Education of Tibetan Buddhist Nuns* (Tibetan Nuns Project, Interviewer) [Interview]. https://tnp.org/the-education-of-tibetan-buddhist-nuns/?mc_cid=74edd6d899&mc_eid=987b3f085d

Nhât Hanh, T. (2009). *Interbeing: Fourteen Guidelines for Engaged Buddhism* (Repr). Full Circle.

Norbu, D. (2001). Refugees from Tibet: Structural Causes of Successful Settlements. *The Tibet Journal*, *26*(2), 3–25.

O'Donnell, E. L. (2018). At the Intersection of the Sacred and the Legal: Rights for Nature in Uttarakhand, India. *Journal of Environmental Law*, *30*(1), 135–144. doi:10.1093/jel/eqx026

Padma'tsho (Baimacuo), & Jacoby, S. (2020). Gender Equality in and on Tibetan Buddhist Nuns' Terms. *Religions*, *11*(10). doi:10.3390/rel11100543

Palmo, D. (2019, November 1). Tibet's Rivers Will Determine Asia's Future. *The Diplomat*. https://thediplomat.com/2019/11/tibets-rivers-will-determine-asias-future/

Price-Wallace, D. (2023). Buddhist Pro-Woman Attitudes Towards Full Ordination: Tibetan and Himalayan Monastics' Views. *Journal of Global Buddhism*, *24*(1), 1–24. doi:10.26034/lu.jgb.2023.3140

Roloff, C. (2020). *The Buddhist Nun's Ordination in the Tibetan Canon: Possibilities of the Revival of the Mūlasarvāstivāda Bhikṣuṇī Lineage*. Numata Center for Buddhist Studies.

Rome, D. (2009, August 5). The Green Buddha. *Tricycle: The Buddhist Review*. https://tricycle.org/trikedaily/the-green-buddha/

Schneider, N. (2022). A Revolution in Red Robes: Tibetan Nuns Obtaining the Doctoral Degree in Buddhist Studies (Geshema). *Religions*, *13*(9), 838. doi:10.3390/rel13090838

Shakya, T. (1994). The Development of Modern Tibetan Studies. In R. Barnett (Ed.), *Resistance and Reform in Tibet* (pp. 1–14). University of Indiana Press.

Sponberg, A. (1998). Green Buddhism and the Hierarchy of Compassion. In M. E. Tucker & D. R. Williams (Eds.), *Buddhism and Ecology: The Interconnection of Dharma and Deeds* (pp. 351–376). Harvard University Center for the Study of World Religions, Harvard University Press.

The Fourteenth Dalai Lama of Tibet. (1987, September 21). *Five Point Peace Plan* [Address]. U.S. Congressional Human Rights Caucus, Washington, D.C. https://www.dalailama.com/messages/tibet/five-point-peace-plan

The Fourteenth Dalai Lama of Tibet. (1988, June 15). *Strasbourg Proposal* [Address]. European Parliament, Strasbourg, France. https://www.dalailama.com/messages/tibet/strasbourg-proposal-1988

Toh, S.-H. (2012). Engaged Buddhism and its Contributions to Sustainable Development and ESD. In R. Clugston & S. Holt (Eds.), *Exploring Synergies between Faith Values and Education for Sustainable Development* (pp. 62–66). Earth Charter International and Earth Charter Center for Education for Sustainable Development at University of Peace. https://www.iau-hesd.net/sites/default/files/documents/febr_2_2012.pdf

Tucker, M. E., & Williams, D. R. (Eds.). (1998). *Buddhism and Ecology: The Interconnection of Dharma and Deeds*. Harvard University Center for the Study of World Religions, Harvard University Press.

Waldau, P. (2003). The Specter of Speciesism: Buddhist and Christian Views of Animals. Oxford University Press. doi:10.1093/0195145712.001.0001

ENDNOTE

1 Earlier versions of this work have appeared in my unpublished Masters- and M.Phil. theses (Chawla, 2010, 2015). I gratefully acknowledge Anurag Sharma and Esha Lee, who interned with me during the winter break in 2023-24, for assistance with primary research for this chapter.

Chapter 5
Re-Defining Tree Worship as Artistic Becoming for Environmental Education

Shruti Jain
https://orcid.org/0000-0002-9239-7062
O.P. Jindal Global University, India

ABSTRACT

Trees play a profound role in the spiritual rituals of the Shakta Tantric tradition, acting as a catalyst for the practitioner's spiritual transformation. Yet, the practice of tree worship often seems obscure to the modern mind, necessitating innovative approaches to integrate its essence into contemporary educational frameworks. This chapter endeavors to redefine worship as an artistic process that could potentially cultivate empathy, highlighting its capacity to deepen human connections with the environment by exploring the implications of adopting a tree-centric or arboromorphic perspective within educational frameworks.

WORSHIPING TREE AS GODDESS

Worshipping the Tree
No, I'll never sign a contract on your feet
I'll love you as an independent girl
The way love between the hunter and the woodland
happens, the way a penguin loves a glacier
No, you've never asked me to sign a contract.
I thank, bow to you, beloved tree.
Steadfast I worship you in love and fight,
forest and hunter joined in our blood's stream.[1]

The worship of trees has been practiced in India since times immemorial. The early Hindu sacred texts, the Vedas and Upanishads, make frequent reference to sacred trees, referring to them as the most im-

DOI: 10.4018/979-8-3693-2577-3.ch005

portant living forms on earth. (Fowler-Smith, 2009, p.43) Within the living Goddess traditions of India too, nature (prakriti) considered to be the ultimate source of life, has been held in profound reverence. The introductory section explores the diverse constellations where trees hold significance in Hindu traditions, particularly within the context of Shakta tantric practices.

A central concept in Shakta cosmology is what Johnsen (1999) calls "the world tree". Johnsen describes the world tree as a symbol of the human spine comprising of chakras associated with primal instincts in the lower region: self-preservation, reproduction, and aggression and with the capacity for selfless love and genuine connection with others signifying divine potential: creativity, intellect/intuition, and spiritual absorption in the human realm or middle region. Shiva is situated at the highest chakra, symbolizing our Higher Self, observing, guiding, protecting, and testing us. Enlightenment occurs when Shakti, represented by the kundalini, reunites with Shiva. (Johnsen, 1999, p.21) Apart from the symbolic use of the tree in kundalini yoga, the forest, embodying the pinnacle of the Earth's fertility and productivity, finds representation in various forms, notably as the Earth Mother or the Tree Goddess. Aranyani, the Goddess of the Forest, holds a unique position within Hindu mythology, notably honored in one of the most vivid hymns of the Rigveda. Found in the *Aranyani Suktam* (Hymn 146, Book 10), she is portrayed as enigmatic, delighting in the tranquility of woodland glades and fearlessly venturing into remote wilderness. The supplicant in the hymn beseeches her to unveil the secret of her unyielding courage and solitude amidst the fringes of civilization. Adorned with bell anklets, her elusive presence is often signaled by the gentle tinkling of these ornaments, hinting at her ethereal dance:

GODDESS of wild and forest who seemest to vanish from the sight, How is it that thou seekest not the village? Art thou not afraid?

2 What time the grasshopper replies and swells the shrill cicala's voice,
Seeming to sound with tinkling bells, the Lady of the Wood exults.

3 And yonder, cattle seem to graze, what seems a dwelling -place appears:
Or else at eve the Lady of the Forest seems to free the wains.

4 Here one is calling to his cow, another there hath felled a tree: At eve the dweller in the wood fancies that somebody hath screamed.

5 The Goddess never slays, unless some murderous enemy approach,
Man eats savory fruit and then takes, even as he wills, his rest.

6 Now have I praised the Forest Queen, sweet-scented redolent of balm,
The Mother of all sylvan things, who tills not but hath stores of food.

(The Rig Veda, Book 10, HYMN 146, Verses 1-6)

In Bengal, Aranyani is closely linked with revered species like the sal (Shorea robusta) and asvattha (Ficus religiosa). Similarly, in Comilla, she is revered as Bamani, while in Assam, she takes the form of Rupeswari. (Shiva, 1988, p. 54) The sacred tree, acting as a microcosm of the cosmos, symbolizes the boundless wellspring of cosmic fertility, offering a tangible connection to the divine order. Notably, the Aranya Devi Temple in Arrah, Bihar, stands as a rare testament to her enduring reverence. The Srimad Devi-Bhagavatam mentions *Goddess Shatakshi,* the goddess with innumerable eyes— *later known as Shakambari (the goddess of vegetation)* —who incessantly shed tears for nine days creating rivers that renourished the world undergoing a calamity caused by the usurping of the knowledge of the Vedas by the demon *Durgamasura* (Srimad Devi-Bhagavatam, Book 7, Chapter 28). In *Kaula Tantra*, a Tantric tradition which is characterised by distinctive rituals and symbolism connected with the worship of Shakti and Shiva that is associated with cremation-ground asceticism (Flood, 1996, p.166), tree spirits merge with Yoginīs, and the trees they inhabit as Yakṣinīs are revered as Kula trees. According to the

Kulachudamani Tantra the "Sadhaka should rise in the morning, make his Pranama to the Kula trees (Kulavriksha), and contemplate upon the Kula (Shakti) from the Muladhara to the Brahmarandhra and meditate on the Guru. The Kulavrikshas according to the Kameshvara Tantra are Shlesmataka, Karanja, Nimba, Asvattha, Kadamba, Vilva, Vata, Ashoka."[2](Vedantatirtha, 1918, p.7-8) Dyczkowski, in his exploration of the *Kubjika cult*[3], unveils that within the multifaceted Newar tantric system, Goddess Kubjika holds the role of a forest goddess among her numerous other aspects. "The mandala of the goddess is the tree within which she resides and over which she presides in the form of a Yaksini, an ancient India folk deity of vegetation and the earth. The whole tradition and its teachings are represented as the Tree of Consciousness which grows out of the Divine Current (divyogha) of the Divine Kaula tradition. [...] This tree is especially represented as a tamarind (cinca) that grows on the Island of the Moon, the goddess's mandala, from the middle of the triangle. Thus, in the later Kubjika Tantras, where this imagery is elaborated, the goddess is frequently called the Lady of the Tamarind (Cincini or Cinca). Under the shade of this tree, which is said to be the pure bliss present within all beings, one attains the supreme liberating repose beyond pleasure and pain. One of the founders of the Kubjika cult was called Vrksanatha ('the Lord of the Tree') because he achieved the perfect repose (visrama) of liberation under this tree when the goddess appeared to him. There, in the shade of the tree, she transmitted the enlightening Command (ajna) which gave him the authority (adhikara) that empowered him both to attain this realization and to transmit the teachings which lead others to it." (Dycskowski, p.70-71)

Another noteworthy tradition that integrates tree worship with the worship of the Goddess is the autumnal celebration of Goddess Durga – the *Durga Puja*. The sixth (ṣaṣṭhī) and seventh (saptamī) lunar days involve specific rites: awakening the goddess within a bilva tree (bodhana), worshiping her manifestations as Cāmuṇḍā and Kālī on the branches, summoning her nine emanations within nine leaves (navapatrapūjā/patrikāpūjā), and animating an unfired clay image of the goddess (prāṇapratiṣṭhā). According to the Skanda Purana (Chapter 250, Verses 1-19, p.1064-1065) the bilva tree grew out of the sweat droplets of Devi Parvati, while she was engaged in intense sadhna on the Mount Mandara. Just as Sri resides in the Tulasi plant so Parvati resides in the Bilva plant. She is said to reside in every part of the bilva tree in her various incarnations. – as *Girija* in the roots, as *Maheshwari* on its branches, as *Daksayani* in its stem, as *Gauri* on its bark, as *Aparna* the middle of the bark, *Durga* in the flower and as *Uma* in the branches and twigs. On the behest of Girija, nine crores of Saktis are stationed on all its thorns for the sake of protecting living beings. (Skanda Purana, Ch. 246, Verse 36, p.1057)

Navapatrika is a collection or sheaf of nine plants which together form one entity, the Kalabau or banana-wife (it is also called the Nava-Durga or nine Durgas). It is organized around a young banana or plantain plant, which is wrapped by an aparajita vine with a variety of other plants (usually bilva, turmeric, arum, pomegranate, barley or manaka, rice, jayanti, and ashoka) and dressed as a married woman. Some informants say that the plants represent different parts of Durga's body: head, hair, breast, nose, and so on. The leaves of the banana or plantain are twisted like a woman bowing her head, with turmeric as Durga's skin when she was incarnate as the golden Uma, and paddy rice representing Durga's aspect as fortune or Lakshmi. Pratapachandra Ghosha mentions the various goddesses worshiped in Navapatrika: Kalika in the arum plant, Durga in the turmeric plant, Brahmani in the plantain, Karttiki in the sesbania, Shiva (the goddess) in the bilva tree, Raktadantika in the pomegranate, Sokarahita in the ashoka tree, Chamunda in the mana tree, Lakshmi in the rice plant. (McDaniel, 2004, p. 227-8) In the autumnal worship of the goddess as Lakshmi, the Navapatrika is taken to in some parts of Bengal, as the best representative of the goddess. (Das Gupta, 1953, p.54)

Bengalis also venerate the Goddess "*Sasthi*, who is a guardian deity of child-birth. They found a relation between *Vata* and *Sasthi*, hence, plant *Vata* during *vrksapratistha* ceremony. No image is needed for the worship of Sasthi, she is represented by Vata." (More & Valhe, 2020, p.155) Shashti, the goddess who rules over women and the protector of children, is regularly worshipped at six seasons of the year and at special seasons also. At child-birth, and even before, her blessing is invoked; and after birth until the child attains to manhood or womanhood, it is supposed to be under her special care. In the sickness of children her aid is first invoked to effect their recovery. At this festival it is not common to set up an image of the goddess; but as the banyan tree is regarded as specially sacred to her, she is represented by it on this occasion. On the day of worship the women of the neighbourhood march with music playing to the tree; the mothers of sons with joy written on their faces, those as yet unblessed with disappointment in their looks and fervent prayers for this blessing in their hearts and even on their lips. The mantras are recited by the priests; the honoured mothers who have sons give presents to those who have none, and the procession wends its way back to the village. (Murdoch, 1991, p. 46)

David Haberman writes, "While most of the neem tree worshipers I spoke with identified the tree as Shitala, for them Shitala is usually considered to be one of the many forms of the Great Goddess, the Mother of Life, who assumes a variety of manifestations to nourish all aspects of life. She is frequently addressed as Ma, Mata, Mai, or Maiya, all words that mean 'Mother' and that are commonly used to address a goddess within Hinduism. I met a man worshiping a neem tree at a temple on the pilgrimage circuit that encircles Bañaras who informed me that the neem is a 'goddess-tree' (devi-vriksha)". (2010, p. 180)

Haberman (2013) explains how watering trees, wrapping them with colorful cotton string, circumambulating them clockwise, and offering them food and flowers are some of the important ways trees are honored. They are also addressed, touched, and embraced in intimate ways. In this context, trees are clearly sentient beings with whom humans can have mutually beneficial relationships. "Sometimes trees are depicted in literature as expressing desires for contact with people.[…] They are described as craving mainly for touching (sparśa), embracing (āliṅgana), or stroking and kicking with the foot (pādaghāta). The range of their cravings is partly determined by a tree species, e.g. in the case of priyaṅgu creeper touching is desired most, but sprinkling with wine from the mouths of young women is the whim of the bakula tree (also known as vakula), the red-flowering aśoka tree – before it blossoms again – craves for foot of a maiden or young woman's heel kicking, whereas glancing is required by the tilaka tree, the kumbaraka tree expects embracing above all, while the mandāra tree needs pleasurable talk and the cūṭa tree's desire is just feeling human's breathing upon itself, whereas the campaka, nameru and karṇikāra trees enjoy laughing, singing and dancing respectively."[4] (Jakubczak, 2017, p.81)

It is clear from the various instances cited above that the reverence for trees transcends mere symbolism, fostering a profound connection with nature and the divine. Even in contemporary times, individuals continue to worship trees. But only few are able to transcend the mere ritual aspect of tree worship and contemplate sustainable methods of cohabiting with trees like Yogesh Kesharwani, a resident of Jabalpur who constructed a house around a 150-year-old peepal tree, ensuring no harm befalls the tree in the process.[5] The question then arises - what exactly does the practice of tree worship entail? The next section deals with the act of worship and how it can be re-invented in the modern times to bring a more meaningful engagement with trees.

WORSHIP AS BECOMING-OTHER

The practice of worship has often devolved into a mere display of unquestioning reverence towards the deity. People engage in worship primarily with the expectation that the deity, capable of alleviating their suffering, will grant them boons. Consequently, the deity transforms into a mere means for fulfilling desires, losing its deeper significance.

Worship is a symbolic process of transformation. The gesture of folding hands in worship symbolizes this transformation, as one hand joins the other, implying a unity where one becomes the other. In the ritual of *prana-pratishtha* – life is infused into a stone which is the *murti-rupa*. But in addition, there is also the *prakriti-rupa*—referred to the tree as a 'natural form' of divinity that appears without the aid of any human intervention. (Haberman, 2010, 182). An important technique in the Shakta tantric tradition is the technique of *Jivanyasa* (infusion). In this technique, "the sadhaka infuses his body with the life of the Devi, the Mother of All" (Woodroffe, 2009, p. 517). The *mantras* that are chanted during the sadhana are physical descriptions of the Goddesses. Through the repetition of the sounds and the focus on the Yantra, the infusion is achieved. Whether looking inward in meditation or gazing outward upon the world, the fully realized *sadhaka* sees the same thing: the blissful projection of the Self—as—Goddess. (Lidke, 2017, p. 30). The worshipper utters the ritual words and with his body performs the ritual acts, such as the gestures (Mudra), the giving of offerings, and so forth. And the reason is, as the Gandharva Tantra says: "By meditating on anything as oneself, man becomes that." The mind assumes the form of that object – [...] So the worshipper is enjoined constantly to think: "I am the Devi and none other". By meditating on *Devi*, man becomes *Devi*. (Woodroffe, 2009, p. 419)

This can be understood through the Deleuzian concept of Becoming-other. Becoming-other is established via 'diversity, multiplicity [and] the destruction of identity.' (Deleuze, 1995, p. 44); it presupposes breaking out of our old outlived habits and attitudes so as to creatively 'bring into being that which does not yet 'exist' (Deleuze, 1994, p. 147). This makes the process of becoming-other a process of "creative learning" that pays "attention to places and spaces, to retrospective as well as untimely memories of actual and potential actions, and to dynamic forces that are capable of affecting and effecting changes, thus contesting the very identity of subjects participating in the process." (Semetsky, 2011, p. 139)

Becoming the other, be it the deity or the tree, hence represents a form of creative learning that prompts us to recognize the external world as an integral part of ourselves. This process has the potential to instigate changes both within and outside the individual, fostering empathetic connectedness. Sherma while defining Tantra as a "theology of identification" emphasizes that it is not only a source of "inspiration" for "personal spiritual empowerment." (1998, p.124), but also includes "a radical empathetic connectedness with the human community or the ecosystem." (ibid.)

It is worth contemplating on the reasons why ancient scriptures have established connections between Goddesses and plant life. From a theological standpoint, that which supports and nourishes humanity, has always been attributed a divine status in various texts. From an anthropological perspective, the veneration of trees in India can be comprehended as acknowledging the crucial role of forests in facilitating rainfall, vital for sustaining a primarily agricultural economy. Several trees also have medicinal value. The neem tree, for e.g. "is honored particularly for its healing abilities in early Puranic literature. The medicinal value of the neem is so considerable that it is often called the 'village pharmacy' and has accordingly been used in many health-related situations. Renowned for its antiseptic and disinfectant properties, the tree is thought to be particularly protective of women and children." (Haberman, 2010, p.178-9) However, as Jakubczak rightly remarks: "What is worth emphasizing, it is not the tree that was worshipped but the spirit residing in them." (2017, p. 80) Even Kesharwani's statement underscores this, when he states the reason for refraining from cutting the peepal tree: "We believe that 350 million gods and goddesses reside in one peepal tree." Does tree worship then only amount to primitive animism? According to Haberman, the anthropomorphic activity of attributing Goddess like qualities or personhood to trees need not be disparaged on account of its anthropomorphism, but "can be best understood as an intentional strategy for establishing more intimate relationships with the nonhuman world." (2010, p. 173) The question then arises, who do we worship while worshiping a tree – the goddess or the tree or both? Mukhopadhyay reminds us that "we need to understand that the Goddess is neither wholly metaphysical nor wholly material; she may be seen as a mediatrix between materiality and metaphysicality." (2023, p. 28).[6] Haberman explains this dual nature of goddess as tree worship through the term *svarupa*, which "is a technical one with a double meaning. It literally means 'own-form', and frequently refers to 'the form assumed by a deity'. Therefore, it can mean the specific form a deity takes in the world. Divinity within Hinduism is typically understood to be infinite and all pervasive, but manifests in particular concrete forms. Accordingly, divinity—although unified at the unmanifest level—can manifest as a multitude of individual beings. A particular tree being, then, is both specific and nondifferent from the Supreme Being." (2010, p.181) From this follows that Tantric traditions acknowledge the intrinsic value of trees. This stems from the tantric "perception as disclosures of sacred in the ordinary". (Kochetkova, 2022, p. 93) In his interpretations of the *Vigyan Bhairav Tantra*, Osho elevates the tree's significance, aligning it with the Tantra tradition's principles of becoming. By emphasizing the intrinsic value of nature objects, he imparts a lesson in empathy and inclusion through his understanding of spiritual transformation. While describing the techniques of becoming one with the whole, Osho remarks: "Sitting under a tree, look at the tree, then close your eyes and feel that the tree is within you. [...] When you feel that the tree is within you, immediately you will feel more young, more fresh. And this is not imagination, because the tree and you both belong to the earth. You both are rooted in the same earth and ultimately rooted in the same existence. So when you feel that the tree is within you, the tree IS within you – this is not an imagination – and immediately you will feel the effect. The tree's aliveness, the greenery, the freshness, the breeze passing through it, will be felt within you in your heart. Include more and more existence and don't exclude." (Osho, 1991, p.290) The revitalizing impact of trees, underscored by Osho, has also been scientifically validated. In their findings, Singh and Hageneder reveal, that "trees don't only partake in their own bioelectrical processes, which mainly take place in the living layer of cells beneath the bark and in the leaves; they also constantly feed the positive charge of the ionosphere into the earth's crust, which is negatively charged. One of the simple laws of physics tells us that a live conductor of electricity gives off its own electromagnetic field. The dispute over whether trees possess their own aura can now be laid to rest. Because the electrically active layers can be found just under the bark, hugging trees has a deeply vitalising effect on us. The same is true for sitting with your back resting on a tree. There is no doubt that doing this leads to reciprocal stimulation for both human and tree. Doing this means that the flow of energy going through the tree can directly interact with the spine and energetic processes of the human brain." (2007, p.21)

This elucidates the scientific underpinnings of the concept of the Shakta world tree, as described by Johnsen (1999). It also provides insight into the rationale behind the preference of numerous spiritual practitioners to engage in meditation while seated beneath a tree. The Om Swami Samartha Sadhana Kendra in Ranavali, Maharashtra, India, for instance, exemplifies one such locus for tree meditation, with over 5000 trees where individuals convene to acquire and refine their tree meditation techniques.

Common practices within this setting include tactile interaction, verbal communication, praying, embracing, and contemplation in the proximity of trees.[7]

In his discourses, Osho also points toward the possibility of cultivating empathy through mindful interaction with trees: "As you are changing every moment, the tree is also changing – if you can feel a deep affinity with it, what they call empathy. Empathy means you have become so sympathetic that really you become one. The moods of the tree become your moods. And then, if this goes deeper and deeper and deeper, you can talk, you can have a communication with the tree. Once you know its moods you start understanding its language, and the tree will share its mind with you. It will share its agonies and ecstasies." (Osho, 1991, p.495) In the interpretations of Osho, one can observe a transition from deity-centric worship to tree-centric or nature-centric meditation. This shift reminds one of what Cooper-Smithson terms Arboromorphism – "the literal, literary, or psychological process of becoming a tree." (2022, p. 233) Arboromorphism, focusing on the idea of "becoming a tree," encourages individuals to identify with and emulate the qualities of trees, such as resilience, interconnectedness, and rootedness. This perspective can inspire behaviors that prioritize environmental conservation, such as tree planting, nurturing green spaces, and recognizing the intrinsic value of ecosystems. It can foster empathy and emotional connection with the environment. Viewing nature as possessing agency and consciousness can lead to actions aimed at protecting and preserving natural habitats, as well as promoting ethical treatment of non-human beings.

BECOMING TREE AS RECONNECTING WITH NATURE

In the discourse of ecological psychology, the act of becoming tree can be construed as a process of re-establishing symbiotic ties with the natural world. Michael Cohen posits that within the fabric of our surroundings lies a latent intelligence, perceptible through a myriad of sensory inputs, serving as incontrovertible evidence of our intrinsic interconnectedness. Regrettably, contemporary societal constructs often sever these ties, marginalizing natural phenomena and fostering a cognitive disassociation from them. Consequently, our capacity to address the ramifications stemming from the absence of nature is compromised.[8] This existential discontent also forms the genesis of Sumana Roy's latest book "How I Became a Tree" (2017) where she embarks upon an introspective journey to write about her "experience of metamorphosis", her "spiritual and emotional transformation into a tree" (Roy, 2017, p.26), exploring the commonalities between trees and humans and the lessons trees offer.

Roy's transformation into a tree at the conclusion of the book is not literal; rather, it serves as a metaphor for elevating human consciousness to the nuances of the arboreal realm: "Not until a bird came and sat on my shoulder around sunset one day. I did not move. I do not know about the bird but I was certain that in the thinning margins of that forest in Baikunthapur I was, at last, ready to be a tree." (p.222).

Roy's writing does not seek to provide definitive answers but rather prompts thought-provoking questions. One encounters a deliberate "self-imposed resistance to closure" (Retallack, 2003, p. 13) within her prose, inviting readers to engage in ongoing contemplation and exploration. Hers is the kind of poetry, which in principle, as Retallack writes, engages "in exploratory projects and procedures, it is the work's poethical form of life—what informs its geometries of attention—[…]." (p.19)

Roy questions the impact of her transformation on her understanding of time: "So, when I look back at the reasons for my disaffection with being human, and my desire to become a tree, I can see that at root lay the feeling that I was being bulldozed by time" (p. 3). She then decides to write at a pace called

'Tree Time'" (p. 4). She asks herself whether it was possible to write like a tree. – which meant for her losing her sense of grammar which is a linguistic baton with which to control time. (p. 62).

Roy's exploration delves into the desire to embody the essence of a tree, embracing what she terms "Tree Time," which could serve as a powerful catalyst for environmental education. She writes, "I wanted that confidence of the tree, the complete rejection of all that made humans feel inferior or superior" (Roy, 2017, p.12). She muses about age: "Age, I was certain, was important to trees. The wrinkles on our face and neck, the accumulation of folds around hips and thighs had, civilizationally, become embarrassing to humans. The age of trees was to be found in similar lines, in circles denoting lived years, in the girth of time that gave aged trees a kind of sober dignity" (p.6). Roy further comments on the industrial age and its perception of age: "Our lives in the industrial age, lived bizarrely as an approximation of machines, had made us think of age as ugly—in the way machines rusted, wasted, and gradually became ugly before they fell apart" (p.6), and embraces the present moment: "carpe diem, seize the moment, living in the present—that was tree time, a life without worries for the future or regret for the past" (p.6). She highlights the societal abundance and its disconnection from inner life: "epidemic of abundance around me. Ours was an age of excess—more food and clothes and houses and things than we needed, an extravagant show of wealth and emotions without either being connected to the inner life" (p.11). Reflecting on tree care, she notes, "But I didn't know a single tree which needed to use makeup. [...] Both the hairdresser and the gardener seemed to believe in the value of snips and cuts for a better future, and in this I began to see the kinship of my undisciplined hair with the wayward branches of trees. I had begun to feel the violence of seasonal pruning and cutting that was inflicted on plants and trees" (p.10). She ponders on affection and attraction towards trees: "Was I being adulterous in loving more than one tree?" (p. 12). She reflects on the ahistorical nature of plants: "plants continue to remain ahistorical creatures" (p. 16) and notes the congruence between being and becoming: "there was no gap between who they were and who they wanted to be" (p. 16). She contrasts human noise with tree silence: "There were two things about this—one was the noise of humans, the other was the vocabulary of silence of the active life of trees. The opposition was terribly stark— the complaining tone that accompanied human work life contrasted with the near silence of the industriousness of trees" (p.23). Roy examines tree responses to wind and sound: "I began calibrating the reactions of different tree leaves to the wind" (p.24) and "My experiments with the sound recorder had brought about a new realization—that trees shared a natural sound with people. It is the sound of resistance—like protesters 'raising their voice', trees produced a sound that held in it their fight against wind, water, rain, to tearing, cutting and breaking. Like everything else, about sound too, they were economical. Revolution. Rebellion. Resistance. All other sounds were noise" (p. 25).

As is evident, Roy's account prompts a deep reflection on the connection between trees and humans, portraying trees as silent instructors with invaluable wisdom for humanity. Her observations on tree aging, resilience, and resistance to external forces lay a fertile groundwork for developing educational modules aimed at nurturing a stronger bond with nature. Further, by reminiscing on instances from her childhood of "the fascinating play of art imitating life and art imitating that art" (p. 48) Roy writes about how the character from O Henry's short story *The Last Leaf* "Johnsy has begun to believe in her life—and death—being a mimicry of the life of leaves on the tree in winter" (ibid.) and how the leaf her schoolmate Saroj had drawn for the play had gained agency and become an "actor" (p.50). Embodying the essence of trees through acting hence serves as a performative exploration, bridging the gap between the human and non-human realms. Roy's musings point toward this kinesthetic understanding of art which fosters a deeper appreciation for the interconnectedness of all life forms.

Similarly, the famous painter Nandlal leaves a long-lasting impression on Roy, when she realizes that drawing leaves is also in a way an exercise of becoming leaves. Through Nandlal's analogies, Roy discerns the parallels he draws between the anatomy of trees and that of humans: "The tree-becoming human is also on the next page where the illustration of tree bark made me want to scratch a rough patch of skin on my knee, where so many scars of childhood games live." (p. 39)

Roy's arboreal meditations arise out of her critique of modernity. Through this lens, she employs creative becoming as a tool to navigate through human fallibilities. This can prove to be of great significance within the realm of environmental education, as it contains in it the seeds for an environment-sensitive pedagogy. An intriguing facet of this discourse lies in exploring artistic avenues that facilitate a connection between humans and the non-human world. By encouraging individuals to delve into their subjective experiences and emotions related to nature, these art forms become conduits for connecting with the non-human world.

TOWARDS AN ENVIRONMENT-SENSITIVE PEDAGOGY

According to Luce Irigaray, "The vegetal world is also an irreplaceable teacher about the cultivation of life. It tells us that life cannot develop without endurance, patience, and faithfulness to oneself" (Gagliano et al., 2017, p. 22). Undoubtedly, modern education, with its focus on subjects like physics, chemistry, botany, and zoology, prioritizes understanding nature. However, solely studying science may not cultivate empathy in students. What's the value of knowing about trees if it doesn't deter students from harming them? Knowledge of trees should encourage inclusivity and empathy towards our arboreal companions. This is where arboreal humanities, with their emphasis on artistic becoming, may play a significant role. The Indian education system for instance, surely incorporates Yoga in its all-school curriculum. School children are taught asanas (yogic postures) such as *Vrkshasana* (the tree pose) or *Tadasana* (palm pose) in their Yoga classes. But do they learn why they assume these positions? Do they realize, as they stand in the tree pose, that they embody a tree? Are they instructed that, akin to trees, they can endure, display patience, and remain faithful, echoing Irigaray's thoughts? Do educators ever incorporate discussions on tree worship into their curriculum? Is there room for recognizing and appreciating cultural rituals that might be perceived as "religious" or "primitive," yet have the potential to shift perspectives towards our ecosystem?

Sumana Roy's arboreal memories can offer a robust framework for the art classes where students engage in drawing, dancing like or acting as or with a tree – thinking and feeling for trees – understand that the arboreal companions are not silent non sentient beings but are very much alive and an integral part of their ecosystem. Literature classes, apart from presenting stories on conservation of trees, can also incorporate exercises of writing to "tree time" – where students reminisce their associations with trees and in turn just dwell upon the various possibilities of connecting with trees. This suggests that educationists must contemplate ways of creating spaces for imaginative interactions to promote tree conservation and adopting an environment-sensitive pedagogy. This can be accomplished by prioritizing "becoming-other" or "creative learning" in their teaching methods.

Cohen describes one such nature-connecting activity: "Find a natural area that is attractive to you and something that is attractive in it (it is always based on this attraction kind of thing). Then you verbalize ''I like, or I love this sound of the wind in the trees because . . . '' You just say that you love it and you explain what it is that you love about it or why you love it. And then, you take that same sentence and you change it and you say, ''I love myself because . . . '' And you take the same thing you described, that sound of the wind, and describe yourself as it. And all of a sudden you find that what was attracting you to the wind is also part of you—and the wind says, ''You are a beautiful expression of nature in contact with these trees.'' And then all of a sudden, say, ''I am a beautiful expression of nature in contact with trees. I never thought of myself that way. I do not even think I am beautiful.'' (2010, p. 57) By prompting students to identify elements in nature that evoke personal attraction, such as the sound of wind in trees, Cohen initiates a process of verbalizing appreciation and identifying specific qualities that resonate with them. This exercise facilitates a deeper understanding of personal preferences and values in relation to the natural world. Moreover, Cohen's method guides participants to reframe expressions of appreciation for nature into expressions of self-love. By recognizing and internalizing the idea that the qualities admired in nature are inherent within themselves, students may experience an elevation in self-esteem, self-acceptance, and a heightened sense of interconnectedness with the environment. Alternative Art based activities suitable for educational environments could include [9]:

1. **Tree Collage**: Provide students with magazines, newspapers, and colored paper. Ask them to create collages depicting different types of trees, forests, or natural landscapes. Encourage them to explore texture, color, and patterns found in nature.
2. **Leaf Printing**: Collect a variety of leaves from different tree species. Dip the leaves in paint and press them onto paper to create leaf prints. Encourage students to realize the similarities between human anatomy and tree anatomy.
3. **Tree Bark Rubbings**: Take students outdoors and provide them with sheets of paper and crayons. Instruct them to place the paper over tree bark and gently rub the crayon over the paper to create bark rubbings. Encourage them to explore the textures and patterns of different tree barks and perhaps even see the similarities between human finger prints and tree bark patterns.
4. **Tree Sculptures**: Provide students with modeling clay or recycled materials such as cardboard, paper mache, or wire. Ask them to sculpt three-dimensional representations of trees or forest scenes. Encourage them to use their imagination and creativity to bring their sculptures to life.
5. **Seasonal Tree Observations**: Have students observe a particular tree throughout the changing seasons. Encourage them to document their observations through drawings, paintings, or mixed media artwork. Compare and contrast the tree's appearance and surroundings in different seasons.
6. **Environmental Installation Art**: Challenge students to create environmental installation art using found materials from nature. This could include creating temporary sculptures, land art, or installations that reflect their connection to trees and the environment.
7. **Tree of Life Collage**: Ask students to create a "Tree of Life" collage representing the interconnectedness of all living things. Provide them with images of trees, animals, plants, and humans to include in their collage. Discuss the importance of trees as providers of oxygen, habitat, and resources for other species.
8. **Writing to Tree time**: Ask students to maintain a journal and write about their experiences of interaction with trees. Students may also express themselves via poetry or drawing or even just words. Pay attention to the emotions and the content and not to the grammar while evaluating such journals.
9. **Reading, Acting, or Dancing:** Introduce diverse texts from world literature in literature classes. Curate theatrical productions centered around narratives depicting human interactions with trees. When curating plays aimed at fostering empathy towards trees, it is crucial to select stories that

emphasize the intrinsic bond between humans and nature, underscore the significance of environmental stewardship, and highlight the vital role of trees in sustaining life. Texts such as "The Boy and his Tree" by Rabindranath Tagore, "The Last Leaf" by O Henry, "The Giving Tree" by Shel Silverstein, "The Lorax" by Dr. Seuss, "The Secret Garden" by Frances Hodgson Burnett, or even "Avatar", especially the part that focuses on the Na'vi people's profound connection to their natural environment and their struggle to safeguard their homeland, Pandora. Additionally, engaging students with folktales like "The Willow Wife" (Japanese) or "A Flowering Tree" by A.K. Ramanujan can provide rich material for exploration and theatrical interpretation. Dancing recitals based on mythological stories such as the legend of Tulsi or the tale of Daphne too can serve as captivating performances that not only entertain but also educate audiences about the profound connections between humans and nature as depicted in these timeless narratives.

CONCLUSION

In this chapter, an attempt has been made to delve into the veneration of trees within Hindu tradition, particularly in the context of Shakta tantric tradition, to uncover the deep reverence for nature and its intrinsic spiritual connection. It emerged that this worship tradition extends beyond mere symbolic gestures, encompassing a profound understanding of the natural world as a manifestation of divine energy. The paper attempts to re-define the concept of tree worship as a creative process of becoming-other withing the context of more secular educational frameworks. The symbolic transformation of becoming-other in worship can be transferred to an ethics of creative learning and empathetic connectedness. Tantric traditions underscore the idea of Arboromorphism, the process of becoming a tree and accept the intrinsic value of trees advocating for ethical treatment of the environment. Sumana Roy's arboromorphic journey of transformation into a tree in "How I Became a Tree," prompts deep reflection on humanity's relationship with nature. Her introspective prose challenges societal norms and emphasizes interconnectedness between humans and trees. Roy's account serves as a catalyst for developing educational modules aimed at fostering environmental awareness and nurturing a stronger bond with the natural world. The chapter advocates for an environment-sensitive pedagogy that transcends anthropocentrism, fostering empathy and understanding towards the environment through artistic expression and subjective experiences related to nature.

REFERENCES

Chauhan S. & Chauhan S. V. S. (2019). *Worship and trees in India*. Sibirskij Lesnoj Zurnal (Sib. J. For. Sci.). . doi:10.15372/SJFS20190404

Cohen, M. J. (2010). *The Ecopsychology Interview*. Project NatureConnect Friday Harbor, Washington. *Ecopsychology*, *2*(2), 53–57. doi:10.1089/eco.2010.0047

Cooper-Smithson, H. (2022). Arboromorphism. *Environmental Humanities*, *14*(1), 233–236. doi:10.1215/22011919-9481539

Das Gupta, S. (1997). Evolution of the Mother Worship in India. Swami Madhavananda & R. C. Majumdar, (eds.) The Great Women of India. Advaita Ashrama.

Deleuze, G. (1994). *Difference and Repetition* (P. Patton, Trans.). Columbia University Press.

Deleuze, G. (1995). *Negotiations 1972–1990* (M. Joughin, Trans.). Columbia University Press.

Desai, S. & Smith, H. (2018). Kinship across species: Learning to care for non human others. *Feminist Review 118*(1), 41-60. DOI: 41305-018-0104-0. doi:10.1057/s

Dyczkowski, M. S. G. (2001). *The Cult of the Goddess Kubjika. A Preliminary Comparative Textual and Anthropological Survey of a Secret Newar Goddess*. Franz Steiner Verlag.

Flood, G. D. (1996). *An Introduction to Hinduism*. Cambridge University Press.

Fowler-Smith, L. (2009). *Hindu Tree Veneration as of Mode of Environmental Encounter*. The MIT Press Stable. https://www.jstor.org/stable/20532588

Frank, G. (1985, Summer). Becoming the Other [University of Hawai'i Press on behalf of Center for Biographical Research Stable.]. *Empathy and Biographical Interpretation, Biography*, *8*(3), 189–210.

Gagliano, M., Ryan, J. C., & Vieira, P. (Eds.). (2017). *The Language of Plants. Science, Philosophy, Literature*. University of Minnesota Press.

Griffith, R. (Ed.). (1896) *The Rig Veda*. https://sacred-texts.com/hin/rigveda/index.htm

Gupta, S. M. (1971). *Plant Myths and Traditions in India*. E.J Brill. doi:10.1163/9789004611542

Haberman, D. L. (2010). *Faces in the Trees*. The Journal of Religion. *Nature and Culture*, *4*(2), 173–190.

Haberman, D. L. (2013). *People Trees. Worship of Trees in Northern India*. Academic. doi:10.1093/acprof:oso/9780199929177.001.0001

Haraway, D. J. (2016). *Staying with the Trouble: Making Kin in the Chthulucene*. Duke University Press.

Hume, J. (2022). Meeting in Forest: Education in Posthuman Era. *Cultural and Pedagogical Inquiry, 14*(1), 118-127. http://ejournals.library.ualberta.ca/index.php/cpi/index

Jakubczak, M. (2017) The Woman-and-Tree motif in the ancient and contemporary India. Retracing the Past, Historical Continuity in aesthetics from a global perspective. International Association for Aesthetics.

Johnsen, L. (1999). *The Living Goddess. Reclaiming the Tradition of the Mother of the Universe*. Yes International Publishers.

Kochetkova, T. (2022). Environmental Management in Transition: Lessons from Tantra. In S. P. Sahni, T. Bhatnagar, & P. Gupta (Eds.), *Spirituality and Management. From Models to Applications* (pp. 77–97). Springer. doi:10.1007/978-981-19-1025-8_6

Lawlor, L. (2008). Following the Rats: Becoming-Animal in Deleuze and Guattari. *SubStance, 37*(3).

Lidke, J. S. (2016) Towards a Theory of Tantra-Ecology. *Sutra Journal*. http://www.sutrajournal.com/towards-a-theory-of-tantra-ecology-by-jeffrey-s-lidke

Lidke, J. S. (2017). *The Goddess Within and Beyond the Three Cities, Sakta Tantra and the Paradox of Power in Nepala-Mandala*. DK Printworld Pvt. Ltd.

McDaniel, J. (2004). *Offering Flowers, Feeding Skulls. Popular Goddess Worship in West Bengal*. Oxford University Press. doi:10.1093/acprof:oso/9780195167900.001.0001

More, T., & Valhe, V. (2020). Symbolic Significance of Sacred Nyagrodha Tree Worship in Indian Religion and Traditions. *Bulletin of the Deccan College Post-Graduate and Research Institute, 80*(October), 149–157.

Murdoch, J. (1991). Hindu and Muhammadan Festivals. Compiled from Wilson, Wilkins, Crooke, Sell, Hughes and Other Writers. Asian Educational Services.

Nitzke, S., & Braunbeck, H. G. (2021). Arboreal Imaginaries. An Introduction to the Shared Cultures of Trees and Humans. *Green Letters., 25*(4), 341–355. doi:10.1080/14688417.2021.2072633

Osho. (1991). *Vigyan Bhairav Tantra. The Book of Secrets*. Rebel Publishing House.

Paulson, S. (2019). Making Kin: An Interview with Donna Haraway. *Los Angeles Review of Books*. https://lareviewofbooks.org/article/making-kin-an-interview-with-donna-haraway/

Pintchman, T. (Ed.). (2007). *Women's Lives, Women's Rituals in the Hindu Tradition*. Oxford University Press. doi:10.1093/acprof:oso/9780195177060.001.0001

Reid-Bowen, P. (2016). *Goddess as Nature: Towards a Philosophical Thealogy*. Routledge. doi:10.4324/9781315585338

Retallack, J. (2003). *The Poethical Wager*. University of California Press.

Rodrigues, H. P. (2003). *Ritual Worship of the Great Goddess. The Liturgy of Durga Puja with Interpretations*. State University of New York Press. doi:10.1353/book4546

Roy, S. (2017). *How I Became a Tree*. Yale University Press.

Semetsky, I. (2011). Becoming-other: Developing the Ethics of Integration. *Policy Futures in Education, 9*(1), 138–144. doi:10.2304/pfie.2011.9.1.138

Sengupta, M., Banerjee, P., & Wright, C. (2004). Worshipping the Tree. *Indian Literature, 48*(6).

Sherma, R. D. (1998). Sacred Immanence: Reflections of Ecofeminism in Hindu Tantra. In L. E. Nelson (Ed.), *Purifying the Earthly Body of God: Religion & Ecology in Hindu India* (pp. 89–132). State University of New York Press.

Shiva, V. (1988). *Staying Alive. Women, Ecology and Survival in India*. Kali for Women.

Singh, S., & Hageneder, F. (2007). *Tree Yoga. A Workbook, Strengthen Your Personal Yoga Practice Through the Living Wisdom of Trees*. Earthdancer.

Tagare, G. V. (2003). *The Skanda-Purana*, Part 18. In G. P. Bhatt (Ed.), *Ancient Indian Tradition & Mythology* (Vol. 66). Motilal Banarasidass Publishers Pvt. Ltd.

Vedantatirtha, G. C. (Ed.). (1915). Kulachudamani Tantra. Sanskrit Text Depository & Luzac & Co.

Vijnananda, S. (1921-22) *The Srimad Devi Bhagavatam*. https://www.pdfdrive.com/srimad-devi-bhagavata-purana-english-by-sw-vijnanananda-d39957378.html (Last access: 3.02.2024)

Woodroffe, J. (2009). Sakti and Sakta. Essays and Addresses on the Sakta Tantrasastras. Celephaïs Press.

ENDNOTES

1 Translated from Bengali by Paramita Banerjee and Carolyne Wright (Sengupta et.al, 2004, p. 48-49)

2 Here are the English common names of the trees: Shleshmastaka – sebesten, karnja – pongamia, nimba – neem, asvattha – peepal, kadamba – burflower, vilva/ bilva – stone apple, vata – banyan, Ashoka – ashoka

3 Kubjika cult is the western transmission of the Kaula tradition that focusses on the hunch-backed crone named Kubjika. (Flood, 1996, p. 166)

4 The English common names of the trees are as follows: mandara – China rose, bakul – bullet wood, priyangu – perfumed cherry, champaka – champa/ joy perfume tree, karnikara – bayur tree/ dinnerplate tree.

5 https://phys.org/news/2019-06-indian-family-tree-house.html (Last access: 3.2.2024)

6 While engaging in dialogue with Reid-Bowen, Mukhopadhyay further illuminates how not theology but 'thealogy' resonates with atheist secular sensibilities: "The celebration of immanence and the natural in thealogy does square with those atheist sensibilities which reject the supernatural world and celebrate—sensuously—the natural world." (2023, p. 28)

7 Vrikshsadhna (Tree Meditation) at the Om Swami Samarth Sadhna Kendra in Ranavali, Maharashtra, India. https://www.youtube.com/watch?v=4fDwys68VLo&t=35s (Last access: 7.4.2024)

8 Cohen, Michael. Nature Connected Psychology: Creating Moments That Let Earth Teach. https://www.ecopsychology.org/journal/gatherings3/cohen.html (Last access: 7.4.2024)

9 One can find other fascinating ideas about environment sensitive pedagogy with respect to trees here: https://childsplayabc.wordpress.com/2020/06/25/exploring-trees-and-woods-activities-and-ideas/ (Last access: 7.4.2024) While responding to Donna Haraway's challenge of "staying with the trouble", Desai and Smith (2018, p. 42) discuss an intriguing idea: promoting care for trees through mobile game applications. They discuss a case study on a digital campaign for mobile game applications aimed at fundraising for reforestation and afforestation projects while raising awareness about the importance of trees in sustaining ecosystems. These games, categorized under the genre of "serious games," aim to promote positive social and environmental change through engaging gameplay. Such games can serve as informal learning spaces, influencing human environmental understanding and encouraging constructive exploration of human-environmental relations. Using the example of Tilt World, an educational game about carbon and the environment, the case study illustrates how gameplay can support forest conservation efforts, particularly in regions like Madagascar, where deforestation threatens endemic species such as lemurs. (p.48) There are also "games that ask players to care for a digital tree, such as Tree Story Game (TSG), conceived by the Los Angeles-based US company Zig Zag Zoom. In TSG, players 'care for' saplings until they grow into mature trees by being attentive to metres indicating the percentages of thirst/water, hunger/satiety, sunlight and love/life the digital tree is experiencing."

(p.49) However, not all players are convinced about the efficacy of such games: "I really wanted to play to help plant trees.... All you do is water the tree, feed it food that trees do not eat (why are we feeding them burgers and french fries?? Would rather learn about trees and the kinds of fertilizers they need) tap to turn the sun on, and poke it for affection. ... The cause is great, but the game itself is not." (p. 49) Hence, adopting natural art-based methods emerges as a logical approach for fostering a deeper connection with nature.

Chapter 6
Karma and Conservation: Unifying Hindu Philosophy and Environmental Consciousness

Siddharth Kanojia
https://orcid.org/0000-0002-1479-5292
O.P. Jindal Global University, India

ABSTRACT

Hindu philosophy renders a distinctive perspective concerning the way humans interact with the environment, as it is firmly anchored in old scriptures and customs. A belief that all life is interconnected and the understanding that the environment is an essential component of the cosmic order is paramount to this viewpoint. This chapter penetrates the fundamental ideas of Hindu philosophy that form the basis of environmental consciousness. It encompasses concepts like Dharma, an acronym for righteous duty, Karma, which means action and consequences, and Ahimsa, which means non-violence. The chapter inquires about the ethical underpinnings of environmental stewardship in Hindu philosophy while illustrating people's duty to preserve the universe's delicate balance with animals and plants. It gazes at how Vedic philosophies can promote thankfulness, environmental consciousness, and a sense of oneness with the natural world. The chapter further explores the symbolic meaning of customs such as Yajna (sacrificial offerings) and how they can encourage reciprocity with the environment.

INTRODUCTION

The historian Lynn White Ir penned a piece for Science in 1967 discussing the ecological problem' historical causes. White contends that people's actions towards their surroundings are influenced by their perception of their place in nature (White, 1967). He further claimed that the teachings of late medieval Latin Christianity, which saw humans as superior to the rest of God's creation and everything else as made for human use and enjoyment, are to blame for the exploitative viewpoint that has caused much of the environmental crisis, particularly within Europe and North America (White, 1978). He proposed that rejecting the idea that nature exists only to serve humankind is the only way to address the ecologi-

DOI: 10.4018/979-8-3693-2577-3.ch006

cal crisis. Thereby, Scientists, theologians, and environmentalists were forced to discuss White's claim that religion was to blame for the ecological crisis (Muller, 2009). Throughout the discussion, instances from various cultures were used to bolster the argument that environmental exploitation has been ruthless even in nations where there is a strong religious reverence for the natural world, nations where Hinduism is practiced. Practices of Buddhism, Taoism, and Shintoism were mentioned in support of Thomas Derr's criticism (Diwedi, 2001). Here, Derr's claim regarding the contribution of Hinduism to the ecological catastrophe is contested. We must comprehend how a Hindu's perspective on nature has been influenced by his religion's understanding of the universe and creation (Retnaselvam & Singh, 2019). Such an explanation is required to clarify Hindu customs and beliefs, and consequently, the historical significance of Hinduism about environmental degradation by humans. About this, Hinduism shimmers as a beacon of age-old wisdom among the many and varied philosophical traditions by offering a profound comprehension of the complex interrelationship between humans and nature (Mallik, 2022). Hindu philosophy, which has its roots in religious texts, philosophical writings, and spiritual practices, offers a distinctive perspective on how all living things are interdependent with their surroundings. Hindu philosophy revolves around a holistic worldview, which is the conviction that the universe functions as a dynamic, interconnected web. This interconnectedness proposes an illustration of humanity as merely one strand in the enormous artwork of existence by extending beyond human relationships to encompass the natural environment as a whole. Significant principles like Dharma, Karma, and Ahimsa operate as an ethical compass, leading people in the direction of peaceful coexistence with the natural world.

Dharma, which is often translated as "righteous duty," encompasses everyone's obligation to promote harmony and order in the universe at large (Muthuswamy, 2018). The concept also entails sustainable practices in the context of environmental consciousness by appreciating the intrinsic value of all living forms in cosmic dance and their interconnectedness. Whereas, the notion that every human action has implications for both individuals and the ecosystem is furthered by the concept of karma, which is referred to as "*action and consequence*" (Ghose, 2007). This concept entails a thoughtful way of living that promotes environmental stewardship by coordinating actions with the greater good. Similarly, the sect of ethics in Hinduism centers on the belief of ahimsa, or non-violence. The concept of ahimsa draws close attention to the need to treat all living things with compassion and reduce environmental harm while traversing lightly on Earth (Godrej, 2010).

The vibrant fabric of Hinduism is not woven only by its theological precepts, but also by diversified customs and celebrations that act as an entry point to elevated environmental awareness (Kapoor, 2016). The rituals, like the symbolic Yajna (sacrificial offerings), become expressions of oneness with the natural world by encouraging thankfulness and a sense of solidarity (Suara et al., 2018). These customs imbibe a heartfelt regard for the environment by transforming prosaic activities into conspicuous relationships with the surroundings. Meanwhile, renowned festivals like Diwali lead to awakening the consciousness of the ecological balance that supports life on Earth (Shah, 2019).

The upcoming part of this paper reviews the ecological symbolism connected to Hindu deities, such as Goddess Earth (Bhudevi), the embodiment of the environment, and Lord Vishnu, the preserver of the universe. These archetypes underscore the necessity for humans to preserve and protect the Earth and inspire environmental ethics. The other parts discuss the relevance of Hindu environmental ethics in the modern era, highlighting programs and movements that support eco-friendly, sustainable, and conservation efforts within the Hindu community worldwide. Hindu philosophy can promote peaceful coexistence between humans and the environment as it examines the difficulties and possibilities of fusing traditional wisdom with contemporary environmental concerns (Tomalin, 2016).

The Sacred Essence in Hindu Philosophy

The Hindu religion is unquestionably deeply rooted in the idea of the sanctity of life. Humans have no control over their own lives or the lives of non-human creatures because only God possesses absolute sovereignty over all living things (Ottuh, 2020). As a result, neither can humanity govern the planet in God's stead or give other species varying degrees of relative value. The text of Yajurveda masterfully conveys the concept that the Divine Being is the single fundamental force of unity:

"The loving sage beholds that Being, hidden in mystery, wherein the universe comes to have one home; Therein unites and therefrom emanates the whole; The Omnipresent One pervades souls and matters like warp and weft in created beings" (Yajurveda, 2.8)

Given that God's creation is sacred, no harm can be done to other species without a good reason. Consequently, every life human and non-human has equal value and the same right to exist. The Atharvaveda states that other animals are also meant to inhabit this planet, not just humans.

"Born of Thee, on Thee move mortal creatures; Thou barest them - the biped and the quadruped; Thine, 0 earth, are the five races of men, for whom Surya (Sun), as he rises spreads with his rays the light that is immortal" (Atharvaveda, 12.1-15)

Furthermore, Hindu scriptures all bear witness to the idea that the Supreme Will is ultimately responsible for the universe's creation, upkeep, and destruction (Narayanan, 2001). For instance, in the Gita, Lord Krishna says to Arjuna "*Of all that is material and all that is spiritual in this world, know for certain that I am both its origin and dissolution*" (Gita, 7.6) And the Lord says: *"The whole cosmic order is under me. By my will it is manifested again and again and by my will, it is annihilated at the end"* (Gita, 7.6). For that reason, God and Prakriti, or nature, were considered to be the same by the ancient Hindus. The Prajapati is the creator of the earth, sky, oceans, and all other species; however, he also serves as both their defender and ultimate destroyer (Kara, 2008). The only Lord of creation is him. It can be further derived that humans have more responsibilities and duties than other creatures, but they also have no particular authority or privilege over them.

In an analogy between spirituality and ecological stewardship, Hinduism's rituals and symbolic practices play an essential part in forming environmental consciousness. For instance, Yajna is a symbolic ritual with deep Hindu roots that involves sacrificial offerings. However, the symbolism of Yajna as a balanced exchange between humans and the environment may be overshadowed by its ritualistic nature. The dilemma arises when a literal interpretation of the ritual eclipses the symbolic meaning of Yajna as a reciprocal gesture. If materialistic offerings are not approached with a conscious awareness of ecological impact, they may contribute to environmental degradation (Huang & Rust, 2011). Factually, Yajna ceremonies frequently make use of wood, ghee, or clarified butter, as well as other offerings. Herein, sustainability is a concern when natural resources are being depleted, especially if they are not being replaced in an environmentally responsible or ethical manner (Hackett & Dissanayake, 2014). Further, some Yajna ceremonies produce an abundance of waste, like burnt offerings or materials hurled away. The integrity of the ritual's environmental symbolism depends on the proper disposal of these leftovers to avoid polluting the environment. However, over time, there has been a commendable shift within Hinduism towards the adoption of eco-friendly rituals and sustainable practices in response to growing environmental concerns (Cairns, 2020). This entails employing biodegradable supplies, trimming down waste, and introducing eco-friendly components into ceremonies. Although these initiatives are an appreciated shift from traditional methods, the obstacle is making sure they are widely adopted and maintained. In light of deeply embedded traditional norms and the availability of sustainable alternatives, the shift to eco-friendly alternatives may encounter resistance; consequently, a careful balancing act between tradition and progressive environmental values is required (Boström & Klintman, 2008).

Ethical Responsibilities towards Flora and Fauna

The primary principle of Hindu theology concerning the treatment of animals is the conviction that the Almighty Himself took on the forms of diverse species. According to the religious texts, God has taken on many forms over the ages. A fish was his first incarnation, followed by a tortoise, a boar, and a dwarf. He was a man-lion in his fifth incarnation. When he was Rama, he was closely related to monkeys, and when he was Krishna, cows were all around him (Leeming & Page, 1996). Furthermore, these species are revered and respected by Hindus who believe in the cycle of birth and rebirth, in which a person may reincarnate as an animal or a bird (Okafor & Stella, 2018). This offers a strong basis for the concept of ahimsa, which is the non-violence towards both humans and animals. Accordingly, Hindus hold a strong belief in the nonviolent philosophy. The idea that one can obtain God's grace by not harming or killing his creation is strongly emphasized in almost all Hindu scriptures (Howard, 2023). For example, Visnupurrana (3, 8, 15) states that "*God, Kesava, is pleased with a person who does not harm or destroy other non-speaking creatures or animals.*" In addition, abstaining from meat is both a duty and an appropriate behavior. Yajnavalkya Smriti threatens those who murder domesticated and protected animals with hellfire (Ghora Naraka): "*The evildoer who takes the lives of protected animals must spend as many days in hell as there are hairs on that animal's body*" (Yajnavalkyasmriti, v. 180; Acaradhyayah).

Whereas, Buddhism and Jainism appeared by the end of the Vedic and Upanishadic eras, and the numerous kings who followed these religions strengthened their protection of plants, animals, and birds (Kent, 2013). These religions upheld environmental preservation principles and developed partly as a protest against the Hindu religious orthodoxy and rituals. With his public declarations, the Buddhist emperor Ashoka encouraged the planting and preservation of plants and animals (Gupta, 2013). Whereas, tree worship was widespread and quite common as early as the Rigveda era. For the Regvedic seers, the tree represented God's many attributes. Regveda believed that plants possessed supernatural abilities; in fact, a whole hymn was dedicated to praising them, mostly about their therapeutic qualities (Singh, 1997). The admiration for the natural world grew during the era of the great epics and Puranas. It was once believed that trees were sentient beings with feelings of joy and sorrow. The widespread belief persists that each tree possesses a Vriksa-devata, also known as a "tree deity," which is revered with prayers and offerings of water, flowers, candies, and sacred threads. Additionally, planting trees is still considered a religious obligation for Hindus. Subsequently, Hinduism has long worshipped plants and trees, primarily for their mythological and religious significance but also for their practical uses (Haberman, 2013). The Hindu forefathers believed it was their responsibility to preserve trees and to do so, they gave each tree a sacred status in their religion.

Vedic Underpinnings

It is believed that Vedic culture is a culture of nature. The God of Rain, the protagonist of The Rig Veda, is a celebration of nature. This Rig Vedic hymn beautifully describes nature: "Rivers throb and ripple at the first touch of His hand." *The sun shines, the moon shimmers, the stars twinkle, and the flowers bloom when He smiles. The universe is stirred by the first light of the rising sun; shimmering gold is scattered across the smiling rose buds; the fragrant air is filled with the lovely melodies of singing birds; dawn is the imaginative dream of God.*"

(Rig Veda, 1.6.3). There are multiple suktas in the Rigveda alone where the Rishi attributes deities to the natural elements, highlighting the significance of these objects in human life or the lives of all living things. For example, the Rigveda's Vayu-sukta asserts that Vayu, or the air, is the fundamental nature of all living things (Rig., 10-186.2). This fact was established in the Vedic era by the fur-fathered sages through their mantras, in which the Rishi expresses his profound gratitude to Vayu by referring to it as his father, brother, friend, etc. According to Tai, Sikhsavalli, and Anuvaka (XII), "*jalat prasastamam lokopakarakam kimapi nasty*" refers to the water as the essence of life. There is ample evidence of the interaction between man and nature in the Aranayakand of the Ramayana and the AryanakaParva of the Mahabharata. In the Ramayana, after Ravana kidnapped Sita, the plants and animals expressed their sympathy for Rama and cleared his path in their native tongues: "*Evamukta Narendra te mrgah sahasotthitah I Daksinabhimukhah sarve darsayanto nabhah sthalam*II *Maithili hliyamano sa disam yamabhyapadyat* I *Tena margena gacchanto niriksante naradhipam* II *Darsayanti ksitim caiva daksinam ca disam mrigah* I *Sadhu gcchavahe deva disametam ca nairrtim* II (Ibid, Aranyanda, 17.19–22). Thus, the pages of ancient Indian scriptures bear witness to the close relationships that exist between man and nature. "*It brings great sin if the fire is not offered purifying material (havan); if trees are cut down on auspicious days, if human waste, spit, and cough, etc., are dropped in water,*" states ShantiParva of Mahabharat. (Parva Shatrani). Arjuna begs Lord Krishna in Bhagavad Gita chapter eleven to reveal his universal form, which encompasses all of existence. Hinduism preaches not to pollute the environment, but to preserve and protect it (Dwivedi 2003). Further, According to the Padma Purana, "*A person who is engaged in killing creatures, polluting wells, and ponds and tanks, and destroying gardens, certainly goes to hell.*" Padmapurana and Bhoomikhanda 96.7-8. A hymn in the Rig Veda emphasizes the value of trees in mitigating pollution, or Pradushana. "*Trees absorb pollution, so don't cut them down.*"(Rig Veda 6:48:17). At that time, nobody could have predicted that there would be an issue such as ozone depletion, but Our Yajurveda emphasized how critical it was to preserve the environment (Diya 2023)." *Avoid disturbing the sky and contaminating the surrounding area.*" Yajur Veda 5:43. The Mahabharata warns people about the importance of protecting and caring for trees: "Puspam puspam.

The ancient Hindu scriptures place great emphasis on the idea that humans are inextricably linked to the natural world, with Earth and man sharing a similar bond as a mother does with her child. One of the most exquisite hymns, BhoomiSuktam, or Hymn to Mother Earth, praises the beauty of Mother Earth and teaches us important lessons about the attitude we should all have towards the environment. The first verse begins with a strong declaration outlining how the Earth and its surroundings should be understood (Tomislav, 2018). Besides, Indian seers believe that the elements of the environment are fire, water, earth, sky, and wind. Therefore, the environment is greatly affected by these five observable gross elements, without which human survival is virtually impossible (Kanojia, 2024). The idea that God is present in every aspect of the universe is supported by several Indian philosophers. Thus, these are not the typical elements that Sanskrit literature has deified. The significance of these elements being deified is that people will naturally and psychologically be eager to grow to respect these elements. The ancient Indian seers saw nature in this way, viewing it as both a creator and a creation. Both living things and inanimate objects are interconnected throughout the entire world. There is a close bond between humans, animals, and plants. Since God is greater than all of these, it is everyone's responsibility to preserve nature's beauty and avoid upsetting its balance (Sagoff, 1974). Thus, Hinduism cannot persuade its adherents to harm or violate the environment for petty, self-serving ambitions. Every Hindu religious text and cultural practice teaches us to honor the Earth and all of its blessings. In verses 3.14–16 of the Bhagavad Gita, Krishna explains the ecological cycle to Arjuna: "*All living bodies subsist on food grains, which are produced from rains.*" Rain is the result of performing yajna, or sacrifice, and yajna is the result of following rules.

Practical Discourse

The Hindu philosophy, as previously mentioned, upholds the values of respect for nature and the interdependence of all living things; however, there are situations in which these principles are compromised by actions that worsen the environment. A typical instance of a contradiction in religious practice is the use of materials from endangered species or environmentally hazardous substances in certain rituals and ceremonies (Apffel-Marglin, 2011). In particular, it is against the concept of ahimsa towards all living things to use sandalwood, which is derived from sandalwood trees, in religious rites. While, some plants, like *peepal* (sacred fig) and *tulsi* (holy basil), are overused in religious ceremonies and for medicinal and commercial purposes. These species' ability to survive in their native environments may be threatened by overharvesting, which can also cause wild populations to decline (Barrow, 2019). Other than that, pilgrimage is an imperative cultural and religious aspect of Hinduism, and it frequently causes environmental stress. Renowned pilgrimage locations, like the Ganges River banks, have been damaged by pollution from pilgrims' waste, offerings, and industrial effluents (Kala, 2018). Even though these places are considered sacred, conservation efforts have been hindered by the negative environmental effects of large crowds and the absence of sustainable management techniques. Furthermore, conventional views can influence agricultural practices that put short-term profits ahead of sustainability over the long run. Driven by the need for higher yields, the careless application of chemical pesticides and fertilizers deteriorates soil fertility and contaminates water sources, endangering biodiversity and human health. On the same tandem, as part of religious rituals, gigantic idols made of plaster of Paris and other non-biodegradable materials are submerged in rivers, lakes, and oceans during festivals like Ganesh Chaturthi and Durga Puja. considering that the idols contain hazardous materials like lead, mercury, and cadmium that seep into the water and endanger aquatic life, this practice pollutes the water (Bhattacharya et al., 2014).

Given that this is the Hindu religion's tradition, philosophy, and ideology, why is the current environmental crisis occurring? As demonstrated by observations. The interactions with all living things, including plants and animals, are influenced by our religious and ethical convictions. The public's faith in the preceding cultural tradition is lost if, for whatever reason, these admirable ideals are replaced by ideas that are either imposed upon society or invaded by another culture (Michaels, 1992). As the foreign language, culture, and administrative system gradually take hold. It permeates all societal levels, and since religious leaders and Brahmans are unable to provide suitable leadership or answers, it is only natural for the masses to grow more egocentric and inward-looking (Chu, 2014). Because people are preoccupied with ensuring their survival and freedom, religious values that serve as sanctions against environmental destruction do not remain highly valued in such circumstances. As a result, economic factors take precedence over respect for nature. That appears to be the outcome of 700 years of foreign cultural hegemony in India. The situation became more complicated when the British introduced Christianity and Western secular institutions and values in addition to Muslim culture (Soper & Fetzer, 2007). Even though it is far too simple to hold these outside forces responsible for the shift in Hindus' views toward the natural world, the truth remains that they seriously impeded the religion's ability to uphold its core principles of encouraging reverence and respect for God's creation (Kellert & Farnham, 2013). However, it is not conceivable to fully attribute the complexity of the current environmental crisis within the Hindu tradition to outside factors like colonialism or the imposition of foreign ideologies. Even though these elements had considerable effects on how India's heritage was disrupted and how society's values were shaped, internal problems also contributed to the diminution of environmental ethics in Hinduism. Beyond that, the dispersion of leadership among Hindu communities and the absence of coherent direction from religious authorities have made environmental stewardship even less adhered to. As a consequence, despite the rich spiritual legacy that emphasizes harmony with nature, the interaction of external pressures and internal dynamics has created a precarious equilibrium where economic imperatives frequently outweigh ecological concerns, contributing to the ongoing degradation of the environment.

Pedagogical Application for Environment Education

Multiple research studies have indicated that Indian philosophy has important pedagogical applications for environmental education. The Indian philosophy places a strong emphasis on the relationship between nature and humanity, having been shaped by the perspectives of thinkers such as Swami Vivekananda, Mahatma Gandhi, Rabindranath Tagore, and Sri Aurobindo. From the teaching and pedagogical standpoint, the notion of interconnectedness by employing Buddhist philosophy i.e., Indra's Net. This metaphor juxtaposes the universe to a massive net, in which each node stands for a distinct entity and is interconnected with all other nodes (Cambray, 2009). This can assist students in comprehending how decisions made in one area of the ecosystem can have implications for the entire system. In addition to this, lessons on how to minimize harm to all living things and how to treat the environment with respect and compassion can be learned by applying the Jain principle of ahimsa (non-violence) to environmental conservation. This may spark conversations about ethical considerations in resource consumption as well as sustainable living techniques. At the same time, the Hindu concept of dharma, which includes one's obligation and responsibility to the environment and future generations, can be explored to inspire students to consider their role as environmental stewards and the significance of carrying out their environmental responsibilities (Rogayan, 2019). Similar to this, the concept of simple living from Mahatma Gandhi's teachings can be introduced in a classroom setting. He pushed for a way of life that is in balance with the natural world and places an emphasis on self-sufficiency and minimal consumption (Devika & Arulmani, 2014). Thereby, pupils can be encouraged to adopt sustainable habits like cutting back on waste, using less energy, and adjusting their lifestyle to the cycles of the natural world. On the other hand, an exchange of ideas about the Samkhya philosophy's views on the connection between Purusha (consciousness) and Prakriti (nature) can begin (Deshmukh, 2019). It will assist students in realizing that nature is a reflection of our inner consciousness as well as an external environment. Encourage them to engage in mindfulness and meditation in natural environments to foster a closer bond with the natural world. Similarly, pedagogical lessons can make use of the Hindu and Buddhist concepts of samsara, or the cycle of birth and rebirth, to highlight the cyclical nature of ecological processes like the carbon cycle, water cycle, and nutrient cycle. It will help pupils comprehend how crucial it is to maintain ecological balance to guarantee the continuation of life on Earth. Ultimately, teaching students about the Buddhist idea of anicca (impermanence) can help them understand how dynamic and constantly changing ecosystems are. Therefore, acknowledging the need for resilience and adaptability can be facilitated by talking about how ecosystems constantly fluctuate and adapt to changes in the environment and human activity. Accordingly, Educators may inspire students to adopt sustainable practices in their daily lives and establish a stronger connection with nature by introducing these ideas and concepts from Indian philosophy into environmental education.

CONCLUSION

It is readily apparent from considering the convoluted relationships among Hindu customs, outside influences, and the present environmental crisis that changes in cultural and religious beliefs have played a major role in exacerbating the problems facing the environment. The observations demonstrate the centuries-long deterioration of Hinduism's innate reverence for nature, especially as a result of the introduction of Western institutions and values and the impact of foreign cultural hegemony. The ongoing conflict between upholding traditional values and giving in to the demands of a changing global environment must be acknowledged, along with the complexity of the environmental crisis. The narrative is further complicated by the incorporation of economic factors and the preference for survival over environmental sanctity. We must evaluate cultural, religious, and societal values to achieve a harmonious coexistence between tradition and the demands of a changing world, as we work to address the environmental crisis. Hinduism's tenets of reverence and respect for nature can be revived, offering hope for a sustainable future where the environment is recognized as an essential component of our shared history.

REFERENCES

Apffel-Marglin, F. (2011). *Subversive spiritualities: How rituals enact the world*. Oxford University Press.

Barrow, E. (2019). *Our Future in Nature: Trees, Spirituality, and Ecology*. Balboa Press.

Bhattacharya, S., Bera, A., Dutta, A., & Ghosh, U. C. (2014). Effects of idol immersion on the water quality parameters of Indian water bodies: Environmental health perspectives. International Letters of Chemistry. *Physics and Astronomy*, *20*(2), 234–263.

Boström, M., & Klintman, M. (2008). *Eco-standards, product labeling and green consumerism*. Palgrave Macmillan. doi:10.1057/9780230584006

Cairns, R. (2020). *Dharmic environmentalism: Hindu traditions and ecological care* [Doctoral dissertation, Memorial University of Newfoundland].

Cambray, J. (2009). *Synchronicity: Nature and psyche in an interconnected universe* (Vol. 15). Texas A&M University Press.

Chu, T. (2014). *Human purpose and transhuman potential: a cosmic vision of our future evolution*. Red Wheel/Weiser.

Deshmukh, V. D. (2019). Nava-Samkhya: New understanding, interpretation, and proposal for the classical Samkhya philosophy. *Journal of Alternative Medical Research, 11*(1), 53–59.

Devika, V. R., & Arulmani, G. (2014). Mahatma Gandhi's ideas for work, career, and life. In *Handbook of career development: International perspectives* (pp. 105–117). Springer New York. doi:10.1007/978-1-4614-9460-7_7

Diya, V. S. (2023). *Environmental Consciousness and the Nine Schools of Indian Philosophy*. Cambridge Scholars Publishing.

Dwivedi, O. P. (2001). *Classical India. A companion to environmental philosophy*, 37-51.

Ghose, L. (2007). Karma and the possibility of purification: An ethical and psychological analysis of the doctrine of karma in Buddhism. *The Journal of Religious Ethics, 35*(2), 259–290. doi:10.1111/j.1467-9795.2007.00306.x

Godrej, F. (2010, September). *Gandhi's Body: Asceticism, Pain and Suffering in Environmental Political Discourse*. 2010 Conference of the American Political Science Association, Washington, DC.

Gupta, A. (2013). Altruism in Indian religions: embracing the biosphere. In *Altruism in cross-cultural perspective* (pp. 101–112). Springer New York. doi:10.1007/978-1-4614-6952-0_8

Haberman, D. L. (2013). *People trees: worship of trees in northern India*. Oxford University Press. doi:10.1093/acprof:oso/9780199929177.001.0001

Hackett, S., & Dissanayake, S. T. (2014). *Environmental and natural resources economics: Theory, policy, and the sustainable society*. Routledge. doi:10.4324/9781315704586

Howard, V. R. (2023). The Nonviolence Conundrum: Political Peace and Personal Karma in Jain and Hindu Traditions. *Religions, 14*(2), 178. doi:10.3390/rel14020178

Huang, M. H., & Rust, R. T. (2011). Sustainability and consumption. *Journal of the Academy of Marketing Science, 39*(1), 40–54. doi:10.1007/s11747-010-0193-6

Kala, C. P. (2018). Cultural significance and current conservation practices of the Ganga's ecosystem and environment. *Applied Ecology and Environmental Sciences, 6*(4), 128–136.

Kanojia, S. (2024). Guardians of Atolls: Examining the Paris agreement's role in climate change mitigation and resilience in the Maldives. In Bridging Health, Environment, and Legalities: A Holistic Approach (pp. 271-291). IGI Global.

Kapoor, S. K. (2016). Hinduism: the faith eternal. Advaita Ashrama (A publication branch of Ramakrishna Math, Belur Math).

Kara, A. (2008). *And the Thunder Said DA: Stories of Dharma in the Mythology, Philosophy, and Religion of India*. iUniverse.

Kellert, S. R., & Farnham, T. (Eds.). (2013). *The good in nature and humanity: connecting science, religion, and spirituality with the natural world*. Island Press.

Kent, E. F. (2013). *Sacred groves and local gods: religion and environmentalism in South India.* Oxford university press. doi:10.1093/acprof:oso/9780199895465.001.0001

Leeming, D. A., & Page, J. (1996). *God: myths of the male divine.* Oxford University Press.

Mallik, B. (2022). Beauty, Wonder, and Sustainability: Reconnecting Humans to Nature for "Durable Harmony". In *Legends in Gandhian Social Activism: Mira Behn and Sarala Behn: Addressing Environmental Issues By Dissolving Gender And Colonial Barriers* (pp. 437–488). Springer International Publishing. doi:10.1007/978-3-030-95431-4_9

Michaels, W. B. (1992). Race into culture: A critical genealogy of cultural identity. *Critical Inquiry, 18*(4), 655–685. doi:10.1086/448651

Mueller, M. P. (2009). Educational reflections on the "ecological crisis": Ecojustice, environmentalism, and sustainability. *Science & Education, 18*(8), 1031–1056. doi:10.1007/s11191-008-9179-x

Muthuswamy, V. (2018). Hinduism and Social Responsibility. *Religious Perspectives on Social Responsibility in Health: Towards a Dialogical Approach*, 91-105.

Narayanan, V. (2001). Water, wood, and wisdom: Ecological perspectives from the Hindu traditions. *Daedalus, 130*(4), 179–206.

Okafor, J. O., & Stella, O. (2018). *Hinduism and ecology: Its relevance and importance.*

Ottuh, P. O. (2020). Religious Approach to Non-Anthropocentric Ethics in Environmental Philosophy. *Cogito-Multidisciplinary research Journal*, (1), 7-24.

Retnaselvam, J., & Singh, P. (2019). *Sustainability and the sacred: a comparative study of Indian religious environmentalism with special reference to Christian and indigenous communities in the State of Kerala.* Research Gate.

Rogayan, D. V. Jr. (2019). I heart nature: Perspectives of university students on environmental stewardship. *International Journal of Engineering Science and Technology, 1*(1), 10–16.

Sagoff, M. (1974). On preserving the natural environment. *The Yale Law Journal, 84*(2), 205–267. doi:10.2307/795401

Shah, V. (2019). *Hindu Culture and Lifestyle: Living Indian traditions in the age of artificial intelligence.* Notion Press.

Singh, N. K. (1997). *Vedic mythology.* APH Publishing.

Soper, J. C., & Fetzer, J. S. (2007). Religious institutions, church–state history and Muslim mobilisation in Britain, France and Germany. *Journal of Ethnic and Migration Studies, 33*(6), 933–944. doi:10.1080/13691830701432780

Suara, I. D. M., Gelgel, I. P., & Yasa, I. W. S. (2018). Pitra yadnya implementation for Hindu heroes. *International Journal of Social Science and Humanity, 2*(3), 136–148.

Tomalin, E. (2016). *Biodivinity and biodiversity: The limits to religious environmentalism.* Routledge. doi:10.4324/9781315569352

Tomislav, K. (2018). The concept of sustainable development: From its beginning to the contemporary issues. *Zagreb International Review of Economics & Business*, *21*(1), 67–94. doi:10.2478/zireb-2018-0005

White, L. Jr. (1967). The historical roots of our ecologic crisis. *Science*, *155*(3767), 1203–1207. doi:10.1126/science.155.3767.1203 PMID:17847526

White, L. T. (1978). *Medieval religion and technology: collected essays* (Vol. 13). Univ of California Press.

Chapter 7
Yoga and Environmental Education

Pranjal Khare
O.P. Jindal Global University, India

Anny Gabriela Molina Ochoa
O.P. Jindal Global University, India

ABSTRACT

This chapter deals with the interplay of practice of yoga and environmental education. It analyzes in detail the intricacies of both the disciplines and puts forward the connection of yoga with the environment. It dives deeper into the different schools of yoga and how they are related in promoting environmental awareness. This chapter also discusses the philosophical foundations of yoga and explains how the core principles and beliefs of yoga relate to one's understanding of the natural world. It further explores the integration of yoga with environment education, discussing in detail the various educational approaches to it along with numerous initiatives running worldwide to bridge the gap between yoga and environmental education. This chapter emphasizes the need to utilize the combined potential of yoga and environmental education towards a sustainable future.

INTRODUCTION

The origins of yoga can be traced back as far as 3000 BC to the Indus Valley Civilization, where its presence is found in the early texts like the Vedas and Upanishads. These texts have mentioned yoga's philosophical underpinnings, emphasizing the interconnectedness of all beings and the pursuit of liberation from suffering. Later, yoga was then further codified in the Yoga Sutras of Patanjali during the classical period around 500 BC. In the Yoga Sutras of Patanjali, eight limbs of yoga have been outlined which includes ethical principles (*Yamas* and *Niyamas*), physical postures (*Asanas*), and meditation practices. Yoga as a practice and philosophy, which has found its roots in India's cultural and traditional history has evolved and crossed all cultural barriers and has become known to all across the world through fa-

DOI: 10.4018/979-8-3693-2577-3.ch007

mous figures like Swami Vivekananda, who was a well known Indian philosopher and author and Bellur Krishnamachar Sundararaja Iyengar, a renowned yoga teacher and author.

The word Yoga is derived from the Sanskrit word 'yuj' which means 'to unite'. So, its literal meaning is 'union' and can be defined as the process of bridging the gap between individual self and the supreme self, that is the universal spirit. Yoga's main philosophy lies in the profound belief of interconnectedness, whereby everything in the universe, including our individual selves, vibrates in harmonious unity. It further propagates that one's mind, body and spirit is no longer seen as separate entities and is intertwined as inseparable facets of the whole. In his acclaimed book 'Yoga for a World Out of Balance', an esteemed yoga teacher Michael Stone puts forth his views that, "The techniques of yoga, including body practices, working with the breath, and discovering the natural east of the mind, reorient practitioners to the very deep continuity that runs through every aspect of life until they realize that the mind, body, and breath are situated in the world and not apart from worldly life in any way." Thus, it fosters the feeling of harmony, balance and interconnectedness. Furthermore, extensive research investigating yoga in clinical settings has demonstrated encouraging results for both physical and mental well-being of the practitioner (Chakravorty, 2021).

Traditionally, yoga thrived in the embrace of nature and was practiced by the yoga practitioners in mountains, forests, or beside rivers and seas whereby they imbibed the virtue of self sufficiency through gardening, animal care, and wilderness survival. This intense connection with nature was not merely incidental but, as a matter of fact, it fueled the growth of yogic wisdom within them. However, in recent times, the natural roots of yoga have often become tangled, if not severed, amidst the part and parcel of modern life. The modern day yoga is being attributed to the physical postures (or *asanas*) and breathing exercises. It is perceived by the people primarily as a physical exercise activity in order to manage or treat any health condition but that is only one aspect of yoga and it has more to it (Ross et al., 2013).

Amidst the rising global environmental crisis, it is a stark reminder that yoga's ancient wisdom about nature's importance is not a relic of the past, but a requisite component for the future. In the recent era, there is rising concern about the constant environmental degradation which includes deforestation, environmental pollution, loss of biodiversity, etc. These natural disasters have a significant impact on the physical and mental health of those affected. People might become anxious and depressed and easily start manifesting conditions such as aggression, hyper vigilance and sleep disturbance. Various research studies show that climate change is adding to the mental health problems of people who rely heavily on nature for their livelihood. However, there have also been case studies where they have done an assessment of the interplay between yogic practices, connectivity with nature and climate change. The results from these studies demonstrate that yoga cultivates a deeper connection with the environment which naturally converts into more mindful and responsible choices for the planet. It fosters positive emotions like happiness and interconnectedness, empowering practitioners to become active advocates of environmental conservation. The practitioners are then more inclined in making efforts to mitigate climate change. Yoga offers a powerful tool for mobilizing individuals towards environmental action (Kanchibhotla et al., 2021).

Across the globe, the intricacies of yoga unfold in diverse expressions. The practice of yoga is perceived and applied differently in different countries. In the United States, yoga has experienced a surge in popularity, with millions of practitioners embracing its physical and mental benefits. However, the American perspective on yoga often emphasizes its role in personal wellness rather than its cultural or spiritual dimensions. Environmental awareness within the American yoga community has led to eco-friendly practices and initiatives. The Chinese traditional practices like Tai Chi and Qigong often integrate with yoga and have welcomed yoga as a complementary discipline as Yoga's emphasis on holistic well-being resonates with the Chinese ethos of balance and harmony. In Japan, the practice of yoga takes on a unique color, a fascinating blend of Eastern tradition and Western influence (Jakovich & Murray, 2019). Yoga blends seamlessly with Zen meditation, emphasizing mindfulness and inner peace. This synergy creates a practice that fosters mental clarity and inner peace, aligning with Zen's focus on meditation and self-awareness. The Japanese do not follow the dynamic *Vinyasa* yoga (fast paced yoga) which is popular in the West but instead focuses on slower and more introspective movements which naturally helps in cultivating stillness and introspection within the flow of poses. In Australia, many yoga studios and practitioners emphasize the connection between yoga and nature, offering outdoor classes, beach-side retreats, and bush yoga experiences. The Australian yoga community often celebrates the symbiotic relationship between physical well-being and environmental appreciation, fostering a sense of gratitude for the country's diverse ecosystems. The yogic lifestyle involves a plant based diet and a deeper connection with the environment, a slow but steady shift in the choices is observed and preference is given to the products which are manufactured using sustainable means (Cramer et al., 2017). Australia's rich Indigenous cultures contribute a unique layer to the practice of yoga in the country. Some yoga practitioners in Australia seek to incorporate indigenous wisdom, acknowledging the deep connection to the land that is intrinsic to Aboriginal and Torres Strait Islander cultures. This synergy of ancient Indigenous knowledge with the principles of yoga creates a distinctive Australian approach that honors both the traditional custodians of the land and the global heritage of yoga. Further, in the Scandinavian countries, where nature and sustainability are deeply ingrained in the culture, yoga finds resonance in its connection to the environment. Scandinavians view yoga not only as a means of physical exercise but also as a way to connect with nature and promote mindfulness. Yoga retreats in natural settings are popular, emphasizing the holistic relationship between personal well-being and environmental consciousness.

Furthermore, not just yoga, but environmental education also holds essential value when the Earth is facing unprecedented environmental challenges. There is an emerging need for a generation of informed and engaged citizens. The role of environmental education steps in, not merely as a subject in textbooks but also as an enlightening source that empowers individuals to become active stewards of our planet. Environmental education is not just learning about the environment or its issues. It is about instilling a deeper understanding of the complex ecosystem, the intricate connections between human actions and ecological consequences, and the responsibility people hold towards the environment. It helps in equipping individuals with the knowledge, skills, and values to make informed and better decisions, be a proponent of environmental protection, and participate in creating a sustainable future. The pedagogical approach can encompass diverse settings to include teaching about the environment from early childhood classrooms to community workshops and outdoor adventures. Students are encouraged to engage in activities like planting trees, monitoring water quality, or participating in science projects, analyzing data, identifying environmental issues, and exploring potential solutions (Emma & Sheila, 2016). All these efforts promote a sense of wonder and awe for the natural world and help in developing a deeper connection with the Earth and its myriad wonders. However, conventional environmental education methods are often associated with rote memorization of facts and figures about ecosystems and fail to encourage critical thinking, problem solving, and decision-making skills. Additionally, environmental education is often treated as a standalone isolated subject, distinct from other disciplines. This fragmented and isolated approach puts limitations on the ability to foster holistic understanding of the interconnectedness between environmental issues and societal systems.

Therefore, there is an urgency in seeking ways and methods to move towards achieving a sustainable environment and future. In light of these circumstances, though conventional education methods play an important role, it would not be sufficient and there is a requirement of unconventional approaches that would promote a deeper and intense connection with the natural world. Thus among these unconventional approaches, the role of yoga emerges as an impactful mechanism in environmental education whereby it encourages environmental awareness and sustainability during the time when our planet is facing grave environmental issues (Betal, 2008). The environmental awareness fostered by yoga is more relevant than ever. It holds the crucial key to securing the future of our planet and our very species. Though yoga and environmental education may seemingly be two separate and distinct disciplines, yet they converge on a very powerful foundation whereby both emphasize the importance of personal health, connection, and coexisting peacefully with the environment. The integration of yoga and environmental education has the combined potential to instill a sense of connection and belongingness with the environment and promote eco-consciousness and sustainable living.

This chapter deals with the interplay of practice of yoga and environmental education. It analyzes in detail the intricacies of both the disciplines and puts forward the connection of yoga with the environment. It dives deeper into the different schools of yoga and how they are related in promoting environmental awareness. This chapter further discusses the philosophical foundations of yoga and explains how the core principles and beliefs of yoga relate to one's understanding of the natural world. It further explores the integration of yoga with environment education, discussing in detail the various educational approaches to it along with numerous initiatives running worldwide to bridge the gap between yoga and environmental education. The authors emphasize the need to utilize the combined potential of yoga and environmental education towards a sustainable future.

UNDERSTANDING ENVIRONMENTAL DIMENSIONS OF DIFFERENT SCHOOLS OF YOGA

There exist numerous forms, divisions, or approaches within yoga. Each route focuses on the betterment of various facets of the human personality at different phases of a practitioner's growth throughout the practice of yoga (Newlyn, n.d.). Although yoga, in its essence, tries to create a connection with the universe and all living things, each school of yoga approaches environmentalism through its peculiar methods.

1. Hatha Yoga

The literal meaning of Hatha is 'yoga using force' which focuses on its transformative power. Many yogic scholars associate 'Ha' with the sun and 'Tha' with the moon, which represents the practice's focus on balancing opposing energies within our being. However, Hatha yoga is not just about physical postures and goes beyond to actively use the body as a tool for experimentation, movement, and ultimately, profound change in both body and mind. This school of yoga is associated with purification of body and mind. Although it is considered to be the modern branch of yoga, it finds its roots in Tantra and is attributed to the legendary yogi Gorakshanath. It is through the dedicated practice of Hatha yoga that an individual can cultivate a yogic body, a vessel which is not only free from disease, but also liberated from the restrictions that often bind us all. This yoga also focuses on building strength and flexibility through the practice of asanas making the body more fit and capable of sustainable activities like gardening, planting or volunteering for environmental causes. It tries to instill a deeper appreciation for the natural world and its intricate details (Zafeiroudi, 2018).

2. Raja Yoga

The word Raja denotes terms like 'King', 'Ruler' and 'Emperor'. The Raja yoga is associated with noble lineage and meticulous approach and is considered to be the supreme form of yoga. This school of yoga, which is also called Classical yoga, is connected with eight limbs of yoga as mentioned by Patanjali. It puts emphasis on doing away with ego and carving a path towards union with the universal self. Through meditation and *swadhyaya* (or self inquiry), it calms down the thoughts and manages the intellect of an individual and naturally helps in promoting pro-environmental behavior (Riordan et al., 2022). It also promotes inner peace and stillness which inspire compassion for all living beings, ultimately motivating individuals to protect the environment. It fosters a deeper understanding and appreciation of the natural world when a person disconnects from distractions and immerses oneself in nature.

3. Jnana Yoga

The word Jnana basically means 'wisdom' or 'knowledge'. This school of yoga puts emphasis on acknowledging and recognizing the truth about oneself. It signifies the transformation of theoretical knowledge to practical wisdom and involves uncovering one's human purpose in life in connection to both the universe and nature. Jnana Yoga isn't just centered on mere intellectual understanding but rather delves into intuitive or radiant knowledge that arises from the profound depths of one's inner self (Jain, 2023). Its fundamental features include the ability to discern between what is genuine and what is illusory, the self from the non-self, and having virtues of detachment, leading to a state of mind devoid of cravings. Additionally, recognizing the interconnectedness of oneself and nature fosters a deep sense of connection and care for the planet. This school also encourages questioning and reflection, inspiring individuals to critically analyze their environmental impact and seek solutions to problems which persist in the ecosystem.

4. Bhakti Yoga

The literal meaning of Bhakti is 'devotion'. This school is also known as devotional yoga or having complete faith. Bhakti yoga is not about one's intellectual beliefs, but instead puts emphasis on heartfelt devotion to whatever a person sees as ultimate reality, which can either be God, or a Higher Power, or even a deeply cherished ideal (Healthline, n.d.). It offers a different path whereby it thrives on a powerful emotional connection with that object of faith, a connection that allows to channelize your energy and devotion towards personal growth. This school of yoga sees nature as a manifestation of the divine which inspires a sense of responsibility and high regard towards the environment. The practitioners of this school often embrace simplicity, reducing their consumption and minimizing their ecological footprint.

5. Karma Yoga

Known as yoga of action, this school of yoga puts emphasis on the spiritual and enlightened truth that true fulfillment comes not from personal gain, but from selfless service. It inspires individuals to immerse themselves in the present moment of mindful action and to let go of the craving for rewards. This is the real journey which leads to enlightenment. By performing one's duties without attachment to the results, leads individuals to engage in environmental activism and volunteer for the pro-environment action. Accordingly, they consider that every act, be it big or small, is seen as a contribution to the collective well-being, motivating individuals to make environmentally conscious choices (Jain, 2016). A renowned Indian yoga guru and a spiritual leader Sri Sri Ravi Shankar have written on this that, "Nature is a living temple, and it is our duty to protect and preserve it. Let us live in harmony with nature and experience the joy and peace that it brings."

6. Tantra Yoga

This school of yoga is associated with yoga of realization. Tantra yoga isn't about rules or asceticism but dives deep into the body's inner landscape, exploring the vibrant tapestry of sensations and energies which weaves through an individual's mind and body. The yoga exercises focus on innumerous subtle points in the body (Healthline, n.d.). It embraces the body as a temple, a map to spiritual awakening. Additionally, Tantra yoga practices can be used to foster a deep connection with the earth's energy and channel this in instilling a sense of respect and connection with the natural world. It puts emphasis on creating balance and harmony and advises practitioners to seek stability within themselves and with the environment.

7. Kundalini Yoga

Kundalini yoga is considered as yoga of energy. This school of yoga puts emphasis on unlocking the dormant potential of the seven luminous chakras which lies within every person. The lowest of chakras is associated with primal instincts, while the highest echoes with divine potential. Each chakra holds a key, and through deep concentration and dedicated practices like asanas and mantras, we activate these chakras, unleashing the hidden energies and abilities they contain (Healthline, n.d.). The awakened Kundalini energy encourages individuals to actively participate in environmental initiatives and take prompt action for the betterment of the environment. Amongst the practitioners of this school, there is a belief that this Kundalini energy is connected to the earth's energy encourages practitioners to care for the planet to nurture their own spiritual growth.

PHILOSOPHICAL FOUNDATIONS OF YOGA FOR ENVIRONMENTAL STEWARDSHIP

One of the important features of yogic philosophy is the concept of *Brahman*. It is basically the universal consciousness, or in other words, it is the interconnectedness of all things. This concept of interconnectedness is not just theoretical but, as a matter of fact, it is experienced, felt, and understood through the practice of yoga. The practitioner through yoga fosters the virtue of stillness which allows us to perceive the subtle threads that bind us all to every facet of existence. We acknowledge and appreciate the breath we share with plants, the energy we exchange with the sun, and the interconnectedness of our thoughts, emotions, and actions. This profound realization promotes the feeling of responsibility and compassion, in a sense that by harming others, we are ultimately harming ourselves.

The recognition of *Duality* is another core principle of yogic philosophy. In life, an individual experiences life through pairs of opposites, that is, joy and sorrow, light and darkness, effort and surrender. The role of yoga becomes crucial in this as it teaches us to navigate these polarities with equanimity, accepting their impermanence and appreciating the dynamic interplay between them. The goal is not to strive to eradicate negativity, but instead, witness and acknowledge it with mindful awareness, perceiving it as a transitory aspect of the ever changing feature of existence. This principle of duality promotes the feeling of resilience and inner peace where an individual can very well observe the ups and downs of life without being swept away by its currents.

It is pertinent to note that at the heart of yogic philosophy lies the pursuit for *self-mastery*. The aim is not to control or conquer the self but rather befriend it, understand its complexities, and foster its highest potential. Thus, this is not about taming the wild beasts within ourselves, but it is more about peeling back layers like an onion. An individual can gradually peel back the layers of conditioning and ego, revealing a deeper, more authentic self by just following the practices like meditation and self-inquiry. However, this path requires both self-discipline and self-compassion. This enlightened path ultimately leads to a life lived with greater clarity, purpose, and inner peace. The yogic philosophy is not a destination, but rather a lifelong journey and individual takes of exploring and discovering oneself. By diving deeper into the sea of yogic wisdom, one can unveil the truth that binds us all, accept and adapt to the push and pull of duality, and learn to sail through the path of self-mastery. Yet the most profound revelation is the thought that the greatest teacher lies nowhere but within ourselves.

Furthermore, this lifelong journey of yogic philosophy is not a linear path but is intertwined with environmentalism. The philosophy of yoga offers a web of ethical principles, guidelines, meditation techniques, and mindful practices that instills a sense of harmonious coexistence with the Earth, its precious natural resources and all living beings in it. The concept of interconnectedness is deeply embedded in yoga's origins in ancient India. This can be found in *Yamas* and *Niyamas*, which offer a mechanism for peaceful living in harmony with the environment, others and oneself (Malhotra, 2001). Through an exploration of yogic philosophy, environmentally resonant principles come to the forefront and can be an effective tool to foster a connection of the individual with nature and commitment to sustainability. Amongst these, *Ahimsa* (or non-violence) is the cardinal principle which emphasizes on the concept of non-violence and restricts us from harming any living being, including our ecosystems as well, in order to not disrupt the delicate balance of the environment. According to the esteemed Indian yogi Paramahansa Yogananda, "If humanity will consciously observe non-violence, kindness, and harmlessness to nature, it will begin to recognize itself as part of the great ecology of the universe." *Aparigraha* (non-possessiveness) is the principle which focuses attention on renouncing worldly attachments and desires and urges to consume only that which is necessary and refrain from exploiting natural resources. The principle of *Asteya* (non-stealing) puts emphasis on the virtue of integrity which simply means to not take anything which doesn't belong to you which includes the idea that whatever you take from nature, has to be returned back to it. In yoga philosophy, this goes beyond the concept of physical theft and encompasses values of fair exchange and avoids exploiting nature or acting unethically. The principle of *Saucha* (cleanliness) highlights the need to extend maintaining cleanliness and proper hygiene beyond one's sphere to encompass environmental integrity, inspiring us to treat nature with respect and care. Another crucial principle is *Santosha* (contentment) which puts stress on appreciating and accepting what we are and what we have already and thus to be content and gradually moving further from there. On this, BKS Iyengar has written that, "the yogi feels the lack of nothing and so is naturally content."

These yogic principles illuminate the path towards choices that benefit the planet and contribute to its well being and serve as a moral compass (Jain, 2022).

It is also pertinent to note that yoga philosophy also draws attention to the concept of interdependence. Through *Asanas* (or physical postures) one mimics natural elements and phenomena. For example, *Trikonasana* (Triangle Pose) mirrors the flowing of a river, *Tadasana* (Mountain Pose) mirrors the stability of a mountain and *Yoga Nidra* (yogic sleep) fosters a deeper connection to the vastness of nature, promotes humility and gratitude for the intricate web of life. Furthermore, *Pranayama* (breathwork) helps us in connecting with the rhythm of the Earth, thereby mimicking the ebb and flow of the tides which is mostly done by breathing through an alternate nostril. Additionally, there are various meditation practices like *Dharana* (concentration) and *Dhyana* (meditation) which ultimately imbibes a sense of connectedness with nature, thereby diminishing the perceived separation between oneself and the environment. Therefore, when an individual truly understands this intricate web of life, a profound sense of responsibility towards our planet would be instilled. This understanding drives the individual to make choices and take actions that benefit not just oneself, but the entire ecosystem and human community as a whole (Iyengar, 2012).

The effect of yoga on one's environmental consciousness is quite impactful. The yoga practitioners cultivate virtues of mindfulness and self-awareness which helps them to develop a deeper connection to their inner landscapes. This naturally helps them extend the connection to encompass the world around them. There is greater empathy and heightened sensitivity towards all living beings in the ecosystem (Amel et al., 2007). Additionally, yoga tends to shift perceptions and values in a sense that its emphasis on interconnectedness counters the anthropocentric perspective of the world at large, stressing on the bigger picture that humans are not different from, but rather an essential segment of nature's intricate web of life. This shift in perceptions and values promotes pro-environmental virtues like sustainability, respect for the ecosystem and everyone in it, faithful and responsible use of the natural resources. As per the research done by Macrothink Institute in 2021, embracing a yogic lifestyle naturally minimizes our footprint on the planet. The yogic practice of adapting to a vegetarian diet, stressed upon by the principle of *ahimsa* (non-violence), brings down animal suffering to a large extent which aligns with environmentally responsible behavior. Jennifer Summer through her work '*Eating Our Way to Sustainability? Leisure, Food and Community Economic Development*' goes on to apply new meaning to the terms: sustainable leisure, sustainable communities and sustainable community economic development. The study supports the argument that we can indeed eat our way to sustainability if we choose food-related leisure activities that enable others to eat as well (Summer, 2018). To explore the connection between sustainable behavior and mindfulness, a research was done on 100 adults attending a *Midwestern sustainability expo* on two factors a) acting with awareness and b) observing sensations. It was observed that there is a significant positive co-relation between acting with awareness and self-reported sustainable behavior (Amel 2009).

Likewise, yogic virtues of simplicity and self-regulation disconnects us from the clutches of consumerism and its environmental impact. Thus, yoga is not merely a passive practice, which has become synonymous with the physical aspect of *asanas* (or postures) these days, but also it helps in cultivating self-discipline, resilience, strength, and empowering individuals towards pro-environmental action. This positive impact of yoga on environmental awareness is not just merely on papers but very much practical and is supported by various research groups and studies. In particular, a study was conducted in 2021 on 195 people, including men and women, to conduct an inquiry into the possible effects of practice of Yoga on a person's eco-consciousness and sustainability. The results statistically demonstrated a remarkable increase in environmental awareness and positive attitude towards pro-environmental behaviors in comparison with a control group. The virtues taught by principles of yoga had an impact on the practitioner with respect to higher levels of ecological concern, responsible consumption habits, and engagement in green initiatives. As a matter of fact, the results stipulated yoga to be a promising supplement of bolstering the environment friendly programmes handled by various social and environmental organizations and groups to contribute to the ongoing pursuit of sustainable practices (Zafeiroudi et al., 2021)

ASSESSING EDUCATIONAL APPROACHES AND INITIATIVES

Witnessing the destruction of the environment can be spine chilling for many, especially the young minds. However, while some like Greta Thunberg can turn their fear into fuel for action, others often find themselves paralyzed by a 'fight' or 'flight' or just simply shutting down. Thus, there is also a need for incorporating yoga with education in the modern schooling system. Extensive research on the issue highlights that yoga in education can be a powerful tool for nurturing students' physical well-being, cognitive abilities, emotional resilience, interconnectedness with nature and promoting environmental behavior and sustainability and therefore makes a compelling case for the inclusion of yoga in the school curriculum (Sinha & Kumari, 2021).

However, to truly care about the environment and to foster a deep connection with the ecosystem, students need more than just facts. The environmental challenges we face call for a shift in perspective. With PM Modi declaring June 21st as the International Yoga Day and making it compulsory for all the state and center affiliated schools to celebrate it by conducting mandatory yoga sessions, India has taken a step forward to impart yoga education through schools and colleges (Omkar, 2019). As a country that propounded yogic philosophy, India in comparison to first world countries is failing to include the yoga education in its school and college curriculum. Despite being a practical subject there is need to impart theoretical aspects of yoga by making it a part of routine material education. It was with National Education Policy 2020, yoga or yog acquired a significant role in holistic education system (Ranade & Nikam, 2023). Further NEP 2023 seeks to provide for playing sports and yoga education compulsory in all educational institutions right from primary school level to senior secondary level in order to make it a part of school curriculum (Bill No. 41 of 2023).

There is a need to promote awareness in fostering eco-consciousness which necessitates the provision of effective tools to dismantle educational barriers. A paradigm shift is required whereby there is integration of holistic education's emphasis on interconnectedness with yoga's focus on mindfulness and connection. It becomes essential to integrate conventional monodisciplinary education with extensive interdisciplinary education because environmental education makes it necessary to incorporate distinct pedagogical approaches to comprehend and teach the intricacies of environmental issues. It is observed that mere adherence to well-structured curriculum or teaching to inculcate virtues with positive attributes is insufficient. The fostering of environmental consciousness requires the implementation of efficacious tools that facilitate intense engagement with values like positivity and optimism (Bhatnagar, 2023). Those who engage genuinely can experience a permanent shift in their perspective and values which will leave a lasting impression and inspire environmental action.

In light of these needs, the Pratipaksha Bhavana method has become one of the most important pedagogical approaches to integrate yogic philosophies with environmental education. The word Pratipaksha Bhavana, which finds its root in the Sanskrit language, means 'cultivation of the opposite'. This ancient yoga practice is mentioned in the Yoga Sutras of Patanjali which emphasizes on transforming negatives into positives, in a sense that the students should focus on empathy instead of aggression. By focusing on nurturing strengths, students can transform challenges into opportunities and build a future where environmental well-being thrives. This method can be applied to achieve environmental sustainability, where stakeholder participation, collaboration, and capacity building are crucial for success. A key contributor to environmental degradation is the evils of greed which can be toned down by encouraging contentment and fostering strong human connections. These virtues will gradually prove invaluable in shaping the students' future roles as managers, scientists, collaborative team members, and ultimately, custodians of our delicate ecosystems (Ganatsios et al., 2021). This holistic-yoga fusion holds the key to unlocking a generation of environmentally conscious learners and leaders.

Various research also indicates that the benefits of integrating yoga in education extend beyond student populations, demonstrating its effectiveness in enhancing teacher education as well, thereby imbibing the sense of sustainability. The synergy of mindfulness practices of yoga such as *Pranayama*, Hatha yoga, and meditation into the routines of practicing teachers have the effect of cultivating a sustainable sense of self, long-lasting happiness, and enhanced well-being which positively impacts both established educators and those who are in pre-service training. These mindfulness practices in new teaching and learning approaches are further contributing to the overall transformation of teachers, students, and education as a whole. Though this whole process promotes individual serenity and well-being, its impact goes beyond the personal sphere of an individual, transforming classrooms and school communities through the cultivation of sustainable happiness and opportunities for collective growth, ultimately creating a sense of harmony in the ecosystem (Kyte, 2017).

With these evolving educational landscapes, there are also various innovative educational programs which incorporate yoga to foster affection, knowledge, and action for the environment. For example, Yoga Calm Earth Warriors is one of the organizations which is pioneering a unique approach whereby they use yoga, mindfulness, and ecopsychology to deepen their understanding and awareness about the environment, through physical and emotional connections. It is pertinent to note that these activities are in compliance with the educational standards and can be integrated with other existing curricular frameworks. The aim of this curriculum is to bridge the gap between young minds and the natural world and empower them to become responsible stewards of the environment (Gillen, 2021). Another such educational initiative is run by the Breathing Space organization which follows the philosophy of yoga and practice and implements the yogic principles of *ahimsa*, *aparigraha*, *saucha* and *santosha* (Breathing Space, 2023). They attempt to foster empathy and awareness towards the environment in their yoga class for kids through activities like nature walks, scavenger hunts, recycled crafts and outdoor yoga which includes practices like sun salutations and tree pose forming a deeper connection with nature (Zafeiroudi & Hatzigeorgiadis, 2014).

It is also pertinent to bring forth the noteworthy efforts of Oceanic Yoga International School in Goa, which is directly affiliated by the Yoga Alliance, USA. It is one of the most prominent organizations providing yoga teacher training programmes in India, particularly Goa and Rishikesh. These teacher training programmes play a crucial role in shaping the landscape of yoga in India. By inculcating eco-consciousness along with the technical know-how about the yoga postures and principles, they empower a generation of teachers and yoga practitioners who will ultimately steer the yoga practice towards a more sustainable and connected future. This is mostly achieved through adopting sustainable practices in yoga studios, including sustainable nutrition choices and practicing mindful consumption habits along with applying yogic principles with respect to environment ethics. These programs seek feedback from trainees and modify their curriculum accordingly to cater to the dynamic needs of sustainability, which helps them stay relevant and efficacious in their effort to promote environmental awareness.

Furthermore, another organization which is working towards the synergy between yoga and environmental education is Green Karma Yoga, which is located in studios in New York City and San Francisco. Their main focus is to integrate social and environmental responsibility into yoga practice and studio operations. To achieve this objective, they promote various eco-friendly yoga classes. Their primary activities also include partnering with local environmental organizations for fundraising and awareness campaigns, host interactive educational workshops and seminars on sustainability topics, and inspiring the young minds to participate in pro-environmental action. Their immense impact naturally led to widespread recognition of their leadership in such green business practices and has inspired other yoga studios as well to adopt similar initiatives.

These pioneering initiatives are merely the tip of the iceberg, hinting at the boundless potential of yoga as a catalyst for environmental action. There are countless initiatives sprouting around the world which are adopting this multidisciplinary approach to integrate yoga and the environment. As more people and communities embrace this approach, its cumulative impact has the potential to generate a transformative tidal wave of positive change for our planet. The integration of yoga and environmental education is proving to be a efficacious method for promoting personal transformation, inspiring action, and building a more sustainable future.

FINAL REFLECTIONS AND FUTURE CONSIDERATIONS

The philosophy of yoga puts emphasis on the intricate relationship between human beings and the environment. The concepts of *Prakriti* (or nature) and *Purusha* (or pure consciousness) when united together give rise to life. The practice of yoga helps in improving the quality of life, inculcates in people healthy and positive lifestyle habits and naturally reduces the carbon footprint. These yogic teachings inspire profound respect for the natural world striving to live in harmony with it (Somarathne et al., 2023). Likewise, the objective of environmental education is to raise awareness about the environment, encourage pro-environmental action and develop a sense of responsibility towards the planet. There is a greater sense of interconnectedness with the natural world and deeper understanding of the ecosystem when educators incorporate yoga sessions, nature walks, and meditation into the environmental education curriculum. Thus, there is an undeniably great potential of integrating yoga with environmental education, yet there are still many significant challenges and hurdles to be addressed and overcome.

Accessibility is one of the biggest barriers to yoga and environmental education programs. The first aspect of this is the socioeconomic limitations whereby insufficient financial resources can restrict access to yoga classes and environmental education programs. There may be a requirement of additional resources, such as trained yoga instructors, appropriate spaces, and materials to efficiently implement yoga and educational programs. In light of these needs, many schools or organizations with limited resources might find it difficult to tap into the potential of yoga and environmental education and ultimately might not incorporate yoga into their environmental education curriculum. However, offering scholarships, sliding scale fees, partnerships with community centers, and utilizing free online resources provide a possible solution to it. Even seeking sponsorship from government, pro-yoga organizations and activists can offer an opportunity to overcome this hurdle. The second aspect of accessibility is the physical limitations whereby some individuals may be physically unable to perform certain yoga postures.

Yoga is not a simple practice and perfecting many of yoga *asanas* takes months or even years, which becomes impossible for individuals with physical disabilities or limitations. Thus, it becomes even more difficult to integrate yoga and environment education programs for such individuals. It is challenging to ensure that yoga sessions are inclusive and adaptable to various physical abilities and ages. To cater to this issue, offering modified poses, using adaptive equipment, and promoting alternative practices like meditation or gentle movement exercises can prove to be a fruitful solution for anyone having such physical limitations.

The practice of yoga finds its origin in ancient Indian philosophy and culture and naturally the obstacle of cultural sensitivity comes into picture when integrating yoga into environmental education. It becomes important to have a culturally sensitive approach which necessitates understanding and respecting the cultural origins of yoga and presenting it in a way that is inclusive and respectful for diverse cultural communities and perspectives. To navigate these concerns, an approach should be adopted encompassing careful consideration and consultation with diverse communities. These approaches may also include incorporating local traditions and practices, utilizing diverse instructors, and fostering community ownership of programs.

Additionally, apart from the issue of accessibility and cultural sensitivities, there may exist limitations in secular context. Traditionally, the practice of yoga has spiritual dimensions as well. In order to integrate yoga with a secular environment education curriculum there is need to incorporate the practices of secular spirituality. In a world where yoga adaption being commercialized are losing their depth and connection to the spiritual core. Secular spirituality is nonadherence to any religion but to spiritual philosophy. It focuses on building the relation with inner self and personal growth rather than connecting with any divine source (Owls & Indigo, n.d.). Tantra Yoga thus through its psycho-somatic methodology can provide an appropriate eco-spiritual practice to achieve secular spirituality thereby achieving secular environment education. As a type of eco-spiritual practice, Tantra yoga can very well cater to issues of mental health such as depression and burnout, frequent in hectic urban lifestyles resulted by disconnection from nature. (Kochetkova & Segalerba, 2024). As discussed where spiritual secularity is an option to have secular environment education, there are scholars like Rita Dasgupta Sherma, who has dwelled into the intersection of ecofeminism and Hindu Tantra. She has explored how Hindu ritual practices and belief systems interact with environmental concerns, highlighting conflicts between religious environmentalism and nature worship (Sherma, 1998).

Another hurdle that needs to be addressed is the issue of integration itself. There exist many pedagogical challenges whereby merging yoga and environmental education necessitates innovative ideas and methods that goes beyond simply adding yoga to existing curriculum. It can be quite challenging to include yoga within an already existing environmental education curriculum without compromising the part and parcel of environmental content itself. It becomes extremely important to find an approach which can seamlessly integrate the two without diluting the core environmental education objectives. One of the approaches can include thematic integration, where there is connecting particular yoga practices to environmental concepts. For example, practicing *Vrksasana* (or tree pose) while discussing the concepts of deforestation or afforestation. Another approach may include experiential learning whereby environment stewards are inspired to imbibe eco-consciousness by combining outdoor yoga sessions with nature observation, service projects, or environmental art activities. These approaches can also be merged with inquiry-based learning where students are motivated to explore environmental issues through yoga practices and reflection. Additionally, there is imperativeness in training educators whereby they need training in both environmental education and yoga to effectively integrate both of them. To ensure that educators are equipped with skills and knowledge about this interdisciplinary approach and can effectively create a harmonious learning experience requires a lot of investment in specialized training programs and professional development.

Integrating yoga and environmental education is an essential step, but it's only part of the journey. The impact of these initiatives remains incomplete without ensuring a way to effectively evaluate and assess them. Developing appropriate mechanisms for a robust evaluation and assessment of the impact of this integrated approach is the need of the hour. To measure the holistic impact of combining yoga with environmental education, traditional assessment methods may need to be adopted. There is a requirement of evaluation tools that can represent the multifaceted impact on environmental attitudes, values, behaviors, and even emotional responses. In this context, gathering observational data and qualitative feedback becomes requisite. This rigorous evaluation will give true accounts of these initiatives' effectiveness in nurturing environmental awareness, responsibility, and action. Through this evaluation and assessment, it will further strengthen the case for securing funding, advocating for wider adoption, and continuously refining the design of these initiatives. Thus, it makes more space for continuously improving these initiatives, tailoring them to specific contexts, and ultimately maximizing the potential of this integration to cultivate a generation of environmentally conscious and responsible citizens.

Yoga and environmental education, when merged together, can reap greater benefits with respect to cultivating eco-conscious individuals who are ready to navigate the complexities of this dynamic world. By integrating the practice and wisdom of yoga into environmental education, there is immense respect for nature, nurturing a generation dedicated to protecting and preserving our ecosystem. In the spirit of Mahatma Gandhi's words, "The future depends on what we do today", we weave the two disciplines of yoga and environmental education together in order to take meaningful action for a greener and more sustainable tomorrow. This combined approach paves the path towards fostering an ecological shift.

As we reflect upon the insights gained, it prompts us to further explore the intricate linkage of yoga and environment education. What are the potential long term benefits of widespread adoption of environmental education based on yoga practices? What are the ethical considerations and best practices for incorporating yoga into environmental education programs? What research gaps remain in understanding the intersection of yoga and environmentalism?

REFERENCES

Amel, E. L., Manning, C. M., & Scott, B. A. (2009). Mindfulness and Sustainable Behavior: Pondering Attention and Awareness as Means for Increasing Green Behavior. *Ecopsychology*, *1*(1), 14–25. doi:10.1089/eco.2008.0005

Betal, C. (2008). Conservation of ecology and environment through yogic lifestyle. *Journal of Environmental Research and Development*, *2*(4), 905–912. https://scholar.google.com/citations?view_op=view_citation&hl=en&user=SvszxRkAAAAJ&citation_for_view=SvszxRkAAAAJ:u5HHmVD_uO8C

Bhatnagar, D. (2023). *Environmental Consciousness,* Medium. https://medium.com/@devanshi.art.bhatnagar/environmental-consciousness-63070dfd4d14

Breathing Space. (2023). *Yoga and Environmental Stewardship: Fostering a Connection with Nature in Children.* Breathing Space. https://breathingspacedc.com/yoga-and-environmental-stewardship-fostering-a-connection-with-nature-in-children/

Chakravorty, S. (2021). Green Yoga and the Art of Living Harmoniously with Nature. *Livemint.* https://www.livemint.com/opinion/online-views/green-yoga-and-the-art-of-living-harmoniously-with-nature-11624164695671.html

Cramer, H., Sibbritt, D., Park, C. L., Adams, J., & Lauche, R. (2017). Is the practice of yoga or meditation associated with a healthy lifestyle? Results of a national cross-sectional survey of 28,695 Australian women. *Journal of Psychosomatic Research, 101*, 104–109. doi:10.1016/j.jpsychores.2017.07.013 PMID:28867414

Dattatray, O. (2019). Yoga Education in India. *Northlines.* https://thenorthlines.com/yoga-education-in-india/

Emma, P., & Sheila, D. (2016). Innovative pedagogical approaches in early childhood care and education (ECCE) in the Asia-Pacific region: a resource pack. *Asia-Pacific Regional Network for Early Childhood (Singapore).* https://unesdoc.unesco.org/ark:/48223/pf0000246050

Ganatsios, H. P., Filippou, A. F., Mpekiri, S., & Danahy, K. E. (2021). Education for cultivation of awareness as an important factor of raising environmental consciousness. *International Journal of Higher Education and Sustainability, 3*(3), 231–265. https://www.inderscienceonline.com/doi/pdf/10.1504/IJHES.2021.117900. doi:10.1504/IJHES.2021.117900

Gillen, J. (2021). A New Approach to Environmental Education: Yoga Calm Earth Warriors. *Yoga Calm.* https://www.yogacalm.org/a-new-approach-to-environmental-education-yoga-calm-earth-warriors/

Healthline. (n.d.). Everything You Need to Know About Tantric Yoga. https://www.healthline.com/health/tantric-yoga

Healthline. (n.d.). What Is Bhakti Yoga? https://www.healthline.com/health/fitness/bhakti-yoga

Healthline. (n.d.). What Is Kundalini Yoga and What Are the Benefits? https://www.healthline.com/health/kundalini-yoga

Iyengar, B. K. S. (2012). *Light on the Yoga Sutras of Patanjali* [ebook edition]. HarperCollins UK. https://www.everand.com/book/234816673/Light-on-the-Yoga-Sutras-of-Patanjali

Jain, R. (2022). How to Practice the Yamas – Applied Yoga Philosophy. *Arhanta Yoga.* Retrieved from https://www.arhantayoga.org/blog/how-to-practice-the-yamas/

Jain, R. (2023). *What Is Jnana Yoga? Meaning, Benefits & How to Practice the Yoga of Knowledge,* Arhanta Yoga. https://www.arhantayoga.org/blog/jnana-yoga-guide/

Jain, R. (2023). *What Is Karma Yoga? Key Principles, Benefits, & Tips for Daily Practice,* Arhanta Yoga. https://www.arhantayoga.org/blog/four-principles-of-karma-yoga/

Jakovich, K., & Murray, J. A. (2019). Yoga in China. *The Asia Dialogue.* Retrieved from https://theasiadialogue.com/2019/12/18/yoga-in-china/

Kanchibhotla, D., Nair, A., & Sangle, P. Gupte & P., Kulkarni, S. (2021). The Correlation between Yoga Practice, Connectivity with Nature and Climate Change Awareness in Indian Population. *International Journal of Innovative Research in Science, Engineering and Technology, 2*(11), 001-005. https://www.iomcworld.org/articles/the-correlation-between-yoga-practice-connectivity-with-nature-and-climate-change-awareness-in-indian-population.pdf

Kochetkova, T., & Segalerba, G. (2024). Environmental Personhood: Philosophical Foundations in Tantric Ecology. Bridging Health, Environment, and Legalities: A Holistic Approach. 1-24.

Kyte, D. G. (2017). Toward a Sustainable Sense of Self in Teaching and Teacher Education: Sustainable Happiness and Well-Being Through Mindfulness. *McGill Journal of Education, 51*(3), 1143–1162. https://mje.mcgill.ca/article/view/9321. doi:10.7202/1039632ar

Malhotra, A. K. (2001). *An Introduction to Yoga Philosophy: An annotated translation of theYoga Sutras*. Routledge.

Newlyn, E. (n.d.). The 6 branches of yoga. *EkhartYoga*. Retrieved from https://www.ekhartyoga.com/articles/practice/the-6-branches-of-yoga

Owls & Indigo. (n.d.). What Is Secular Spirituality? https://owlsandindigo.com/day-15-21daysofyule-what-is-secular-spirituality/

Riordan, K. M., MacCoon, D. G., Barrett, B., Rosenkranz, M. A., Chungyalpa, D., Lam, S. U., Davidson, R. J., & Goldberg, S. B. (2022). Does meditation training promote pro-environmental behavior? A cross-sectional comparison and a randomized controlled trial. *Journal of Environmental Psychology, 84*, 101900. Advance online publication. doi:10.1016/j.jenvp.2022.101900 PMID:36969767

Ross, A., Friedmann, E., Bevans, M., & Thomas, S. (2013). National survey of yoga practitioners: Mental and physical health benefits. *Complementary Therapies in Medicine, 21*(4), 313–323. doi:10.1016/j.ctim.2013.04.001 PMID:23876562

Sherma, R. D. (1998). Sacred Immanence: Reflections of Ecofeminism in Hindu Tantra. In L. E. Nelson (Ed.), *Purifying the Earthly Body of God. Religion and Ecology in Hindu India* (pp. 89–132). SUNY Press.

Sinha, A., & Kumari, S. (2021). Integrating yoga with education in the modern schooling system. *Yoga Mimamsa*. 53(1), 46–58. https://journals.lww.com/yomi/fulltext/2021/53010/integrating_yoga_with_education_in_the_modern.9.aspx

SomarathneE. A. S. K.GunathungaM. W.LokupitiyaE. (2023). How Does Meditation Relate to Quality of Life, Positive Lifestyle Habits and Carbon Footprint? SSRN. https://papers.ssrn.com/sol3/papers.cfm?abstract_id=4549372 doi:10.2139/ssrn.4549372

Summer, J. (2018). Eating Our Way to Sustainability? Leisure, Food and Community Economic Development. *Sustainability*. https://www.mdpi.com/2071-1050/10/5/1422

Zafeiroudi, A. (2018). Hatha Yoga: The importance of Yamas and their application from theory to practice. *Journal of Sport & Recreation Management, 15*(1), 64–76. https://www.researchgate.net/publication/343452538_Hatha_Yoga_The_importance_of_Yamas_and_their_application_from_theory_to_practice

Zafeiroudi, A., & Hatzigeorgiadis, A. (2014). The effects of an outdoor pursuit's intervention program on adolescents' environmental beliefs. *International Journal on Advances in Education Research, 1*(3), 106–118. https://www.researchgate.net/publication/343452414_The_effects_of_an_outdoor_pursuit's_intervention_program_on_adolescents'_environmental_beliefs

Zafeiroudi, A., Pipinia, M., Yfantidou, G., & Georgomanos, S. (2021). The Effects of Yoga Practice on Practitioners' Environmental Behaviours & Sustainability. *Environmental Management and Sustainable Development, 10*(3), 76. Advance online publication. doi:10.5296/emsd.v10i3.18894

KEY TERMS AND DEFINITIONS

Interconnectedness: It is a state where all things are connected to each other within the ecosystem.

Mindfulness: It is the human ability or mental state of being completely aware of one's body, mind and surrounding.

Meditation: It is an exercise where an individual practices mindfulness and improves the concentration and awareness which gives a sense of calm and composure to the practitioner.

Chakras: These are the body's seven main energy points which can be found in the spine.

Asanas: These are the physical body postures or poses practiced in yoga which focuses on physical and mental well-being.

Eco-consciousness: It is the awareness and understanding of the environment whereby an individual shows care and concern towards it.

Ecopsychology: It is the interdisciplinary science which integrates ecology with psychology, exploring the linkage between a person's mental well-being and the natural world.

Section 3

Indigenous and Traditional Environmental Education

Chapter 8
The Indigenous Knowledge Systems-Based Ethnoscientific Model of Integrated Eco-Education (IEE):
Multicultural Prospects of Environmental Education

L. Jan Slikkerveer
Leiden Ethnoscience and Development Foundation, The Netherlands

Sander L. Gellaerts
Leiden Ethnoscience and Development Foundation, The Netherlands

ABSTRACT

This chapter delves into the pivotal role of Indigenous Knowledge Systems (IKS) under the framework of ethnoscience in transforming global environmental education. Scholars document the ongoing sixth mass extinction of biodiversity, necessitating urgent collaborative action. Ethnoscience, emerging from joint efforts of pedagogy and UN-led organizations, offers promise in addressing environmental crises. Since the 1990s, ethnoscience studies have highlighted sustainable indigenous management systems. Traditional Environmental Knowledge (TEK), a subset of IKS, imparts sustainable practices and holistic perspectives across generations. TEK fosters global conservation awareness and ecosystem restoration. Building on indigenous cosmovisions like "use the fruits, don't use the wood," a Model of Integrated Eco-Education (IEE) is proposed. This transdisciplinary model bridges the "knowledge gap" between science and ethnoscience aiming to cultivate eco-conscious global citizens. Efforts target multicultural environmental education in both developing and developed nations, emphasizing students' potential to drive ecological transformation.

DOI: 10.4018/979-8-3693-2577-3.ch008

INTRODUCTION

"The role of local, traditional knowledge and indigenous approaches to learning are critical for modern education systems"

Dr. Balakrishna Pisupati (2020).

As in other international debates, the growing concern of biodiversity loss and climate change threatening life on Planet Earth has gained increased priority in environmental education. Although students, teachers and the general public appear to acknowledge that there seems to be a looming crisis, and efforts are made to an understanding of the magnitude of the situation, the complexities and pace of the world-wide threat remain undervalued. The complicated threats of biodiversity, accelerating the evolutionary process towards the "sixth mass extinction of life", are expected to worsen if urgent and targeted action is not taken. There are, however, sparks of hope to turn the tide, albeit largely at the local level.

In view of the global crisis, the discipline of environmental pedagogy has been exploring new avenues to cope with the situation by transforming its framework in theory and practice by linking up with declarations and goals of international organisations, such as the first UN-UNESCO *Intergovernmental Conference* on environmental education at Tbilisi (USSR) in 1977, introducing the concept of sustainable development in environmental education. Thereafter, the first UN *Earth Summit on Environment and Development*, held in Rio de Janeiro, Brazil in 1992, recommending new methods of education, was followed by the introduction by UNESCO (2014) of the concept of Education for Sustainable Development (ESD) to promote the inclusion of the key sustainable development issues into teaching and learning. The 2030 *Agenda for Sustainable Development*, adopted by all United Nations Member Sates in 2015, created 17 Sustainable Development Goals (SDGs), including SDG 4: Quality Education to "Ensure inclusive and equitable quality education and promote lifelong learning opportunities for all.

Meanwhile, in the practical setting of the international debate on global sustainable development, the role of Indigenous Knowledge Systems (IKS) gained a non-negligible position since the 1980's. The shift in the "development paradigm", supported by the discussion on the interaction between science and ethnoscience, had evolved from the discipline of cognitive anthropology. In environmental research, Traditional Ecological Knowledge (TEK) underscored the paradigm shift, showing to provide a substantial contribution to sustainable community development. Likewise, in environmental education, several programmes in ethno-education on the environment proved rather effective, albeit confined to local and regional settings.

Following the process of de-colonization, the world-wide emergence of multicultural societies called for restructuring the educational systems according to the expectations and demands of sub-cultures, retaining their traditional culture and ways of life, including the transfer of knowledge to their younger generations, often in their local language. A key role in local education is the holistic concept of their traditional cosmovision and philosophy of nature, fostering a ethic of respect and care for the environment and its resources.

Embarking of the positive experiences in Sub-Saharan Africa and South-East Asia with the introduction of models in Integrated Community Health Development (ICHD), Integrated Microfinance Management (IMM), Integrated Community-Managed Development (ICMD), all seeming to bridge the "knowledge gap" between science and ethnoscience, this Chapter provides a IKS-based Ethnoscientific Model of Integrated Eco-Education (IEE) designed for effective Environmental Education in Multicultural settings around the globe.

Today's Unprecedented Global Ecological Crisis

Since the dawn of humanity, Planet Earth and its plant and animal species have been confronted with an unprecedented all-embracing global ecological crisis, permeating in loss of biodiversity, linked with climate change, air, and water pollution, and the depletion of natural resources for food, energy, and livelihood of humankind. Such mass extinctions of plants and animals, caused by both natural and non-natural events, are part of the evolutionary process, albeit recently they evolved into the human-led *Sixth Extinction* (*cf.* Leakey & Lewin 1996; Wagler 2011; Walsh 2022).

Today's Holocene or Anthropocene extinction refers to the ongoing extinction event of numerous species of plants and animals during the Holocene epoch, where humans cause climatic changes that result in the loss of other species.

As the current global predicament eventually threatens human life is eventually threatening human life itself, concerned scientists, conservationists, and teachers have lately drawn peoples' urgent attention to the current existential menace. The famous anthropologist Richard Leakey (1996: 77) warned that: "*Like all other past species that once called Earth home, we too would likely go extinct*". However, Leakey added to his prediction of the *Sixth Extinction* his conviction, that humans could delay such a catastrophic event if the crises are timely recognized and addressed properly, underscoring his important contributions to the protection of the environment through his lifelong dedication to nature conservation and education (Leakey & Lewin 1996).

The planet's biodiversity includes all living species, genes, and ecosystems, meticulously described and assessed by Heywood & Watson, known as the *Global Biodiversity Assessment* (UNEP 1995) as a follow-up to the previous *World Conservation Strategy* (IUCN 1980), *Our Common Future* (WCED 1987) *Caring for the Earth: A strategy for Sustainable Living* ((IUCN/NEWP/WWF 1991) and the *Global Biodiversity Strategy* (WRI/IUCN/UNEP 1992). These important studies not only document the great diversity of approximately 9 million species of plants, animals, protists, and fungi, and nearly 8 billion people presently inhabiting the Earth, but also warn that human actions were dismantling the Earth's ecosystems causing the loss of biodiversity at an alarming rate. Cardinale *et al.* (2012) point out this precarious situation by arguing that human intervention has threatened biodiversity, while paradoxically, humans depend on it most. Following the warnings of the first *Earth Summit on Environment and Development*, held in Rio de Janeiro, Brazil in 1992, their observation has led to the key question of how the growing loss of biological diversity will alter the functioning of ecosystems and their ability to provide humanity with the goods and services needed to survive.

More recently, the *UN Global Assessment Report on Biodiversity and Ecosystem Services* (2019) documents that the present decline of biodiversity is faster than at any other point in human history, with a million species facing extinction across the globe, many even within decades. The Report reaffirms Leakey's warning of more than a quarter of a century ago, that life on the Planet is now facing the world's *Sixth Extinction* (*cf.* Leakey & Lewin 1996).

Soon thereafter, the G*lobal Biodiversity Outlook 5* (GBO-5), published by the UN Convention on Biological Diversity (CBD 2020: 1), came to the harsh conclusion that "As a planet, we are failing to meet all the targets set for slowing down biodiversity destruction by 2020." The document not only warns of the alarming degradation of nature but also points out that biodiversity loss increases the risk of future pandemics.

Such a bleak perspective is also transpiring through the projects of the *Millennium Ecosystem Assessment* (MA) in 2005, indicating that biodiversity loss and transformation of habitats are likely to continue

Figure 1. Graphic representation of the five major threats to biodiversity, including the biggest threat being the change in land and sea use.
Source: Acuity (2023).

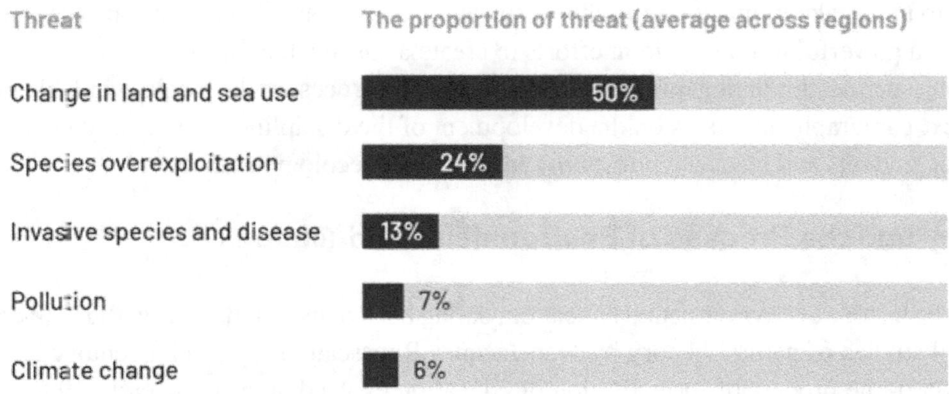

for the foreseeable future. Among the reasons mentioned is the fact that most of the direct drivers of biodiversity loss, including climate change, habitat transformation, and overexploitation, are predicted to either remain constant or increase shortly. Even more worrisome facts have recently been presented by the *Sixth Assessment Report* (AR6) of the *Intergovernmental Panel on Climate Change* (IPCC) (1923), documenting that the human-induced rise of global warming of 1.1 degrees C, unparalleled over past centuries to millennia, is now occurring everywhere around the world, resulting in an unprecedented loss of biodiversity in recent human history.

The biggest threat is the change in land and sea use, while on average, the world loses approximately one acre of rainforest every *two seconds*. The five major threats are listed below, as is shown in Figure 1.

Despite such worrying figures, there are, also reports documenting the results of recent efforts of reversing the ecological crisis, albeit in a modest way, under the motto: *Global problems, local solutions*. An inspiring example of the early education of young people in nature conservation is provided by Jane Goodall, the celebrated naturalist who launched her *Roots & Shoots Programme* (Goodall 1992). The Programme aims to encourage young people around the world to make a positive change by providing teachers and team leaders with free resources and a place to share their stories with the world, has evolved to today's global action-oriented community of more than 700.000 members. As a UN Messenger of Hope, Goodall (1921) explores in her *Book of Hope* that among the four reasons for hope in the present challenge - human intellect, the resilience of nature, and the strong human spirit - the power of young people will help humanity to find hope for the future.

Avery (2012) documents in *Our Vanishing Flora*, that although Warwickshire, U.K. loses one plant species every three years, there is also hope. The *Warwickshire Wildlife Trust* worked with farmers and landowners across the county on successful habitat restoration projects helping to support populations of otters, voles, and different species of farmland birds. Similar projects have been reported on the successful re-introduction of beavers in The Netherlands, where these mammals had become extinct in 1826. Also, under the protection of the *European Biodiversity Programme Natura 2000* (2009/147/EG), projects for the re-introduction of endangered species elsewhere in Europe include popular fieldtrips, courses, and workshops to educate the youth in conservation. In this way, Member State Governments (MSG's), International Organisations (IO's), and Non-Government Organisations (NGO's) have become major actors in local initiatives to try to reverse the ecological crisis at the grassroots level, which deserves a follow-up at the international level.

Recently, conservationist David Attenborough (2021) issued a serious wake-up call to all MSGs, IOs, and NGOs to undertake joint action to achieve sustained conservation, underscoring the importance of education as a powerful tool in the joint efforts to create a sustainable future.

Following such accumulating information on the ongoing process of the barely avoidable mass extinction, the next paragraph elaborates on the development of the discipline of pedagogy in environmental education of students and teachers in response to the current ecological crises.

Pedagogy and the Process of Environmental Education

Following the history of environmental education, dating back to its foundations in the 18th century with the classical studies of natural history by Jean-Jacques Rousseau (1754) and a century later by Louis Agassiz (1859), the process of formalisation of education evolved later in the early 20th century with the nature study movement in the United States. The pedagogy-oriented movement combined scientific investigation with personal and spiritual experiences from human interaction with the natural world. The underlying hypothesis of this movement, known as *biophilia,* provided the educational framework of human attraction to nature and to interact with other forms of life (*cf.* Turner 2003; Louv 2008). It was, however, Anna Botsford Comstock (1911) who in her *Handbook for Nature Study* introduced the popular nature study movement across the United States as a process to educate children and students about the cultural values of nature.

For pedagogy, defined in contemporary academic discourse as a discipline representing the theoretical framework of education which enables teachers to know the objective, process, and outcomes of the teaching process, the emerging changes in society in the course of the 20th century necessitated adaptations in the implementation of pedagogy in the practice of the education process. With the emerging socio-economic and environmental problems in the Western world, a new form of environmental education developed, known as *conservation education*. Conservation education soon evolved into a research-based education of human utilisation of the natural environment and its resources through skilled management, preservation, and restoration of natural resources intending to meet the material, cultural and philosophical expectations of present and future generations.

Since the end of World War II, the growing critical global situation of pollution, climate change, resource depletion, and habitat destruction began to cause an increased general awareness that the environmental problems were transcending regional and national boundaries, rendering such environmental concerns a particular focus for international institutions and organisations. Kaiser & Meyer (2016) argue, that most environmental problems, including biodiversity loss and climate change, require transnational and global coordination, and support the growing conviction that national and international funding agencies should address the increasing environmental "knowledge gaps" which had been impeding the development of the effective management and conservation of the environment on a global scale. In the practical setting of the growing worldwide ecological challenges, the *UN Conference on the Human Environment* (1972) became a new point of embarkation to address the global need for concerted actions to protect the carrying capacity of the globe, bringing together MSG's, IO's and NGO's with the aim to coordinate global action for environmental protection.

In the meantime, the key concept of sustainable development was introduced in 1987 by Brundtland (1987, p.41) in her UN Report *Our Common Future* as: "*development that meets the needs of the present without compromising the ability of future generations to meet their own needs."* In December 1987, the UN General Assembly passed a resolution adopting sustainable development as a central guiding principle of the UN, governments, organisations, and private institutions.

Less than a decade later, the *International Environmental Education Programme* (1975-1983) (IEEP 1975) was established in 1975 by UNESCO-UNEP with a view to implement their objectives to provide a new structural vision for a worldwide mobilisation of education for environmental awareness. IEEP implemented its mission by developing the capacity of environmental education through programmes of training, technical cooperation, policy research and knowledge sharing for protecting nature and conserving natural resources. According to IEEP, the objective of the school curriculum is that it becomes a fluid conversation between teachers and learners based on reliable sources of information.

In his review of IEEP's contribution to international environmental education over the past decades, Dorn (2020) observes a *New Global Ethic* in the move to fostering an environmental ethic through a problem-based, interdisciplinary study of the ecology, focused on the adoption of the concept of sustainable development. The author documents that UNESCO-UNEP increasingly integrated sustainable development as a fundamental element of environmental education, focused on teaching students the rational use of natural resources.

However, the ambitions of IEEP met with mixed success, also underlined by the global gateway *Connect* of the *UN Office for Partnerships* to deliver the SDGs (2016). On the one hand, the societal processes, not directly involved but adjacent to the field of environmental education, were confronted with increasing problems of globalisation and conservation, causing a negative effect on the development of environmental education. On the other hand, as concluded by Dorn (2020), IEEP's substantial input through international conferences, declarations and agreements did not always succeed in meeting the challenges to teaching and learning embedded in the local contexts. His conclusion links up with the assessment of McCormick (1991) on the *Global Environmental Movement,* that ultimately, education is a national matter, rendering the work of UN specialised agencies less productive in environmental education than the input of national NGO's. Another aspect of the discourse on environmental challenges dealt with by UNEP and its supporting organisations such as the *International Union for the Conservation of Nature* (IUCN) and the *World Wildlife Fund* (WWF) is mentioned by Macekura (2015) as drawing more attention for the underlying themes of poverty and inequality. In his publication *Of Limits and Growth: The Rise of Global Sustainable Development in the Twentieth Century*, he underscores the success of developing countries in influencing IOs and NGOs to conceptualise conservation in partnership with economic development. For environmental education, this consideration means to UNEP, IUCN, and WWF that a global strategy of conservation could only be successful if all nations of both the developing and the developed world would jointly engage in collaborative strategies in which economic development would only be successful if that would be implemented in an ecologically sound environment.

Under the impetus of the shift towards global action for environmental protection by UNEP and its supporting organisations, followed by IEEP's vision of environmental awareness for a world-wide education movement, the theoretical framework of environmental pedagogy began to focus on the integration of environmental education in curriculum development at various levels of national schooling systems, designed to foster the students' awareness of the need of conservation of the environment. Moreover, IEEP's movement to a *New Global Ethic* to foster an environmental ethic and adopt the concept of sustainable development in education opened up a new perspective for development-oriented environmental education. This perspective gave a new meaning to the previous recommendations from the *Intergovernmental EE Conference* at Tbilisi (USSR) in 1977, underscoring the crucial role of education in the new global challenge to attain a form of sustainable development which would fit into an educational framework of sound environmental management.

In 1994, the representatives of ninety-two governments and twenty-five international organisations came together in Spain for the *World Conference on Special Needs Education* (UNESCO 1995) to reaffirm their commitment to *Education for All*, aimed at the provision of basic education for everybody and urge a radical changeover from the exclusion of disable participants being a direct violation of human rights. Mitchell (2015, p.1) points at the world-wide impetus of the *Convention on the Rights of People with Disabilities* (UN 2006), noting that: "*With the impetus of the UN Convention on the Rights of People with Disabilities, inclusive education is an idea whose time has arrived around the world*". The author describes a model of inclusive education which broadens its scope to include vision, curriculum development, teaching, access, support, resources and leadership. In practice, inclusive pedagogy involves the ways in which courses, classroom activities, curricula, and assessments consider issues of diversity in an effort to engage all students in meaningful, relevant, and accessible education.

In a recent article, Loreman (2017) elaborates further on the different pedagogical approaches of the late 20th century to assist teachers who are working with students in inclusive learning environments. These approaches have been based on models of special education, still relevant in classrooms around the world today. Several examples are provided of these approaches derived from the basic principles of inclusive pedagogy, including the models of *Differentiated Instruction, Universal Design for Learning,* and *Inclusive Pedagogical Approach in Action* (*cf.* Tomlinson 2001; Rose & Meyer 2002; Florian 2015).

The *Chicago Center for Teaching and Learning* (CCTL 2023) links pedagogy with diversity as a means to describe inclusive pedagogy as the way in which courses, classroom activities, curricula, and assessments consider issues of diversity in an effort to engage all students in meaningful learning. However, the yet inadequate body of environmental knowledge and practice in education not only necessitates a substantial input from the science of ecology, but also from supporting MSG's, IO's and NGO's regarding the coordination of the development of the environmental education process on a global level.

Linking up with the recent effort in pedagogy on environmental education, the next paragraph seeks o provide an overview of the joint efforts of the MSG's, IO's and NGO's to change course.

Joint International Efforts from MSG's, IO's and NGO's

In the international setting of the growing ecological problems and the early conceptualisation of rational use and conservation of the biosphere in the UNESCO *Biosphere Conference* (1968) in Paris, preceding the UN *Conference on the Human Environment* (1972) held in Stockholm, Sweden, as the first world conference making the human environment a major topic, substantiates a milestone of MSG's, IO's and NGO's to address the urgent need of concerted efforts to protect the carrying capacity of the globe.

In its *Stockholm Declaration and Action Plan for the Human Environment* (1972:24), an important recommendation was made to all organisations of the United Nations system: "*to take the necessary steps to establish an international programme in environmental education, interdisciplinary in approach, in school and out of school, encompassing all levels of education and directed towards the general public...*" In its *Principle 19* regarding *Education in Environmental Matters*, the Conference declared environmental education essential to broaden the basis for a responsible conduct for the younger generation as well as for adults in protecting and improving the environment in its full human dimension.

Inspired by the key concept of *sustainable development*, a re-appraisal of the *Second United Nations Development Decade* (1971-1980) transpired through the debate, focusing on the concept of socio-economic development. The dynamic, development-oriented approach towards the educational, informational, social and cultural aspects of environmental issues triggered the Conference to particularly recommend developing new materials and methods for all types and levels of environmental education.

Soon after the establishment of the *International Environmental Education Programme* (IEEP) in 1975 by UNESCO-UNEP with a view to implement its objectives to provide a new structural vision for a world-wide mobilisation of education for environmental awareness, a new environmental ethic developed through the community-oriented, problem-based, interdisciplinary study of the total environment. Two years later, the world's first intergovernmental conference on environmental education was organized by UNESCO-UNEP in Tbilisi, Georgia (USSR) in 1977. The *Tbilisi Declaration,* considered a major breakthrough in environmental education, laid out the principles and goals of global environmental education. The "environment" was regarded in its totality, encompassing natural, technological and social components. Consequently, the *Tbilisi Declaration* (1977, p.1) defined goals for environmental education in the curriculum beyond "ecology" to include the development of: *"a clear awareness of, and concern about, economic, social, political, and ecological interdependence in urban and rural areas"* The wider conceptualisation of environmental education became the major basis for environmental education, known as *Education for Sustainable Development* (ESD) (*cf.* UNESCO 2018).

Noting that the concept of "sustainable development" had re-energised the development of education, IEEP (1975) stressed that environmental education would remain effective only if it used an integrated approach. As similar UN international efforts were undertaken to become more sustainable, sustainable development soon became the new global ethic in environmental education.

Twenty-years following the establishment of IEEP, the *UN Office for Partnerships* established in 2016 its *Global Gateway* to create partnerships to help to realise the UN *Sustainable Development Goals* (SDGs) (2015). Aimed at elevating education as a win-win strategy for the SDGs, the Gateway explores priorities to progress the action for education everywhere to enhance the urgently needed transformation of education by 2030. It links up with the call for SDG No. 4 on quality education at the UN *Office for Partnerships* (2023:16) for commitment and investment in education which seeks to avoid the widespread twin crisis of equality and relevance, where: "*84 million children will be out of school, and 300 million more who attend school will still lack basic numeracy and literacy skills to thrive in a rapidly changing world"*.

Despite such international endeavour, the critical biodiversity loss and climate change are having their effect on the direction of environmental education. Challenges continue to exist, not only in convincing the general public and their leaders that humans are well underway to the *Sixth Extinction*, but also in determining and agreeing on the appropriate actions to be taken in order to avoid further ecological destruction.

The conclusion of Dorn (2020:105) that: *"from the perspective of the twenty-first century IEEP's urgent calls for a coherent plan of action in developing and implementing international environmental education were prescient*" and that: "*these challenges have educational components"* underscores the failure of the international community to effectively implement the proposals launched by *IEEP* (1975), today nearly half a century ago. Indeed, challenges continue to persist, not only in the need to create an *a priori* awareness among teachers, students and the general public and their leaders, but also to determine and agree on the appropriate actions to be taken to avoid further ecological destruction. While both challenges have educational components, it becomes obvious that there is a world-wide discrepancy between what people know about the environment and how they act upon that knowledge. The next paragraph will elaborate on this fundamental educational challenge within the context of the "knowledge gap" between scientific and ethnoscientific knowledge in relation with environmental education.

It is difficult, if not impossible, to comprehend the state of environmental education in the present without recognising the transformation which environmental education underwent during the second half of the twentieth century, as described above. The IEEP Conference, declarations, agreements, workshops, seminars, and pilot projects were all necessary elements of goal of UNESCO-UNEP to encourage the international community to commit to environmental education both nationally and globally. Even these efforts, however, prove yet insufficient to overcome the practical routines, attitudes, and approaches to education and training in the classroom, as they remain largely grounded in the local contexts of the formal education systems. Although IEEP ultimately met with modest success, it remains the task assigned to the UNESCO-UNEP Programmes to formulate and implement environmental education globally, which would reach all members of the community locally to foster a critical shift in human behaviour towards sustainable use and conservation of the environment in time. In line with the above-mentioned international efforts from various organisations to step-up nature conservation and transform the degradation of resources, the next paragraph takes a closer ethnoscientific look at the contribution of indigenous peoples' TEK to sustainable management systems around the globe.

Traditional Ecological Knowledge: Ethnoscience and the Recognition of IKS

In view of the fact that an estimated 476 million indigenous people live on the planet today, dispersed over 90 countries, representing around 5,000 different cultures, and are making their voices, that there are today an estimated 476 million indigenous people are living on the planet dispersed over 90 countries, representing around 5.000 different cultures, who are making their voice.

more and more heard, a growing international recognition of their IKS has emerged, specifically regarding their indigenous rights.

Their in-depth knowledge of their ecological community and conservation-based indigenous management systems of their natural resources are of particular interest. In particular, their in-depth knowledge of their natural habitat and their conservation-based indigenous management systems of their natural resources is particularly extensive and conservation-oriented. The relevant component of IKS, known as Traditional Ecological Knowledge (TEK), is rooted in their centuries-long wisdom and experience with nature, which is also reflected in their cultural tradition of educating their younger generations in the balanced use and management of natural resources.

The scientific interest in the indigenous peoples' cultures, their way of life and their use of natural resources had evolved from the early work of the American anthropologist Franz Boas (1858–1942), the "Father of American Anthropology" who paved the way to the development of *ethnoscience* by introducing his concept of *cultural relativism* as an approach to understand indigenous peoples' scientific practices in their own right. His work was complemented by Goodenough (1957:171) who introduced ethnoscience, also known as the "new ethnography" in the 1950s by defining local peoples' knowledge systems from their own perspective: "*How local people perceive and interact with their environment as reflected in their own words and actions*".

The emergence of the discipline of *ethnoscience* from cognitive anthropology in the late 1950s, substantiated the development of a new study field which became concerned with the knowledge and belief of local peoples, and how these cognitive factors determine their own perception of the world around them from an *emic* – insiders - point of view. The ethnoscientific approach evolved in contrast with the age-long conventional study of 'other peoples' living outside the Western World. Such *etic* – outsiders – point of view had been established in the descriptions of 'primitive people' and 'natives' in Ethnography and Anthropology of since the 18th century, describing their observations from a Western perspective, characteristic for the era of colonialism.

In the ongoing search for a global solution to the critical loss of biodiversity, one of the critical issues concerns the age-long misconception of "Humans place *above* Nature" – instead of "Humans place *within* Nature" - which dominated Western philosophies since the Renaissance (*cf.* Leakey & Slikkerveer 1993). While the related Western ecological theories have tended towards *"Stewardship of the Earth"*, justifying global control over and exploitation of natural resources, most indigenous ecological principles are concerned with peoples' position in a wider comprehensive cosmological order of their universe. These principles are often embedded in local institutions and include values, norms and beliefs regarding the maintenance of the "balance of nature" which evolved over generations, and which encapsulate specific conservation beliefs and practices. In this way, the role of ethnoscience and the related "new" perspectives in TEK could assist in transdisciplinary research of the practical behavioural patterns of indigenous management systems and the guiding belief systems of a broader concept of environmental education for development.

The theoretical orientation of *ethnoscience* refers to the study of indigenous knowledge manifest in various sub-fields, where as complementary to *science*, a perspective of *ethnoscience* has been introduced on the basis of indigenous people's own knowledge, belief and practices. The use of the local peoples' language, their indigenous classification of plants, animals and ecosystems, their religion and indigenous cosmologies and philosophies of nature and the environment are among the major subjects of ethnoscience research. This interest led to the development of a number of specialist sub-fields of ethnoscience, such as ethnolinguistics, ethnohistory, ethnobotany, ethnomedicine, ethno-economics, ethno-ecology, ethnopsychology, and – last but not least - ethno-education. Although it became very interesting to learn how indigenous peoples interpret their own culture and their environment themselves differently, providing new insights into the evolution and development of humans on Planet Earth, the accumulated body of indigenous knowledge remained initially rather static.

Not long thereafter, the potential of IKS became truly relevant in the practical setting of community development, and in 1980 it was the American ethnoscientist Michael Warren, who introduced IKS as an essential key factor into the dynamic process of socio-economic development and change. Soon thereafter, interest in indigenous peoples' sustainable conservation and management practices emerged from IKS-based studies of local farmers' knowledge and skills in how to select, breed and utilise their floral and faunal resources. These indigenous peoples' knowledge systems have amply been described as "traditional knowledge", "folk knowledge", "farmers' knowledge", or "rural people's science" by Warren, Slikkerveer & Titilola (1989); Chambers, Pacey & Thrupp (1989); Barker, Oguntoyinbo & Richards (1992); Slikkerveer & Slikkerveer (1995); Posey (1999).

The publication of the classical work by Warren, Slikkerveer & Brokensha (1995) on *Indigenous Knowledge Systems: The Cultural Dimension of Development,* substantiated the paradigm shift in the development debate of the 1990s from ignoring and neglecting IKS to including them in the theory and practice of sustainable community development. In policy planning and implementation, the shift from an etic "top-down" approach to an emic "bottom-up" approach introduced the concept of the *Participants View* into local decision-making processes, rendering development projects and programmes in different community sectors rather successful. The *Participants View* became an important component of the special ethnoscience methodology of the "Leiden Ethnosystems Approach" introduced by Slikkerveer (1995).

The above-mentioned official recognition by the UN *Conference on Environment and Development* (UNCED) in Rio de Janeiro in 1992 of the critical contribution of indigenous peoples and their ecological knowledge, established in the *Principles* encouraged further empirical research on sustainable practices among the indigenous communities around the globe.

At that time, the *Indigenous Knowledge & Development Monitor* (IKDM) was launched in 1990 by the Center for Indigenous Knowledge and Agricultural Development (CIKARD), the Leiden Ethnosystems And Development Programme (LEAD), and the Center for International Research and Advisory Networks (CIRAN, followed by a series of interesting studies on IKS&D conducted at the *Leiden Ethnosystems and Development Programme*, established in 1987 at Leiden University in The Netherlands.

These studies document important IKS-research in subfields such as ethnomedicine, ethnobotany, ethnopharmacology, ethno-economics, ethno-education, and ethno-ecology, predominantly in Sub-Saharan Africa by Gheneti (2007) in Ghana; Ibui (2007) in Kenya; Chirangi (2013) and De Bekker (2020) in Tanzania; the Mediterranean Region by Aiglsperger (2014) in Crete and South-East Asia by Agung (2005) and Leurs (2010) in Bali; Angerler (2009) in Sumatra; Djen Amar (2010), Ambaretnani (2012), Erwina (2019), Saefullah (2019), Febriyanti (2021) and Diah Susanti (2023) in West Java, and Williandri (2024) in East Java.

According to Berkes (1993: 3), Traditional Ecological Knowledge (TEK) is: "…*a cumulative body of knowledge and beliefs, handed down through generations by cultural transmission, about the relationship of living beings (including humans) with one another and with their environment process between local and global knowledge systems*." The definition underscores that TEK also includes the important holistic cosmovisions, so far largely excluded from "scientific" ecology. Although there are both similarities and differences between science and ethnoscience, Bronowski (978:9) considers that the practice of science, including magic are: "*fundamental characteristics of human societies*", underscoring that science and ethnoscience are the result of the same human intellectual process of creating order out of disorder.

The concept of cosmovision - or cosmology - refers to the way in which indigenous peoples perceive their central position in the universe or cosmos. Cosmovisions comprehend their assumed interrelationships with the three natural, spiritual and human realms or worlds, which should be maintained in a harmonious way in order to safeguard humans' physical and spiritual well-being throughout their lifetime. The cosmovision guides the participants in their thoughts, belief and behaviour, including their interactions with the universe to respect and care of the use of the natural world and its resources in a sustainable way. Figure 2 provides a schematic representation of the Sundanese cosmovision of *Tri Tangtu*, comparable with the traditional Balinese cosmovision of *Tri Hita Karana* and the Javanese cosmovision of Tri Loka (Agung, 2005; Saefullah, 2019).

Meanwhile, impressive ethnobotanical research data have been collected by Schultes & Raffauf (1990); Balée (1994) and Posey (1999) of the Ka'apor people in the Brazilian Amazon, who recognise at least 768 species of plants from seed to reproductive adult stages. These and similar ethnoscience studies from South America, Sub-Saharan Africa and Asia document an enormous body of indigenous knowledge about the biodiversity living in their habitat. While Sponsel (2013) argues that such rich TEK also provides biologists and conservationists with urgently needed data for the inventory of local biodiversity, it should not be forgotten that most "scientific" knowledge – particularly of medicinal plants and fruits – originates from the indigenous knowledge and practice of local healers, TBA's and sages (*cf.* Wheelright 1974; Dharma 1981; Shultes, Hofmann & Ratsch 2001; Slikkerveer 2005).

The important, albeit long-ignored position of indigenous peoples in the international conservation debate has recently been strengthened by the growing worldwide recognition of their traditional practices

Figure 2. Schematic representation of the Sundanese cosmology of Tri Tangtu
Source: Saefullah (2019).

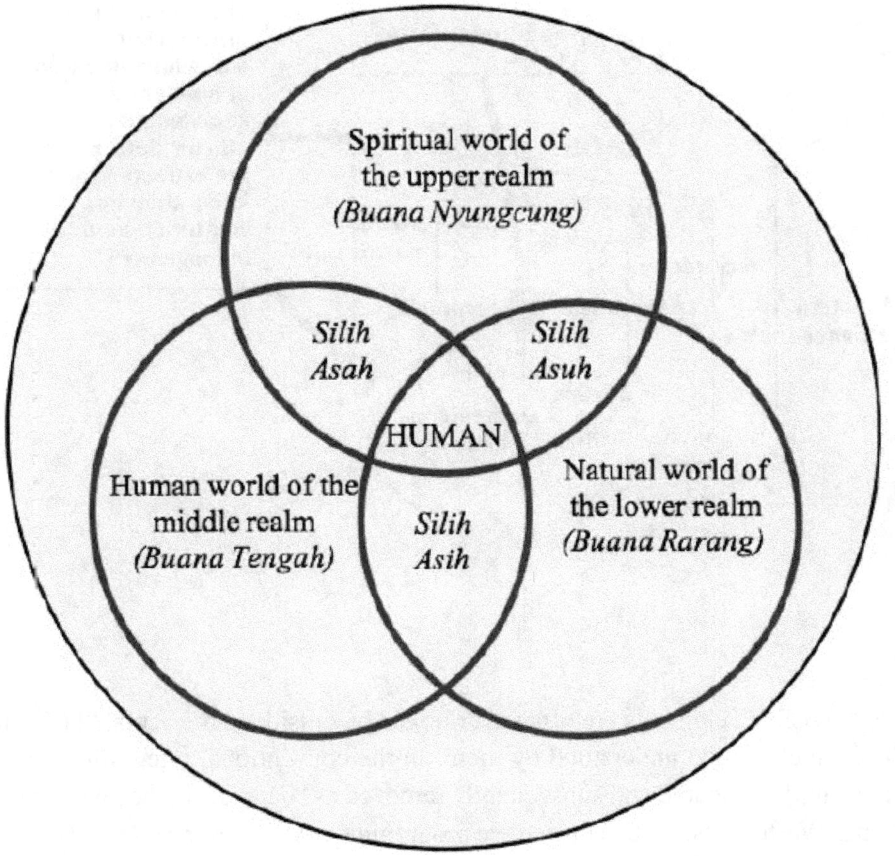

and worldviews, centered on a subsistence-oriented use of resources. Their relevance for the development of the *New Global Ethic* in the move towards fostering general environmental awareness among students and staff elsewhere around the globe is significant for transforming the attitude of the economic philosophy of commercially-based surplus-building, characteristic for most Western neo-liberal approaches.

Because of TEK's recent recognition for its contribution to a general awareness of the value of indigenous management of natural resources and conservation of bio-cultural diversity, recent research has particularly focused on the cultural embeddedness of components of practical knowledge, which are based on the central spiritual basis of the peoples' cosmovision concerned. Figure 3 shows that the various components of factual knowledge of TEK are embedded in the center of the web of IKS, where the "spirituality core" of the local culture is located. The representation of the embeddedness of TEK in a larger "web" of IKS also identifies the "invisible" factors of wisdom, beliefs, norms, and values, as well as their indigenous institutions, which guide indigenous peoples' behaviour as the key elements in their overarching, holistic worldview.

Figure 3. Conceptualisation of Traditional Ecological Knowledge (TEK) within the indigenous knowledge web
Source: Casimirri (2003)

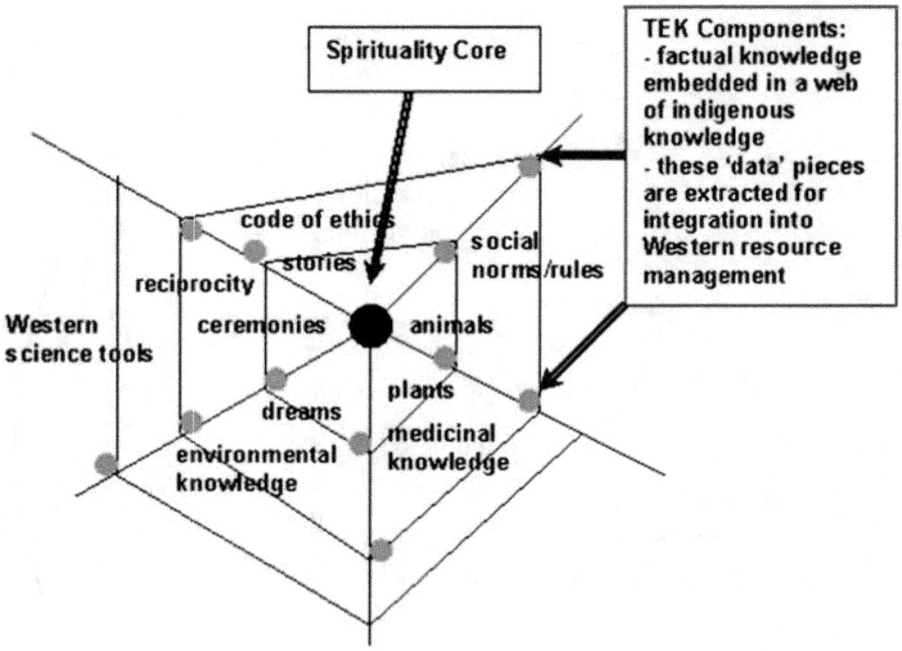

While these "invisible" elements are often overlooked by outsiders, the actual "facts" form the components of TEK which can be understood by them. In the conventional integration process, however, these "facts" are simply validated and subsequently removed as "data" from the "web" to be applied in a isolated position in Western resource management programmes. As Casimirri (2003: 3) argues, it becomes possible that: "*removing these data points would weaken the structural integrity of the web. Likewise, the data points are interconnected to the web and cannot be fully understood when they are removed from their context*". In other words, TEK has to be considered as interconnected with the worldview concerned, forming one whole system. As consequence, TEK integration models designed to contribute to sustainable development should be interactive and holistic in their approach and include the invisible factors represented in connection with cosmologies, such as spirituality, sanctity and morality, highly relevant for local peoples' perceptions, activities and behaviour.

Alongside biodiversity, indigenous knowledge as part of cultural diversity is also at risk of extinction. In addition to the growing international concern over the loss of *biodiversity*, a similar apprehension has emerged at the beginning of the 1990s among anthropologists, ethnoscientists and ecologists. Their concern relates to the ongoing loss of the related *cultural diversity* in terms of indigenous peoples' systems of knowledge, belief, and practice concerning the institutionalized customs, norms, and values related to genes, species and ecosystems at the community level.

In this context, Slikkerveer (1992) introduced the concept of *bio-cultural diversity* in his study of conservation of bio-cultural diversity in East Africa, stressing their crucial complementarity for achieving an alternative, less exploitative philosophy of nature and the environment. The concept links up with the coincidence of areas of high *cultural diversity* with areas of high *biological diversity*, or – analogously to Vavilov's classical assessment of the world's diversity centres (1951) – reflecting the coherence of "Centers of Biodiversity" with "Centers of Cultural Diversity". Such *Centers of Bio-Cultural Diversity* would not only serve as most valuable research areas for multidisciplinary teams, but could also be designated as sites where indigenous and global knowledge systems could integrate based on shared power and self-determination, and as such serve as a knowledge center for sustainable development (*cf.* Slikkerveer 2018). Such center should also build up a global database, which could bridge the "knowledge gap" between scientific and ethnoscientific knowledge in many sectors of society, particularly in ecology and environmental education.

The ongoing rapid loss of bio-cultural diversity in many developing countries is largely the result of unbalanced historical processes of commercially-oriented acculturation and transculturation during the colonial era, in which indigenous societies have largely been dominated by knowledge and technology from the West. Kothari & Das (1999) show that Western science and resource use have eroded most indigenous knowledge systems in India, as happens in the rest of the world. They describe several efforts that are being made to revive these systems for enhanced biodiversity conservation in India. By the 1990s, the unidirectional "Transfer-of-Technology" (ToT) development approach from developed towards developing countries had been replaced by the equality-based "Participatory Development" (PD) approach to involve the indigenous people in development projects by including them in the decisions affecting their way of life.

The momentous UNEP publication *Cultural and Spiritual Value of Biodiversity – A Complementary to the Biodiversity Assessment*, edited by Posey (1999) also highlights the crucial role of cosmologies in the sustainable use and management of resources among indigenous peoples in different settings. So, Pierotti & Wildcat (1999) compare and analyse local beliefs about the close relationship with nature among the indigenous peoples in North America as opposed to the concepts of the environment of Western European immigrants. Hay-Edie (1999) demonstrates that local landscape perception and sensory emplacement are important factors which should be included in decision-making for environmental strategies. Similarly, the role of local belief systems in the management of resources is an important issue. Gonzales, Chambi & Machaca (1999) describe in this context the central role of seeds in the cosmovision in contemporary Andean agriculture, while Adimihardja (1999) provides an account of the Sundanese cosmology in traditional agriculture in West Java (Indonesia). Cosmovisions prevalent in the "Great Tradition" and "Little Tradition" of India are documented by Mahale & Soree (1999), who conclude that traditional and modern knowledge systems can be integrated for sustainable management and conservation.

While TEK's practical significance has recently been demonstrated in a growing number of conservation projects, its significant contribution to environmental education remained confined to practical *ad hoc* examples in curriculum development, rather devoid of any reference to its holistic approaches to sustainable protection and conservation of the environment. In the LEAD Programme of Leiden University – later transformed in 2022 into the independent LEAD Foundation - several studies focused on the research of the role of traditional cosmovisions in local education, pertaining to sustainable community development. In Indonesia, such important role of *Tri Hita Karana* in Bali has been described by Agung (2005), *Tri Tangtu* in Subang, West Java by Saefulla (2021), *Kearifan Kesehatan* in the Sunda Region of West-Java (2023), and *Bakti* and *Sayan* in the Tengger Region, East-Java by Wiliandri (2024).

Although the former colonial domination of the field of international education has received ample attention in the present debate, the "knowledge gap" between scientific and ethnoscientific knowledge in relation with environmental education is now in need of due consideration. As in the classroom, such a "gap" is only visible between indigenous *i.e.*, local and global knowledge; it is still continuing, mainly due to the uneven position of both systems in the societies worldwide. As Agrawal (1999) argues, empowerment is needed to reach a system of power-sharing with indigenous peoples, enabling mutual understanding and integration between both knowledge systems on an equal basis.

Initial steps have recently been taken to explore a balanced contribution from indigenous peoples' knowledge and experience of traditional use and management of the natural environment to environmental education, but are reporting several obstacles to reach a systematic global programme of comprehensive learning and training in ecological pedagogy. The study by Korteweg & Root (2016) among the indigenous peoples belonging to the various First Nations in Canada document that traces of the effect of rather negative attitudes from the colonial past have not disappeared from the public education system. By consequence, indigenous peoples' younger generations are still suffering from a misplaced feeling of inferiority of their cultures, which hinder the process of reconciliation between settlers and local communities needed for mutual understanding and learning of environmental knowledge and belief among teachers and students. Korteweg & Russell (2012) and Tuck, McKenzie, & McCoy (2014) elaborate on the endeavour of both indigenous an non-indigenous scientists and teachers to decolonise and indigenise environmental education through special land-based education programmes, most exemplary for non-Canadian regions elsewhere on the globe.

The wider positive impact of involving indigenous peoples in environmental education is also described by Nesterova (2020) based on the *Facing the Mountain Education* programme among the *Bunun* indigenous people in Taiwan, in which environmental learning and reconnection with the natural world of both indigenous and non-indigenous groups have participated. Although the aim of the programme was learning with and from the natural environment, its additional objectives include, as Nesterova (2020, p.1) notes: *'re-building and strengthening indigenous identities, cultures and ways of life, and potentially contributing to decolonisation of settler societies and reconciliation between groups'*. Core principles of the programme could also be beneficial to education in other comparable situations, such as the need to forge substantial interconnection between the individual and communal identities with the land and the culture. In addition, the focus on collaborative, shared decision-making processes between generations, as well as concentration on localisation of land-based knowledge have similar applicability in environmental education elsewhere.

As a follow-up to the growing potential of TEK in advanced, sustainable development-based environmental education, the changing socio-cultural embedding of such educational systems in societies is further highlighted in the next paragraph.

The Multicultural Perspective on Ethno-Education

The contemporary development and expansion of environmental pedagogy and its sub-fields are not only the result of a growing worldwide awareness of the urgent need to protect and preserve the environment, but are also driven by the imperative to explore new approaches to environmental education as a means to reach a rigorous cultural reorientation towards human respect and appreciation of nature in conjunction with sustainable use and management of its resources by present and future generations. The compelling transformative change needs a transdisciplinary approach which goes far beyond the conventional forms and processes of knowledge development and innovation.

By consequence, the critical reflection on the participants, the context, content, and process in the development of international educational curriculums concerning the environment requires a transformative perspective on the integration of science and ethnoscience in order to gain "the best of both worlds" with a view to achieve a global system of sustainable environmental education.

Today's rise of a general realisation that the critical environmental problems are transcending regional and national boundaries is clearly reflected in the special focus on the development of a world-wide education system for present and future generations at schools, universities, organisations and government agencies. With such rising socio-economic and environmental problems in the Western world, a new form of environmental education was introduced in 2020 by the *IUCN Commission on Education and Communications* (IUCN-CEC) recommending a dedicated, focused and transformational strategy of *Education for Conservation* (EfC), dealing with future needs of education in conservation action.

The global context of the implementation of EfC, however, has in the meantime also changed after both the Western nations, as well as the newly independent states, experienced a fundamental transformation in their socio-cultural composition as the result of massive re-migration and settlement after the conclusion of the former colonial era to become modern multicultural societies. These societies are characterized by a configuration of population groups from different ethno-cultural backgrounds living and working together in society. Such multicultural societies have different customs, beliefs, languages, art, traditions, and behaviours and are found both in developed and the developing countries. The related concept of *multiculturalism* refers to recognition and respect for the diversity of opinions and lifestyles of the minority groups, without demanding their assimilation into the dominant culture. As consequence, the education systems had to develop and adapt to the multicultural composition of the population of contemporary societies and attune their education system to the expectations and needs of the different ethnocultural population groups.

In the process, a division between, on the one hand the dominant, and on the other hand, the minority groups both tended to adhere to their own educational system – coined as formal *versus* informal - where particularly in the former colonies after independence, the minority groups resorted to their educational traditions, often in their own local language. In this way, the process of ethno-education emerged as an important sub-field of applied ethnoscience in conjunction with ethno-pedagogy, in which environmental education gained a central position. The process of ethno-education developed soon to become a world-wide relevant approach to overcoming ethnocultural differences in the classroom, strongly supported by international organisations, such as UNESCO-UNEP.

In several countries in Latin America, the early development of ethno-education took place, initially influenced by the emphasis on a process of political empowerment and preservation of indigenous groups and their socio-cultural traditions in a multicultural society. This process links up with the above-mentioned attention for the need of decolonization and reconciliation in environmental education among the indigenous First Nations and the settlers in Canada.

In these former colonial regions, where different communities and their often parallel knowledge systems and institutions practiced their own education and training of their youngsters, the often dominant Western education system—implemented in a non-indigenous language—created tensions and opposition that transpired through the pursuit of empowerment and emancipation of local communities.

Soon, ethno-education aimed to solve the educational problems from such discrepancies between formal and informal education systems in multicultural societies in developing countries. Other implications of the new orientation of ethno-education in terms of education of indigenous peoples and their diverse cultural background became manifest in the political arena of the multicultural countries in Latin America. In 1965, the tone for the early development of ethno-education was set by *the International Convention on the Elimination of all Forms of Racial Discrimination* (UN-ICERD 1965). Thereupon, countries including Peru, Ecuador, Chile, Venezuela, Brazil, Colombia and Bolivia introduced and implemented ethno-educational policies with the objective not only to provide a system of bilingual education, but also advance intercultural relations, rescue and preserve the cultural identity and make the needs of ethnic groups visible (*cf.* Flores & Mena 2018).

UNESCO (2008:8) underscores that: *'the most significant contribution of ethno-educational experiences lies in the recognition and appreciation of the ethnic and cultural diversity of its students and communities, as a starting point for the development of new learning and for the affirmation of its identity; also, that life in the educational centers promotes intercultural relations of respect and fraternity in a local and subregional socio-geographical context characterized by multiculturalism.*

The Omprakash Project in Barranquilla, Colombia is a striking example of a formal ethno-education programme for ethnic minorities which provides key features defining Colombia as a multicultural nation. As the Omprakash Team (2007-2021) reports, formal ethno-education means for the state the materialisation of the drive for inclusion for those groups which have been marginalised in a multicultural nation.

In health education in Colombia, Hernández-Martínez *et al.* (2021) document the experience of ethno-education as a special form of health education based on cultural diversity as a means to attain improved health status of the participants. The authors conclude that although local, national and international guidelines have been established, ethno-education in Colombia continues to be a challenge, it is essential that: *'the planned activities be integrated into the communities' cultural perceptions.* In the area of worldwide environmental ethno-education, Tretyakova *et al.* (2021, p.1371) provide an overview of various educational systems, which seek to implement the ethno-pedagogical bases into a modern educational process of ethno-pedagogical bases in the modern innovative educational process of a transition from the educational environment to the educational ecosystem. The study shows that current efforts of upgrading and development of education requires the inclusion of folk pedagogy elements in the system of personal formation and the educational process, underscoring that: *"It is important to continue the traditional ethnic trajectories of love for the motherland, native nature, and education of citizenship in spiritual formation"*. The progress in development studies of the recognition and integration of IKS into the socio-economic process of development and change of other multicultural societies in developing countries since the 1990s has further substantiated the sub-field of ethno-education, in which the overriding role of TEK became prominent in the context of multicultural educational programmes.

However, although the origins of ethno-education emerged in several independent countries in Latin America during the second half of the past century, the world-wide recognition of the indigenous peoples' right to take their life into their own hands prompted similar groups in other developing countries to put political pressure to adopt such educational process to safeguard their way of life and their traditions, including their ways of transfer of knowledge and experience to the young generations. Epelle & Alafonye (2010) describe several strategies implemented by the Government of Nigeria to bring about the desired national unity and integration in order to overcome the cultural problems of education being the result of the different cultural backgrounds of the people, and the differences in their way of life.

A similar complex situation has been described for the recent development of ethno-education in Indonesia in an interesting publication by Malihah *et al.* (2018) who elaborate not only on the understanding of what equality and inclusivity entail, but also on the social vision to achieve social justice as a precondition to realize one of the important United Nations Sustainable Development Goals (SDGs). In particular, the integration of social justice into research in education is strongly recommended by Maftuh & Adriany (2018) in order to address inequities and inequalities in the society through the analysis of the imbalance of power structures.

In the same publication, Slikkerveer (2018) underscores the need of an ethnoscience-based approach to integrate local languages into nationwide education management in Indonesia in order to develop a sustainable system of ethno-education for all future generations.

In addition, Sulistyani & Rahrdjo (2019) analyse the contradictions of the traditional educational system in the Sikep society in Java with the formal government system of education, resulting in a discrepancy where most of the Sikep people avoid joining the formal school education. The main aim of the study is to design a specific local educational model which can be implemented to increase the participation of the Sikep people in the formal school system. In this project, the theory of the field of *Ethnography of Communication* is used for exploring how everyday communication in the Sikep communities create a local specific educational discourse. The authors describe the way in which the ethno-educational approach can successfully be applied for the preparation of national education in the Governments Plan of the 'Society 5.0 Era of Indonesia' (2023), where the combined cultural values and technological skills can become very useful tools in education. As this 'Society 5.0 Plan' is based on the concept of integration of humans and technology, the authors' suggestion for the integration of cultural values is most promising as they encompass not only indigenous environmental knowledge and practice, but also traditional cosmovisions and respect for the natural environment.

The above-mentioned multicultural setting affects an appropriate design and implementation of an integrated model for the envisaged eco-education approach on a global scale, as described in the next paragraph.

Towards IEE: An Ethno-Scientific Model of Integrated Eco-Education

Spurred on by the above-described recent acceleration of the increasing loss of biodiversity and climate change, not only scientists, MSG's, IO's and NGO's but also teachers, students and the general public have recently become rather concerned about the ongoing deterioration inflicted by humanity on the natural environment of the planet.

While effective environmental education is imperative to train students and staff to reduce harmful processes of exploitation of natural resources and restore destroyed natural areas, it is also critical to educate people in general to adopt ecological awareness and an environment-friendly philosophy of nature, operationalised in a holistic approach towards sustainable behaviour.

In the present situation, more than ever before, innovative lifelong environmental education and training is urgently needed. Many efforts are being made to foster a general awareness among all generations, particularly the youth, of the threat of ultimate extinction of all life – including humans – by ongoing unlimited, non-sustainable processes of profit-oriented industrial, technological and agricultural exploitation of resources. In a recent enlightening review of the current state of Environmental Education and Management, Akhmetova *et al.* (2020) elaborate on the various doctrines of nature conservation, underlying the development of different environmental education models. The authors' assessment of the centralist exclusive doctrines of nature protection by the State such as the former USSR, where the state dominates the exploitation and protection of resources, indicate that these have not been able to create an environmental culture among the population concerned.

Following the introduction of the system of national parks and other protected areas, where the indigenous peoples living in these protected areas and their traditional management systems have become the focal point of conservation, an environmental education model was developed, based on the indigenous participatory doctrine. This model encompasses several opinions, ideas and visions about the individual and communal responsibilities towards nature. Ecological education is largely implemented by the indigenous communities and is reflected in the locally-based training of students, with special attention for TEK, rendering this doctrine highly conservation- oriented.

The intermediate phase of the transition from one doctrine to another is reflected in the model of Inclusive Education, in which all parties concerned are expected to care about everything. The inclusive doctrine promotes the development of an ecological culture of all individuals, groups and organisations. In line with the above-mentioned *Declaration of the World Conference on Human Rights*, organised by the United Nations in 1993 in Vienna, Austria, the model aims to ensure that all students have equal opportunities without discrimination, learning together in one environment. The relations with some groups of the 'Green Movement' have become complicated since some green parties have brought their pragmatic ideas and actions for the protection of nature into the political debate.

In a recent article, Sadruddin (2023) provides an interesting case study of the design and pilot of an inclusive environmental education course in Pakistan, where a community is living under extreme environmental conditions. In addition to effective training in environmental literacy and strategies to support the communities during environmental catastrophes, the focus of the course is on peace-building approaches, such as collaborative problem-solving case studies and development of conflict management skills. The project recommendations include translating of the course into a regional language, and the training of local educators. As mentioned by Akhmetoval *et al.* (2020), three types of models of Environmental Education (EE) have been identified in the debate by authors such as Ramsaroop & van Rooyen (2013) and Mironov (2020).

In addition to the above-mentioned comprehensive *Inclusive Model of EE*, elaborated by Ramsaroop & Van Rooyen (2013), the *Multidisciplinary Model of EE* involves the greening of the conventional subjects of both the natural sciences and the social and humanitarian sciences. The *Single-Subject Model of EE* focuses on the study of ecology as an independent subject in the curriculum of each educational level. The *Mixed Model of EE* embarks on the previous two types by introducing environmental knowledge based on a combination of the conventional academic subjects and holistically in independent integrated subjects.

Meanwhile, a new model has been introduced at the *World Summit on Sustainable Development* in 2002 in Johannesburg,, known as *Education for Sustainable Development* (ESD). The World Summit adopted a *Political Declaration and Implementation Plan* which included provisions covering a set of activities and measures to be taken in order to achieve development which takes respect for the environment into account. The provisions also include the creation of a global solidarity fund for the eradication of poverty and the launch of ten-year programmes to support regional and national initiatives aimed at accelerating the transition to viable production and consumption patterns.

Under such impetus of the *World Summit on Sustainable Development* (2002), a growing number of studies were conducted in support of the interdisciplinary approaches to Environmental Education (EE), some of which focus on the development of a systemic approach to curriculum design of environmental education. As already mentioned in Paragraph 4, Ramsaroop & Van Rooyen (2013) document such systemic interdisciplinary approach implemented in education planning in South Africa. The study provides a follow-up to the model of *Education for Sustainable Environments*, in which Van Rooyen (2002) identifies seven interactive dimensions of constituents of the concept 'environment', *i.e.* the biophysical, social, economic, political-juridical, scientific technological, personal and contextual, linking up with the Tbilisi principles of EE (UNESCO 1978). The outcome of the study shows the build-up of the current nature of knowledge, skills and attitudes of teachers implemented in their curriculum to teach EE in a holistic and systemic approach. Furthermore, Ramsaroop & Van Rooyen (2013, p.595) underscore their conclusion that: '*The findings of this investigation emphasize the need to empower teachers with in-depth knowledge of EE so as to be able to teach it in an integrated manner.*' In November 2019, the 40th session of *UNESCO General Conference* adopted a new global framework on *Education for Sustainable Development: Towards achieving the SDG's* or *"ESD for 2030"*.

The new framework was officially launched at the *UNESCO World Conference on Education for Sustainable Development* in Berlin in 2021. The aim is to contribute to more peaceful and sustainable societies and develop a sense of responsibility for our planet in line with the 17 *Sustainable Development Goals* of the UN (2015). Acknowledged by the UN General Assembly through the *Resolution on ESD*, the framework linked up with the UN *Decade of Education for Sustainable Development* (2005 - 2014) and the *Global Action Programme* (GAP) on ESD (2015 – 2019).

The new orientation of UNESCO on sustainable development in education underscores the need of knowledge integration models to be designed for providing a contribution to sustainable development as agreed upon in the SDGs of the UN (2015). Moreover, these models should be holistic in their approach and also include the invisible factors represented in connection with indigenous cosmologies, and include spirituality, sanctity and morality, highly relevant for local peoples' perceptions, activities and behaviour.

In line with the classical concept of *Education for Sustainable Development* (ESD of IEEP (1975), stressing that environmental education would remain effective only if it made use of an integrated approach, the newly-developed model of *Integrated Eco-Education* (IEE) represents a systemic model of integrating major scientific, ethno-scientific and expert knowledge systems, operational in the dynamic arena of sustainable development.

As shown in Figure 4, it aims to provide a contribution to the imperative ecological shift in human awareness and behaviour regarding the urgent conservation and protection of nature and the environment on the planet. Following Fernández (2011) useful distinction, the model applies a differentiation between *integration* and *inclusion*. While the former aims at the normalisation of students' lives and their integration into regular schools, the second has no particular goal as it just embodies a basic human right. Integration promotes adaptation processes in curriculum development, while inclusion safeguards the participation of disabled students.

Analogously to the Model of *Integrated Community-Managed Development* (ICMD) introduced by Slikkerveer (2016), this model is based on the doctrine of multicultural indigenous participatory development. It represents the dynamic process of integration between the three major knowledge systems operational in the environmental education framework, plotted against the background of the natural environment. The effect of the processual integration of knowledge, belief and practices involved in environmental education for sustainable development is represented in the interaction of these systems within the multicultural societies, manifest in both the developed and the developing countries (*cf.* Figure 4).

The IEE model locates the system of integrated eco-education in the center, being the interactive outcome of the ongoing evolution of scientific, ethnoscientific and expert knowledge into a body of comprehensive knowledge of all dimensions involved in the human interactions with the natural environment, rendering the dynamic, interactive relationships between the different systems both inter-dimensional and holistic towards effective environmental education for sustainable development around the globe.

Figure 4. A comprehensive model for Integrated Eco-Education (IEE), based on the multicultural indigenous participatory development doctrine.

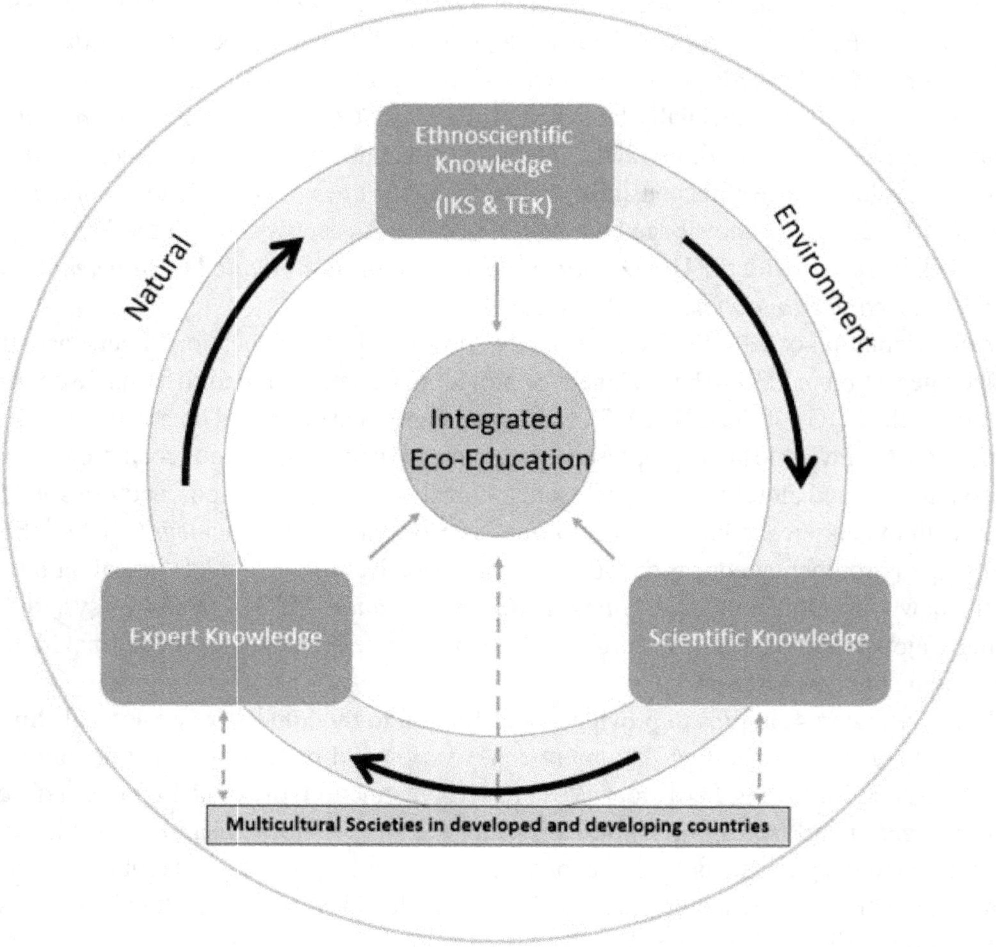

CONCLUSION

The Chapter embarks on the growing concern of biodiversity loss and climate change threatening life on Planet Earth which has gained increased priority in environmental education. While over the past decades, a general recognition has emerged among students, teachers and the general public that humanity is confronted with a hardly avoidable ecological crisis, and joint efforts are being made to cope with the situation, the complexities and pace of the world-wide threat remain undervalued. Although the current evolutionary process towards the human-induced "sixth mass extinction of life" may worsen if appropriate action is not taken, there are, however, sparks of hope to turn the tide, albeit largely at the local level.

It is well-documented that the discipline of environmental pedagogy has put much effort in exploring new avenues to cope with the critical situation. The transformation of its framework in theory and practice linked up with the various declarations and goals of international organisations, such as UN,-UNESCO, UNEP, WRI and IPBES, pertaining to the adoption of the concept of sustainable development in environmental education. The introduction of the concept of Education for Sustainable Development

(ESD) by UNESO (2014) further substantiated the inclusion of this key concept in teaching and learning. The Sustainable Development Goal No. 4 (SDG-4) of the UN (2015) strengthened the promotion of the objective of quality education to: "Ensure inclusive and equitable quality education and promote lifelong learning opportunities for all.". The recommendations seemed effective at the theoretical level of environmental education, but met with mixed success in the implementation at the classroom level in the field.

Meanwhile, in the practical setting of the international debate on global sustainable development, the role of Indigenous Knowledge Systems (IKS) had reclaimed their position to provide a valid contribution to participatory socio-economic development since the 1980's. The reorientation and revitalization of IKS in the development process prompted the shift of the "development paradigm" from a "top-down" to a "bottom-up" approach of the 1990s, supported by the discussion on the interaction between science and ethnoscience.

As highlighted above, the contribution from Traditional Ecological Knowledge (TEK) to sustainable environmental conservation pertained to a growing interest in its role in the transformational strategy of *Education for Conservation* (EfC), recently introduced by IUCN-CEC (2020) to deal with future needs of education in conservation efforts around the globe.

Special importance for the timely transformation of environmental education is attached to the complementary principles of TEK of practical knowledge and experience with the holistic conceptualisation of the traditional cosmovisions showing great respect and care f the natural environment. In this way, it is shown that in environmental education, several TEK-based programmes in ethno-education on the environment have proved rather effective, albeit confined to local and regional settings.

In view of the urgent need for effective environmental education in order to foster an ecological shift to overcome socio-economic and cultural l challenges and prepare for the future, it is important to take into account that the world-wide emergence of multicultural societies call for restructuring the educational systems according to the expectations and demands of sub-cultures, retaining their traditional culture, including the transfer of knowledge to their younger generations, often in their local language. Apart from the practical factors of TEK, its key role in environmental education is the provision of the holistic concept of cosmovisions and philosophy of nature, fostering a ethic of respect and care of teachers and students for the environment and its resources.

The Chapter concludes with the positive experiences gained in Sub-Saharan Africa and South-East Asia with the introduction of IKS-based integrating models bridging the "knowledge gap" between science and enthnoscience, including the models in Integrated Community Health Development (ICHD), Integrated Microfinance Management (IMM), and Integrated Community-Managed Development (ICMD).

The resulting TEK-based Ethnoscientific Model of Integrated Eco-Education (IEE) is designed for effective environmental education in multicultural settings around the globe, rendering the dynamic, interactive relationships between the different systems both inter-dimensional and holistic towards effective environmental education for sustainable development with the objective to contribute to foster an ecological shift to overcome socio-economic and cultural challenges and prepare for the future of Planet Earth.

ACKNOWLEDGMENT

The authors wish to express their gratitude for the great inspiration and encouragement of H.R.H. Prof. Dr. Anak Agung Gde Agung, Mrs. Dr. Martha Tilaar, Mrs. Prof. Dr. Vina Adriany, and Dr. Kurniawan Saefullah in Indonesia. Special thanks go to Mrs. Drs. Mady K.L. Slikkerveer MA for her unflagging support.

REFERENCES

Acuity (2023). *Whitepaper: Preserving biodiversity: The Next Big Challenge after Climate Change*. Acuity.

Adimihardja, K. (1999). Cosmology and Biodiversity of the Kasepuhan Community in the Mount Halimun Area of West Java, Indonesia. In D. A. Posey (Ed.), *Cultural and Spiritual Values of Biodiversity: A Complementary Contribution to the Global Biodiversity Assessment*. Intermediate Technology Publications Ltd.

Agung, A. A. G. (2005). *Bali Endangered Paradise? Tri Hita Karana and the Conservation of the Island's Biocultural Diversity*, [PhD Dissertation], Leiden Ethnosystems And Development Programme (LEAD) Studies, No. 1, Leiden: Leiden University.

Aiglsperger, J. (2014). *Yiatrosofia yia ton Anthropo: Indigenous Knowledge and Utilisation of MAC Plants in Pirgos and Pretoria, Rural Crete: A Community Perspective on the Plural Medical System in Greece*. [Thesis, PhD], Leiden Ethnosystems And Development Programme (LEAD) Studies, No. 8, Leiden, Leiden University.

Akhmetova, A. I., & Kolomiyets, O. M., Arpentieva, M. R., & Golubchikova, M. G.(2020). Environmental Education and Management: Exclusive, Inclusive, and Indigenous Doctrines, IOP Conference Series: Earth and Environmental Science. *Sci. 459 052086*. DOI 10.1088/1755

Ambaretnani, P. (2012). *Paraji and Bidan in Rancaekek: Integrated Medicine for Advanced Partnerships among Traditional Birth Attendants and Community Midwives in the Sunda Area of West Java, Indonesia*. PhD Dissertation. *Leiden Ethnosystems And Development Programme (LEAD) Studies*, No. 7. Leiden: Leiden University.

Attenborough, D. F. (2021). *A Life on Our Planet: My Witness Statement and a Vision for the Future*. BBC Documentary.

Avery, M. (2012). *Our Vanishing Flora*. HarperCollins Publishers.

Balakrishna, P. (2020). *Strategy on Education for Conservation (EfC): A Contribution from IUCN CEC and FLEDGE in Support of Realizing the Future Conservation Goals*. IUCN CEC and FLEDGE.

Balée, W. (1994). *Footprints of the Forest: Ka'apor Ethnobotany—The Historical Ecology of Plant Utilization by an Amazonian People*. Columbia University Press.

Barker, D., Oguntoyinbo, J. S., & Richards, P. (1992). The Utility of the Nigerian Peasant Farmer's Knowledge in the Monitoring of Agricultural Resources. *General Report Series No. 4*, London, UK.

Berkes, F. (1993). Traditional Ecological Knowledge in Perspective. In J. T. Inglis (Ed.), *Traditional Ecological Knowledge: Concepts and Cases*. Canadian Museum of Nature and International Development Research Centre.

Bhatiani, G. (2022). *Sustainable Urban Networks for Dynamic and Resilient (Sundar)*. RTI International. DOI., doi:10.13140/RG.2.2.24783.76964

Boas, F. (1928). *Anthropology and Modern Life*. Routledge.

Bronowski, J. (1967). *The Common Sense of Science*. Harvard University Press.

Brundtland, G. H. (1987). Our Common Future, World Commission on Environment and Development (WECD). Oxford UK: Oxford University Press.

Cardinale, B. J., Duffy, J. E., Gonzalez, A., Hooper, D. U., Perrings, C., Venail, P., & Naeem, S. (2012). Biodiversity Loss and its Impact on Humanity. *Nature*, *486*(7401), 59–67. https://doai.org/10.1038/nature11148. doi:10.1038/nature11148 PMID:22678280

Casimirri, G. (2003). *Problems with Integrating Traditional Ecological Knowledge into Contemporary Resource Management*. In *Congress Proceedings of the III World Congress*. Quebec City, Canada.

CCTL. (2023). *Mission Statement of the Chicago Center for Teaching and Learning (CCTL)*. University of Chicago, Chicago Center for Teaching and Learning.

Chambers, R., Pacey, A., & Thrupp, L. A. (1989). *Farmer First: Farmer Innovation and Agricultural Research*. Intermediate Technology Publications. doi:10.3362/9781780440149

Chirangi, M. M. (2013). '*Afya Jumuishi*': Towards Interprofessional Collaboration between Traditional and Modern Medical Practitioners in the Mara Region of Tanzania. PhD Dissertation. *Leiden Ethnosystems And Development Programme (LEAD) Studies* No. 8. Leiden: Leiden University.

CIKARD. LEAD & CIRAN(1990). Indigenous Knowledge and Development Monitor (IKDM), The Hague: CIRAN.

Comstock, A. B. (1911). *Handbook for Nature Study*. Comstock Publishing Company.

De Bekker, J. C. M. (2020). *Transcultural Health Care Utilisation in Serengeti of Tanzania: Towards Applied Ethnoscience in Public Health Management*. PhD Dissertation. PhD Dissertation. *Leiden Ethnosystems And Development Programme (LEAD) Studies* No. 12. Leiden: Leiden University.

Dharma, A. P. (1981). *Indonesische Geneeskrachtige Planten*. De Driehoek.

Djen Amar, S. C. (2010). *Gunem Catur in the Sunda Region of West Java: Indigenous Communication on the MAC Plant Knowledge and Practice within the Arisan in Lembang*. PhD Dissertation. *Leiden Ethnosystems And Development Programme (LEAD) Studies* No. 6. Leiden: Leiden University.

Dorn (2020). "A New Global Ethic»: A History of the United Nations International Environmental Education Program, 1975-1995". *Foro de Educación 18*(2), 83-108. DOI:. doi:10.14516/fde.808

Epelle, A., & Alafonye, U. O. (2010). Enhancing the Classroom Environment for Effective Teaching, Learning, *International Journal of Educational Development (IJOED)* (Jan 2010). https://www.researchgate.net/publication/362114552

Erwina, W. (2019). *Iber Kasehatan* in *Tatar Sunda*, Indonesia: Utilisation of Traditional and Modern Health Information and Communication System in Sukamiskin, Bandung. PhD Dissertation. *Leiden Ethnosystem And Development Programme (LEAD) Studies* No. 10. Leiden: Leiden University.

European Union (2000). *European Biodiversity Programme Natura 2000* (2009/147/EG), Brussels: EU

Febriyanti, M. (2021). *Ubar Kampung:* Indigenous Knowledge and Practice of Medicinal, Aromatic and Cosmetic (MAC) Plants Used for the Treatment of Diabetes Mellitus in the Tatar Sunda Region of West Java, Indonesia. PhD. Dissertation. *Leiden Ethnosystems And Development Programme (LEAD) Studies* No. 13. Leiden University.

Flores, A., & Mena, P. (2018). Cultural and Intercultural Education: Experiences of Ethno-Educational Teachers in Colombia [on line]. *The Australian Journal of Teacher Education, 43*(7), 62–81. doi:10.14221/ajte.2018v43n7.4

Florian, L. (2015). Conceptualising Inclusive Pedagogy: The Inclusive Pedagogical Approach in Action: Inclusive Pedagogy Across the Curriculum. In *International Perspectives on Inclusive Education* (Vol. 7). Emerald Group Publishing Limited., doi:10.1108/S147936362015000007001

Gheneti, Y. (2007). *Microcredit Management in Ghana: Development of Co-operative Credit Unions among the Dagaaba*, PhD dissertation, *Leiden Ethnosystems And Development Programme LEAD Studies* No. 2. Leiden: Leiden University.

Gonzales, T., Chambi, N., & Machaca, M. (1999). Agriculture and Cosmovision in the Contemporary Andes: The Nurturing of the Seeds. In D. A. Posey (Ed.), *Cultural and Spiritual Values of Biodiversity: A Complementary Contribution to the Global Biodiversity Assessment* (pp. 167–260). Intermediate Technology Publications Ltd.

Goodall, J. (1992). Roots and Shoots Programme, Washington D.C., USA: Jane Goodall Institute (JGI).

Goodall, J. (1999). *In the Shadow of Man.* Mariner Books.

Goodenough, W. H. (1957). "Cultural Anthropology and Linguistics"., in Garvin, P.C. (Ed.), *Report of the Seventh Annual Roundtable Meeting on Linguistics and Language Study*, 167-173. Washington, D.C: Georgetown University.

Hay-Edie, T. (1999). Landscape perception and sensory emplacement in decision-making. In D. A. Posey (Ed.), Cultural and Spiritual Values of Biodiversity (pp. 65–74). Intermediate Technology Publications Ltd.

Hernández-Martínez, A., et al. (2021). Ethno-education: Health Education Based on Cultural Diversity, *City Diversity Research* (2022). Volume 3 Issue 2, .1–10.

Heywood, V. H., & Watson, R. T. (Eds.). (1995). Global Biodiversity Assessment, United Nations Environment Programme. Cambridge, UK: Cambridge University Press.

Holzhauer, R. W., & Gellaerts, S. L. (2022). Ideas and the Law. In G. Enthoven, S. Rudnicki, & R. Sneller (Eds.), *Towards a Science of Ideas: About the Emergence, Evolution and Expansion of Ideas and Their Translation into Action*. Vernon Press.

Ibui, A. K. (2007). *Indigenous Knowledge, Belief and Practice of Wild Plants among the Meru of Kenya.* PhD Dissertation. *Leiden Ethnosystems And Development Programme (LEAD) Studies,* No. 3. Leiden: Leiden University.

IPBS. (2020). *Global Assessment Report on Biodiversity and Ecosystem Services*, Intergovernmental Science-Policy Platform on Biodiversity and Ecosystem Services (IPBES), Bonn, Germany, IPBES: www.ipbes.net

IPCC (1923). *Sixth Assessment Synthesis Report* (AR6) *of the Intergovernmental Panel on Climate Change* (IPCC). Geneva, Switzerland: ipcc-media@wmo.int.

IUCN. (1980). *World Conservation Strategy: Living Resource Conservation for Sustainable Development*. IUCN.

IUCN. (2023). Commission on Education and Communications (IUCN-CEC). Gland, Switzerland: IUCN

IUCN/NEWP/WWF. (1991). Caring for the Earth: A strategy for Sustainable Living. North American Wetlands Conservation Council (NEWP), International Union for Conservation of Nature (IUCN) and World Wildlife Fund (WWF). Gland, Switzerland: IUCN.

Kaiser, W. K., & Meyer, J. H. (2016). *International Organizations and Environmental Protection: Conservation and Globalization in the Twentieth Century,* New York, USA. *Berghahn., 2017*. Advance online publication. doi:10.3167/9781785333620

Knapp, C. N., Fernandez-Gimenez, M., Kachergis, E., & Rudeen, A. (2011, March). Using Participatory Workshops to Integrate State-and-Transition Models Created with Local Knowledge and Ecological Data. *Rangeland Ecology and Management*, *64*(2), 158–170. doi:10.2111/REM-D-10-00047.1

Korteweg, L., & Russell, C. (2012, May). Decolonizing + Indigenizing = Moving Environmental Education Towards Reconciliation. *Canadian Journal of Environmental Education*, 17.

Kothari, A., & Das, P. (1999). "Local Community Knowledge and Practice"s. In D. A. Posey (Ed.), *Cultural and Spiritual Values of Biodiversity: A Complementary Contribution to the Global Biodiversity Assessment*. Intermediate Technology Publications Ltd.

Leakey, R. E., & Lewin, R. (1996). *The Sixth Extinction: Patterns of Life and the Future of Humankind.* Bantam Double Dell Publishing Group, Inc.

Leakey, R. E., & Slikkerveer, L. J. (1993). *Man-Ape Ape-Man: The Quest for Human's Place in Nature and Dubois' 'Missing Link.* Kenya Wildlife Service Foundation.

Leurs, L. N. (2010). *Medicinal, Aromatic and Cosmetic (MAC) Plants for Community Health and Bio-Cultural Diversity Conservation in Bali, Indonesia.* PhD Dissertation, *Leiden Ethnosystems And Development Programme (LEAD) Studies,* No. 5. Leiden: Leiden University.

Loreman, T. (2017). *Pedagogy for Inclusive Education, Oxford Research Encyclopedia of Education*. Oxford University Press.

Macekura, S. J. (2017). *Of Limits and Growth: The Rise of Global Sustainable Development in the Twentieth Century*. Cambridge University Press.

Maftuh, B., & Adriany, V. (2018). Bringing Social Justice into Research in Education. In E. Malihah, T. Aryanti, V. Adriany, H. Yulindrasari, & A. Izharuddin (Eds.), *Research for Social Justice*. Routledge Taylor and Francis Group.

Mahale, P., & Sorée, H. (1999). Cosmovisions and Agriculture in India. In D. A. Posey (Ed.), *Cultural and Spiritual Values of Biodiversity: A Complementary Contribution to the Global Biodiversity Assessment*. Intermediate Technology Publications Ltd.

Malihah, E., Aryanti, T., Adriany, V., Yulindrasari, H., & Izharuddin, A. (Eds.). (2018). *Research for Social Justice*. Routledge Taylor and Francis Group.

McCormick, J. (1991). *Reclaiming Paradise: The Global Environmental Movement*. Indiana University Press.

Millennium Ecosystem Assessment (MA). (2005). *Ecosystems and Human Well-being: A Framework for Assessment, MA* (Vol. 1). Island Press.

Mironov, A. V. (2020). The Technique of Studying the World in Primary School. Moscow. *Pedagogy*. Advance online publication. doi:10.1088/1755-1315/459/5/052086

Mitchell, D. (2015). Inclusive Education is a Multi-Faceted Concept. *Center for Educational Policy Studies Journal*, *5*(1), 9–28. doi:10.26529/cepsj.151

Nesterova, I. A. (2020). Ethno-education as a Means of Enhancing the Health of the Population of Colombia. *Russian Journal of Humanities and Social Sciences*, *13*(4), 1602–1606.

Omprakash Team (2007-2021). *Ethno-Education in Barranquilla*. Colombia, Seattle, WA, USA: Omprakas

Pierotti, R., & Wildcat, D. R. (1999). Traditional knowledge, culturally-based World-Views and Western Science. In D. A. Posey (Ed.), *Cultural and Spiritual Values of Biodiversity: A Complementary Contribution to the Global Biodiversity Assessment*. Intermediate Technology Publications Ltd.

Posey, D. A. (1999). Cultural Landscapes, Chronological Ecotones and Kayapó Resource Management. In D. A. Posey (Ed.), *Cultural and Spiritual Values of Biodiversity: A Complementary Contribution to the Global Biodiversity Assessment*. Intermediate Technology Publications Ltd.

Posey, D. A. (Ed.). (1999). *Cultural and Spiritual Values of Biodiversity: A Complementary Contribution to the Global Biodiversity Assessment*. Intermediate Technology Publications Ltd.

Ramsaroop, S., & Van Rooyen, H. (2013). Exploring Educator Competence in Teaching Environmental Education in Schools in Gauteng. *Africa Education Review*, *10*(3), 595–613. doi:10.1080/18146627.2013.853555

Rose, D., & Meyer, A. (2002). *Teaching Every Student in the Digital Age: Universal Design for Learning*. Association for Supervision and Curriculum Development.

Sadruddin, M. M. (2023). Design and Pilot of an Inclusive Environmental Education Course for a Community Living Under the Unprecedented Environmental Emergencies in Pakistan- A Case Study. *Journal of Peace Education*. Advance online publication. doi:10.1080/17400201.2024.2314302

Saefullah, K. (2019). 'Gintingan in Subang': An Indigenous Institution for Sustainable Community-Based Development in the Sunda Region of West Java, Indonesia. PhD Dissertation. *Leiden Ethnosystems And Development Programme (LEAD) Studies* No. 11. Leiden: Leiden University.

Schultes, R. E., & Raffauf, R. F. (1990). *The Healing Forest: Medicinal and Toxic Plants of the Northwest Amazonia*. Dioscorides Press.

Shultes, R. E., Hofmann, A., & Ratsch, C. (2001). *Plants of the Gods: Their Sacred, Healing, and Hallucinogenic Powers*. Healing Arts Press.

Slikkerveer, L. J. (1989). Changing Values Attitudes of Social and Natural Scientist towards Indigenous Peoples and Their Knowledge Systems. In D. M. Warren, L. J. Slikkerveer, & S. O. Titilola (Eds.), *Indigenous Knowledge Systems: Implications for Agriculture and International Development. Studies in Technology and Social Change No. 11*. Iowa State University.

Slikkerveer, L. J. (1990). *Plural Medical Systems in The Horn of Africa: The Legacy of 'Sheikh' Hippocrates*. Kegan Paul International.

Slikkerveer, L. J. (1992). Indigenous Agricultural Knowledge Systems in Kenya: Towards Conservation of Bio-Cultural Diversity in East Africa. In W. M. Adams & L. J. Slikkerveer (Eds.), *Indigenous Knowledge an Change in African Agriculture. Studies in Technology and Social Change No. 9 Ames*. Iowa State University.

Slikkerveer, L. J. (1999). Ethnoscience, 'TEK', and its Application to Conservation. In D. A. Posey (Ed.), *Cultural and Spiritual Values of Biodiversity: A Complementary Contribution to the Global Biodiversity Assessment* (pp. 167–260). Intermediate Technology Publications Ltd. doi:10.3362/9781780445434.005

Slikkerveer, L. J. (2006). The Challenge of Non-Experimental Validation of MAC Plants: Towards a Multivariate Model of Transcultural Utilization of Medicinal, Aromatic and Cosmetic Plants. In R.J. Bogers, L.E. Craker & D. Lange Medicinal and Aromatic Plants: Agricultural, Commercial, Ecological, Legal, Pharmacological and Social Aspects (pp. 1-28). Springer.

Slikkerveer, L. J. (2016). *Integrated Community-Managed Development: Reflections on the Ethnosystems Approach*. LEAD Programme of Leiden University.

Slikkerveer, L. J. (2016). LEAD: Leiden Ethnosystems And Development Programme. Verleden, Heden en Toekomst van het Leidse Speerpunt. Lokale Kennis Systemen en Ontwikkeling 1993-2016-2020. Internal Report, Leiden: LEAD Programme of Leiden University.

Slikkerveer, L. J. (2018). Towards the Development of Ethno-Education: an Ethnoscience-Based Approach to Integrated Education Management in Indonesia. In E. Malihah, T. Aryanti, V. Adriany, H. Yulindrasari, & A. Izharuddin (Eds.), *Research for Social Justice*. Routledge Taylor & Francis Group.

Slikkerveer, L. J. (2019). Towards a Model of Integrated Community-Managed Development. In Slikkerveer, L.J., Baourakis, G. & Saefullah, K. (Eds.) *Integrated Community-Managed Development* (ICMD)*: Strategizing Indigenous Knowledge and Institutions for Poverty Reduction and Sustainable Community Development in Indonesia - A Community-Based Contribution to the United Nations 2030 Agenda for Sustainable Development.* Spinger Nature Switzerland AG. doi:10.1007/978-3-030-05423-6_4

Slikkerveer, L. J., & Dechering, W. H. J. C. (1995). LEAD: The Leiden Ethnosystems And Development Programme at Leiden University in The Netherlands. Warren, D. M., Slikkerveer, L. J. & Brokensha, D. (Eds.) (1995) The Cultural Dimension of Development: Indigenous Knowledge Systems. IT Studies on Indigenous Knowledge and Development, London, UK: Intermediate Technology Publications Ltd.

Slikkerveer, L. J., & Slikkerveer, M. K. L. (1995)."Taman Obat Keluarga (TOGA) Indigenous Indonesian Medicine for Self-reliance. In D. M. Warren, L.J. Slikkerveer, & D. Brokensha, (Eds.) The Cultural Dimension of Development: Indigenous Knowledge Systems.Intermediate Technology Publications Ltd.

Sponsel, L. E. (2013). *Spiritual Ecology: A Quiet Revolution.* Praeger.

Sulistyani, H. D., & Rahrdjo, T. (2019). Ethno-Pedagogy Perspective on Ethnic Minority Discourse of Education, *Proceedings of the 2nd International Conference on Educational Sciences Series: Advances in Social Science, Education and Humanities Research.* 10.2991/ices-18.2019.53

Susanti, R. D. (2023). '*Kearifan Kesehatan Lokal'*: Indigenous Medical Knowledge Practice for Integrated Nursing of the Elderly with Cardiovascular Disease in Sumedang, West Java: Towards Transcultural Nursing in Indonesia. PhD Dissertation. *Leiden Ethnosystems And Development Programme (LEAD) Studies* No. 12. Leiden: Leiden University.

Suyanto. (2023). Digital Innovation in the Era of Society 5.0 Encourages Indonesia's Economic Recovery in Post-Pandemic, *International Journal of Integrative Sciences*, *2*(7), 1141-1148.

Tomlinson, C. A. (2001). *How to Differentiate Instruction in Mixed-Ability Classrooms* (2nd ed.). Association for Supervision and Curriculum Development.

Tuck, E., McKenzie, M., & McCoy, K. (2014). Land Education: Indigenous, Post-Colonial, and Decolonizing Perspective on Place and Environmental Education Research. *Environmental Education Research*, *20*(1), 1–23. doi:10.1080/13504622.2013.877708

UN (1971). *Second United Nations Development Decade* (1971-1980). New York, USA: UN.

UN (1972). *Stockholm Declaration and Action Plan for the Human Environment*. New York, USA: UN.

UN (1992). *Earth Summit on Environment and Development* (Rio de Janeiro, Brasil). New York, USA: UN.

UN (2002). *World Summit on Sustainable Development* (Johannesburg). New York, USA: UN.

UN (2006). *Convention on the Rights of People with Disabilities*. New York, USA: UN.

UN (2014). *Decade of Education for Sustainable Development* (2005-2014). New York, USA: UN.

UN (2015). *2030 Agenda for Sustainable Development (SDG).* New York. USA: UN.

UN (2016). *Global Gateway of IEEP: UN Office for Partnerships to Deliver the SDGs*. New York, USA: UN.

UN. (2019). Global Assessment Report on Biodiversity and Ecosystem Services. New York, USA: UN.

UN. (2020). *Global Biodiversity Outlook 5* (GBO-5), UN Convention on Biological Diversity. New York, USA: UN.

UN-ICERD. (1965). *International Convention on the Elimination of all Forms of Racial Discrimination*. New York, USA: UN-ICERD.

UNEP. (2017). *Conservation through Cultural Survival: Indigenous Peoples and Protected Areas*. UNEP.

UNESCO. (1968). *Biosphere Conference*. Paris: UNESCO.

UNESCO. (1995). *World Conference on Special Needs Education*. Paris: UNESCO.

UNESCO. (2021). *World Conference on Education for Sustainable Development (Berlin)*. Paris: UNESCO.

UNESCO-UNEP. (1975). Education for Sustainable Development (ESD of IEEP). Paris/Nairobi, Kenya: UNEP.

UNESCO-UNEP. (1975). International Environmental Education Programme (IEEP) (1975-1983). Paris/Nairobi, Kenya: UNEP.

UNESCO-UNEP. (1975). *International Environmental Education Programme (IEEP)*. UNEP.

UNESCO-UNEP. (1977). *The Tbilisi Declaration, Tbilisi, Georgia (USSR)*. UNEP.

Van Rooyen, H. G. (2002). Education for a Sustainable Environment: A Potential Model. Johannesburg, SA:_ Rand Afrikaans University.

Vavilov, N. I. (1951). *The Origin, Variation, Immunity and Breeding of Cultivated Plants*. Chronica Botanica Company. doi:10.1097/00010694-195112000-00018

Wagler, R. (2011). *Human Extinction and the End of the World: An Analysis of the World's End Scenarios and Related Topics*. GRIN Verlag.

Walsh, A. (2022). What to Expect from the World's Sixth Mass Extinction. Deutsche Welle., Retrieved February 5, 2022, from.

Warren, D. M., Slikkerveer, L. J., & Brokensha, D. (Eds.). (1995). *The Cultural Dimension of Development. Indigenous Knowledge Systems. IT Studies on Indigenous Knowledge and Development*. Intermediate Technology Publications Ltd. doi:10.3362/9781780444734

Wheelright, E. G. (1974). *Vintage Medicinal Plants and Their History*. Dover Publications.

Williandri, R. (2024). 'Bakti and Sayan' Traditions among the Tenggerese People in East Java:The Role of Indigenous Institutions in Integrated Elderly Care Development in Indonesia. PhD Dissertation, *Leiden Ethnosystems And Development Programme (LEAD) Studies* No. 1. Leiden: Leiden University.

WRI/IUCN/UNEP (1992). *Global Biodiversity Strategy*. New York, USA: UN.

Chapter 9
The "Ecosophy" Concept Incubated in the Indigenous Schools of the Sierra Nevada de Santa Marta, Colombia

Sara Acevedo Garzon
Jindal Global University, India

ABSTRACT

The interaction dynamics upheld by Colombian indigenous communities with nature have been grounded in environmental principles derived from their ancestral knowledge and worldview. In the present day, these principles, which fostered the preservation and attentive guardianship of their natural surroundings, also permeate indigenous schools and their educational methods by teaching their myths, rituals, and agricultural heritage. This chapter examines the environmental principles and the pedagogical techniques upheld by the Arhuaco indigenous schools in the Sierra Nevada de Santa Marta in Colombia. It will delve into the Arhuaco worldview and the potential parallels between Arhuaco's environmental principles and the philosophy "Ecosophy," which promotes harmonious and wise ecological behavior. Along the same line, integrating experiential activities in nature into classrooms will also be highlighted as a reasonable means for nurturing a sense of stewardship and empathy towards the natural world in children.

INTRODUCTION

Colombia's highest mountain, the Sierra Nevada de Santa Marta, is home to four indigenous communities, descendants of the extinct Tairona's native people. The Tairona civilization flourished in northeast Colombia from approximately the 7th to the 16th century A.D. and was known for its complex social, political, and economic systems. After their disappearance in the 16th century A.D., caused by the arrival of the Spanish colonizers, numerous indigenous groups successfully survived in isolation from Spanish influence due to the rugged conditions of access to the high mountains and their biological

DOI: 10.4018/979-8-3693-2577-3.ch009

richness (Izquierdo, 2017). This text aims to explore one of the four remaining communities carrying on the legacy of the Tayrona civilization: the Arhuaco people.

The Arhuaco are renowned, at national and international levels, for their environmental preservation endeavors, rooted in their sacred responsibility as guardians of the Sierra Nevada. According to the Arhuaco worldview, the Sierra Nevada de Santa Marta embodies the world's heart because the universe was created on its peaks. Therefore, as inhabitants of this holy territory, they consider themselves in charge of maintaining its natural balance (Dolmatoff, 1982). The Arhuacos' role as guardians of nature has influenced their economic, social, and cultural spheres and distinctly defined their management of natural resources. In this regard, the Arhuaco worldview comprises some environmental principles governing their relationship with nature, which prescribe rituals and ceremonies to sacralize various activities of their daily life, such as cultivation, harvesting, pest control, and response to seasonal variations and natural disasters, among others (Jaramillo, 2010).

The paper herein is dedicated to analyzing the Arhuaco self-conception as guardians of nature, the intergenerational transmission of this role in the schools, and the environmental repercussions of such a conception. For this purpose, the narrative will initially delve into the Arhuaco worldview and explore the stories, beliefs, and rituals on which their way of life is founded. Subsequently, the educational approach employed in indigenous schools will be examined, encompassing the delineation of school curricula, community involvement, course offerings, and the challenges inherent in Arhuaco identity pedagogy. Lastly, the paper will present the philosophical perspectives of "ecosophy" to explore its parallels with the environmental principles observed by the Arhuaco and their positive environmental impacts.

Mapping the Geographical Features of the Arhuaco Homeland

The mountainous system of the Sierra Nevada de Santa Marta covers an area of 21158 km2 to the Colombian northeast, rises to a height of 5775 m, and is located 42 km from the Caribbean coast, making this the highest coastal mountain in the world (Jaramillo, 2010). With its range of altitude levels, the Sierra Nevada supports diverse ecosystems, promoting high levels of endemism and serving as a significant water source for the region. Since 1964, 383.000 hectares of this territory have been safeguarded under the designation of "The Sierra Nevada de Santa Marta National Natural Park," the highest environmental protection level, which mandates that no activities other than scientific research, education, and protection endeavors be carried out in this area. This protected ecosystem, underscored in 1979 by UNESCO as a Biosphere Reserve, is part of the territory currently inhabited by over 85,000 indigenous people divided into four indigenous groups: Arhuacos, Wiwa, Kankuamos, and Kogi (Botero, 2005).

The Arhuaco people are settled in the Nabusímake city on the western slopes of the Sierra Nevada and are composed of more than 34,000 members who speak their dialect, iku (D.A.N.E., 2018). The economic livelihood of the Nabusímake population rests upon subsistence agriculture. Typically, families inhabiting the region own land across various thermal levels, allowing them to practice rotational horticulture. In the lowlands, crops like corn, plantain, yucca, sugarcane, and avocado are cultivated, while colder climates support the growth of vegetables, fruits, potatoes, and sweet potatoes. Moreover, some households participate in livestock rearing, which involves raising cattle, sheep, pigs, and poultry (Niño, 2011). The last population census conducted in 2018 showed that 54% of its members live in indigenous reservations, a collective property granted by the government to ancestral dwellers, and that a substantial 69,4% of the population resides in traditional indigenous dwellings crafted predominantly from wet clay soil without a fixed structure or mud-layered onto interwoven wood (D.A.N.E., 2018).

The Arhuaco Designation of Spiritual Mountain Guardians

In the community's organizational structure of Arhuaco, myths play an essential role in explaining the origin of Arhuaco, in assigning their role in the community, and in the behavioral guidelines they must follow as its members (Dolmatoff, 1977). For this reason, to understand the Arhuaco attribution of nature's guardians, it is necessary to return to their ancestral stories transmitted by oral tradition. According to their origin myths, the dawn of time witnessed a divine matrimony between the forces of the infinite and the material world, culminating in the genesis of life on Earth. This sacred union unfolded through a celestial explosion, eventually settling in the Sierra Nevada de Santa Marta, from which the mountain peaks were created (Bunchanawim, 1997). Thus, it is believed that the Sierra Nevada mountains represent the children of the universal mother and that the ancestors of all humanity live there. Under this understanding, mountains embody the divine ancestors from which the Arhuaco stems, and it is revered as a holy sanctuary that enables humans to contemplate and admire the sacred qualities of divinity (Petitpierre, 1975).

These original forces or divine ancestors are called by the Arhuaco "Elder Siblings," they are in charge of sustaining the fabric of the universe's cosmic. These entities are perceived as the spirits of nature or the subtle energies that lie beneath matter, and the bond established with them is compared to that of a mother and child. The figure of "Elder Siblings" fulfills several functions in the collective imagination of the Arhuaco; they are simultaneously conceived as the origin, the rational basis of moral norms of conduct, and the ultimate purpose of all human experience (Mamo, 1997).

On the other hand, Arhuaco's position is that of the "Younger Brothers" in charge of their "Elder Siblings" camouflaged in the natural forces of the material world. Because the Arhuaco appraises themselves and all humans as descendants of the "Elder Siblings," they consider humans to be, in turn, one of their extensions or essential parts (Mamo, 1997). The belief in humans' divine lineage not only attributes them with divine essence but also positions them as conduits on earth through which the original divinity of the "Elder Siblings" can be accessed. Hence, one of the primary purposes of the Arhuaco people is to acknowledge themselves as intermediaries between heaven and earth, assuming the role given by their ancestors of safeguarding the balance of existence on earth (Plan de Salvaguarda del Pueblo Arhuaco, 2015).

After comprehending the divine origin attributed to the Arhuaco by their myths of origin, the subsequent section will delve into the three fundamental pillars derived from their myths that serve as the foundation of their societal structure: the "Law of origin," the sacred territory, and the original dialect.

The "Law of Origin": The Law of Life's Permanence

The "Law of Origin" is conceived by the Arhuaco people as a system of social and spiritual governance whose primary purpose is to preserve the balance of nature's cycles by regulating the relationship between humans and the environment (Arias & Zuleta, 2020). These conduct guidelines constitute a compendium of moral principles that profess humility and promote imitating nature. It was inherited by the "Elder Siblings" to reach the vitality of individuals and other living beings and thereby ensure Mother Earth's vitality. For this reason, the "Law of Origin" is also known as the law of life's permanence (Jaramillo, 2010).

The implementation of the "Law of Origin" is significantly steered by the Arhuaco leaders, known as Mamo. They are pivotal in guiding the community, serving as political leaders, shamanic healers, spiritual guides, and judges, and fulfilling various other administrative functions (Arias & Zuleta, 2020). Grounded in the Arhuaco conviction that destiny is inscribed in their navel from the day of birth, the role of Mamo is predetermined from infancy, and their entire upbringing is oriented toward molding them as leaders (Mamo, 1997). Mamos are attributed with spiritual attributes that enable them to establish direct communication with the "Elder Siblings." Thus, the divine source from which their authority and wisdom emanate grants them legitimacy in the eyes of the community.

Environmental Principles of Reciprocity and Complementarity Unfolded in the Sierra Nevada Sacred Territory

As mentioned in previous paragraphs, the Sierra Nevada de Santa Marta mountains are conceived as the materialization of the original divine forces. This conception prompts the Arhuaco to regard mountains as a holy source from which they can draw divine attributes through contemplation and worship (Petitpierre, 1975). Thus, the approach of the Arhuaco with nature is characterized by reverence for its sacred symbolism and the pursuit of wisdom and purification. Also, from the Arhuaco vision, the Earth is correlated with the feminine energy of creation, as the Earth is fertilized by the masculine energy of the sun. Therefore, the Earth is the mother of all living creatures because it has been assigned the task of gestation and creation of life (Dolmatoff, 1977). In this light, it is because of this mother-child approach that the Arhuaco have with the Earth that their relationship with it is mainly based on respect.

Another essential element that strengthens the Arhuaco notion of respect for nature is their understanding of the scarcity and limit of natural resources. According to their worldview, the energy that makes up the material world is finite and limited. Therefore, energy's multiple expressions and forms belong to a circuit of transformation, where energy is not created but constantly recreated (Dolmatoff, 1977). This notion suggests that while energy is limited because it remains constant but capable of assuming diverse forms, even more importantly, it is always borrowed from the circuit. The Arhuaco then assumes the use of natural resources from a lending perspective, in which they, as borrowers, accept the duty to retribute nature for the fruits taken from it to maintain the balance of the energy flow in the circuit (Guerrero, 2005).

Therefore, the balance of nature's circuit becomes the main objective of the Arhuaco people, leading the environmental principles of reciprocity and complementarity to assume pivotal roles in the Arhuaco organizational and administrative frameworks. These important principles advocate "giving back" the gifts given by nature so as not to disturb their biological cycles. This behavior is recognized as a cosmic duty aligned with the law of cause and effect, where each action is expected to yield a reciprocal and complementary contribution (Guerrero, 2005). Reciprocity is, therefore, a symbol of justice that requires humans to reintegrate all the outputs into the circuit. This permanent task of balancing the energy circuit gives humans active participation in safeguarding the biological processes of nature, making them not mere spectators of the ecosystem but playing a vital role in sustaining it (Dolmatoff, 1977).

In this context, the Arhuaco role as guardians of the mountain is accentuated by the principles of reciprocity and complementarity, which makes them key agents in sustaining nature. As a derivative of this belief, the Arhuaco consider that elemental occurrences like sunrise and harvest are intrinsically linked to their faithful engagement in sacred rituals and ethical conduct. For instance, the Arhuaco people celebrate rituals to welcome the sowing and harvesting periods, make offerings to thanks and ask nature for certain favors, and offer tributes to stop plagues and natural disasters (Plan de Salvaguarda del Pueblo Arhuaco, 2015).

Because the Arhuaco whole reality unfolds according to the goodness of their intentions and actions, each activity is preceded by rituals and symbolisms that re-signify and sacralize them. Cultivation is, for example, a form of reverence for Mother Earth's power of creation; the pilgrimages to the mountain, beyond reaching its peaks, represent the opportunity to ascend to the origin; diseases among the people or the land are a warning sign, signaling an imbalance in the energy circuit that calls for remedial offerings, among other. The territory, therefore, functions as a canvas rich with spiritual meanings, where the cultural identity of the Arhuaco people is unfolded (Mamo, 1997).

The Ancestral Dialect as the Keeper of Ancestral Knowledge

The Arhuaco wisdom is not the result of an intellectual effort recorded in books but the product of a widely lived experience within the community. The diverse spaces for conversation that emerge during activities such as weaving, planting, and cooking, among others, foster stronger bonds among community members and facilitate the transmission of knowledge through dialogue. Because of this, the primary source of the Arhuaco people's ancestral knowledge lay in the verbal transmission of information, known as oral tradition. In this light, due to the importance of disseminating their knowledge, stories, and rituals through oral tradition, the Arhuaco people's native dialect, iku, is attributed great importance in faithfully conveying his ancestral vision of the world. (Guerrero, 2005).

Given that language and words exist to delineate or signify preconceived notions in the human mind, words are usually analogous to a finger pointing to the moon; in serving as a tool for pointing out ideas, words are separate from the ideas themselves and could never entirely capture them (Tauste, 1992). A similar idea is shared by the philosophy of the Tao in its popular dilemma, which says that the Tao that can be named is not the true Tao (Schuon, 2001). With this perspective, it can be said that the Arhuaco people conceive their dialect as the symbolic bridge that allows them to reach their deities, principles, and truths with the highest possible authenticity. Hence, its native dialect, iku, is regarded as an authentic medium that preserves the community's original beliefs.

As language functions as the custodian of ancestral knowledge, ensuring that the original essence of their worldview remains intact, the education of Arhuaco children in rural schools is centered around the instruction of their dialect, which is the central axis in teaching their worldview (Castro, 2013). Nevertheless, imparting their traditional ancestral knowledge to the new generations is comprehensively designed by teaching experiential activities that intimately connect children with the intricate dynamics of their territory.

Education in Arhuaco Schools: Preserving Tradition, Nurturing Ecological Wisdom

This section explores the structure of Arhuaco schools, community involvement in children's education, curriculum content, and related aspects. While it offers an overview of children's activities, specific games, and exercises developed in the schools are unknown due to a lack of secondary sources. However, the section provides a detailed examination of the approach and practices of Arhuaco schools, offering insight into the effective integration of the Arhuaco worldview into educational institutions.

The duty of imparting cultural education among native peoples has historically rested with their families, making them the principal responsible for preserving their cultural heritage. This dynamic has shaped the foundational concept of education in indigenous communities, rooted in self-directed learning acquired through everyday activities of rural life within the family context. Thus, education has been an experiential activity for indigenous communities, demanding active involvement in the territory and a practical, learn-by-doing methodology (Castro, 2013).

Following this trend, the Arhuaco people retain the family's central role as the principal provider of cultural education nowadays. However, due to the disintegration of families due to migration and divorce or due to the increasing loss of customs among its members, the gross of Arhuaco's educational task has passed into the hands of state schools (Bunchanawin, 1997). As a result, the five elementary schools and the one agricultural high school in Nabusímake have been challenged with the need to embrace Arhuaco's principles to offer Arhuaco children educational experiences of intimate interaction with the environment (Castro, 2013).

From a legal perspective, incorporating Arhuaco principles within the regular education designed by the central government has been feasible due to the extensive regulations in Colombia that empower indigenous schools to promote their ancestral knowledge. In this regard, the Colombian Constitution of 1991 affirms the autonomy of indigenous communities to design their educational programs and prioritize the imparting of their ancestral knowledge: "Education must be linked to the environment, productive, social and cultural processes, with due respect for their beliefs and traditions" (Constitution, 1991, Art. 55). More specifically, Law 115 of 1994, which regulates the education of indigenous communities, enshrines the cultural orientation that their education must have: "Its purpose will be to strengthen the processes of identity, knowledge, socialization, protection and adequate use of nature, community systems and practices of organization, use of vernacular languages, teacher training and research in all areas of culture." (Law 115, 1994, Art. 56).

However, the autonomy granted by these regulations to indigenous to design their educational projects is simultaneously a requirement to move beyond their oral tradition and establish a formal educational model. Consequently, the directive to impart their ancestral knowledge in the classrooms under modern methodologies has transformed the process of Arhuaco knowledge transmission, forcing them to introspect the most effective means for doing so. Based on this requirement, the Arhuaco schools have proposed a community education model wherein the primary goal of education is community service and where key community figures such as families and spiritual leaders are integral participants in guiding children's educational journey (Castro, 2013).

For instance, the academic curriculum of the Arhuaco schools is shaped through a collaborative effort of community members such as spiritual leaders, professors from the educational committee, and the parents committee (Castro, 2013). Every contribution stemming from their individual life experiences, aspirations, and principles is given equal weight and consideration. By consensus, they are in charge of delineating the core curricular principles grounded in their cultural and ethnic identity, which will be later complemented by guidelines from the central government through the Ministry of National Education (Castro, 2013).

Elsewhere, to ensure parents are actively involved in education, they are used to teaching traditional skills in class, like weaving and agriculture. The parents' close relationship with their children's studies becomes even more meaningful when considering that it is the age or the maturity of the Arhuaco children, and not the school levels, which determines their ability to grasp certain subjects and acquire specific skills. For this reason, parents must provide consent for their children to be instructed on particular topics, as they may have certain reservations about the age-appropriateness of some ancestral information. As well as the choice of whether children can receive contemporaneous knowledge beyond the Arhuaco culture rests with their corresponding parents (Mindiola, 2018). For the explained schema, the program of study for each child is always subject to determination by the child's parents, professors, and Mamos.

In turn, the role of Arhuaco professors goes beyond the classroom. Their guidance doesn't solely pertain to academic matters but also extends into students' personal lives during times and spaces outside their official working hours. Professors are seen as social and cultural managers, functioning as dedicated activists committed to the continuous service of the community (Castro, 2013). Then, it is essential that the professor has a solid understanding of Arhuaco principles and is attuned to their aspirations. For this reason, there is a preference for hiring professors from the community, prioritizing those who, despite lacking university degrees, have demonstrated their capabilities as community leaders and experts in the territory dynamics. Nevertheless, numerous Arhuaco professors are community members who have completed both undergraduate and graduate degrees (Mindiola, 2018).

On the other hand, the Mamos serve as psychological and philosophical advisors of important consideration for the schools in Nabusímake. Because they are considered the "Elder Siblings" representatives, they are responsible for directing the younger generations toward constructive pursuits that enhance the community's welfare (Jaramillo, 2010). The involvement and credibility of the Mamos in the educational process are facilitated by the community's intimate size and by the bonds of fraternity among its members, who regard child-rearing as a joint community responsibility for the benefit of all.

Moving on to the process of formulating the academic curriculum, the Arhuaco people prioritized the instruction of the physical elements of their environment, emphasizing topics such as the geography and botany of its territory over other information deemed part of general culture. For this reason, specific courses related to global history, state formation, and even territorial organization of Colombia, among others, are excluded. The Arhuaco primary focus is stringent; it consists of exhaustively knowing the natural resources of their territory, such as their rivers, mountains, lakes, and valleys, as well as the economic activities of these zones and their administrative organization (Castro, 2013).

The curricular program is structured in four thematic axes: community and territory, mathematics, agricultural work, conservation and protection of the environment, and community production. In the course on community and territory, students are instructed in subjects such as administrative community organization, political participation, ancestral history, and indigenous worldview. The dimension associated with culture and identity is conveyed through the practice of traditions and rituals, as well as through recreational activities such as dances, music, and crafts. There is also an exclusive course for teaching their mother tongue, "iku," and another for Spanish as a secondary language (Castro, 2013).

The Arhuaco children learn a crucial lesson in their agricultural curses: the practice of intercropping. Rejecting monoculture for its detrimental effects on soil biodiversity and nutrition, the Arhuaco people advocate for intercropping method, which involves cultivating multiple crops on a single plot of land, fostering synergistic relationships that yield mutual benefits. Alongside imparting the benefits of intercropping, children are taught the essential rituals that precede its implementation. Moreover, children are instructed to offer to Mother Earth, which is believed to secure a bountiful harvest (Mamo, 1997). Through these rituals, children cultivate a profound reverence for the Earth's gifts, viewing them as manifestations of its benevolence, which responds to their prayers and needs. Engaging in the harvest, the Arhuaco reaffirm their reciprocal bond with the Earth, feeling cherished and safeguarded by its provisions.

Another pivotal class in Arhuaco schools is weaving bags, a practice rooted in Arhuaco tradition. It has served as a platform for coming together, engaging in discussions about significant aspects of life, and strengthening their bonds. The act of weaving and creating different drawings on the bags is compared with the creation of the earth and all its living beings, involving a profound sense of creativity and communion with life. Additionally, selling these products represents one of Arhuaco's primary means of subsistence. Hence, teaching this skill is highly valuable, contributing to community cultural preservation and economic viability (Arias & Zuleta, 2020).

Curiously, in the Arhuaco community, some recreational activities have a negative connotation because they are considered leisure activities. The Arhuaco people believe that in order to rest and refresh the mind is sufficient to switch activities, thereby eliminating the need for leisure time. For this reason, the music, dance, and weaving classes bear primarily a cultural significance as they strive to impart the essence of their ancestral rituals. While the physical education course is integrated as a regular ecological walk, intertwined with community tasks (Castro, 2013).

How to Become an Arhuaco?

After reviewing the subjects covered in Arhuaco schools, oriented to the preservation of their cultural identity, a fundamental question emerges: Can these teachings truly mold the children into authentic Arhuaco? And, maybe more significantly, what does being an Arhuaco mean? Despite the elusive nature of these questions, the following discussion will delve into the Arhuaco people's considerations regarding the notion of being an Arhuaco.

The essence of being Arhuaco encompasses several elements whose significance and priority differ from case to case. However, the ones they consider most important are as follows: firstly, having Arhuaco lineage; secondly, familiarity with and adherence to customs and traditions; and thirdly, knowing the native dialect. Furthermore, additional considerations like the adoption of traditional attire, engagement in rituals, meal preparation, and place of residence further contribute to determining one's Arhuaco identity (Mindiola, 2018). In this regard, while there are defining features of Arhuaco's identity, the list is not exhaustive, and its significance is always subjective.

Nevertheless, one essential factor that holds paramount importance in Arhuaco's identity is the profound sense of "conviction" regarding one of their maximum truths: the interdependent connection that human beings have with the environment (Mamo, 1997). In turn, this recognition fosters in the Arhuaco the belief that nature deserves both value and respect, sentiments that ultimately shape the Arhuaco's environmental behavior and identity as guardians of nature.

As per the Arhuaco tradition, acquiring an Arhuaco identity also involves the ceremonial celebration of critical life stages, such as birth, the transition to adulthood, marriage, and death (Mindiola, 2018). In the christening ceremony, the baby is introduced to the "Elder Siblings" for the assignation of its spiritual name and its purpose in life. On the other hand, the ceremony of transition from childhood to adulthood signifies that the Arhuaco children are prepared to start a family. For girls, this transition is marked by their first menstruation, during which they are secluded in a sacred space to deposit their first blood on the Earth. Meanwhile, boys undergo physical changes such as deepening voices and the emergence of facial hair, which serve as signs of their passage into adulthood. Another essential stage is the marriage ceremony, which is preceded by a spiritual cleansing to purify the engaged couple from their past actions and prepare them to begin a life together. Finally, the mortuary ceremony aims to commemorate the deceased's body and spirit reintegration into the energy circuit mentioned earlier (Mindiola, 2018).

One must consider that, according to the Arhuaco worldview, the concept of youth did not initially exist. In their traditional belief, individuals transition directly from childhood to adulthood, with the immediate opportunity to marry and start a family. Nevertheless, a significant shift has occurred, as the new generations of Arhuaco have delayed the age of marriage for various reasons (Mindiola, 2018). These motivations include the opportunity to engage in work or education and the increasingly validated decision to attain greater maturity before entering into matrimony. In this new scenario, the emergence of new occupations for men, such as motorcycle cab drivers, entrepreneurs, and tour guides, has played a crucial role in influencing this decision. Likewise, the economic horizon of women has expanded with the emergence of additional avenues to support their families by taking on positions such as housekeepers and nannies in urban areas (Mindiola, 2018).

In the same manner, the integration of new technological tools like cell phones, televisions, motorcycles, and scythes has induced a fundamental change in the daily activities of the Arhuaco people. For instance, in earlier times, obtaining guidance from the Mamo on personal and spiritual concerns meant the Arhuaco had to embark on prolonged visits to the Mamo place, often extending into the next day due to transportation limitations. Nevertheless, in the present day, the frequency of visits has significantly diminished as numerous consultations are conducted remotely via cell phone. Moreover, the duration of visits has also shortened due to the ease of mobility provided by motorcycle cabs (Mindiola, 2018).

Even more significant is the Arhuaco people's concern that the excessive use of cell phones might diminish the interest of its members in cultivating the land, engaging in weaving, and interacting with the leaders to acquire wisdom. From the Arhuaco perspective, the constant use of cell phones not only supplanted the space once dedicated to reflection but also diminished the innate capacity for reflection, as there is less need to generate original thoughts when information is already readily available (Mindiola, 2018). While these new economic and recreational alternatives offer the Arhuaco opportunity to integrate into Colombian society's prevailing economic and social dynamics, they also denote a rupture with the traditional Arhuaco lifestyle, which is characterized and structured by their cultural and economic activities.

However, despite the abovementioned circumstances, Arhuaco embraces technology as it has also proven beneficial for gaining insights into the external world and exchanging knowledge. Indeed, including all these external dynamics, which have modified elements previously foundational to Arhuaco's identity, is perceived as a chance to redefine the concept of what it means to be Arhuaco (Mindiola, 2018). This opportunity brings us back to the initial question of this subtitle, suggesting that Arhuaco's identity is a fluid concept constantly shaped by the evolving context, which cannot be confined to a rigid list of traits. Nonetheless, one might argue that Arhuaco's identity increasingly moves beyond external attributes, such as dress and rituals. Instead, there is a stronger preference for acknowledging non-physical elements to define itself, such as the conviction of their fundamental belief: the intrinsic connection between humans and nature.

Ecosophy: The Personal Philosophy in Harmony With Nature

Having delved into the beliefs and imaginaries shaping the Arhuaco's identity and their interactions with habitat, it is pertinent to assess how the environmental principles transmitted through their educational institutions share essential similarities with "ecosophy," the philosophy that advocates a wise relationship with the environment. This analysis explores how Arhuaco's engagement with nature has cultivated a unique, subjective, eco-friendly philosophy of life within the community, which may also be called "Arhuaco Ecosophy" and which can offer valuable insights in addressing the sustainable development objectives advocated by modern environmental movements.

The term "ecosophy," coined by Arne Naess, derives from the fusion of two Greek roots: "eco," signifying "place at home," and "Sophia," denoting "wisdom." Together, they encapsulate the notion of ecological wisdom or wisdom of place (Drengson, Devall, & Schroll, 2011). "Ecosophy" underscores a holistic understanding of environmental systems and promotes harmonious interactions between humans and the natural world. It fosters sustainable practices, reverence for biodiversity, and recognition of the interconnectedness of all living beings (Comins, 2016).

The philosopher Arne Naess introduced the "ecosophy" concept as a derivative of his "deep ecology" movement, which aimed to transcend the commonly held "shallow ecology" or materialistic perspectives that had previously prevailed in addressing environmental problems (Naess, 1990). The "deep ecology" gained prominence in the environmental movements of the late 20th century because of its ideals to challenge outdated conceptions of the human-nature relationship that conceived humans as superior species with the legitimacy to dominate other life forms (Drengson, 1995). Part of the importance of the "deep ecology" movement lies in its principles, which recognize the intrinsic value in all living beings and the inherent value of biological diversity, providing the groundwork for the emergence of subsequent ecocentric environmental currents (Devall & Sessions, 1985).

Naess explains that in the endeavor to cultivate wisdom or "sophia" in our interaction with our natural environment or "eco," the attributes, conditions, and dynamics of that "eco" were indispensable. This suggests that the construction of such wisdom, defined as a set of guidelines directing particular behaviors, is intricately tied to the unique and fluctuating conditions of each individual's environment. (Naess, 1990). For this reason, Naess emphasized that "ecosophy" is not a fully formed and complete philosophy but one to be constructed based on each individual's circumstances. Therefore, countless "ecosophies" can arise, each representing an articulated personal philosophy of life (Drengson, Devall, & Schroll, 2011).

Based on this concept, Naess introduced his personal philosophy called "Ecosophy T" (with T denoting the Tvergasein mountains in Norway, where he lived) to exemplify the formulation of a personal philosophy that harmonizes with the well-being of nature. In defining his philosophy, Naess must first establish his ultimate value or highest norm of life, which he identified as his Self-realization (N). From there, he derived a series of hypotheses regarding the nature of the world (H), which in turn led to the formulation of new norms and hypotheses (Drengson, Devall, & Schroll, 2011). Here is the Ecosophy T as presented by Naess:

"N1: Self-realization!
H1: The higher the Self-realization attained by anyone, the broader and deeper the identification with others.
H2: The higher the level of Self-realization attained by anyone, the more its further increase depends upon the Self-realization of others.
H3: Complete Self-realization of anyone depends on that of all.
N2: Self-realization for all living beings!"
"H4: Diversity of life increases Self-realization potentials.
N3: Diversity of Life!
H5: Complexity of life increases Self-realization potentials.
N4: Complexity!
H6: Life resources of the Earth are limited.
H7: Symbiosis maximizes Self-realization potentials under conditions of limited resources.

N5: Symbiosis!"

(Naess, 1990; Naess, 1992)

Instead of imposing ecological guidelines or a unique ecosophy for everyone, through this illustration, Naess encourages others to establish their own ecosophy by first determining their ultimate value of life (N) and, secondly, their objective or subjective interpretations of their surrounding nature (H). One could suggest that the catalog of ecological norms delineated in Naess's "Ecology T" was only the result of his profound aspiration for Self-realization, which was already articulated with harmonious relations with other life forms. However, it is also conceivable that different ultimate values of life, which may be considered materialistic or self-oriented, could produce results like Naess, as the environment's vitality is indispensable for the success of any endeavor undertaken by the entities that inhabit it. Naess' ecosophy can also be seen as an introspective practice that prompts individuals to inquire about the extent to which their natural or material environment influences their ultimate values in life. Therefore, the symbiotic relationship between the satisfaction of the individual and that of the environment becomes evident (Drengson, Devall, & Schroll, 2011).

To better understand the implications of acknowledging the interconnectedness of all living beings, the concept of "animism" can offer us valuable insights. The original definition of the term "animism," as suggested by Edward Tylor, was constrained to its literal sense to describe the religions that "believe in spirits" (Tylor, 1871). Given this initial interpretation, "animism" was deemed to attribute human-like capabilities to non-human entities or phenomena. Similarly, a secondary common usage of this term has been to describe the beliefs held by indigenous communities, wherein elements of nature are assimilated as ancestors, guardians, intermediary deities, and other spiritual entities (Harvey, 2019). However, a third and last definition emerged by the anthropologist Nurit Bird-David in 1999, who broke with the subordination of personhood to "human" qualities such as mind or spirit to propose that the personhood or intrinsic value of all beings arises from the web of interactions they engage in with other entities (Harvey, 2019).

The latter interpretation of animism is of particular interest, as it prioritizes the interactions between beings as the fundamental basis for assigning values and rights to them. Animism diverges from the notion that humans hold the central or ultimate status regarding personhood or relational significance. Instead, humans are perceived as just one variation among numerous beings entwined within a network of interrelations (Harvey, 2019). While emphasizing the interactions between living beings may appear more complex than focusing on their individual attributes, this new "animism," as well as the mentioned "ecosophy," directs our attention to a fundamental element of environmental ethics: the relationship of interdependence among all living beings.

Likewise, the philosopher Spinoza, recognized for being a proponent of pantheism, described human beings as composite entities comprising lower and higher systems that simultaneously constitute and transcend them (Bula, 2008). Spinoza argued that since bodies are shaped and defined by their interactions with other bodies, these interactions have the capacity to define each body's essence. In this regard, the term "Conatus" was coined by Spinoza to denote the inherent nature or way of being of an individual, characterized by the individual endeavors with itself and with other beings to endure and maintain one's existence (Valera & Vidal, 2022). Within this understanding, we find again how all living beings' individuality and intrinsic value are reshaped by the interdependence dynamics held with others.

In associating Spinoza's notion of "Conatus" with Naess's "Ecosophy T," it becomes evident that external relations can serve as natural means to pursue personal goals of Self-realization. The prospect of attaining our Self-realization through an ecological ethic that also contributes to the realization of nature prompts a shift in the question regarding human responsibility towards the environment, reformulating the inquiry as follows: considering my Self-realization, how ought I engage with my environment? This alternative viewpoint may contribute to advancing the dual objective outlined by Val Plumwood: "The first is to re-situate the human in ecological terms, and the second is to re-situate the nonhuman in ethical terms" (Plumwood, 2002).

Crafting Arhuaco Environmental Principles and Worldview Into an "Arhuaco Ecosophy"

Having reviewed the preceding ecological philosophical currents, let's turn our attention to the purpose of this last section: formulating the Arhuaco ecological principles as their distinct "Arhuaco Ecosophy." Guided by Naess's approach to constructing personal ecosophies (Naess, 1990), the initial step to formulating the "Arhuaco Ecosophy" is to define the ultimate value or purpose in the Arhuaco worldview. As extensively discussed at the outset of this text, the purpose of Arhuaco's life is to serve as guardians of the Sierra Nevada de Santa Marta. In this framework, the Self-realization or ultimate value-norm of the Arhuaco people rests upon their recognition and acceptance as "Young Siblings," mediators between the celestial and terrestrial realms (N1), because only through this recognition do they acquire their function as protectors of nature. After that, the second step for the "Arhuaco Ecosophy" construct is to outline the Arhuaco perceptions and beliefs about the world. As for this, various hypotheses could be suggested, such as the Arhuaco creation myths (H) and the concept that energy is a finite substance that is not created but continuously regenerated within a transformation circuit (H). Based on these three beliefs, a proposal of the "Arhuaco Ecosophy" can be summarized as follows:

N1: Recognition of themselves as "Young Siblings" guardians of nature.
H1: Energy is a finite substance continuously regenerated within a transformation circuit.
H2: All its outputs must be reintegrated with the respective inputs to keep the circuit balanced.
H3: The inherent nature of all beings is to contribute to the balance of the ecosystem in which they belong.
N2: The inherent nature of all humans is to contribute to and safeguard the balance of nature.
H4: Because the world was created in the Sierra Nevada, their mountains embodied the divine ancestors or "Elder Siblings."
H5: The divine ancestors merit both respect and worship.
N3: The mountains and their inhabitants merit both respect and worship.

The outlined framework is merely one example of how Arhuaco's fundamental values, coupled with hypotheses about world creation and energy flows, can evolve into an ecological philosophy that fosters balanced relationships with the environment. Therefore, the "Arhuaco Ecosophy" is a prospectus of philosophy formulated from the distinctive perceptions and experiences of the Arhuaco people with their environment, primarily based on their relationships of respect and interdependence. Naess' proposed scheme for designing our own ecosophy also holds promise for the schools of the Arhuaco people, who currently confront the challenge of imparting their ecological wisdom to new liberal generations. In this sense, by considering their numerous and enriching personal experiences with the environment, Arhuaco students could shape their ecological philosophies authentically, free from the constraints imposed by adult authorities or educational systems.

CONCLUSION

This text has explored various approaches to understanding the environmental principles rooted in the Arhuaco worldview and their integration into educational institutions while, in parallel, examining these principles through the lens of ecosophy. Upon reflection on the ecological and communal approach embraced by the Arhuaco schools, it is evident that they serve as vital pillars for preserving their environmental principles, which we call the "Arhuaco Ecosophy." Therefore, it becomes clearer why the Arhuaco have perceived schools as shells wherein the seeds of their ancestral knowledge germinate, and where, as was proposed in this text, it is possible to nurture the emergence of diverse potential ecosophies among future Arhuaco adults (Castro, 2013).

From the perspective of ecosophy, the Arhuaco educational approach emerges as highly rational for cultivating a personal ecological philosophy. By immersing children in nature and facilitating a range of experiences, from worship and protection to cultivation and construction, schools are nurturing a profound connection with the environment that will be essential in constructing their own harmonious and balanced relationships with the environment. Relying on their personal experiences and not exclusively on the guidance of their professors and elders, students will be able to form their own ecosophy in the years to come. Perhaps, with this meaningful encounter, they could find the determination to navigate the obstacles imposed by technology and modern society, thereby reaffirming their role as guardians of the Sierra Nevada de Santa Marta.

The rising trend of facilitating nature encounters in schools is increasingly approved by academics and scientists who have determined the correlation between individuals' connection with nature and their pro-environmental actions. These studies indicate that experiences fostering a connection with nature, locally or globally, can lead to equal positive impacts (Klaniecki, Leventon, & Abson, 2018). Because more and more researchers express that childhood exposure to nature is strongly and positively linked to pro-environmental behaviors in adulthood (Wells & Lekies, 2006), experts consider that hands-on experiences facilitate a profound connection with the environment and promote practices of ecological awareness.

That's why the approach utilized by Arhuaco schools is profoundly impactful; besides imparting environmental principles, it also empowers children to formulate their own ecosophy grounded in their direct encounters with nature. Because establishing a connection with nature is increasingly recognized as one of the most pivotal strategies for promoting sustainable behaviors, our efforts should be directed toward fostering this reconnection in everyone, especially children. Within this pursuit, may the present and upcoming generations of humanity also find fertile ground, as the Arhuaco schools found, to cultivate future ecosophies harmonious with life on Earth.

REFERENCES

Arias Fuentes, H. J., & Zuleta Cotes, M. K. (2020). *Propuesta de lineamientos para la enseñanza de la filosofía arhuaca en las Instituciones Educativas De Nabusimake Sierra Nevada De Santa Marta* [Doctoral dissertation, Universidad Santo Tomás].

Asamblea Nacional Constituyente. (1991). *Constitución Política de Colombia. Bogotá, D.C.: Recuperado.* ANC. https://www.constitucioncolombia.com/indice.php

Botero, E. U. (2005). *Natural Resource Conservation and Management in the Sierra Nevada of Santa Marta: Case Study. Universidad de los Andes, Facultad de Economía*. CEDE.

Bula, G. (2008, May). Spinoza y el pensamiento ecológico. In *Ética y filosofía política, filosofía de la religión e historia de la filosofía: I Congreso Colombiano de Filosofía.* (p. 395). Universidad Jorge Tadeo Lozano. 10.2307/j.ctv23dxch6.23

Castro, C. M. (2001). Pedagogía indígena: una visión diferente de hacer educación en Colombia. *Nodos y Nudos, 2*(10).

Comins Mingol, I. (2016). La Filosofía del Cuidado de la Tierra como Ecosofía. *Daimon, 67*, 133.

Confederacion Indigena Tairona CIT, O. d. (2015). *Plan de Salvaguarda del Pueblo Arhuaco* . Nabusimake.

de la República de Colombia, C. (1994). Ley 115 del 8 de febrero. Recuperado de http://www.mineducacion.gov.co/1621/articles-85906_archivo_pdf.pdf

Decreto 804 de mayo 18 De 1995, Ministerio del interior. Recuperado de: Departamento Administrativo Nacional de Estadística (DANE). (2018). Censo Nacional de Población y Vivienda 2018.

Drengson, A. (1995). The deep ecology movement. *The Trumpeter, 12*(3).

Drengson, A., Devall, B., & Schroll, M. A. (2011). The deep ecology movement: Origins, development, and future prospects (toward a transpersonal ecosophy). *International Journal of Transpersonal Studies, 30*(1), 11.

Giraldo Jaramillo, N. (2010). Camino en espiral. Territorio sagrado y autoridades tradicionales en la comunidad indígena iku de la Sierra Nevada de Santa Marta, Colombia. *Revista Pueblos y fronteras digital, 5*(9), 180-222.

Gómez, F. F. G. (2013). Learning and adaptation as conservation practices in resilient traditional socio-ecological systems: The Elder Brothers of Sierra Nevada de Santa Marta. *Revista de Tecnología, 12*(1), 99–109.

Guerrero, O. M. B. (2005). *El diálogo de saberes y la educación ambiental* (Vol. 7). Univ. Nacional de Colombia.

Harvey, G. (2019). Animism and ecology: Participating in the world community. *The Ecological Citizen, 3,* 79-84. https://www.mininterior.gov.co/la-institucion/normatividad/decreto-804-de-1995

Izquierdo, D. A. N. (2017). El Pueblo Arhuaco, Defensor del Territorio. *Americanía: Revista de Estudios Latinoamericanos*, 86-102.

Klaniecki, K., Leventon, J., & Abson, D. J. (2018). Human–nature connectedness as a 'treatment'for pro-environmental behavior: Making the case for spatial considerations. *Sustainability Science, 13*(5), 1375–1388. doi:10.1007/s11625-018-0578-x

MAMO. (1997). *Universo arhuaco No. 10.* Antioquia, Medellín, junio.

Mindiola Romo, B. (2018). *El ser Arhuaco: transformaciones en las formas de enseñar-aprender los patrones culturales en la Comunidad de Nabusimake.*

Naess, A. (1990). Ecology, community & lifestyle. London, UK: Cambridge.

Naess, A. (1992). The encouraging richness and diversity of ultimate premises in environmental philosophy. The Trumpeter. *Journal of Ecosophy*, *9*(2), 53–60.

Niño Izquierdo, R. C. (2011). *Organización y uso del territorio por la comunidad Indígena Arhuaca de Nabusímake Sierra Nevada de Santa Marta.*

Petitpierre, F. (1975). The Symbolic Landscape of the Muiscas. *Studies of Comparative Religion*, (Winter), 36–51.

Plumwood, V. (2002). *Environmental Culture: The ecological crisis of reason*. Routledge.

Reichel-Dolmatoff, G. (1977). Cosmología como análisis ecológico: una perspectiva desde la selva pluvial. *Estudios antropológicos*, 355-375.

Reichel-Dolmatoff, G. (1982). Cultural Change and Environmental Awareness: A Case Study of the Sierra Nevada de Santa Marta, Colombia. *Mountain Research and Development*, *2*(3), 289–298. doi:10.2307/3673093

Schuon, F. (2001). *Imágenes del espíritu: shinto, budismo, yoga*. Olañeta.

Sessions, G., & Devall, B. (1985). Deep ecology. Salt Lake City: Peregrine Smith Books.

Tauste, A. M. V. (1992). Función metalingüística y uso del lenguaje. *Epos: Revista de filología*, (8), 123-123.

Tylor, E. B. (1871). *Primitive culture: Researches into the development of mythology, philosophy, religion, art and custom* (Vol. 2). J. Murray.

Valera, L., & Vidal, G. (2022). Pantheism, Panentheism, and Ecosophy: Getting Back to Spinoza? *Zygon*, *57*(3), 545–563. doi:10.1111/zygo.12800

Wells, N. M., & Lekies, K. S. (2006). Nature and the life course: Pathways from childhood nature experiences to adult environmentalism. *Children, Youth and Environments*, *16*(1), 1–24. doi:10.1353/cye.2006.0031

Chapter 10
Expanding Environmental Education in Honduras:
A Holistic Approach for Sustainable Development

Anny Gabriela Molina Ochoa
O.P. Jindal Global University, India

Pranjal Khare
O.P. Jindal Global University, India

Sunidhi Setia
O.P. Jindal Global University, India

ABSTRACT

This chapter delves into the importance of environmental education in Honduras, a country with vast biodiversity facing many environmental challenges. It explains the historical background of environmental education and recognizes the significance of indigenous knowledge and its relationship with the environment. Additionally, it explores existing initiatives towards environmental education. The chapter highlights some difficulties when implementing environmental education, such as limited funding, inadequate infrastructure, and language or cultural barriers. Despite these challenges, it emphasizes the importance of a holistic approach to the adaptation and reformation of the educational system. Finally, the conclusion emphasizes the need for collaborative efforts between government agencies, educational institutions, and community organizations to collectively take part in effectively implementing environmental education initiatives to build a greener and more resilient future for Honduras.

INTRODUCTION

"Every passing car gives him hope. Juan listens carefully to every sound, ready to thumb a lift whenever a car passes by. Sometimes, he is lucky, and he can jump in the back of a pickup. Most of the time, he

DOI: 10.4018/979-8-3693-2577-3.ch010

has to keep walking. Three hours to go and another three to come back. It's a long way for a thirteen-year-old child to walk to school, but it's his way of life." (Cotza, 2013)

Juan's example of resilience and determination highlights how environmental education positively impacts future generations and can empower individuals to take action and create positive change in their communities.

This book chapter delves into the fundamental role of environmental education in developing countries, with particular attention to Honduras. It offers a broad overview of the topic, traces the historical roots of environmental education in the country, and highlights its potential as a catalyst for a sustainable future.

The chapter begins by laying the foundation with a quote that tells the story of Juan, a thirteen-year-old boy whose daily six-hour trip to school personifies the resistance and determination of people who fight for education in countries in which development. This story is a moving reminder of the transformative power of education, particularly environmental education, which empowers people to make positive changes in their communities. In sustainable development, environmental education plays a fundamental role in fostering awareness and understanding of environmental problems and promoting the adoption of environmentally responsible practices and behaviors.

At the community level, environmental education can play a crucial role in developing communities. By providing knowledge and skills related to the environment, environmental education enables people to make informed decisions and actively participate in solving local environmental problems. This includes promoting sustainable practices, such as the efficient use of natural resources, proper waste management, and biodiversity conservation.

Under the title "Understanding the context of environmental education in Honduras," the chapter's analysis of Honduras, a country rich in biodiversity but facing numerous environmental challenges, serves as the primary case study for this exploration. The chapter recognizes the country's pressing environmental problems, such as climate change, deforestation, pollution, and climate-related natural disasters. It underlines the potential of environmental education as a tool to raise awareness and foster a sense of responsibility among people, thus contributing to achieving sustainable development goals.

Following the title "Historical roots of environmental education in Honduras," the chapter also delves into the historical roots of environmental education in Honduras, shedding light on indigenous knowledge of the Mayan civilization and its relevance to contemporary environmental education. Explores the role of modern indigenous communities, such as the Lenca, Miskito, and Garifuna peoples, in promoting biodiversity conservation and sustainable practices.

However, the path towards widespread environmental education is not without difficulties. The "Initiatives/programs towards environmental education" title in this chapter outlines the obstacles that must be addressed, such as lack of funding, shortage of trained professionals, inadequate infrastructure, and cultural barriers. It proposes a holistic approach to overcome these challenges, emphasizing the need for international partnerships, private sector participation, improved infrastructure, and access to new technological resources.

Finally, the title "The Way Forward: Recognizing the Need for a Holistic Approach to Environmental Education" explores the thesis that promoting environmental education in developing countries, especially in Honduras, can serve as a catalyst to envision a sustainable future for future generations. It underscores the importance of integrating indigenous knowledge into general environmental education and highlights the potential of environmental education to promote sustainability and conservation efforts in Honduras. By addressing systemic problems and investing in specific interventions, we can overcome environmental education challenges in the country, fostering a culture of environmental management.

Understanding the Context of Environmental Education in Honduras

Environmental problems in developing countries are not different from those in developed countries; they all suffer from transboundary environmental risks, except for the fact that developing countries do not have the resources to stand against or endure the consequences of environmental degradation, which includes all kinds of pollution, principally air and water pollution, waste and resources management, deforestation, and land degradation (Khan & Naqvi, 2000). Honduras is a Latin American country rich in biodiversity, flora, and fauna. However, given its geographical location, this area is prone to numerous environmental challenges, including climate change, deforestation, pollution, and weather-related natural calamities (Morello, 1997). This chapter will explore fostering environmental education in developing countries, especially in Honduras, as a catalyst for envisioning a sustainable future for future generations.

The need for environmental education in developing countries is related to these nations' different ecological challenges. These countries face numerous environmental challenges, including deforestation, pollution, overconsumption, and climate change. (UN, 2024). These problems have far-reaching consequences for the local ecosystems, human health, and socio-economic stability. It is essential to understand the root causes of these challenges and the role environmental education can play in addressing them.

Raising awareness and fostering a sense of responsibility for developing countries, but most notably for the people and the different societies that live in these areas. Environmental education will serve as a powerful tool to raise awareness about the environmental issues developing countries face. Educating individuals about the impacts of their actions on the environment can foster a sense of responsibility and empower them to make informed choices that contribute to sustainability (UN, 2024).

In 2015, the UN adopted Sustainable development goals as a universal call to the nations to act to protect the environment and achieve the goals by 2030. The 17 SDGs are integrated and framed to achieve social, economic, and environmental sustainability (UNDP, n.d.). Environmental Education plays a vital role in the sustainable development goal, as it equips individuals with the knowledge and skills necessary to understand complex environmental issues and develop sustainable solutions (UN, 2023). Through the incorporation of environmental education within the educational framework, we have the potential to cultivate a cohort of environmentally knowledgeable individuals who possess the capacity to contribute to the advancement of sustainable development actively. Education for Sustainable Development (ESD) is one of the important 2030 Agenda for Sustainable Development that is considered a driver for achieving all 17 SDGs. ESD empowers countries to make informed decisions in favor of environmental integrity, economic viability, and a just society for present and future generations (UNESCO, n.d.).

The plan for integrating environmental education encourages active participation and engagement in environmental conservation efforts. We can instill a sense of ownership and inspire collective action toward a more sustainable future by involving students and community members in hands-on activities, such as tree planting, waste management initiatives, and sustainable agriculture practices.

Developing countries face numerous challenges, including overcoming resource limitations, especially when implementing environmental education programs. Collaboration between government agencies, educational institutions, civil society organizations, and local communities is crucial for effectively implementing environmental education initiatives.

Historical Roots of Environmental Education in Honduras

To understand the context of environmental education in Honduras, this chapter will delve into the historical roots that native civilizations and cultures in this country share among their communities. Like most Latin American cultures, Honduran communities have a rich history and deep connection with the environment, understanding the importance of their indigenous knowledge and how they provide valuable insights into their unique cultural practices and traditions. Their indigenous knowledge has great potential to play an integral role in environmental education (Recio, 2015).

To be able to talk about environmental education in Honduras. Or rather, in the panorama of Latin America, because most countries share similar experiences, we will begin by talking about the legacy of the Mayas as a pre-Columbian civilization (Thompson, 1927). An ancient culture that existed long before the current Lenca community existed. To understand its history, we must start with the geographical location of these civilizations; they were in the different countries that make up the Latin American territory. And to be more precise, in this chapter of the book, we will relate to Honduras as a country that is part of Central America.

It is worth mentioning that one of the oldest legacies, counted as a physical form of the Mayan literature, is called. *Popol Vuh*, from all the ancient communities settled in this region, the Mayan-Quichés of Guatemala have left the most valuable mythological legacy that explains in some way the aboriginal thought that the environment had concerning different spheres of life (Knapp, 1997).

The different ethnicities, communities, or peoples that currently exist in the American continent, to be more precise in Latin America, were not in the current state they are structured socially and culturally; before them, other previous cultures existed in that land, and they were isolated from the rest of the world. This can be followed by the three great civilizations that developed in this geographical space, better known as Mesoamerica, among which we can mention the Incas in Peru and the Aztecs in Mexico. The Mayas cover the regions of Mexico, Guatemala, and Honduras (Bernal, 1994).

By studying the Mayan civilization, we intend to explain how environmental education expanded from a holistic approach. As mentioned above, the Mayans were better known as a civilization that possessed special knowledge about the movements of the stars, thus studying astrology and creating a perfect calendar. Similarly, their knowledge encompassed or expanded to different literary and artistic works, unfortunately, at the time of the conquest of the American territory. The documents pertaining to the culture of indigenous people in these areas were destroyed by the first colonizers or Christian missionaries who were sent. Given through the Spanish crown and forced to conquer the lands. In this way, these cultures were subject to abandoning their old religious beliefs. However, different authors such as Sahagun, Las Casas, Torquemada, and others have preserved some literary works.

According to the historian Bernal Diaz del Castillo, the indigenous people of Mexico, as mentioned before, the Aztecs, had a civilization very similar and related to the Mayas. They had "booklets of tree bark paper called amate, and in these booklets, they left a record of their stories in relation to time or past things." Here, we can see the intrinsic relationship between indigenous communities or cultures and the environment without leaving aside the cultural development and historical legacy.

The *Popol Vuh* is an antique literary and historical work of art based on the legends of the Mayan civilization. It deals with the world's creation, man's origin, the Mayan civilization's origin, and its different groups. Regarding the environment, *Popol Vuh* recounts the creation and growth of trees and vegetation, which precede man. In addition, it describes an ancient world with incredible biodiversity in an imaginary time, suggesting a connection with the natural environment. The text also presents a relationship between man and the Earth, articulated around images or stories from the indigenous world, reflecting an ecological golden age and a possible and desired future society (Knapp, 1997).

In summary, the *Popol Vuh* deals with themes related to the environment, the creation and growth of vegetation, and the relationship between man and the Earth, which makes it a relevant work for understanding the Mayan vision of the natural environment. It is extracted from this literary work *Popol Vuh* the recognition of a deity, which they called *The Creator* and The Modeler, *"the Maker, the Modeler, "mother-father of life, of humankind, giver of breath, giver of heart, bearer, upbringer in the light that lasts of those born in the light, begotten in the light" (*Popol Vuh Tedlock, 72). This abstract from the *Popol Vuh* illustrates how the cultural relationship between religion and the environment was essential to the Mayan civilization. It has always stood out with an intrinsic relationship since the deity that this indigenous group recognized, respected, and admired for being part of what we now consider the environment conformed by the different elements of water, Earth, air, and everything that surrounds us, even including flora and fauna.

The Mayan civilization in Honduras collapsed in 822, just two years after the death of its sixteenth ruler, Yax Pasah. The collapse of the Mayan civilization in the southern lowlands of Honduras and other regions is attributed to various factors, including overpopulation, environmental degradation, warfare, shifting trade routes, and extended drought. After the collapse, the Mayans did not disappear; they established cities in other regions, such as the northern lowlands and the highlands. Descendants of the Maya still live in modern-day Honduras, along with other Central American countries (Zamora, 1978).

Indigenous Knowledge and its Role in Environmental Education in Honduras

From the example of the Mayan civilization, it is proven that societies and communities without proper environmental education suffer from the consequences of environmental degradation, droughts, and, of course, the continuous population growth that leads to food and resource depletion. This scenario is very similar to what is happening worldwide because of climate change and other environmental challenges.

The joint effort being made internationally and at a local level is improving in some way the various impacts and risks of environmental degradation. For example, modern indigenous knowledge includes all the cultures and communities after colonization, especially about Honduras, including the Lenca people that came after the Mayan civilization. Now, in modern times, they are the largest indigenous community in Honduras, making up over 60% of the indigenous population. The aboriginal culture of the Lenca has virtually disappeared and is not well known. However, there are efforts to revive and preserve their cultural heritage. Indigenous movements are attempting to restore the Lenca language, and there have been reports of elementary school textbooks in the original language being distributed to public schools (Metri, 2018). This culture proves how indigenous knowledge about protecting the environment can be transmitted even if the actual culture is disappearing because they are teaching sustainable agriculture to their children and people outside their communities. Their traditional practices and knowledge reflect the Lenca people's connection to nature. They deeply understand the local ecosystems, plants, and animals and have developed sustainable agricultural practices harmonizing with the environment. Traditional herbal therapy is a popular treatment method among the Lenca, highlighting their reliance on natural remedies (Weeks & Black, 1992). The Lenca people have a deep conscience about environmental knowledge, preservation, and activism due to their deep connection to the environment. They have been actively involved in environmental activism and the defense of their ancestral lands. They have faced significant challenges and threats, particularly from the business sector, including mining and energy corporations (Valenzuela, 2020).

Donatila Girón Calix, a leader of the Lenca Indigenous Movement of La Paz Honduras (MILPAH), has emphasized the importance of recovering indigenous identities and practices as central elements in defending Honduras' environment and human rights. Her leadership and activism have focused on the intersection of indigenous knowledge and environmental protection. However, the Lencas are not the only indigenous people from Honduras; several cultures and groups were established in the Honduran territory after Spain's colonization.

Similarly, the Miskitos are using their indigenous knowledge to spread awareness about the importance of biodiversity conservation. The Miskito people are an indigenous group that primarily inhabits the northeastern region of Honduras known as La Moskitia, which runs along the border with Nicaragua and the Caribbean (Helms, 1969). They have a rich cultural heritage and a historical connection to the land. The Miskito people have a strong relationship with their natural environment and possess traditional knowledge and practices closely tied to their agricultural and fishing activities. They rely on sustainable farming techniques and climate-smart agricultural actions to adapt their traditional practices to changing environmental conditions. Their staple crop is cassava; they keep poultry, cattle, and other farm animals (Galeana & Pantoja, 2013).

The Miskito people have been actively involved in environmental activism and the defense of their ancestral lands. They have faced challenges related to deforestation and land ownership. Honduras has one of the highest deforestation rates in Latin America, and the Miskito people have been affected by this loss of forestland. Efforts have been made to grant the Miskito people legal ownership of their homeland. In 2013, the Honduran government gave more than 1.6 million acres of coastal land to the Indigenous Miskito people in La Moskitia. This recognition of their customary rights and ownership has been seen as a step towards a new development agenda in Honduras, guided by Indigenous Peoples (Cultural Survival, 2013).

Projects have been implemented to promote the traditional knowledge of the Miskito people and strengthen their administrative, organizational, and financial capabilities. The goal is to increase the economic benefit from their agricultural and fishing activities while incorporating climate-smart agriculture practices, gender participation, and food nutrition. These initiatives aim to improve the livelihoods of the Miskito people while ensuring the sustainability of their practices and respect for their traditional indigenous knowledge. It is important to note that the Miskito people's traditional ecological practices may vary among different communities and individuals. Additionally, external factors such as globalization and changes in land use patterns have influenced their traditional practices to some extent (World Bank, 2023).

Additionally, the Garifuna people in Honduras are one of two Afro-Honduran communities recognized as distinct ethnic groups within the country. They have preserved their ethnic and cultural identity apart from the mestizo mainstream. (UNHCR, 2018). The Garifuna are the descendants of African-Carib populations from the Caribbean Island of St. Vincent who were exiled to the Honduran coast in the 18th century. Their unique cultural heritage includes language, dance, music, and ancestral knowledge. The Garifuna people in Honduras have a rich cultural heritage and possess traditional ecological knowledge closely tied to their way of life. Here are some aspects of their environmental knowledge and practices: fishing is a primary subsistence strategy for the Garifuna, providing them with food and economic livelihood (Bellamy et al., 2023). However, their fishing practices and reliance on marine resources have been undermined by environmentally conscious policies prioritizing tourism and viewing fishing activities as a threat to protected areas.)

As part of their culture, the Garifuna people collectively understand land ownership and use. They do not adhere to the concept of private property and instead view land rights as collective. However, they have faced challenges related to land dispossession and encroachment by external forces, such as tourism, African Palm cultivation, and drug trafficking. Efforts have been made to secure their ancestral lands and protect their rights. They have also preserved their distinct cultural identity, including their language, customs, beliefs, and rituals. Their cultural practices are intertwined with their relationship with the environment and reflect their deep connection to the coastal regions where they reside. As a tradition, they practice sustainable agriculture, including cultivating crops like cassava. They have developed techniques adapted to their local environment, promoting long-term soil fertility and productivity. The Garifuna people's practices are related to a food security policy and food sovereignty (MacNeill, 2020).

These Honduran indigenous communities exemplify how the intergenerational equity principle from sustainable development works. All the ethnic groups mentioned before have a deep connection with their culture and ancestral knowledge, which is related to their environmental knowledge and transmitted from generation to generation. These are great examples of informal environmental education in Honduras, learning that occurs outside the traditional classroom setting rather than is more like the knowledge and education gained through various daily life experiences, work, family, and culture. The importance of indigenous knowledge in environmental education and its role in fostering a sustainable future is noticeable. Indigenous communities like the Lenca, Miskito, and Garifuna have developed sustainable practices that harmonize with the environment. These practices, passed down through generations, are examples of informal environmental education. By integrating this indigenous knowledge into mainstream education, we can enrich our understanding of the environment and promote sustainable practices on a broader scale.

Highlighting the Importance of Preserving and Integrating Indigenous Knowledge Into Mainstream Environmental Education

Indigenous knowledge plays a vital role in preserving indigenous communities' heritage. Still, it is necessary to highlight the importance of maintaining and integrating indigenous knowledge into mainstream and formal education, given that most indigenous knowledge is passed down between the different cultures and communities in an informal way of expertise by instructions given orally and the various daily activities. It can be seen how these cultures want to preserve ancestral knowledge within their communities; in this way, it has been protected and maintained all this time. However, it is essential to highlight the distrust on the indigenous peoples' part and the long adaptation process that needs to be implemented to integrate these cultures and their knowledge into Honduran society.

With the concern of preserving indigenous knowledge and moving from distrust to trust, it is crucial to acknowledge and respect the rights of indigenous communities. This includes their right to their land, their cultural heritage, and their traditional knowledge. Building trust requires open dialogue, mutual respect, and recognition of indigenous rights. It also involves addressing these communities' challenges, such as land dispossession and the impact of external forces like tourism and commercial activities. By working collaboratively with indigenous communities and respecting their rights and knowledge, we can foster a relationship of trust and mutual learning. It is essential to emphasize the shared goal of environmental conservation and the mutual understanding that can occur between indigenous communities and the broader society.

It is relevant to point out that access to education is one of the issues that have made this integration of indigenous people and their knowledge hard into the formal educational system in Honduras. (Gavin, 2017) According to Coombs and Ahmed (1974), informal education is *"the lifelong process by which every person acquires and accumulates knowledge, skills, attitudes, and insights from daily experiences and exposure to the environment"* (La Belle, 1982) on the other hand, formal education refers to structured and organized learning within educational institutions such as schools, colleges, and universities. It follows a set curriculum and is typically led by qualified teachers or instructors.

In Honduras, environmental education is not essential to the education system. Still, it should be implemented across schools and colleges to inculcate environmental awareness, promote responsible behavior, and foster individuals to make positive environmental contributions. In most schools, anything related to the environment is included under the extracurricular activities, in which institutions organize different activities like environmental clubs, eco-brigades, and community guidance, but this is limited to NGOs and private institutions or schools because since it is not mandatory in the curriculum, it requires more interest. In addition to the economic factor, organizations are needed locally to sustain and support environmental and conservation activities, which limits even more environmental education and awareness.

Initiatives/Programs Towards Environmental Education

Environmental education can embed sustainable practices and environmental conservation in individual behavior. Environmental education helps alter an individual's behavior and inculcates environmental responsibility, crucial in promoting sustainable environmental practices. Honduras, a country rich in biodiversity but facing environmental challenges, has recognized the importance of integrating environmental education into its national agenda to deal with these challenges. Various environmental educational programs and initiatives are implemented at national, regional, and local levels to raise awareness of environmental issues. At the national level, the Honduran Ministry of Environment outlines the goals and objectives of environmental education. For implementing environmental programs at the regional level, the Honduran Environmental Protection Agency has a network of regional offices that work with communities to promote sustainable practices. There are government-led programs, which include the National Environmental Education Program, the Citizen Monitoring of Water Quality Project, and the Community Forest Management Program.

Furthermore, at the local level, there are several non-governmental organizations (NGOs) working toward environmental education, which includes the Honduran Foundation for Environment, the Foundation for the Protection of Lancansan, the Association of Friends of Pico Bonito National Park, and the Tropical Agricultural Research and Higher Education Centre. The chapter's principal aim is to expand and enhance environmental education in Honduras towards the common objective of sustainable development. It highlights the advancements made in the country so far and discusses the significant initiatives taken by government and non-government organizations to integrate and promote environmental education.

Recognizing the importance of early exposure to environmental concepts, the Secretaría de Educación de Honduras (Ministry of Education of Honduras) has taken significant steps to integrate the environmental curriculum into the national education system across all grade levels. This has mandated the inclusion of environmental education in subjects like biology, geography, and civics, incorporating modules on topics such as biodiversity conservation, climate change, and sustainable development. As per the United

Nations Educational, Scientific and Cultural Organization (UNESCO) report, such incorporation helps students connect with economic and social factors by developing an encompassing understanding of environment-related issues (UNESCO, 2019). One such example is the schools of District Lancaster, where science teachers have integrated environmental literacy into their new curriculum. The project aims to nurture native plants and attract wildlife to school playgrounds, helping students learn by practice (LancasterOnline, 2024). With such mainstreaming environmental education in schools' curriculums, Honduras will have a generation of environmentally conscious citizens equipped with the skills and knowledge to address critical environmental challenges.

Well-trained educators are the backbone of any education system. Realizing the importance of trained educators, the government of Honduras has invested in teacher training programs for the effective implementation of environmental education in schools. To equip teachers with the skills and required knowledge to engage with children and provide informative environmental lessons, the Environmental Education Directorate within the Ministry of Education collaborates with national/international academic institutions and NGOs to organize training sessions, workshops, and seminars for teachers. These sessions cover various topics, including environmental science, pedagogy, and curriculum integration techniques. These programs provide educators with the necessary resources, pedagogical tools, and professional development opportunities to effectively teach environmental concepts. One such example is the '*Honduras Teachers Workshop*' conducted by Pure Water for the World (PWW) in rural schools in Honduras. The workshop was conducted in 10 rural schools of Trojes to sensitize the students about environmental hygiene and help teachers develop a continuous teaching plan that includes environmental hygiene in the standard class curriculum (Pure Water for the World, 2016).

Another such collaboration is the Global Environmental Education Partnership (GEEP), initially launched in April 2014 as a collaborative effort between US & Taiwan. Over 15 countries, including Honduras, have now joined the partnership. Through demand-driven and result-oriented programs, GEEP provides a platform for countries to share knowledge regarding global environmental education (International Environmental Partnership, n.d.). To strengthen the environmental literacy and education standards, the member countries implement and share the experiences from the programs from their home countries.

In addition to formal training and education settings, the government of Honduras has established environmental education centers across the country. These centers are hubs for experiential learning, community engagement, and environmental advocacy. One notable example is the community reforestation project by Tree, Water, and People (TWP) and Asociación Hondureña para El Desarrollo (AHDESA) at the National University of Agriculture (UNA) in Olancho, Honduras (Claudia, 2019). This UNA worked with the local communities and offered interactive exhibits, nature trails, and educational programs designed for ten elementary schools in the area. Through learning by participation method, the project focused on raising awareness among students about local ecosystems, deforestation, and forest degradation. The program taught students exercises like land mapping, explained concepts like topography and hydrology, etc., and helped them learn the solution by planting trees.

Further, the Honduran government conducts community outreach and awareness campaigns to reach a wider audience and promote environmental awareness at the grassroots level. These initiatives involve collaborating with local communities, NGOs, and civil society organizations to deliver targeted environmental conservation and sustainability messages. The Ministry of Environment and Natural Resources, in partnership with the Honduran Ecological Movement, organizes environmental fairs and community workshops in rural and urban areas. These events provide platforms for knowledge sharing, skill-building, and dialogue on waste management, reforestation, and water conservation. Honduras seeks to empower individuals to take collective action toward environmental stewardship by engaging directly with communities. One such example is Communities Organizing for Watersheds, Honduras. It is a women-led community participation program to help adapt communities to the changing climate in areas threatened by water shortage (United Nations Climate Change, 2023).

The government of Honduras, recognizing the integral linkage between indigenous knowledge and environmental conservation, has tried to integrate traditional ecological knowledge into its environmental education initiatives. Indigenous communities possess valuable insights into local ecosystems and sustainable practices and are invited to participate in curriculum development, teacher training, and environmental governance processes. For instance, the Lenca Indigenous Council collaborates with the Ministry of Culture, Arts, and Sports to develop educational materials incorporating traditional ecological knowledge into school curricula. Honduras honors cultural diversity by incorporating indigenous perspectives while enriching environmental education with age-old wisdom. Another example is the Saving the Planet initiative by the Inter-American Development Bank. It's an education project on climate change for the indigenous and Afro-Honduran communities to support the teaching of climate change and include the same in the formal primary education curriculum (Inter-American Development Bank, n.d.).

The aim of a sustainable environment can be achieved with the collaborative efforts of the government and other organizations in any nation. Recognizing the efforts of various national organizations, this chapter highlights the examples and efforts of multiple NGOS in promoting a holistic environmental education approach in the country. Mesoamerican Permaculture Institute (IMAP), founded in 2000, is a regional organization that works in Honduras to promote permaculture principles and practices. It aims to promote and work for biodiversity protection, food sovereignty, and agroecological education. (Mesoamerican Permaculture Institute, (n.d.). They offer workshops, courses, and demonstration projects to educate communities about sustainable land use, organic gardening, and ecological design. The organization educates people on how to share and preserve the knowledge of their ancestors. Asociación de Jóvenes en Movimiento por la Naturaleza (AJUN), founded in March 2005, is a youth-led organization that advocates for environmental conservation and climate action. They organize educational campaigns, reforestation projects, and youth-led initiatives to engage young people in environmental stewardship (Association of Young People United for Nature, n.d.). Un-Mundo is another grassroots, well-established organization that works with local communities to promote environmental protection. It organizes competitions, workshops, and community-led programs to share the best practices and uplift the sustainable environment (Lit World, 2019). Asociación Ecológica De San Marcos De Ocotepeque (AESMO) is a community of people established in 1990 with an aim to find creative solutions for the environment-related issues faced by Honduras's municipality San Marcos Ocotepeque (World Land Trust, n.d.). It works with the World Land Trust (WLT), collaborating with other national and international actors to develop shared governance, conservation, and sustainable development processes. These non-governmental initiatives complement government efforts toward environmental education in Honduras, reach diverse communities, and foster grassroots participation in conservation efforts.

In conclusion, the Honduran government has implemented a range of initiatives to advance environmental education and promote sustainability. From integrating environmental curriculum in schools to fostering community engagement and incorporating indigenous knowledge, these efforts reflect a commitment to building a more environmentally literate and resilient society. Continuing investment in environmental education is essential for addressing pressing environmental challenges and fostering a culture of environmental stewardship in Honduras.

Challenges of Environmental Education in Honduras

Despite the progressive measures the Honduran government took to expand and develop environmental education in this country, its implementation faces several challenges that hinder its effectiveness and impact. Therefore, recognizing the need for a holistic approach to environmental education cannot be dispensed with. This may require international partnerships, private sector engagement, improved infrastructure, access to new technological resources, etc. This part of the chapter explores and analyzes the critical challenges faced by the environmental education system or programs launched in Honduras. It points out the plausible social, ecological, and cultural reasons hindering the proper functioning of the growth of the respective government initiatives to provide for an integrated environmental education approach. There are areas for improvement, which include limited funding, scarcity of trained professionals in environmental education, limited access to technology and resources, inadequate infrastructure, language and cultural barriers, etc.

As a developing country, Honduras faces constraints in allocating sufficient financial resources to education, including environmental education. According to a report by the World Bank, Honduras has historically given a relatively low percentage of its GDP to education, limiting funding available for environmental education initiatives (World Bank, 2019). This lack of financial resources hampers the development of comprehensive programs, provision of adequate materials, and training for educators. This ultimately pulls the environmental education-related programs down and is a significant challenge for developing nations like Honduran. Further, rural areas in Honduras often lack basic educational infrastructure, including schools and educational facilities. This lack of infrastructure, compounded by poor transportation networks, hinders many communities' access to formal environmental education. A study published in the International Journal of Educational Development highlights the disparities in educational access between urban and rural areas in Honduras, with rural populations facing more significant challenges in accessing quality education (Carr-Hill et al., 2016). Such limited access to environmental education exacerbates disparities in environmental awareness and participation in conservation efforts.

Along with resource scarcity, another major challenge faced by the programs supporting environmental education is course integration. While there have been efforts to integrate environmental education into the national curriculum, implementation remains inconsistent. Challenges include educators' resistance, competing curriculum priorities, and a lack of standardized guidelines for incorporating environmental themes into different subjects. A study published in Environmental Education Research emphasizes the need for comprehensive policy frameworks and teacher training programs to support effective curriculum integration (Gough & Scott, 2016). Educators may struggle integrating environmental education into their teaching practices without clear guidance and support.

Further, many teachers in Honduras lack the training and resources to teach environmental concepts effectively. While there are initiatives to provide professional development opportunities, such as workshops and training programs, these efforts are often limited in scope and reach. A United Nations Development Program (UNDP) report highlights the importance of investing in teacher training and capacity building to strengthen environmental education (UNDP, 2020). However, ongoing challenges, such as high teacher turnover rates and a shortage of qualified environmental educators, hinder efforts to build and sustain capacity in this area. Honduras, a multicultural and multilingual country with diverse indigenous communities speaking various languages and dialects, faces language barriers and cultural differences that hinder the delivery of environmental education initiatives, particularly in reaching indigenous populations with culturally relevant content and approaches. A study published in the

International Journal of Environmental & Science Education emphasizes the importance of culturally responsive pedagogy in environmental education to engage diverse learners (Monroe & Watkins, 2017). Environmental education initiatives may fail to resonate with local communities without addressing language and cultural barriers.

In conclusion, environmental education in Honduras faces numerous challenges that hinder its effectiveness in promoting sustainability and conservation efforts. These challenges include limited financial resources, inadequate infrastructure and access, curriculum integration challenges, teacher training and capacity-building needs, and language and cultural barriers. Addressing these challenges requires concerted efforts from government agencies, civil society organizations, educators, and community stakeholders. By addressing systemic issues and investing in targeted interventions, Honduras can overcome these challenges and strengthen environmental education to build a more sustainable future.

THE WAY FORWARD: RECOGNIZING THE NEED FOR A HOLISTIC APPROACH TO ENVIRONMENTAL EDUCATION

Environment education of children is said to be one of the essential components of conservation efforts worldwide. It is important to note that to preserve and promote sustainable environmental development, there is a need to inculcate knowledge about environmental issues and behavioral and social sensitivity. Thus, children's environmental education is based on the idea of intergenerational knowledge and value transfer for future generations. However, for a holistic approach, it is essential to focus on the environmental education of adults who have the power to act now. After discussing in detail the projects initiated and challenges in the Honduran environmental education system, this chapter highlights suggestions and a way forward to achieve a wholesome and integrated environmental education system.

Judy Braus and David Wood emphasize the importance of environmental education in schools in their book "Environmental Education in the Schools: Creating a Program that Works! Volume 2" and provide insights into creating effective programs. They discuss the significance of integrating environmental education into the curriculum, engaging students in practical environmental actions, and fostering a sense of responsibility towards the environment.

Analyzing the social, economic, political, and cultural aspects of Honduran, the authors suggest increasing youth participation in environmental development programs and making the projects based on the country's needs. It is pertinent to note that the existing infrastructure and resource capability should thoroughly access any country's education model. Thus, the authors suggest having a tailor-made environmental education system to promote environmental literacy in Honduran and not just applying any successful education model from the developed nations. For example, a considerable portion of the teacher population in rural Honduran has no more than a high school education, the dropout rates of the students increase with the age of children, and the student-faculty ratio is also significantly high (Ham & Castillo, 1990). These factors are critical in formulating any policy for its successful implementation and functioning. Thus, understanding the disparities and highlighting the importance of understanding the local educational context requires adapting educational materials to the local context and providing insights into the challenges and potential solutions for effectively introducing environmental education in rural Honduran schools.

Further, the chapter proposes an Education for Sustainable Development (ESD) in higher education. It offers ESD in teaching learning and assessment approaches to have a holistic framework to equip learners with values and skills that help them sustain a vision to promote economic prosperity, environmental integrity, and social justice (Quality Assurance Agency, n.d.). Keeping in mind the European Union's Recommendation on learning for the green transition and sustainable development, the authors emphasize the need for a transformative and interdisciplinary approach to learning for the green transition, focusing on cognitive, practical, and socio-emotional skills (Education & Training Foundation, n.d.). As a developing nation, the Honduran can also seek support from UNESCO in introducing and expanding educational activities focusing on sustainability issues such as climate change and biodiversity. One of the essential factors for successful environmental education to focus on along with the learner and educator is the parent, who largely mold a child's behavioral aspects and value system. It is suggested that the teaching-learning pedagogies be drafted to involve and educate the parents for a further wholesome value integration for social sensitivity towards environmental challenges. It is further proposed to have action-based learning with the inclusion of parental activities in the curriculum of natural science classes or new subjects in connection to ecological literacy. Thus, the involvement of educators, policymakers, students, and parents is vital in successfully implementing these recommendations and fostering a culture of sustainability within educational institutions (Jelavic, 2014).

Lessons can be learned from environmental education programs and initiatives, such as Eco-Schools, one of the most extensive global sustainable school programs. It is designed to be pupil-led and aims to engage the next generation in action-based learning, starting in the classroom and expanding to the community. It focuses on integrating environmental work with the curriculum, helping schools align with the principles of effective environmental education, and involving the communities. Additionally, several other programs empowering young people to be environmentally sensitive are offered by the Foundation for Environmental Education, which provides several educational programs, including Eco-Schools, LEAF, and Young Reporters for the Environment. These programs emphasize the importance of hands-on experiences, outdoor learning, and advocating for sustainable business practices.

Furthermore, the teaching pedagogues and curriculum drafting can be done by learning through the teacher training programs and strategies evolved by organizations like SEED and Better Planet Schools that actively support schools and engage students in environmental education and sustainability initiatives. These projects emphasize the need for collaborative efforts by young and adult change-makers to work together and build a resilient and sustainable world.

Having discussed different successful programs nationally and internationally, the authors suggest that the following critical aspects of environmental education in schools can be focused on while framing such programs or projects in Honduras.

1. There is a need to focus on Whole-School Policy and Staff Motivation. This will help establish a comprehensive environmental education policy at the school level and motivate staff to integrate environmental education into the curriculum organically.
2. To have the collaborative efforts of all the change actors, there is a need to include after-School Program Integration. This will provide insight into integrating environmental education into after-school programs, including parents' and educators' high interest levels and motivations, such as fostering an appreciation for nature and addressing environmental problems.
3. There is a need to integrate programs that encourage young people to engage in their environment by allowing them the opportunity to protect it actively, starting in the classroom and expanding to the school and eventually fostering change in the community at large.

4. While drafting the frameworks for environmental education, the Guidelines for Excellence by The North American Association for Environmental Education (N.A.A.E.E.) can be considered for framing a successfully evaluated project. These guidelines provide a framework for developing, implementing, and evaluating environmental education programs, emphasizing key characteristics such as gathering information, designing instruction, and delivering and evaluating programs.
5. There is also a need to focus on pupil-led environmental actions like Eco-Schools programs, which are pupil-led initiatives that empower students to take impactful environmental actions and address climate change, showcasing the potential for student involvement in environmental education. This can be achieved by incorporating environmental education into classroom activities, such as reading and writing, and using the environment as a laboratory for science and math experiments.
6. Lastly, the authors suggest evaluating environmental education programs to determine their effectiveness and efficiency and using evaluation to improve the program.

Concludingly, while suggesting a holistic approach as a way forward, the chapter intends to pave the way for an environmental education system that needs to move students beyond awareness to environmentally responsible behavior. This continuum begins with environmental awareness and knowledge and ends with students becoming actively dedicated to improving and maintaining environmental quality. The chapter stresses the need to understand that environmental education is not tied solely to the science curriculum but cuts across all subject areas, including business, economics, language arts, history, social studies, and the humanities.

These suggestions align with the principles of effective environmental education. They can help the schools in Honduras and the educators to create impactful programs that empower students to become environmentally conscious individuals and thus promote and achieve sustainable education goals.

REFERENCES

Bellamy, K., Garcia, V., Ahmed, S., & Archila, A. (2023). Ethnic Difference at the Center of Land Struggles in the Americas: A Complex History of Marginalization and Multidimensional Challenges among the Garifuna in the Northern Honduras. *Journal of Poverty, 27*(6-7), 474-494. https://www.tandfonline.com/doi/full/10.1080/10875549.2023.2173702

Bernal, I. (1994). Formación y desarrollo de Mesoamérica. In D. Cosío Villegas (Ed.), Historia general de México: volumen I (4th, reimpresión ed., pp. 125–164). El Colegio de Mexico. doi:10.2307/j.ctv47w8sq.6

Carr-Hill, R. (2016). The rural-urban gap in education in Honduras. *International Journal of Educational Development*, *46*, 52–61.

Claudia, M. (2012, April 9). *Notes from the Field: Environmental Education in Honduras.* Sustainable Development, Sustainable Livelihoods. https://treeswaterpeople.wordpress.com/2012/04/09/notes-from-the-field-environmental-education-in-honduras/

Cotza, L. (2013, June 10). Improving education standards in Honduras: A long road ahead. *The Guardian.* https://www.theguardian.com/global-development-professionals-network/2013/jun/06/education-standards-honduras

Cultural Survival. (2013). Campaign Update: Honduras Gives Title to Lands to Miskito People. Cultural Survival. https://www.culturalsurvival.org/news/campaign-update-honduras-gives-title-lands-miskito-people

Education & Training Foundation. (n.d.). *Education for sustainable development (ESD)*. https://www.et-foundation.co.uk/resources/esd/

Galeana, F. & Pantoja, E (2013). *Strengthening Indigenous Peoples' Land Tenure Governance: the Case of the Miskito People in Honduras*. Academia. https://www.academia.edu/4314700/Strengthening_Indigenous_Peoples_Land_Tenure_Governance_the_Case_of_the_Miskito_People_in_Honduras

Gavin, M. (2017). *Accountability: Education in Honduras.* UNESCO. https://unesdoc.unesco.org/ark:/48223/pf0000259546

Gough, A., & Scott, W. (2016). Environmental education policy research: A systematic review. *Environmental Education Research, 22*(6), 800–813.

Ham, H. S., & Catillo, L. (1990). Elementary Schools in Rural Honduras Problems in Exporting Environmental Education Models &om the United States. *The Journal of Environmental Education, 21*(4), 27–32. doi:10.1080/00958964.1990.9941935

Helms, M. W. (1969). The Cultural Ecology of a Colonial Tribe. *Ethnology, 8*(1), 76–84. doi:10.2307/3772938

Inter-American Development Bank. (n.d.). *Save the Planet Initiative: An Education Pilot Project on Climate Change for Indigenous and Afro-Honduran communities*. IADB. https://www.iadb.org/en/whats-our-impact/HO-T1245

International Environmental Partnership. (n.d.). *Global Environmental Education Partnership*. IEP. https://www.iep-global.org/programs/global-environmental-education-partnership-geep/

Jelavic, H. M. E. (2014). *Education for sustainable development: assessing the development impact of environmental education (EE) on Roatan, Honduras*. Universitet i Agder / University of Agder. http://hdl.handle.net/11250/222386

Khan, S. R., & Naqvi, A. (2000). Root Causes of Environmental Degradation. In *The Environment-Poverty Nexus: An Institutional Analysis* (pp. 1–7). Sustainable Development Policy Institute., https://www.jstor.org/stable/resrep00631.4

Knapp, B. L. (1997). The Popol Vuh: Primordial Mother Participates in the Creation. *Confluencia, 12*(2), 31–48. https://www.jstor.org/stable/27922440

La Belle, T. J. (1982). Formal, Nonformal and Informal Education: A Holistic Perspective on Lifelong Learning. *International Review of Education / Internationale Zeitschrift Für Erziehungswissenschaft / Revue Internationale de l'Education, 28*(2), 159–175. https://www.jstor.org/stable/3443930

LancasterOnline. [LNP]. (2024, February 2). *This week's good things: Environmental education takes root, helping hands in Honduras, and F&M College volunteerism.* Lancaster Online. https://lancasteronline.com/opinion/editorials/this-weeks-good-things-environmental-education-takes-root-helping-hands-in-honduras-and-f-m/article_d81ef58c-c15c-11ee-9da0-7be190c33228.html

LitWorld. (2019, May 23). *Growing Environmental Awareness in Honduras*. LitWorld. https://www.litworld.org/blog/2019/5/23/un-mundo

MacNeill, T. (2020). Indigenous food sovereignty in a captured state: The Garifuna in Honduras. *Third World Quarterly*, *41*(9), 1537–1555. doi:10.1080/01436597.2020.1768840

Mesoamerican Permaculture Institute. (n.d.). *Sowing the seeds helping it grow*. Mesoamerican Permaculture Institute. https://imapermaculture.org/

Metri, N. (2018). *Internet Society: How the Lenca Are Restoring the Past to Build Their Future*. Internet Society. https://www.internetsociety.org/blog/2018/04/lenca-people-restoring-past-build-future/

Monroe, M. C., & Watkins, K. B. (2017). Culturally responsive environmental education: An indigenous perspective. *International Journal of Environmental and Science Education*, *12*(3), 429–444.

Morello, D. (1997). *Honduras and Deforestation*. American. http://www1.american.edu/TED/honduras.html

Pure Water for the World. (2016, May). *Honduras teacher Workshop*. Pure Water for the World. https://www.purewaterfortheworld.org/wpcontent/uploads/2015/10/PWW_Honduras_HygieneTeacherTraining-v2.pdf

Quality Assurance Agency. (n.d.). *Education for Sustainable Development*. QAA. https://www.qaa.ac.uk/the-quality-code/education-for-sustainable-development

Recio, M. E. (2015). *Honduras Resource Tenure and Sustainable Landscapes Assessment*. USAID Tenure and Global Climate Change Program. https://www.land-links.org/wpcontent/uploads/2016/09/USAID_Land_Tenure_TGCC_Honduras_Assessment_En.pdf

The World Bank. (2023). *Betting on sustainability in the Moskitia, in Honduras, to create new livelihoods*. The World Bank. https://www.worldbank.org/en/news/feature/2023/04/24/betting-on-sustainability-in-the-moskitia-honduras-to-create-new-livelihoods

Thompson, J. E. S. (1927). The Civilization of the Mayas. *Anthropology Leaflet, 25*, 1–96. https://www.jstor.org/stable/41444125

UNDP. (2020). *Environmental Education for Sustainable Development in Honduras: Challenges and Opportunities*. United Nations Development Programme.

UNDP. (n.d.). *The SDGs in Action*. UNDP. https://www.undp.org/sustainable-development-goals

UNESCO. (2019). *Education for Sustainable Development in Latin America and the Caribbean: A Stocktaking Report on the Decade of Education for Sustainable Development (2005-2014)*. UNESCO Regional Bureau for Education in Latin America and the Caribbean.

UNESCO. (n.d.). *Sustainable Development Goals - Resources for educators*. UNESCO. https://en.unesco.org/themes/education/sdgs/material#:~:text=Its%20aims%20form%20one%20of,for%20present%20and%20future%20generations

UNHCR. (2018). *World Directory of Minorities and Indigenous Peoples*. Minority Rights Group International. https://www.refworld.org/reference/countryrep/mrgi/2018/en/65136

United Nations. (2023). *Communities Organizing for Watersheds – Honduras.* UN. https://unfccc.int/climate-action/momentum-for-change/activity-database/communities-organizing-for-watersheds

United Nations. (2023). *Global Sustainable Development Report 2023: Times of crisis, times of change: Science for accelerating transformations to sustainable development.* UN. https://sdgs.un.org/gsdr/gsdr2023

United Nations. (2024). *Climate Action, Education is key to addressing climate change.* UN. https://www.un.org/en/climatechange/climate-solutions/education-key-addressing-climate-change

United Nations. (2024). *Department of Economic and Social Affairs, Population growth, environmental degradation and climate change.* UN. https://www.un.org/en/desa/population-growth-environmental-degradation-and-climate-change

Valenzuela Pérez, L. (2020). *A Lenca Voice of Resistance Against a Culture of Extractivism: Donatila Girón Calix. Cultural Survival Quarterly.* https://www.culturalsurvival.org/publications/cultural-survival-quarterly/lenca-voice-resistance-against-culture-extractivism

Weeks, J. M., & Black, N. J. (1992). Notes on the Ethnopharmacology of the Lenca Indians of Western Honduras and Eastern El Salvador. *Mexico, 14*(4), 71–74. https://www.jstor.org/stable/23758612

World Bank. (2019). *Education Sector Public Expenditure Review for Honduras.* World Bank Group.

World Land Trust. (n.d.). *Asociación ecológica de san marcos de ocotepeque (aesmo).* World Land Trust. https://www.worldlandtrust.org/who-we-are-2/partners/aesmo/

Zamora, O. M. F. (1978). EL COLAPSO MAYA. *Anuario de Estudios Centroamericanos*, *4*, 489–507. https://www.jstor.org/stable/25661662

KEY TERMS DEFINITIONS

Action-Based Learning: The methodology related to optimizing the learning process and creating an environment that fosters healthy and effective learning outcomes.

Ecological Literacy: Based on the ecocentric perspective, it is the knowledge and application of ecological principles to promote sustainable development.

Environmental Education: Educational process related to protecting the environment, understanding the issues from an ecocentric perspective, and working on restoring and preserving ecosystems.

Global Environmental Education Partnership: Also known and referred to as GEEP, this partnership focuses on integrating environmental education initiatives and promotes ecological literacy.

Honduran Education Policy: This policy is defined by the Ministry of Education of Honduras and is focused on improving its curriculum so it can be according to the guidelines established by the United Nations and the Sustainable Development Goals.

Indigenous Knowledge: Refers to the teachings, practices, and wisdom that belong to indigenous people and that are passed down from generation to generation. It is focused on understanding humans and their relationship with the environment, their way of living, and preserving their culture.

Sustainable Development: The goal established by the United Nations focuses on finding a balance between development and the preservation of the environment; it emphasizes meeting the needs of the present generation without compromising future generations.

Chapter 11
Forgotten Realms:
Mainstreaming Indigenous Knowledge Into Environmental Education

Sunidhi Setia
O.P. Jindal Global University, India

Darshna Gupta
Mahindra University, India

Samanvi Narang
O.P. Jindal Global University, India

ABSTRACT

The chapter argues that human life is fundamentally attached to nature and that the latter's language plays a crucial role in defining our identity. Indigenous communities, in particular, have deep knowledge of their environment rooted in their traditions and cultural identity. They have a reciprocal relationship with nature, operating in tandem with it and never otherwise. This knowledge reflects a delicate balance required to maintain life on this planet. Thus, the chapter questions the current education system, which primarily focuses on scientific principles, solutions, and figures, and highlights how indigenous knowledge can contribute to our environmental education. Lastly, it proposes to further open the horizon to new research and adaptation in environmental education at various levels, scholarly accounts, and governmental policies.

INTRODUCTION

Finding the Realms

The crucial feature of human life is its fundamentally dialogical character. We became full human agents, capable of understanding ourselves, and hence of defining our identity, through our acquisition of rich

DOI: 10.4018/979-8-3693-2577-3.ch011

human languages of expression—where the latter include, in addition to spoken and written language, also the "languages" of art, of gesture, of love, and the like." (Fred Dallmayr 2002, 60).

One understandable perspective in these choices of words by Fred Dallmayr points towards the continuous efforts of humans to collect and preserve knowledge irrespective of the places or communities to which they belonged. The mode in which this knowledge has been collected is varied and not necessarily written, but collectively, it can be referred to as science in various formats.

Most of this science focuses on the interactions between human beings and nature and the former's observations about the latter. Though the contribution has come from all spectrums, when we intend to engage with our ecology, the knowledge of indigenous communities has captured the attention of scientists since times immemorial. Indigenous communities refers to networks of individuals who have undergone colonization of their territories and cultures, and whose cultural identities have been influenced and transformed by boundary violations. For indigenous communities, their connection to the land is a crucial element of their identity, distinguishing them from those who colonize the land and hinder them from living in accordance with their own traditions (Aikman, 1999). The idea encompasses the teaching and learning of traditions by elders, parents, and neighbors, which emphasizes the importance of staying rooted and enhancing the knowledge and skills required to sustain and be sustained by their own communities. Buried in ancient wisdom indigenous knowledge is an integrated system that reflects deep understanding of nature and biodiversity required to sustain the balance of the ecosystem.

Their approach highlights the deep respect these communities have always held for the biodiversity around them and have operated in tandem with it and never otherwise. These communities, with their intricate understanding of the relationship between human beings and the environment based on their centuries-old wisdom, observations, and experiences, reflect the delicate balance required to maintain life on this planet.

Distinct techniques have been developed by indigenous communities to encode valuable data within frameworks of thoughts that are associated with specific landscapes. This data comprises geographical, genealogical, biological, and other types of evidence that chart human connections to flora and fauna, land and water.

Most recently, the acclaimed Netflix series "Kaala Pani" abundantly highlights this simple fact about indigenous communities. The plot revolves around an epidemic that has erupted on an island and how the entire population is grappling with it, except the indigenous community that resides on the island. The latter are somehow immune to the virus, while the administration deals with a record number of deaths amongst the civilian population. It is eventually shown that the deep ecological knowledge of the community about the island has helped them in evading the virus. While this series is a depiction of how valued their contribution is, it also raises some fundamental questions. Do we really need such epidemics to know the significance of this community? Do these communities deserve the segregation we are treating them with? Do they always need to struggle to show their existence?

Our environmental education has primarily been earmarked by scientific principles, solutions, and figures. However, our scientific understanding has not aided us in effectively understanding the mysteries of our natural environment. And, primarily Indigenous Knowledge encompass the complex, intergenerational, and "cumulative experiences and teachings of Indigenous peoples" (Battiste, 2002)—are not problems to be overcome or addressed by schooling. The perspective shift that is needed is the approach toward the education system (Battiste, 2002). *As Article 31 of the United Nations Declaration of the Rights of Indigenous Peoples (UNDRIP) states, Indigenous peoples have the right to maintain, control, protect and develop their cultural heritage, traditional knowledge and traditional cultural expressions, as well as the manifestations of their sciences, technologies and cultures, including human and genetic resources, seeds, medicines, knowledge of the properties of fauna and flora, oral traditions, literatures, designs, sports and traditional games and visual and performing arts. They also have the right to maintain, control, protect and develop their intellectual property over such cultural heritage, traditional knowledge, and traditional cultural expressions. (United Nations, 2007)*

Today, as our world deals with many climate-related challenges, the approach and knowledge of indigenous communities are widely discussed in academia for coming up with new-age solutions and looking for insights that might have been missed so far.

It is time that we deeply delve into the realms of traditional ecological knowledge, lying around us for centuries. The reason for this discussion is that we inhibit in time these forgotten realms to prevent such environmental crisis that scientists and environmentalists around the globe are predicting. It is high time that we give them the respect that they have been giving to our planet for ages.

Tracing the Traditional Ecological Knowledge

Traditional Ecological Knowledge (TEK) or Traditional Knowledge (TK) can be defined as community intellect that is transmitted and preserved by generations of the indigenous communities (Cocks, 2006). It is imperative to note that the local TEK management practices can help adapt to socio-ecological and sustainable development (Harisha et al., 20215). Traditional Knowledge is often interchangeably used with 'indigenous knowledge' describing the information or relevant practices ensuring sustainable and conservational use of environmental resources (Jasmine et al., 2015). It is a system resource classification based on the environment observation and self-management by the communities (Emery, 1997). This knowledge system has weaved the ecological, social, and spiritual aspects closely in a thread. In contrast to the western science explanation based on global generalization, the Traditional Knowledge is primarily based on socio-ecological interaction in a small geographical area. The knowledge that humans possess are often tied to a specific place or ecosystem, but they are not necessarily unchangeable or rigid. For instance, religious knowledge can be easily transferred and employed to facilitate human interactions with spirits of ancestors or other non-human entities, no matter where these encounters occur. Similarly, ecological knowledge can be transported, as they require using locally available resources and carefully observing how living beings and natural processes interact within any ecosystem to ensure human survival. (Margaret, 2014).

Additionally, certain types of specialized information are kept by gatekeepers who have a strong connection to important sites. These gatekeepers may include tribal leaders, ritual practitioners, or medicine people. They have a vested interest in preserving this information and have deep experience in doing so. These keepers undergo meticulous training to correlate certain sections of traditional stories to particular events and places and this thorough and continuous repetition enables cultural coherency (Vansina, 1985). A good example of this practice can be seen in the Australian communities, wherein such keepers are referred as "traditional owners" and are held accountable for observing the interactions of the landscape (Smith & Wobst, 2005). The Maori tribe has a similar concept, securing certain special and protected knowledge with only a handful members of the tribe (Smith, 2012). Whereas, in Native American contexts, oral tradition-keepers, historians, and interpreters were marked from their birth (Vizenor, 2008).

Almost 20% of our planet is home to some 370 million indigenous people, described as 'first' or 'original' people, having a historic and cultural bond with their biodiversity. (UNESCO, 2016) Their diverse knowledge is deeply rooted in the environment and cultural cohesion, ensuring sustainable management of natural resources, and environment protection in the changing world. (Magni G, 2017)

Traditional Ecological Knowledge is a 'way of life', it is dynamic wherein through experience of previous generations and current experience, the stable knowledge keeps evolving with the changing environment. It is the accumulation of environmental understandings gained through incremental learning, including trial and error and experimentation. Notably, its richness and depth is contributed from various scientific disciplines, such as biology, botany, geography, and cosmology. (Augustine, 1997; Berkes, 2012).

This knowledge is vital not just for the local communities but can help to understand and conserve the ecological resources better. This is derived from generations of communities observing and living in the environment, acquiring experiences over time. For instance, in preparation for hunting, tribes may prepare their tools in a specific location and use standing stones to predict the appropriate time and season to hunt based on the movements of the stars.

When related to biodiversity conservation, this knowledge has become fundamental to human life, especially in rural and remote areas across the globe. (Rhea et. al, 2005) Thus, the livelihoods of millions of people referred to as indigenous people means being involved in the management of diverse ecosystems, natural resources, and conservation of biodiversity and a number of success stories on resource management have been reported globally. (Maikhuri et. al 2013) The efficacy of traditional ecological knowledge can be witnessed through the example of the Inuit hunters (Arctic), who learnt to survive the unforgiving cold environment and the efficient usage of specific instruments or the medicine practitioners in the jungles of Amazons who became skilled at distinguishing varied potentially poisonous plants. While this traditional knowledge aided these people in their survival during those times, at present, the same knowledge opens the window for the present generations to understand the people of the past.

Essentially, traditional ecological knowledge is a valuable set of data based on the functioning of our natural environment, supported by the learning, skill set and undertakings of localized people. If given adequate attention, this data can aid in the overall collective well-being of the human society and ecosystems.

Protecting Indigenous Knowledge through Domestic Initiatives: Traditional Knowledge has been the backbone for food security and health care for poor people in developing and least developed countries. For example, documentation of traditional biodiversity-related knowledge on the status, uses, and management of biological resources constitutes the People's Biodiversity Registers (PBRs) in India. (Rhea et. al, 2005) The documentation is, therefore vital for the appreciation and protection of this valuable knowledge. According to a study by Brayboy and Castagno (2009), Indigenous Knowledge has been found to have a positive impact on education. However, the study also found that the integration of Indigenous Knowledge with mainstream culture in a multilingual and multicultural learning environment can be challenging due to existing educational practices and implicit racism. The authors suggest that culturally responsive schooling can be achieved by leveraging local community and knowledge to develop a curriculum that aligns with academic standards, as demonstrated by several case studies in schools serving Indigenous students.

With International treaty obligations like Convention on Biological Diversity (CBD), India and other contracting states have implemented various plans to respect, promote and document TK to make it more accessible.

The chapter will now discuss these policies and plan and will deconstruct them focusing on the need for change in education policy through the case study of India.

India is one of those 17 countries which are extensively rich in biodiversity and its related traditional knowledge. India's diverse cultural richness has been weakened by the impact of colonialism, and as a result, people tend to attribute India's progress and development to Western technology and knowledge systems. This has led to an excessive focus on high-tech infrastructure and technologies as a symbol of modernisation, even though the reality is quite different. Most individuals, including those in positions of political power, often refer to Indian philosophy or traditional knowledge, but they are mainly alluding to Aryan knowledge, which includes literature on medicine, agriculture, and yoga. By using the phrase "Indian culture," they are attempting to homogenize and eliminate cultural diversity in favor of the dominant community's culture, as written and propagated in Sanskritised texts. This use of the term "Indian" to describe culture or knowledge reflects an ethnocentric colonialist attitude.

Despite the presence of several significant constitutional provisions that ensure educational, economic, public employment-related, and political safeguards, there are still numerous instances of indigenous groups being subjected to atrocities by non-indigenous groups (including the modern state). This includes the displacement and dispossession of millions of people from their ancestral lands.

Another example is disappearing of indigenous knowledge and practices, evident in indigenous fishing communities, such as the Kolis of Maharashtra's coast. Many artisanal fishers have abandoned their traditional methods of fishing and now rely on modern technology tools like GPS. Due to a lack of financial support, low subsidies, unsustainable infrastructure projects, and the worsening climate change, the fisherfolk are being pushed out of practice and business. The government is now urging indigenous fishers to switch from open-sea fishing to commercial fish cultivation, which is causing a drastic disruption in the way of life of the Kolis and other fisher communities, leading to changes in their food systems, culture, and future.

But, in recent years, initiatives have been put in place to give due acknowledgement to indigenous knowledge and protect them from exploitation, discussed below:

1. AYUSH: To deal with India's rich Traditional Knowledge related to biological resources, the Ministry of Health and Family welfare has established a separate department, AYUSH which promotes and develop the recognized ancient knowledge related to namely Ayurveda, Yoga, Unnai, Siddha and Homeopathy. With AYUSH and related plans, the government has promoted various projects relating to cultivation of medicinal plants.
2. Wealth of India Series is an initiative by the National Institute of Science Communication and Information Resources (NISCAIR) to document renewable sources and mineral wealth in the Indian Subcontinent. Under this series an encyclopedia has been prepared with the correct identity of each plant with its economic advantages, distribution in the forests.
3. Under the Department of Science and Technology, an autonomous body named the National Innovation Foundation (NIF) was established in 2002 to promote, respect, and reward Traditional Knowledge innovations at the grass root level. The foundation has a specialized team of experts that aid the TEK holders in protecting their intellectual property under the existing intellectual property laws. The expertise is extremely helpful for those who are completely unaware in regards to their intellectual property rights
4. In a collaborative project with the Council of Scientific and Industrial Research, the MoHFW has developed Traditional Knowledge Digital Library to prevent bio-piracy and misappropriation of

Traditional knowledge at various International Patent Offices. Regarding this objective, India has signed the TKDL Access Agreement with countries like the European Union, US (United States), UK, Canada, Germany, and Australia. The library has a collection of more than thirty thousand medicinal formulations and is made available online to provide relevant information to patent and trademark examiners in offices of respective jurisdictions, refraining from a grant. The data is made available in five UN languages, viz., French, German, English, Spanish and Japanese for convenience irrespective of the fact that data originated in languages like Sanskrit, Urdu, Persian and Hindi.

5. Globally Important Agricultural Heritage System is an initiative started by Food and Agricultural Organization of the United Nations in 2002 to recognize, manage and protect the Indigenous Knowledge related to agricultural biodiversity, food-livelihood security.
6. Along with several other initiatives the Government of India has enacted various legislation to conserve, protect and promote the indigenous knowledge through statutory obligation. These statutes include the National Green Tribunal Act, 2010, the Biodiversity Act, 2002, the Protection of Plant Verities and Farmers' Right Act, 2001, the Patent Act, 1970 etc.
7. To advocate, promote and train the indigenous knowledge, the country has established various graduate and post graduate colleges to strengthen the system.

Notably, the Indian civilization was renowned for being far ahead of its times, much before its colonization by the West. Today, the main challenge with applying Western solutions in the global south context is that they are less efficient due to their bias towards problems in the global north. Consequently, traditional and indigenous knowledge systems offer an advantage in this regard. A remarkable instance is the practice of Meghalaya's bamboo drip irrigation system. This system uses abundant bamboo species and natural terrain to meet domestic and agricultural needs while preserving the local ecosystem. As a result, the initiative has contributed to the conservation of India's bamboo forests, which is a significant accomplishment. Conclusively, in the Indian context, modern westernized grey infrastructure is efficient only in resolving the specific issue for which they were designed. On the other hand, traditional indigenous knowledge systems provide a range of holistic benefits, the most significant being resilience, which was evident during the COVID-19 pandemic among communities that still practice these systems. Thus, phase wise reintroduction of indigenous agricultural practices to arrest agrarian distress to both the farmer and the land, is the key for India's progress in the next few decades (Perroni, 2017).

In the field of Indigenous educational research, there is a growing demand for increased participation of Native people in the research process. To achieve this, many Indigenous researchers recommend using qualitative data in addition to quantitative data to gain a deeper understanding of Native students' experiences in mainstream educational settings. By examining mainstream education through the lens of Native students' perspectives, teachers and school leaders can better transform mainstream education (Gilbert 2005; Quinn and Andrews 2004; Wayne et al. 2005).

The discussion in this part of the chapter was to highlight the efforts at national and regional level to conserve, document and understand the indigenous knowledge. These efforts reflect the acknowledgment that Indigenous Knowledge create "a new, balanced centre and fresh vantage point from which to analyze Eurocentric education and its pedagogies" (Battiste, 2002). However, it is pertinent to note that for the optimum use of traditional ecological knowledge for environment sustainability, there is need of incorporating this acquired knowledge in the environment education. Simply, because Indigenous Knowledge can create "new perceptions" and open up "new possibilities" in education (Brayboy &

Maughan, 2009). The vast Traditional Knowledge that has been documented over the period of time if managed well and studied will help to construe the dynamic changes in the ecosystem. This will not only help to understand the traditional ecological system but will also aid the indigenous communities to share and trade their valuable knowledge with other communities.

Deconstructing the Conventional Approach to Indigenous Knowledge

The movement to "de-Indianize" or "civilize" Native people stems from the colonial notion that savages should be transformed from their evil ways. Protestant epistemologies served as the vehicle to assimilate Native peoples into the mainstream culture. After the United States gained its independence, it turned its attention to acquiring more land from Native people in the south and west with minimal cost to the government. Its government's approach to educating Native children was to establish boarding schools run by missionaries with the goal of assimilating them into White American culture. This involved attempting to erase their Native identity and replace it with a Eurocentric one. Their strategy for education was aimed at persuading Native people to sell their land to the government and adopt Eurocentric practices of agriculture, husbandry, and acquiring possessions rather than relying on their traditional relationship with the land (Spring 2008). Another example of the inhumane approach towards natives was when in 1879, Richard Henry Pratt, a military officer, founded the Carlisle Indian School, the first off-reservation Indian boarding school in Carlisle, Pennsylvania. He coined the phrase "Kill the Indian and Save the Man." Although Pratt initially believed that Indians could excel like their white counterparts if given equal opportunities, he later changed his mind and labeled Native students as inferior.

In Tanzania during the colonial period, schools instructed students in English and punished those who used indigenous languages. Students were trained to identify those who spoke languages other than English, thus transforming them into "witch hunters" and teaching them the profitable value of betraying their own community, where loyalty was once highly regarded (Semali and Kincheloe 1999). These stereotypes have continued to impact Native education and research methodology. There were boarding schools that aimed to isolate children from their tribal communities and traditions while assimilating them into American society.

The above-mentioned practices of colonizing are a limited sample, but they demonstrate the principles, convictions, and customs of modern knowledge discourses. Examining the techniques of modernization that have been used on native communities emphasizes that modern knowledge and practices are not universally applicable. Indigenous cultural traditions and territory were under threat from Western practitioners who considered them to be public scientific property. The investigations conducted by archaeologists in the past were carried out in a manner that was destructive and ethically questionable.

It is pertinent to note that due to lack of institutionalized intellectual development the traditional knowledge explanations didn't align with western scientific explanation of the concepts (McGregor, 2004). And, the treatment of indigenous people was more or less close to an object. The environment surrounding them, in countries like North and South America, Australia, and New Zealand was hostile politically crippling and made it difficult for the traditional communities to maintain cohesion. A painful instance relates to the Maori community, wherein, colonization directly interfered with the authority of traditional leaders and traditional knowledge, stripping away Maori control of their land and destiny (Smith, 2012).

Countless Indigenous populations around the world have been devastated by colonialism and still to this day face the oppression of Western society. (Andrew, 2023) Their knowledge for the longest time was considered obsolete and irrelevant. It was also labelled as "difficult" due to its difference from Western knowledge. (Doughlas et. al, 2020). This strengthens the scientific loopholes against the indigenous knowledge system like asymmetry, non-cognitive and nob quantitative. Considering the inductive application of this knowledge, Breidlid (2009) found indigenous knowledge in case of dynamic situations misleading and inadequate due to its glamourized explanations of the scientific phenomenon (Chandra et al., 2014). The mainstream science behind thinking and analyzing is based on an explanation that divides the natural world and products of human experience into disparate parts that do not reflect their places of origin (e.g., the use of Latin names for flora and fauna). In contrast, TEK is inherently integrative and holistic due to being based on sensory awareness and human experience of the complex relationships among multiple organisms in distinct ecosystems (Augustine 1997; Apffel Marglin 2011; Smith 2012).

Science is practised based on natural and environmental resources in any given cultural and socio-economic context. However, the picture of science represented in many textbooks all over the world often neglects its cultural component or restricts it to a Western view on the history of science (e.g., Forawi 2015; Khaddour et al. 2017; Ideland 2018). The mainstream practice of science depends on the natural and environmental resources available in a given cultural and socio-economic context. Indigenous knowledge in science and indigenous views on nature vary among societies and cultures across the globe, depending on the level at which they are observed.

The integration of indigenous knowledge in existing scientific system for sustainable development also hints towards the possible exploitation of these communities owing to visible power imbalance. Language barrier and translation of this knowledge system into English is another crucial issue that could lead to compromised or incorrect understanding of such knowledge. Overall, there are many complexities that have affected the integration of Indigenous knowledge than just the duality of technical and political agendas. Further, Indigenous erasure has been a common practice in settler colonial education, with policies and actions aimed at eliminating Indigenous languages, cultures, and knowledge systems (Wolfe, 2006). This has been achieved through practices such as assimilation, boarding schools, and portraying Indigenous knowledge as primitive. Unfortunately, many educational policies have framed Indigenous knowledge as a problem to be solved through schooling, perpetuating the notion that Western knowledge is superior to Indigenous knowledge. This has led to the erasure of Indigenous knowledge embodied by Native children and a lack of appreciation for the valuable contributions that Indigenous knowledge can make to society. (Lomawaima & McCarty, 2006).

This must change if we want to create a sustainable movement towards more respectful and inclusive education for Native people. Traditional Western research methods will not aid in this transformative change, as they mirror the same obstructive principles as mainstream educational systems. Therefore, many Native scholars are calling for educational research that challenges the inappropriate use of Western methodologies and epistemologies. Researchers in the field of language education have also been exploring ways to utilize a decolonial perspective in order to address the fundamental issues that have contributed to the colonial outlook in language teaching, as well as the disparities in global knowledge production (Kubota, 2020; Macedo, 2019). According to Mignolo and Walsh (2018), it is crucial to adopt a decolonial viewpoint to critique the supposed universality and prominence of Western knowledge, and the belief in the superiority of Western lifestyles over all other cultures and civilizations.

The Keele Manifesto follows the ideology of Mignolo and Walsh and was published in response to a global movement that calls for the decolonization of higher education curriculum. The Manifesto provides guiding principles to decolonize the curriculum, which is still deeply rooted in colonial and Western-centric worldviews. It states that the process of decolonization involves identifying and challenging colonial systems, structures, and relationships. And, it seeks to expose the underlying logics and values that help maintain the dominance of Western knowledge in world knowledge production. After all, decolonization is not a simple integration or a tokenistic inclusion of non-White cultures or non-Western knowledge.

For integration and global acceptance of the Indigenous knowledge system, there is a need for environment education sustainability through transformative actions in schools. While naturalizing Indigenous Knowledge in education will require intentional teacher preparation and systemic change (Kanu, 2005; 2011; Orr, 2004; Ottman & Pritchard, 2010), it is to be maintained that Indigenous Knowledge should be afforded the same respect as mainstream knowledge, have a rightful place within schools, and be regarded as a generative resource and gift that can enrich public education. Thus, this calls for attention to the need for a pedagogical approach for the inclusion of this knowledge in the world through formal or informal education.

Indigenous Knowledge as a Part of Environmental Education

When we speak of education today, we are reminded of a highly competitive and individualistic environment in most schools. But, contrastingly the manner of Indigenous Knowledge Systems focuses on promoting learning as a process that is based on co-operation, holistic values (encompasses the mental, physical, emotional, and spiritual aspects of a person) and an experiential approach advocating for rationality and sustaining of the collective. For many Indigenous communities, the goal of education has always been to nurture the well-being of the whole child, including their emotional, mental, physical and spiritual development. The purpose of education was, and still is, to instill in future generations the skills and knowledge necessary to live a balanced life, a life where individuals can use their unique gifts to contribute to the well-being of their relatives, which includes not only immediate family members, but the entire community and the animals, plants, waterways and land with which we depend on for life itself.

Environment Education can be understood as a process of learning the relationship between the human and the natural habitat around that includes plants, animals, insects, air, water etc. (Whyte, 2018). This part of the chapter focuses on the indigenous knowledge system's role by educating about the sustainable practices and strengthening the ecological sustainability.

Per se, Traditional knowledge is generally associated with biological resources and is invariably an intangible component of such a biological resource. The valuable leads/clues it provides save time, money and investment of modern biotech and other industries into any research and product development. Logically, therefore, a share of such benefits should accrue to the creators and/or holders of such traditional knowledge. In recent times, scientists and researchers have now started to delve into what is referred to as Traditional Ecological Knowledge (TEK) in order to develop a mature grasp of the natural world. Around the globe, researchers are turning to what is known as Traditional Ecological Knowledge (TEK) to fill out an understanding of the natural world. TEK as a branch has been the contribution of those people who have had their generations deal with it for centuries, now, making sense of the environment around them and how they recognize, share and value it. It is a distinguished reflection of each such community having a deep attachment to their natural environment. Previously, TEK has been studied by disciplines under headings such as ethno-biology, ethno-ornithology, and biocultural diversity. However, in the face of climate change and immense loss of biodiversity, scientists have now shifted their focus to this branch.

An example of this is based in the Arctic where global warming is happening at a rapid speed. Scientists with the aid of native people (Skolt Sami) (Jim Robbins, 2018) who have existed there as hunters and fishers, are now collaboratively making note of the minute changes that are happening in the region and learning the adaptation process of the natives.

The Arctic is not the only region wherein scientists have extended a hand towards the natives but around the world, recognition about the uniqueness and usefulness of TEK is on a rise. Researchers have discovered that the Maya people of Mesoamerica preserve an astonishing amount of biodiversity in their forest gardens, in harmony with the surrounding forest. Unsurprisingly, their houses have the most diverse domestic system in the world.

When we discuss the incorporation of TEK into our mainstream education systems, it is not just about teaching of Indigenous content but also about exploring Indigenous ways of knowing and learning. It means acknowledging and respecting Indigenous knowledge systems and ways of life and understanding the inter-connectedness of all things. Much literature states that inquiry-based science instruction can help disadvantaged students achieve academically, but culturally responsive methods are needed to help students embrace their backgrounds, stand against oppressive systems, and prevail over any self-doubt in their abilities. (Brown, 2017) Incorporating culture in science education provides extra meaning to the curriculum and allows students to see their day-to-day lives in their educational experience. (Zidny et. al, 2020).

"A mining giant—euphemistically called "Vedanta" — wanted to extract about 1.8 billion tonnes of high-grade bauxite, the source of aluminium, to feed into giant factories built on this land." The indigenous poor, who were primarily tribal people, had been residing on this hilly terrain for thousands of years until being driven off the plains by the advance of modern civilization. Following years of opposition from them and significant official manipulation and muscle-flexing, funded by the mining aristocracy, the government was forced to call a referendum for twelve carefully chosen villages. (Nirmalangshu Mukherji, 2016)

While struggles like these have shaped our environmental realms and understanding of relations of people with the environment, educators do not find any compelling reasons to highlight such incidents through mainstream education. Imparting knowledge pertaining to instances like these is crucial for inculcating values and rights of these indigenous communities. Moreover, it also inculcates a sense of Emotional Qoutient that helps to empathise with the values of these communities and also with our nature.

The main issue with science education is that students often view the content as irrelevant, boring, and not engaging enough. Researchers have stated that Indigenous knowledge offers rich contexts to initiate learning interest and connect science education to more holistic worldviews that support sustainability. But, literature has also shown that the integration of Indigenous knowledge is a complicated process that has no one solution. Each case of integration deals with its nuances of language culture, geological location, and much more. Locality plays an enormous role in the specifics of each case of integration.

Fullan's (2001) three-dimensional framework is useful for showing that implementing a new educational policy must warrant a change in practice for it to meet its stated outcomes, in three requisites: (1) the possible use of new material; (2) the possible use of new teaching approaches; and (3) the possible alteration of beliefs. Educators should include Indigenous populations in the planning and execution of the integration process, and educators should work toward forming authentic relationships with their local Indigenous tribe. Each instance of Indigenous knowledge integration will be different. It will require its own unique set of interactions and methods, but the notion of authentically working to create a partnership between Western and Indigenous communities is universally applicable.

Activists and practitioners involved in curriculum decolonisation projects should bear in mind that it is crucial to comprehend the roles and viewpoints of teachers. In order to make curriculum decolonisation a success, it is imperative to modify the mindsets and beliefs of teachers. This is because curriculum decolonisation is primarily concerned with bringing about cognitive transformation and breaking down polarised thinking and entrenched belief systems.

For integration to be successful without ontological overlap would require Western science to have intimate familiarity with the epistemic and social goals of Indigenous communities, along with a deep understanding of the local environment. This can be understood with the fictional example of Oraka Tribe in Andaman and Nicobar Islands. The Netflix series 'Kaala Paani' is an account to highlight not just need for including and respecting their community's existence but also makes us appreciates values we have forgotten in this realm. Before the Tsunami hit these islands in 2004, Orakans had moved to a higher ground and notably, without a single casualty. This was because of 'Tinnotu' as they called it. Despite, there is no exact translation of its meaning in any known language, it is closely associated with their innate sense of connect with the nature that they could feel an approaching crisis and safeguarded themselves. Contrastingly, we who depended on the weather forecasts faced prodigious consequences on human lives and resources. While, we are not arguing for a case that modern science is insufficient or inaccurate, we only intend to revitalize their sources and emotional connectivity with the nature that would proffer human population a harmonious lifestyle in resonance with our planet and it's forces.

A land-based approach is an interesting way to merge Indigenous perspectives into education. Land-based learning is an approach to teaching and learning that connects students with the environment and the land. It is a holistic method that incorporates Indigenous knowledge and ways of knowing, and encourages a sense of responsibility and stewardship towards the environment. Land-based learning can take different forms, such as learning from Elders, participating in traditional activities, or engaging in community-based research. To strengthen and deepen Indigenous curricula in areas that have already adopted this type of instructional material, policymakers should take steps to initiate the process. Additionally, they should require school districts in areas that have not yet adopted any curricula to establish relationships with tribal communities.

Language can serve as another way through which indigenous perspectives can be embodied in mainstream education. Language is a crucial aspect of Indigenous cultures, and is essential to the transmission of Indigenous knowledge and ways of life. In order to reconcile and decolonize, revitalizing languages can play a key role in their knowledge promotion. Incorporating Indigenous language into the curriculum, providing language classes for non-Indigenous students, and supporting Indigenous language revitalization initiatives are all ways in which educators can integrate Indigenous perspectives into their teaching practice.

Schools and districts hold significant authority in the implementation of Indigenous knowledge. Districts, in particular, have the potential to initiate genuine relationships with Indigenous communities. This can be achieved through various means such as arranging community talks, organizing visits to tribal reservations by district personnel, or hosting events that concentrate on reconciliation. The key aspect to note here is that Districts possess the power to initiate these efforts, which can act as a driving force for bringing non-indigenous and Indigenous communities closer together. A fine example of this is in Montana (United States), where school and Tribal leaders worked together to create a curriculum called Essential Understandings Regarding Montana Indians, which was a crucial step in implementing the statewide Indian Education for All (IEFA). Although IEFA was initially an unfunded mandate that faced legal challenges, it has since provided significant academic, social, and cultural benefits to both Indigenous and non-Indigenous students in Montana's public schools. Educational resources were developed using Native American knowledge from local schools and Tribal elders, resulting in positive attitudes towards cultural diversity among students when teachers collaborated with Native parents.

Finally, it is important to recognize that Indigenous perspectives are not limited to specific subjects or areas of study. Indigenous knowledge is holistic and interconnected, and can be applied to all aspects of life. It is therefore important for educators to incorporate Indigenous perspectives into all areas of the curriculum, and to recognize the importance of Indigenous knowledge and ways of knowing in shaping our understanding of the world. Teachers have the power to bring more attention to the issue of integrating Indigenous knowledge and to start the process of forming relationships with our local tribal communities. Most importantly, educators need to recognize their internal biases and correct them in order to wholeheartedly mainstream indigenous knowledge. However, most educators will not recognize this internal bias until deep questions are asked about their viewpoint on the Indigenous knowledge and messages they share with their students.

Following Fullan's framework, activists and practitioners involved in curriculum decolonisation projects should bear in mind that it is crucial to comprehend the roles and viewpoints of teachers. In order to make curriculum decolonisation a success, it is imperative to modify the mindsets and beliefs of teachers. This is because curriculum decolonisation is primarily concerned with bringing about cognitive transformation and breaking down polarised thinking and entrenched belief systems.

AN INCLUSIVE APPROACH

Culturally Relevant Education

How does one teach, impart, or include "culture" in education without reducing and potentially distorting its dynamic and holistic nature? Mary Hermes makes the case that culture cannot be "added on" or simply appended to existing school curriculum.

Mainstream educational policies and practices are replete with irrelevant curriculum and pedagogy that have an adverse impact on Native academic achievement. The consequence is an unacceptable rate of failure—a failure not of Native children but of the system to educate Native children and keep them engaged in their schooling. Addressing these shortcomings will require educators to recognize culturally relevant elements that are critical to capturing the interest of Native children in the nation's classrooms.

A good example is, a research project called Hawaiian Cultural Influences in Education is being conducted by Kamehameha Schools, the Hawaii Department of Education, and Na¯ Lei Na 'au'ao. This project aims to identify and share the most effective culturally relevant education practices being used in Hawaii's classrooms. Based on survey responses from teachers, students, and parents/caregivers, the researchers found that using culture-based educational strategies regularly can positively impact students' socio-emotional and educational outcomes.

Another example of culturally relevant education can be seen in the efforts of research centers in New Zealand. The Wilf Malcolm Institute of Educational Research at the University of Waikato's faculty of education focuses on promoting culturally relevant Maori education. They do this through various projects, such as learning journeys from early childhood through school, teaching strategies used in mathematics by Maori students, and teaching and learning under the supervision of Maori doctoral students. The New Zealand Association for Research in Education also sponsors a journal for scholars to disseminate educational research, with the aim of improving the education landscape for Indigenous students across the world.

Thus, the focus of culturally relevant education is to use Indigenous educational research to enhance the learning experiences of Indigenous students in Western educational settings. Scholars have many opportunities for collaborative research initiatives with Indigenous research methodologies at the center of their work, whether they are examining the educational experiences of North American tribes or Aboriginal people in New Zealand. However, sharing knowledge should not be restricted to educational scholars alone.

It is essential to incorporate the knowledge of tribal elders, community stakeholders, and students' voices in these transformational partnerships. Researchers and scholars recognize the challenges a monolithic perspective can create in discussions of educational reform. Hence, diverse stakeholders are vital for these collaborative partnerships in Indigenous educational research to bring about transformative change and create culturally relevant education.

Ideological Beliefs and Practices

In order to enhance education for Native children, it is important to include Indigenous viewpoints in the ideological beliefs and practices. Gaining approval and acknowledgement for indigenous forms of education that are different from the formal schooling system run by the state can be a challenging and gradual process. Such efforts face obstacles in the form of national policies that aim to assimilate and standardize, thereby limiting innovation and diversity in the name of accountability and presumed 'equality'. The individuals holding positions of power at both state and federal levels should recognize the significance of Indigenous knowledge and wisdom. Moreover, it is necessary to impart correct representations and accurate portrayal of Native individuals in the teacher training and administrator certification program curriculums.

Public education gatekeepers such as state departments of education and colleges of education should promote multiculturalism in schools. Unfortunately, many educational leaders and policy makers lack exposure to the wealth of information available in existing Indigenous educational research. Consequently, non-Native perspectives and ideologies persist, hindering school reform efforts for schools serving Native students. Reyhner and Jacobs (2002) criticize the inadequacy of teacher education programs in requiring adequate knowledge of Indian perspectives, despite certification requirements and the new "standards movement" shaping teacher training programs.

Inaccurate depictions of Indigenous cultures and traditions in mainstream education have contributed to the perpetuation of stereotypes in the dominant culture. Textbooks often misrepresent American Indians and Alaska Natives, shaping the dominant culture's perceptions of Indigenous peoples. Unfortunately, mainstream research paradigms form much of the foundation of teacher and administrator education curricula. This approach can lead to ill-informed teacher and administrator candidates who may possess an erroneous knowledge base concerning Indigenous peoples and cultures. Recognizing the crucial role of Indigenous knowledge in preparing educators and school leaders who will work with Native students is the key to transforming these experiences. To be successful, indigenous education professionals must participate freely in all phases of the planning, design, implementation and evaluation of such reforms. As the United Nations Declaration on the Rights of Indigenous Peoples, proclaimed by the General Assembly in 2007, makes special mention of their right to education, 'Article 14. Indigenous Peoples have the right to establish and control their educational systems and institutions providing education in their own languages, in a manner appropriate to their cultural methods of teaching and learning'.

Thus, the next step is to become educated about the genuine history of Native peoples in different countries.

Way Forward

This account has attempted to deconstruct the present educational system by presenting a constructive critique on the colonial and post-colonial lens to environmental education. We have presented a case for acknowledging their presence in tracing the traditional ecological knowledge, how we have ignored the indigenous knowledge by labelling and treating them as people not belonging to this planet. While indigenous knowledge is and should actually be source and core of the current environmental education and it's approaches, we have voluntarily chosen to ignore them and their experiences with the Earth.

Historically, we have attempted to foster a culture to differentiating between indigenous and non-indigenous communities. Examples throughout the chapter have reflected that indigenous lands are where all public education takes place; therefore, respect for these lands, Indigenous peoples, and Tribal sovereignty needs to be fostered. The problem is not with the Indigenous children or their knowledge, but with policies and practices that ignore and demean them. Indigenous students have the right to educational self-determination, to an education that affirms their knowledge and supports their rights as Tribal citizens. We, collectively, as global population are inter-dependent and therefore, the solution is also in inhibiting this interconnectedness.

Therefore, we suggest a multi-pronged inclusive approach to culminate a sense of belonging in both the indigenous and non-indigenous communities. In other words, we must discourage this sense of segregation and value our individual and collective role in educating and saving our environment.

Specifically, we propose changes in curriculum Integration, sensitization and specialised training of our teachers, community collaboration and approach toward policy making in this realm. Culminating Indigenous knowledge within formal curricula, ensuring that traditional ecological wisdom becomes an integral part of environmental education programs at all levels will give us a timely opportunity to understand, respect and save our planet. We advocate for specialized training programs to empower educators with the skills and understanding necessary to effectively incorporate Indigenous perspectives into their teaching methodologies. Also, ways for community collaborations that encourage partnerships between educational institutions and Indigenous communities, fostering a collaborative approach that respects and involves local knowledge holders in the design and delivery of environmental education initiatives.

Policies and drives that help us work for resource proliferation and invest in the creation of educational resources that bridge the gap between Indigenous knowledge and mainstream environmental education, including textbooks, multimedia materials, and interactive tools.

We also recomment policy changes at the institutional and governmental levels to officially recognize and mandate the inclusion of Indigenous knowledge in environmental education frameworks. Governments, NGOs and other institutions must encourage further research to explore and document various

Indigenous perspectives on environmental issues, ensuring a continuous dialogue that enriches the broader understanding of ecological relationships. Government must also launch awareness campaigns to inform the public about the significance of incorporating Indigenous knowledge into environmental education, fostering broader support and understanding.

By implementing these suggestions, we can pave the way for a more inclusive and comprehensive environmental education system that respects and integrates the invaluable wisdom of Indigenous communities.

In conclusion, this chapter has delved into the crucial endeavor of providing a critical account on integration of Indigenous knowledge into mainstream environmental education. Recognizing the value of traditional wisdom and practices from Indigenous communities contributes not only to a more holistic understanding of the environment but also fosters cultural diversity and sustainable coexistence. As we strive for a more inclusive and comprehensive approach to environmental education, acknowledging and incorporating Indigenous perspectives becomes imperative for a harmonious relationship between humanity and the natural world. This endeavor holds the potential to transform not only our educational systems but also our collective commitment to responsible stewardship of the planet we all share.

REFERENCES

Aikman, S. (1999). *Intercultural Education and Literacy: An Ethnographic Study of Indigenous Knowledge and Learning in the Peruvian Amazon*. John Benjamins Publishing Co. doi:10.1075/swll.7

Apffel-Marglin, F. (2011). *Subversive spritualities: how rituals enact the world*. Oxford University Press.

Augustine, S. J. (1997). Traditional Aboriginal knowledge and science versus occidental science. *Paper prepared for the Biodiversity Convention Office of Environment Canada*. Research Gate.

(1994). Austin, Regina. (1994). A nation of thieves: Securing Black people's right to shop and to sell in White America. *Utah Law Review*, (1), 147.

Battiste, M. (2002). *Indigenous knowledge and pedagogy in First Nations education: A literature review with recommendations*. Indian and Northern Affairs Canada.

Berkes, F. (2012). *Sacred ecology*. Routledge Press. doi:10.4324/9780203123843

Biba, J., & ... (2016). Traditional Knowledge systems in India for Biodiversity Conservation. *Indian Journal of Traditional Knowledge, 15*(2), 304–312.

Brayboy, B. M. J., & Maughan, E. (2009). Indigenous knowledge and the story of the bean. *Harvard Educational Review, 79*(1), 1–21. doi:10.17763/haer.79.1.l0u6435086352229

Breidlid, A. (2009). Culture indigenous knowledge systems and sustainable development: Acritical view of education in an African context. *International Journal of Educational Development, 29*(2), 140–148. doi:10.1016/j.ijedudev.2008.09.009

Brown, J. C. (2017). A meta-synthesis of the complementarity of culturally responsive and inquiry-based science education in K-12 settings: Implications for advancing equitable science teaching and learning. *Journal of Research in Science Teaching, 54*(9), 11431173. doi:10.1002/tea.21401

Bruchac, M. (2014). Indigenous knowledge and traditional knowledge. Encyclopedia of global archaeology, 10, 3814-24. doi:10.1007/978-1-4419-0465-2_10

Chakrabarty, S. P., & Kaur, R. (2021). A Primer to Traditional Knowledge Protection in India: The Road Ahead. *The Liverpool Law Review*, *42*(3), 401–427. doi:10.1007/s10991-021-09281-4 PMID:34099944

Chandra, D. V. (2014). Re-examining the Importance of Indigenous Perspectives in the Western Environmental Education for Sustainability: From Tribal to Mainstream Education. *Journal of Teacher Education for Sustainability*, *16*(1), 117–127. doi:10.2478/jtes-2014-0007

Chantiluke, R., Kwoba, B., & Nkopo, A. (Eds.). (2018). *Rhodes must fall: The struggle to decolonise the racist heart of empire*. Zed. doi:10.5040/9781350222380

Douglas, V., Purton, F., & Bascuñán, D. (2020). Possibility not difficulty: Difficult knowledge in K-12 classrooms as opportunities for renegotiating relationships with Indigenous perspectives and knowledge. *The Alberta Journal of Educational Research*, *66*(3), 307–324. doi:10.55016/ojs/ajer.v66i3.56962

Fullan, M. G. (2001). *The new meaning of educational change* (3rd ed.). Teachers College Press. doi:10.4324/9780203986561

Gilbert, L. (2005). What Helps Beginning Teachers? *Educational Leadership*, *62*(8), 36–39.

Harisha, R.P & others. (2015). Traditional ecological knowledge (TEK) and its importance in South India. *Applied Ecology and Environmental Research.*

Harrington, B. (2013). Using Indigenous Educational Research to Transform Mainstream Education: A Guide for P–12 School Leaders. *American Journal of Education*, *119*(4), 487–511. doi:10.1086/670962

Ismail, S. M., & Cazden, C. B. (2005). Struggles for Indigenous Education and Self-Determination: Culture, Context, and Collaboration. *Anthropology & Education Quarterly*, *36*(1), 88–92. doi:10.1525/aeq.2005.36.1.088

Jacob, M., Sabzalian, L., Jansen, J., Tobin, T. J., Vincent, C. G., & LaChance, K. M. (2018). The Gift of Education: How Indigenous Knowledges Can Transform the Future of Public Education. *International Journal of Multicultural Education*, *20*(1), 157–185. doi:10.18251/ijme.v20i1.1534

Kubota, R. (2020). Confronting epistemological racism, decolonizing scholarly knowledge: Race and gender in Applied Linguistics. *Applied Linguistics*, *41*(5), 712–732. doi:10.1093/applin/amz033

Langton, M., & Rhea, Z. M. (2005). Traditional indigenous biodiversity-related knowledge. *Australian Academic and Research Libraries*, *36*(2), 45–69. doi:10.1080/00048623.2005.10721248

Lomawaima, K. T., & McCarty, T. L. (2006). *"To remain an Indian": Lessons in democracy from a century of Native American education*. Teachers College Press.

Macedo, D. (Ed.). (2019). *Decolonizing foreign language education: The misteaching of English and other colonial languages*. Routledge.

Magni, G. (2017). Indigenous knowledge and implications for the sustainable development agenda. *European Journal of Education*, *52*(4), 437–447. doi:10.1111/ejed.12238

May, S., & Aikman, S. (2003). Indigenous Education: Addressing Current Issues and Developments. *Comparative Education*, *39*(2), 139–145. doi:10.1080/03050060302549

Mcgovern, S. M. (2000). Reclaiming Education: Knowledge Practices and Indigenous Communities. *Comparative Education Review*, *44*(4), 523–529. doi:10.1086/447633

McGregor, D. (2004). Traditional ecological knowledge and sustainable development: Towards coexistence. In M. Blaser, H. Feit, & G. McRae (Eds.), *In the way of development, indigenous people, life projects and development* (pp. 72–91). Zed Books. doi:10.5040/9781350220720.ch-005

Mignolo, D. W., & Walsh, E. C. (2018). *On decoloniality: Concepts, analytics, praxis*. Duke University Press.

Negi, V. S., & Maikhuri, R. K. (2013). Socio-ecological and religious perspective of agrobiodiversity conservation: Issues, concern and priority for sustainable agriculture, Central Himalaya. *Journal of Agricultural & Environmental Ethics*, *26*(2), 491–512. doi:10.1007/s10806-012-9386-y

Paquin, A. (2023). The Integration of Indigenous Knowledge in Education. M.Ed. Literature Reviews, 18.

Reyhner, J., & Jacobs, D. T. (2002). Preparing Teachers of American Indian and Alaska Native Students. *Action in Teacher Education*, *24*(2), 85–93. doi:10.1080/01626620.2002.10734422

Ruíz, R. (2010). Reorienting language-as-resource. In J. E. Petrovic (Ed.), *International perspectives on bilingual education: Policy, practice, and controversy* (pp. 155–172). Information Age Publishing.

Smith, C., & Wobst, H. M. (2005). *Indigenous archaeologies: decolonizing theory and practice*. Routledge.

Smith, L. (1999). *Decolonizing Methodologies: Research and Indigenous Peoples*. Zed Books Ltd.

Smith, L. T. (2012). *Decolonizing methodologies: research and indigenous peoples* (2nd ed.). Zed Books.

Spring, J. (2008). *The American School: From the Puritans to No Child Left Behind* (7th ed.). McGraw-Hill.

Stavenhagen, R. (2015). Indigenous Peoples' Rights to Education. *European Journal of Education*, *50*(3), 254–257. doi:10.1111/ejed.12136

Tuck, E. (2009). Suspending damage: A letter to communities. *Harvard Educational Review*, *79*(3), 409–428. doi:10.17763/haer.79.3.n0016675661t3n15

UNESCO. (2016). *Education for people and planet: creating sustainable futures for all. Global education monitoring report*. UNESCO.

United Nations. (2007). *United Nations Declaration of the Rights of Indigenous Peoples*. United Nations.

Vansina, J. (1985). *Oral tradition as history*. University of Wisconsin Press.

Vizenor, G. (2008). Aesthetics of survivance. In G. Vizenor (Ed.), *Survivance: narratives of Native presence: 1-23*. University of Nebraska Press.

Wolfe, P. (2006). Settler colonialism and the elimination of the native. *Journal of Genocide Research*, *8*(4), 387–409. doi:10.1080/14623520601056240

Zidny, R., Sjöström, J., & Eilks, I. (2020). A multi-perspective reflection on how indigenous knowledge and related ideas can improve science education for sustainability. *Science & Education*, *29*(1), 145–185. doi:10.1007/s11191-019-00100-x

Chapter 12
Integrating Traditional Ecological Knowledge (TEK) With Environmental Education in India

Harleen Singh Sandha
O.P. Jindal Global University, India

Pranjal Khare
O.P. Jindal Global University, India

Paridhi Sharma
O.P. Jindal Global University, India

ABSTRACT

Imparting educational awareness and ensuring preservation of traditional ecological knowledge (TEK) encompasses a complex interplay of opportunities and constraints. This chapter seeks to explore the complex terrain that defines TEK conservation and preservation in India. The present chapter seeks to throw light on the concept, meaning and significance of TEK. It discusses the prevalent TEK systems in India in detail. Given the continuous decline of the Indian TEK systems, the chapter stresses upon the need to preserve and protect it. One of the foremost ways of protection is providing adequate legal recognition and protection to Indian TEK. Furthermore, the chapter discusses the role that TEK plays in promoting environmental education through the help of numerous community programs and local initiatives. Lastly, the chapter highlights the importance of integrating TEK within the formal education system to enhance its protection and generate greater awareness.

INTRODUCTION

With extreme heatwaves, droughts, wildfires and rising sea levels, the issue of sustainable development

DOI: 10.4018/979-8-3693-2577-3.ch012

and environment conservation doesn't need an introduction. United Nations Development Programme with the member nations is constantly working on the sustainable development goals to achieve the goals. Without undermining the efforts of the national governments around the world that are looking for scientific solutions in order to achieve a balance in the environment protection and economic growth, the chapter stress upon the importance of indigenous tricks and skills to co-habit with the environment. Beyond a romanticized preservation of indigenous identities, this system offers philosophical anchor for harmonious socio-environmental balance through guiding natural resources as custodians. The traditional knowledge of India preserves a rich yet threatened heritage of communities that is reflected in diverse Traditional Ecological Knowledge ("**TEK**") systems. This system evolved over generations of communities ranging from Eastern Himalayas to the Andaman Islands and provides ways to sustainably manage surrounding flora, fauna, and natural resources. TEK fosters an inherent consciousness towards environmental sensitivity and is expressed through practices like sacred grove conservation, prudent hunting norms and rainfall predictions for agriculture (Singleton et al, 2021). Academics, agency scientists, and policymakers are increasingly looking towards TEK as a possible source of inspiration for new models of ecological restoration, conservation biology, and ecosystem management (Kimmerer, 2002). However, in modern India, livelihood transition induced dislocations, have endangered the inter-generational passage within native rural communities. Reintegrating this fast-eroding traditional knowledge with contemporary sustainable education provides transformative scope for environmental awareness across the world.

This chapter discusses strategies for mainstreaming this knowledge of the land and wildlife reflected in Indian TEK with school curriculum frameworks, policy guidelines and decentralized digitization. Understanding that the key lies in transitioning the perspective on ecological heritage from being 'relics' for selective community revival to being pivotal knowledge partners for ethical environment-centric education model, the chapter stress on need to work on foundation by integrating the TEK in environment education system. Thereby, Authors argues that India needs to leverage its still surviving but threatened wealth of traditional knowledge while also re-instilling care, reverence and balance in the generations choosing to inhabit shared environment (Kimmerer, 2002). The chapter further discusses various examples of TEK System promoting intelligent utilization of resources, which if integrated in education system will help to align with natural regeneration cycle.

In the words of marine ecologist Fikret Berkes, TEK refers to the cumulative body of knowledge, beliefs, traditions, practices, and institutions that guide human communities towards sustainable management of surrounding flora, fauna, and natural resources through generations (Berkes 2018). It encloses the interdependent social, economic, spiritual, ecological, and philosophical worldviews, intrinsic to the cultural heritage of indigenous groups and rural communities. TEK reflects a cooperative human nature which is balanced by embedding conscious conservation ethic. The continuance of TEK systems relies heavily on intergenerational sharing of region-specific knowledge evolved through generations of stability in that specific area (Cajete 2020).

TEK largely can be exhibited through two broad classifications, viz documented TEK and oral/abstract TEK. Documented TEK refers to codified ecological knowledge of biodiversity, plant variety, agricultural practices, forest charts, climate models etc., preserved by specialist knowledge holders within a community. The oral or abstract TEK includes the unwritten traditional wisdom passed on through generations via folklore, taboos, rituals, ceremonies, livelihood practices and customary laws. However, this nuanced and regional nature makes it vulnerable to dilution and extinction (Whyte 2013).

In India, geographical variety, climate dynamism and cultural intricacies have fostered highly diverse contexts for environment-human co-evaluation. This is reflected in the richness as well as accuracy of accumulated TEK systems among the communities living in that geographical area. TEK diversity ranges from documented treatises on agricultural networks in Tamil Nadu, landscapes conserved as community heritage in Meghalaya, extensive soil taxonomy classified by Warli tribes in Maharashtra and sensitive wildlife tracking techniques evolved among Mishing clans from Assam. Nationally, TEK is enriched by environment unique rainwater harvesting in Thar desert, tidal agriculture by Adivasi groups in Andaman and shifting cultivation as practised in the Eastern Himalayan slopes over centuries. Additionaly, it includes shared wisdom across several geographical areas by connecting Kerala to Bihar due to Monsoon forecasting model and thus appreciates not only the plurality but also inter-connectedness of indigenous knowledge forms. Therefore, contextual integration of this dynamism with localized environmental education programs can foster natural adoption pathways for sustainability (Das et al., 2021).

TEK systems include important ecological heritage for community focused preservation in India through Prudent Resource Usages, Soil/Water Conservation, agricultural biodiversity, and sustainable livelihood. TEK enables an intelligent resource utilization that is aligned with natural regeneration cycles. For example, differential harvesting of medicinal plants, seasonal fishing bans, indigenous land and water management practices, such as building check dams, contour bunds, and drainage channels, play a vital role in soil and water conservation (Hatfield et al 2018). Check dams control water flow, preventing erosion and facilitating groundwater recharge. Contour bunds reduce surface runoff, promoting water absorption and preventing soil erosion. Additionally, drainage channels help in effective water distribution and prevent waterlogging. The collective impact of these indigenous practices ensure sustainable agriculture, and safeguarding of vital water resources for communities. By harnessing the knowledge of traditional land-water management, these methods foster resilience in local ecosystem, supporting agricultural productivity and enhancing food security for indigenous populations.

TEK also provides frameworks for environment-friendly rural practices like forest produce harvesting, rain-fed irrigation etc. Therefore, integrating TEK with environmental education creates organic pathways for community mobilization achieving sustainability goals mentioned under National Biodiversity Action Plans or National Water Mission.

INDIAN TEK SYSTEMS

India's cultural diversity has spread diversly from Ladakh's high altitude desert to the wet evergreen forests of Western Ghats. These prominent TEK examples showcase sustainable community guardianship of environment. The Warli tribes inhabiting mountain slopes near Maharashtra's Western Coast have cultivated an extensive ethno-ecological knowledge base. Their ritual paintings also embed cryptic occult references to regional wildlife. The Warli tribe carefully process seeds of the Sahyadri mountains' endemic Waghoba plant during certain lunar phases for germination resilience.

Kani tribes in Kerala's Agasthyamalai hills uniquely practice carving of idols from fragrant resin rich Guy trees without fatally harming and respecting the tree deity. The indigenous Toda community conserved biodiverse shola grasslands in Nilgiri Hills through religious protections for their water temple sites (Haritha, 2019). The Nicobarese islanders in Andaman Sea sustain intricate rain fed coconut palm agriculture through seasons by observing flight paths of Great Nicobar teals.

Similarly, in Nagaland, Ao Naga forestland is governed through indigenous stewardship principles called Genjaobi embedding sustainable rotational agriculture with community hunting rules and ownership rights (Tynsong, 2020). The Apatani tribe of Arunachal Pradesh practices unique paddy fish farming symbiosis where fish cultivation in waterlogged farms enriches soil fertility organically while rearing protein rich small fishes on rice paddies through traditional weirs. Thus, community TEK frameworks, being locally evolved innovations for environmental harmony over generations, provide sustainable development models that balance human needs with environmental integrity. They signify culturally embedded conservation philosophies to draw actionable principles for modern India (Dkhar & Tiwari, 2020).

TEK constitutes the knowledge cultivated by communities through cumulative experiences of interacting with local ecosystem. Honed over generations of co-existing perspective, TEK reveals diverse environmentally conscious worldviews intrinsic to indigenous groups and rural societies across geographies. Therefore, TEK provides working frameworks for ecological stability and sustainability which can be learn through integration in education system to help preserving environmental integrity to secure community wellbeing.

Integrating Environmental Education and Biodiversity Conservation

After understanding how TEK has been dealing with environment to maintain the balance, the chapter now focuses on the question of why there is need to integrate TEK into the environment education. Integrating TEK into environmental education programs promotes the fundamental aspect of sustainable development in India through community's cultural wisdom. TEK through its natural-cycles and generational wisdom if used in education models can be useful to promote organic adoption of conservation behaviour. It can be done through fostering connection with nature, demonstrating sustainable techniques, and enculturing conservation ethics (Alexiades 2021).

TEK based activities like sacred grove walks, heirloom crop trails, etc., develop a spiritual connectivity with specific regional ecology. It demonstrate micro-climate specific innovations like water irrigation facilities, agroforestry models etc., in turn serving as live evidence of sustainable techniques relevant to local environmental challenges and security of livelihood. TEK system practices like transmission through intergenerational storytelling platforms and revived ritual ceremonies promotes indigenous conservation philosophies to influence community consciousness form within. Native biodiversity has been nurtured and sustained in India through niche traditional agricultural ecosystem, community woodlands, and seasonal fishing systems guided by ecological knowledge of indigenous groups over a period of time. Therefore, community TEK frameworks aid environment stability across scales by nurturing regional species and adapting customary practices to achieve ecological balance which is aligned to natural regeneration capacities (Harisha et al., 2016).

This ecological consciousness is intrinsic to TEK. It caters to sustainable community progress pathways in conformity with local environment through optimizing resource use, securing ecological services and fostering eco-friendly enterprises. TEK practices provide for various indigenous resource optimising tactics like seasonal fishing prohibitions, differential wild harvesting, and heirloom seed selection practices. These techniques counter over exploitation while maintaining spiritual ties with natural flow. Further, watershed conservation, sacred woodland protection and wildlife taboos help in securing nature's ecological subsidies like nutrient cycling, carbon sequestration and genetic information storage that have been provided as public goods over generations. TEK provides frameworks for various sustainable rural enterprises related to bamboo handicrafts, herbal dyes extraction, organic agriculture, naturopathy medicine that can meet localized consumption demands without degrading regional ecology.

Thus, viewed through the TEK lens, human ecosystem is spiritually rooted and is based on a system of reciprocal and caring bond with nature. The ethics of balanced co-existence delivers holistic sustainability outcomes across interlinked social, economic and environmental dimensions. Integrating these harmonized models with mainstream development programs and environmental education offers indigenous knowledge for India to localize sustainability (Negi et al., 2023).

TEK Systems Under Threat in India

Despite promoting sustainable community environment balance for generations, the growth of India's diverse TEK systems stands imperilled today due to multiple factors contributing to habitat degradation, decline in agricultural biodiversity, socio-cultural disintegration, gender exclusion, generational gap and legal disenfranchisement.

The rich relevance of land that TEK entails is threatened by expanding resource usage and development footprints, affecting community based ecosystem like sacred groves, village woodlands and pastoral watersheds. Deforestation, wetland drainage, and unmonitored tourism often mobilise communities from TEK evolved niches. The indigenous agricultural diversity that TEK systems protect is declining with agricultural modernization policies targeted towards selective high-yielding agrochemical dependence. Displacement from traditional environment due to urbanization, climate disruptions and conflicts dislocate entire indigenous societies, violently dismantling intergenerational knowledge flow. Even settled communities, and modern institutions introducing standardized curriculum are erasing diverse localized ethics and worldviews engraved in environment. Further Globalized aspirations and neoliberal economies are orienting youth away from familiar livelihood in socio-cultural rupture. Resultantly, elders who safeguarded niche skills around artisanal fishing nets, herbal medicine preparation, and seed selection protocols are dying without transferring contextual TEK legacies rooted in landscapes over generations. Furthermore, national policies imposed without consulting indigenous governance institutions undermine community autonomy in upholding customary resource access frameworks, thereby hindering the continuance of self-determined TEK (Tang & Gavin 2016).

To subside these threats and build TEK management pathways for sustainability, India needs to urgently recognize Community Conserved Areas, operationalize access and benefit sharing of TEK knowledge, democratize biodiversity records through citizen science, and nurture intergenerational TEK dialogue (Singh et al 2020).

OBSTACLES IN THE PROTECTION OF TEK: HIGHLIGHTING THE NEED FOR PRESERVATION

The government's policies and laws, as well as the pressures of economic development, are posing a serious threat to indigenous groups' TEK, raising a serious cause for alarm. The foundation of the identity and cultural heritage of indigenous populations worldwide is Traditional Knowledge ("**TK**") derived from natural resources like medicinal plants. However, indigenous communities are becoming increasingly susceptible to outside interests seeking to monopolize ownership over their traditional resources, causing environmental degradation, and forced relocation (Swiderska et al., 2009). While there exist numerous factors contributing to the erosion of TEK, some of the prominent ones are enumerated below-

- Market Economy: The advent of the market economy has caused a serious decline in the transmission and retention of TEK, as precious knowledge transmitted from one generation to the other, primarily because of the substitution of market-based products in place of local and indigenous products. The market makes food more readily available in local communities, which slows down the dissemination of TK about food production (Kodirekkala, 2017). Similarly, market forces may trigger shifts away from traditionally gathered food and livestock towards highly valuable goods. This phenomenon has in turn been exacerbated by the forces of modernisation and globalisation, putting pressure on the livelihood of many local communities.
- The Developmental Burden: The welfare and rights of indigenous groups have been compromised as a result of rising urbanization and farmland conversion brought on by the pressures of economic growth. The government's developmental ambitions often take precedence over the opinions of the indigenous people (Nepal, 2021). This developmental burden is shared by many developing countries including India.
- Migration and Forced Relocation: In this context, the United Nations' State of World's Indigenous Peoples (SOWIP, 2021) report noted that:

"...it is clear that one of the most significant threats faced by indigenous peoples arises from their displacement, eviction, and separation from their lands, territories, and resources. These issues are expanding and represent significant challenges to the security, health, and survival of indigenous peoples and their cultures."

Thus, the current decline in the transmission of TEK, its channels, beliefs, and livelihood patterns may be the result of voluntary or involuntary non-indigenous immigration and indigenous exodus. For example, the migration of young indigenous people to metropolitan areas in quest of better employment and educational possibilities may have an impact on their interest in studying TEK.

- Imbalances in Environment and Natural Resources: This stems partly from the abovementioned consequence of uneven migration. Changes in TEK may be caused by ecological disturbances, increased population, excessive use, or tensions resulting from immigrants' resource exploitation on native lands. In addition, the forced or voluntary movement of indigenous people or local communities may also have an impact on TEK's ecological foundation.
- Loss of Indigenous Language: Transmission of inter-generation TEK is based primarily on community and local languages and therefore a loss of native language constitutes a loss of TEK. Native dialects are frequently used to transmit TEK, and the loss of linguistic diversity directly affects the loss of information diversity (Nepal, 2021). Thus, TEK is also at risk when traditional languages disappear. More often than not a major aggravating factor for this is that the younger generations of indigenous communities only occasionally visit their communities after spending an extended period of time outside. They consequently frequently lose out on the opportunity to learn from elders and speak their native tongue.
- Lack of Integration in the Formal Education System: It is a well-known fact that the formal education system in India frequently disregards indigenous knowledge and is primarily based on western principles, a direct result of India's legacy of colonialism. The TK system is not highly valued

by the knowledge system founded on Western principles. Except for a few cases, which will be discussed later in detail, awareness about TEK is largely absent in most school curriculums and thus continues to be scanty. The idea that TK is antiquated and unscientific has been cultivated by Western invasions, and thus the system of passing down such knowledge from one generation to the next has collapsed (Honwad, 2018). It is in this manner that the forces of the formalised education system work in complete opposition to the preservation and promotion of TEK, thus affecting not only its existence but also the sustainable livelihood of many traditional communities.

- Lack of Formal and Legal Recognition: Lack of formal legal protection of TEK is another factor which plays a huge role in the loss of preservation and transmission. Except for a few initiatives and legislations which only briefly touch upon TK and its protection, India, like most countries, does not have a standalone legal framework recognising and protecting TK in its entirety.

All these factors and more, together play a significant role in the decline of the transmission of TEK in India today. The younger generation itself needs to take into consideration the necessity of conserving TK in the current context. Inadvertently, the preservation of TK will be pivotal in safeguarding the cultural norms, society ideals, and vital human techniques, so as to further use the knowledge through education.

Preservation and Protection: First Step Towards Promoting Education in TEK

Having discussed the importance, threats and issues relating to TEK system, the chapter now discuss the existing policies in education system and the steps towards promotion of education in TEK. Given this backdrop, it becomes imperative to emphasize on the necessity to retain the dwindling state of indigenous knowledge. One of the important stages in generating greater awareness and promotion of the same is taking effective steps towards first preserving this knowledge by means of appropriate legal and policy framework. Thus, legal governance and recognition of TK will not only help preserve it but also constitutes the key to promoting education and awareness in the field.

In international forums, there is a growing recognition of the significance of safeguarding the customs, knowledge, and innovation of indigenous and local populations. The United Nations Educational, Scientific, and Cultural Organization (UNESCO) and World Intellectual Property Organisation (WIPO) jointly launched the first attempt to protect TK under the IP regime in 1978. This initiative resulted in the 1982 "Protection of Expressions of Folklore against Illicit exploitation and Other Prejudicial Actions". Since then, the Convention on Biological Diversity (CBD), which was adopted in 1992, has brought the protection of TK even more attention (Mohi-ud-din et al., 2019).

Intergovernmental organizations that deal with IP, the environment, and even human rights have made significant efforts to safeguard TK held by indigenous and local communities. These organizations include the World Trade Organization (WTO) and its Council for TRIPS, the WIPO, the Food and Agriculture Organization (FAO), the United Nations Conference on Trade and Development (UNCTAD), and the World Health Organization (WHO) (Gazizova, 2020).

Although, India has not passed any law specifically pertaining to TK, it has taken up steps to fulfill its obligations under the TRIPS agreement, the CBD, and the International Treaty on Plant Genetic Resources for Food and Agriculture 2004. Examples of these actions include, *inter alia*, the Biodiversity Act of 2002, the Protection of Plant Variety and Farmers Right (PPVFR) Act of 2001, and the Patent (Amendment) Act of 2005.

The Biodiversity Act of 2002 mainly deals with foreign people, organizations, or businesses' accessibility to genetic resources and related information in order to guarantee a fair distribution of benefits resulting from the use of these resources. The Act establishes guidelines for the use of TK and biological resources in three different ways, depending upon the case, basis, "Prior approval of National Biodiversity Authority"; "Prior intimation to the State Biodiversity Board"; and lastly, exemption from prior consent or notification for local inhabitants and communities. Further, subject to the Biodiversity Rules', Section 21 and Rule 20, the Act requires that suitable benefit-sharing clauses be included in access agreements and mutually agreed-upon terms pertaining to the transfer and use of biological resources or knowledge acquired from India for commercial purposes, bio-survey, bio-utilization etc.

The PPVFR Act of 2001 and the PPVFR Rules of 2003 categorically provide for the protection of plant breeders' rights over their newly developed varieties, as well as farmers' rights to register new varieties and safeguard, breed, use, exchange, share, or sell plant varieties that they invented and maintained over many generations.

The Patents (Amendment) Act of 2005 includes the following two grounds among the eleven specified grounds for post-grant opposition:

"(j) that the complete specification does not disclose or wrongly mentions the source or geographical origin of biological material used for the invention;

(k) that the invention so far as claimed in any claim of the complete specification is anticipated having regard to the knowledge, oral or otherwise, available within any local or indigenous community in India or elsewhere,"

These additional grounds, in effect, allow the source country to protect the biological material or TK belonging to a local or indigenous community, entitling the holders of such knowledge to be recognized and compensated through suitable benefit-sharing arrangements.

It can be argued that the above-mentioned laws and policy do provide sufficient room to protect TEK along with other indigenous knowledge, but is largely rendered ineffective in the absence of a standalone legislation governing the rights of the indigenous communities in terms of protection and use of their knowledge. Another issue is the lack of recognition by the IPR regime of other countries, thus indicating that the lack of uniformity in international rules acts as a roadblock in the protection of indigenous knowledge both in India and abroad (Forsyth, 2013).

A significant stride made by India in this regard is the creation of a unique mechanism for the protection of indigenous knowledge known as the Traditional Knowledge Digital Library ("**TKDL**"). Notably, India has been one of the first few countries to undertake such an exercise by developing a database of its indigenous knowledge. In order to fully understand the intent and extent of TKDL, it is important to briefly delve into the history and need for the development of TKDL. The global debate on the preservation of TK is based on the issue of bio-piracy, which occurs when TK is appropriated for profit and patents are obtained without getting permission from the owners of that TK and without any compensation in return. India faced the issue of bio-piracy in many cases of patent applications filed by foreign companies for Turmeric, Neem Tree and Basmati Rice and was able to successfully defend its TEK (Swamy, 2014). However, this was not possible without spending huge amount of money to ensure that Indian TEK was protected and such patents were not granted. Thus, the Government of India, via the Ministry of Ayurveda, Unani, Siddha and Homeopathy (AYUSH) and the CSIR, developed the TKDL database in 2001 as a defensive protection mechanism in response to these experiences before overseas patent offices (Mathur, 2003). The TKDL seeks to create a database of previous art references that does not itself constitute prior art by digitizing the body of publicly available literature on TK, particularly that which relates to Ayurveda, Siddha, Unani, and Yoga. The same has been translated into English, French, Spanish, Japanese, and German. When determining whether to grant a patent, prior art plays a critical role since the invention must meet the standards of novelty and non-obviousness. Only if the innovation is not publicly disclosed and is not included in prior art, can the novelty requirement be met (Sen & Chakraborty, 2014).

This makes TKDL a cost-effective strategy that saves the enormous expense that would have been required in order to challenge TK theft through patenting before patent offices worldwide. Furthermore, because it is a controlled access database, the possibility of biopiracy has decreased. The TKDL, as a defensive mechanism, has proven effective in its narrow goal of compiling prior art related to TK. However, TKDL in itself along with other methods of protection under the IPR laws do not sufficiently cover the scope of protection actually required and leaves many doors opened, exposing TK to misuse and misappropriation. TKDL faces many limitations in the form of translation problems, lack of comprehensiveness of the database as TKDL does not include oral TK etc. Thus, a standalone legislation based on a "sui generis" model is required to better protect TK through a system of "access and benefit sharing mechanisms". Such legislations have been inculcated by countries such as Brazil and Peru (WIPO, 2003). It is important to take these matters into consideration seriously as effective education and awareness of TK is not possible in the absence of effective protection by means of a standalone legislation.

TEK IN CLASSROOM AND BEYOND

The synergy between TEK and environmental education has promising avenues for promoting sustainable development and encouraging environmental responsibility. However, there is a widespread concern amongst most theorists and researchers that formal education systems have significantly disregarded cultures, conditions, languages, and knowledge of the indigenous communities throughout the globe (Brady, 1997). Exploring insights from history, figures like Gandhi, Nyerere, and Friere had developed unique educational models based on their own principles in response to the oppressive nature of colonial education. Gandhi formulated his model of 'Basic Education' which was inspired by John Dewey and stressed upon the importance of local knowledge, practical experiences, vernacular languages, and practical experiences. This innovative approach holds the potential to bridge the gap and reconcile both traditional and established scientific knowledge, and formal and informal learning environments. Despite its immense potential, the attempts to its practical application in India has only been limited to specific regions (Agrawal, 1995).

The *Gram Vidyapith* in Gujarat serves as a compelling model and proves the effectiveness of this approach. With their three-year rural studies degree, these *Vidyapiths* (or colleges) represent a significant effort to bring together local, traditional, and practical knowledge and conventional teaching methods. However, despite their commendable model, full integration in student learning remains a strenuous task. This issue can be countered by introducing fieldwork projects which put emphasis on TEK in various areas including animal husbandry, horticulture, agriculture, etc. These projects may include diverse topics like herbal veterinary medicines, livestock ownership partnerships, and drought management.

But the sustainability of these solutions is interdependent on efficiency and activism from teachers and navigating through the university affiliation requirements (Shukla & Gardner, 2006).

Furthermore, these colleges are contributing in making a significant impact as several of them have partnered with NGOs and other external institutions in order to enhance the integration of TEK into their curriculum. Three of these Vidyapiths have partnered with Ahmedabad's Centre for Environmental Education towards a joint effort of launching a project called Samvardhan. This project is aimed at empowering local communities to nurture their environment through TEK and managing their resources sustainably.

In order to protect and promote TEK, it becomes essential to integrate it with environmental education. This integration is crucial for fostering environmental responsibility and promoting sustainable development. There are several inspiring local and community initiatives and programs which are raising awareness about it. Many researchers are undergoing studies to evaluate the impact of integrating environment education with TEK. Although, in India, there exist few nature schools which integrate TEK into formal education, promoting sustainable practices and environmental awareness, the major one being the *Honeybee Network's Barefoot College*. This is a network of indigenous communities spread across India dealing with TEK related to various aspects of the environment including biodiversity conservation, natural resource management, and traditional livelihoods. The network has established a unique school called Barefoot college in the state of Rajasthan in India, which helps students engage with a nature-based curriculum, integrating TEK from the Tharu community into teachings on agriculture, water conservation, and biodiversity. Aranya Eco School is another one of a kind nature school and TK based learning in the state of Assam in India. This educational institution is situated amidst the Kaziranga National Park and is primarily involved in imparting knowledge about medicinal plants, wildlife conservation, and sustainable living practices to the students. Their pedagogical approach involves integrating Karbi Tribal knowledge into its program. This approach is far more impactful than textbook based education.

There is also growing importance of Biodiversity Contests which originally was developed by an NGO called SRISTI, in collaboration with the Honeybee Network. These contests spread across India, Bhutan, and Vietnam engage children in showcasing their TEK. Their primary objective is to nurture a generation of young environmental stewards who are equipped efficiently to protect local resources and biodiversity. In these contests, children demonstrate their plant knowledge by bringing specimens that they've collected. These entries by the children are evaluated by teachers, community elders, and forest staff whereby they interview all the children on the basis of the number of samples of plants and their novelty, their habitat and potential uses. The children are accordingly scored on their knowledge and presentation style, and then suitably rewarded. These competitions inspire further learning through activities like creating a school herb garden and a herbarium, and crafting informative displays (Shukla & Gardner, 2006). It is important to remark that through these initiatives, children demonstrate a deep understanding of plant uses, especially for medicinal purposes, highlighting the wealth of informal knowledge they possess. It shows how TK can be used in schools and help communities manage resources and protect the environment. Colonization stimulated cultural disruption is another criteria that is widely used to challenge the quality or veracity of the local knowledge system.

To bridge the gap between formal school education and community based traditional knowledge, it is essential to highlight the noteworthy efforts made by Uttarakhand Seva Nidhi Paryavaran Shikshan Sanstha, which is an NGO established and running in the Kumaon region of the Himalayas. This educational institution has developed an environment education curriculum, called Our Land Our Life (OLOL), and incorporated it into the school system. This curriculum puts emphasis on local environmental concerns in the Kumaon Himalayas and utilizes embedded pedagogy to address the needs of rural communities.

It addresses livelihood issues pertaining to land, water, and natural resources like fodder, crops, and trees and integrates them into the formal education system. This mechanism helps students prepare for future livelihoods tied to agriculture and environmental sustainability. It helps in connecting the classrooms to real-world problems through a community-based approach, valuing TK and promotes sustainable living. Further, there are many traditional ecological practices which are provided in this curriculum. It is also important to note that this pedagogical approach goes beyond simply teaching TEK and further blends with best modern practices. For example, a few inspiring schools built rainwater harvesting systems through local material and existing TK. Such a unique curriculum, developed in partnership with teachers and villages, goes beyond mere environmental science education and promotes environmental education for a resilient future (Honwad, 2018).

Outside the ambit of formal systems of education, many attempts have been made to integrate TEK with various aspects of environment and resource management. The Medicinal Plant Conservation Centre is one such NGO which is pushing the boundaries of traditional education by seamlessly blending TEK with environment and resource management. The approach of this NGO revolves around their two key programs which includes '*vaidu sammelans*' and 'village botanist' (or barefoot botanist). The term *vaidu* is an indigenous folk term which in common parlance means healers and traditional herbal practitioners (Shukla & Gardner, 2006). In these *vaidu sammelans,* all the local healers and practitioners gather and empower themselves by documenting their plant knowledge, supporting value added ventures, and interacting with other practitioners.

Additionally, in the village botanist program, the *vaidus* and botanists get proper training to strengthen their skills. This multi-pronged community-based approach fosters knowledge sharing, economic opportunities, and collaboration for *vaidus*, ultimately empowering them to contribute to local biodiversity conservation. These programs have made a significant impact by training and spreading knowledge across the local healers and village botanist programs have been successful in generating learning at the local, regional and national levels up to some extent. However, their efforts are more evident at the local level. The only drawback of this initiative is that the impact of this institution might not be long lasting in the long run as most of its efforts are outside the system of formal education. Nonetheless, the success of their community-based approach, involving local indigenous knowledge experts shows great promise for managing resources more effectively in the future. This mechanism works not just by gaining valuable data, but also by giving due respect and legitimacy to the concerns of those who understand their land more closely.

In 2015, India ratified and agreed to adopt the Sustainable Development Goals (SDG) 2030 and in adherence to such goals, it launched its National Education Policy (NEP) 2020 (Singh, 2023). In the policy, there has been recognition of the rights of marginalized and under-represented people and groups. The objective is to give voice to these people and groups by preparing a curriculum which is to be framed by the Ministry of Human Resource and Development, Ministry of Women and Child Development (WCD), Ministry of Health and Family Welfare and Ministry of Tribal Affairs. The policy has included every section of the society to have a diversified curriculum. Notably, it has inculcated TK into the curriculum of the schools. Experiential learning and adopting a pedagogical approach to include learning and teaching in the mother tongue are a few other salient features of the policy. The Indian government has taken a revolutionary initiative to preserve and protect its TK by assimilating it with the formal education system.

The efforts to raise awareness about environmental education and TEK aren't just limited to India, but is snowballing into a global movement with countries and communities worldwide embracing these approaches. For example, in New Zealand, there is integration of native children's culture into their formal educational curriculum. Most of the schools in the country have been following the 'forest schools' programs or 'enviroschools' programs which are imported from other indigenous cultures and traditions and are incorporated into their existing curriculum. In particular, the indigenous views and learnings of the Māori culture gives meaning and authenticity to their forest school programmes which are implemented in early childhood education. These approaches foster an intense connection to the place and naturally forms a strong spiritual relationship with the natural world. The children going to these schools engage themselves in various activities focused on sustainability which includes production of food, observation of water, enhancement of biodiversity and waste minimization (Alcock & Ritchie, 2018).

It is also pertinent to highlight the initiatives taken by Toledo Environmental and Cultural Heritage Alliance (TEACHA) in Ohio, United States. This alliance comprises many prominent activists and researchers and allied community educators whose primary role is to design and implement a framework where environmental and cultural heritage can be easily incorporated into the curriculum of primary schools in the Toledo district. It provides students with such an educational framework with teachings involving present TEK along with other compulsory subjects. This framework of environmental and cultural heritage knowledge helps students maximize their potential within their natural habitat and provide tools necessary to cater to the obstacles that may come in their way of a desired life. Such a curriculum has the future prospect of contributing positively to the health and promotion of social justice of the indigenous communities (Baines & Zarger, 2017).

In Taiwan, scholars have undertaken study to explore the effects of a sustainable development curriculum framed by working with indigenous communities (particularly a small Rukai Indigenous community). The primary objective to undertake this research was a larger concern that formal education methods often ignore the unique experiences of Indigenous students, and this disconnect might gradually make learning less effective and fulfilling for them. The results of the study demonstrate an enhanced sense of place amongst the students. The students were more familiar with their local geographical environment and had further enhanced their connectedness with the place. This place not only includes the geographic location but is also associated with emotional and cultural attachment that residents have with respect to that place (Li & Shein, 2019).

Furthermore, there have been established nature kindergarten in some of the public schools in British Columbia, Canada. These schools stand amidst the Coast Salish First Nations people. Although it is aiming for local roots, the nature kindergarten has been mostly inspired by the forest schools or nature preschools in order to reflect the community's values and beliefs (Elliot & Krusekopf, 2018). This program involves a collaborative design process involving educators, environmentalists, and even children. It promotes fostering a connection with nature, learning from the environment itself as a living teacher, collaborating within a community, and achieving holistic well-being, all of this incorporated with the wholesome understanding and insights of Aboriginal traditions and cultures. Their aim is to build a nature kindergarten experience which focuses on a hands-on approach with the local habitat thereby fostering responsibility towards the environment, honoring Indigenous knowledge and perspective.

All of the efforts and initiatives taken in India and worldwide, review the educational approaches, even those embedded within rigid educational systems and emphasize diverse learning methods. When it comes to effective resource management through traditional community-based approach, every method and knowledge should be utilized to the full potential. Insights from the past show that much of this information and means was mostly disregarded because of the prevalence of formal education systems. Informal knowledge like TEK was denied its due credit but this trend has changed. There is large-scale recognition of the immense potential of TEK and accordingly many initiatives are accessing and integrating it into effective and sustainable resource management and conservation practices. However, there still remain many challenges which are required to be addressed such as overcoming institutional hurdles and developing culturally sensitive knowledge exchange methods. For successful collaboration, environmental educators and activists have to build trust and equitable partnerships with the indigenous communities. It is also essential to recognize and protect the intellectual property rights of indigenous communities to prevent their misappropriation and exploitation.

CONCLUSION

This chapter has briefly discussed the meaning, scope and significance of TK. TK represents the richness and diversity of the Indian society. India has an extensive legacy of interaction between nature and society, particularly between ethnic and indigenous populations and the use of bioresources in the healthcare system. Non-codified indigenous knowledge, such as folk or indigenous medicine without written scriptures, also play a part in the system and serve as a source for richness and novelty in many inventions and ecological resources. This rich TK has been passed down through generations resulting in its precious survival and accumulation.

The various kinds of TEK and their role in promoting biodiversity conservation, sustainable management and environmental education has also been discussed throughout. An important appreciation in this respect pertains to the fact that the system, practices and transmission of TEK is currently under threat due to a variety of reasons such as modernisation, globalisation, gradual shift of the young generation to urban living, environmental degradation, lack of adequate recognition and promotion of TEK etc.

Further, the chapter mentions the importance of protecting TEK through a legislative and institutional framework to conserve it. The current legislative framework for the protection of TK is minimal at best and does not even begin to cover its scope in its entirety. As outlined in the chapter, a significant threat is the issue of bio-piracy, thus necessitating the requirement of a "sui generis" model of protection of TK. A "sui generis" model is based on both defensive (such as TKDL) and positive mechanisms (benefit and access sharing agreements), as has been done by many countries, to preserve India's TK and protect the rights of the traditional holders of such TK. Thus, protection through appropriate law and policy constitutes the first step in the promotion of TEK and its education.

The promotion and dissemination of this knowledge through inclusion in school curriculums, teaching methodologies and on-field participation is a necessity as it allows most importantly the young generation to learn, practice and understand TEK holistically at the local or community level. Thus, community interaction is an important facet when it comes to education of TEK as most of the TK is held by locals and community at the village level. The local and community initiatives, as deliberated, are steps in the right direction. A focused inclusion of TK within school curriculums, though undertaken, exists in a limited manner restricted only to certain parts of the country. It is therefore important that efforts to integrate TK within the formal education systems is undertaken in an organised and uniform manner across the nation and societies. In the long run, the formal education system is nearly unrivalled as a tool of power in promoting TK. It offers a specific type of knowledge and a process for creating knowledge that are dogmatic and exclusive in and of themselves, but they also serve as the foundation for recognition and, thus, credibility in society at large. However, in order to utilize the enormous reservoir of TEK and information for efficient and sustainable resource management and conservation, every effort must be taken to identify and effectively access it.

REFERENCES

Agrawal, A. (1995). Dismantling the Divide Between Indigenous and Scientific Knowledge. *Development and Change*, *26*(3), 413–439. doi:10.1111/j.1467-7660.1995.tb00560.x

Alcock, S., & Ritchie, J. (2018). Early childhood education in the outdoors in Aotearoa New Zealand. *Journal of Outdoor and Environmental Education*, *21*(1), 77–88. doi:10.1007/s42322-017-0009-y

Alexiades, A. V., Haeffner, M. A., Reano, D., Janis, M., Black, J., Sonoda, K., Howard, M., Fiander, C., & Buck, M. (2021). Traditional Ecological Knowledge and Inclusive Pedagogy Increase Retention and Success Outcomes of STEM Students. *Bulletin of the Ecological Society of America*, *102*(4), e01924. doi:10.1002/bes2.1924

Baines, K., & Zarger, R. K. (2017). "It's Good to Learn about the Plants": Promoting social justice and community health through the development of a Maya environmental and cultural heritage curriculum in southern Belize. *Journal of Environmental Studies and Sciences*, *7*(3), 416–424. doi:10.1007/s13412-016-0416-3

Berkes, F. (2018). *Sacred Ecology*. Routledge.

Brady, W. (1997). Indigenous Australian Education and Globalisation. *International Review of Education*, *43*(5), 413–422. https://www.jstor.org/stable/3445057. doi:10.1023/A:1003077718880

Cajete, G. A. (2020). Indigenous science, climate change, and indigenous community building: A framework of foundational perspectives for indigenous community resilience and revitalization. *Sustainability (Basel)*, *12*(22), 9569. doi:10.3390/su12229569

Das, A., Gujre, N., Devi, R. J., & Mitra, S. (2021). A review on traditional ecological knowledge and Its role in natural resources management: North east India, a cultural paradise. *Environmental Management*, 1–22. PMID:34791537

Dkhar, M., & Tiwari, B. K. (2020). Traditional ecological knowledge of tribal communities of North East India. *Biodiversitas (Surakarta)*, *21*(7).

Elliot, E., & Krusekopf, F. (2018). Growing a nature kindergarten that can flourish. *Australian Journal of Environmental Education.*, *34*(2), 115–126. doi:10.1017/aee.2018.27

Forsyth, M. (2013). How Can Traditional Knowledge Best Be Regulated? Comparing a Proprietary Rights Approach with a Regulatory Toolbox Approach. *University of Hawai'i Press, 25*(1), 1-31. https://www.jstor.org/stable/23725729

Gazizova, A. S. (2020). Protection of Traditional Knowledge: The Work and the Role of International Organisations and Conferences. *International Journal of Higher Education*, *9*(8), 95–99. doi:10.5430/ijhe.v9n8p95

Harisha, R. P., Padmavathy, S., & Nagaraja, B. C. (2016). Traditional ecological knowledge (TEK) and its importance in south India: Perspecive from local communities. *Applied Ecology and Environmental Research, 14*(1), 311–326. doi:10.15666/aeer/1401_311326

Haristha, J. (2019). Sacred groves of north Kerala: The last refuge for biodiversity amongst urbanisation. *Mongabay*. https://india.mongabay.com/2019/01/sacred-groves-of-north-kerala-the-last-refuge-for-biodiversity-amongst-urbanisation/

Hatfield, S., Marino, E., & Whyte, K. P. (2018). Indian time: Time, seasonality, and culture in Traditional Ecological Knowledge of climate change. *Ecological Processes, 25*(1), 7. doi:10.1186/s13717-018-0136-6

Honwad, S. (2018). Learning in Schools about Traditional Knowledge Systems in the Kumaon Himalayas. *Journal of Folklore and Education, 5*, 180–193. https://jfepublications.org/wp-content/uploads/2018/09/honwad-JFE20185.2-11.pdf

Kimmerer, R. W. (2002). Weaving traditional ecological knowledge into biological education: A call to action. *Bioscience, 52*(5), 432–438. doi:10.1641/0006-3568(2002)052[0432:WTEKIB]2.0.CO;2

Kodirekkala, K. (2017). Internal and External Factors Affecting Loss of Traditional Knowledge: Evidence from a Horticultural Society in South India. *Journal of Anthropological Research, 73*(1), 22–40. https://www.journals.uchicago.edu/doi/epdf/10.1086/690524. doi:10.1086/690524

Li, W. T., & Shein, P. P. (2023). Developing sense of place through a place-based Indigenous education for sustainable development curriculum. *Environmental Education Research, 29*(5), 692–714. doi:10.1080/13504622.2022.2098933

Mathur, A. (2003). Who Owns Traditional Knowledge? *Economic and Political Weekly, 38*(42), 4471–4481. https://www.jstor.org/stable/4414163

Mohi-ud-din, R., Farooq, S., Majeed, A., Khan, N. A., & Bhat, Z. A. (2019). Legal Framework on Protection of Traditional Knowledge: A Review. *International Journal of Advance Research in Science and Engineering, 8*(1), 100–108. https://www.researchgate.net/publication/331221999

Negi, V. S., Pathak, R., Thakur, S., Joshi, R. K., Bhatt, I. D., & Rawal, R. S. (2023). Scoping the need of Mainstreaming indigenous knowledge for sustainable use of bioresources in the Indian Himalayan region. *Environmental Management, 72*(1), 135–146. doi:10.1007/s00267-021-01510-w PMID:34341866

Nepal, T. K. (2021). An Overview of Threats to Traditional Ecological Knowledge. *Journal of Research in Humanities and Social Science, 9*(11), 01-04. www.questjournals.org

Rao, V. L. N., & Ramana, G. V. (2007). Indigenous knowledge, conservation and management of natural resources among primitive tribal groups of Andhra Pradesh. *Anthropol, 3*(Spec), 129–134.

Sen, S., & Chakraborty, R. (2014). Traditional Knowledge Digital Library: a distinctive approach to protect and promote Indian indigenous medicinal treasure. *Current Science Association, 106*(10), 1340-1343. https://www.jstor.org/stable/24102476

Shukla, S., & Gardner, J. (2006). Local knowledge in community-based approaches to medicinal plant conservation: Lessons from India. *Journal of Ethnobiology and Ethnomedicine, 2*(20), 20. doi:10.1186/1746-4269-2-20 PMID:16603082

Shukla, S., & Gardner, J. S. (2006). *The Role of Traditional Ecological Knowledge in Education for Community-Based Resource Management*. Research Gate. https://www.researchgate.net/publication/42761744_The_Role_of_Traditional_Ecological_Knowledge_in_Education_for_Community-Based_Resource_Management

Singh, B. (2023). Integration of Traditional and Folk Knowledge with Current Education System: Indianizing Education System and Regaining the Status of Vishwa-guru. *AGPE The Royal Gondwana Research Journal, 4*(3), 22-30. https://agpegondwanajournal.co.in/index.php/agpe/article/view/238

Singh, R., Sharma, R. K., Babu, S., & Bhatnagar, Y. V. (2020). Indian Trans-Himalayas. *Pastoralism, 15*(1), 15. doi:10.1186/s13570-020-00169-y

Singleton, B., Gillette, M., & Green, C. (2021). Toward productive complicity: Applying 'traditional ecological knowledge' in environmental science. *The Anthropocene Review, 10*(2), 393–414. Advance online publication. doi:10.1177/20530196211057026

Swamy, R. N. (2014). Protection of Traditional Knowledge in the Present IPR Regime: A Mirage or Reality. *The Indian Journal of Public Administration, 60*(1), 50. doi:10.1177/0019556120140103

Swiderska, K., Argumedo, A., Pant, R., Vedavathy, S., Munyi, P., Mutta, D., Heraclio, H., Song, Y., & Li, J. (2009). *Protecting traditional knowledge from the grassroots up*. International Institute for Environment and Development. http://www.jstor.com/stable/resrep01439

Tang, R., & Gavin, M. C. (2016). A Classification of Threats to Traditional Ecological Knowledge and Conservation Responses. *Conservation & Society, 14*(1), 57–70. https://www.jstor.org/stable/26393228. doi:10.4103/0972-4923.182799

Tynsong, H., Dkhar, M., & Tiwari, B. K. (2020). Review: Traditional ecological knowledge of tribal communities of North East India. *Biodiversitas (Surakarta), 21*(7), 3209–3224. doi:10.13057/biodiv/d210743

Whyte, K. P. (2013). On the role of traditional ecological knowledge as a collaborative concept: A philosophical study. *Ecological Processes, 7*(1), 2. doi:10.1186/2192-1709-2-7

World Intellectual Property Organisation. (2003). *Composite Study on the Protection of Traditional Knowledge* (5th session). Intergovernmental Committee on Intellectual Property and Genetic Resources, Traditional Knowledge and Folklore. https://www.wipo.int/edocs/mdocs/tk/en/wipo_grtkf_ic_5/wipo_grtkf_ic_5_8.pdf

KEY TERMS AND DEFINITIONS

Biodiversity Conservation: Conservation of the rich and diverse flora and fauna.

Indigenous Language: It is the language of local and tribal communities through the medium of which TK/TEK is generally disseminated.

Sustainable Management: The term constitutes the use of Earth's natural resources in such a manner that the future generations are able to enjoy it as well.

Traditional Ecological Knowledge (TEK): Knowledge held by indigenous and local communities related to ecology and nature and is passed from one generation to the other.

Traditional Knowledge (TK): TK is a broad term referring to all kinds of traditional knowledge ranging from agriculture, medicinal plants, water management etc., to any other topic. TEK is a constituent of TK. TK can be both oral and recorded.

Traditional Knowledge Digital Library (TKDL): TKDL is a database maintained by the Indian Government to record India's Traditional Knowledge (TK) including TEK and constitutes prior art.

Section 4
Environmental Pedagogy

Chapter 13
Narratives of Hope:
Teachers' and Students' Roles in Climate Justice Activism

Sibylle Gruber
Northern Arizona University, USA

ABSTRACT

This chapter addresses efforts on how to shift normative narratives focusing on climate catastrophe, environmental disasters, and climate inaction to climate hope, action, and activism. The author discusses how teachers can promote learning communities where current and future teachers can engage successfully with each other and with their students on the complex issue of climate change, and the author encourages educators to become teacher-researchers who can use inquiry-based and critical self-reflection to promote hope and constructive climate activism in students and teachers. The author argues that we need to redefine, blend, and expand existing and limiting boundaries and create an environment where transdisciplinary and intersectional teaching and learning perspectives are foregrounded in order to create a new way of addressing climate change and moving towards climate action.

INTRODUCTION

Recent reports published by the National Oceanic and Atmospheric Administration (Bateman, 2023, "July 2023 brought") show the staggering and devastating impact of the climate crisis on U.S. and global communities. According to Bateman's report titled "July 2023 brought record-high temperatures, devastating floods across the U.S.," , the U.S. experienced "15 individual weather and climate disaster events" from January to July 2023, including flooding, tornados, droughts, wildfires, and hurricanes. These events, Bateman points out, have led to the "highest number of billion-dollar disasters ever recorded for the first seven months of a year since NOAA began tracking these events in 1980" (2023, "July 2023 brought"). On a global scale, Bateman (2023, "Record Shattering") details NOAA's findings that global ocean surface temperatures have risen by 2.02 degrees Fahrenheit, and sea ice coverage has decreased by "about 1 million square miles — roughly the size of Argentina — below the 1991–2020

DOI: 10.4018/979-8-3693-2577-3.ch013

average." The focus on the economic impact of the climate crisis and the many costly climate-driven disasters (Cheng et al., 2024; Smith, 2024; Tingley, 2024; Vahlsing & Liscow, 2023) provides a stark picture of the impact that climate change has on "our economy, health, well-being, security, and quality of life" (Vahlsing & Liscow, 2023).

Most students who sign up for the graduate-level course on environmental rhetoric that I teach at a U.S. university located in the Southwest of the country, are very familiar with the dystopian regional, national, and global outlook on our abilities to slow down the climate crisis or to enact positive change. Many are curious about a course that seems to take on a topic that, based on their lived experiences or their experiences with distressing social media feeds, can only be part of a dystopian philosophy that would focus on problematic environmental practices such as greenwashing, the practices of industrial polluters in lower-income and minority neighborhoods, or the difficulty navigating environmental racism, sexism, homophobia, and classism nationally and globally (Das, 2022; Robinson, 2022; Smiley, et al., 2022; Smith et al., 2022). Students' exposure to climate injustices, their frustrations with local, national, and global attempts to mitigate the climate crisis, the many challenges that differ widely based on specific regional needs, and their worries about their futures in a world of land degradation, forest fires, hurricanes, floods, and tsunamis have led to emotional stress that, according to Mala Rao (2022), has become "an existential threat" (p. 440) reinforced by the perceived "indifference and lack of urgency with which adults and organisations treat climate change" (p. 439).

In this chapter, I bring together many of the current discussion points on how to engage students in environmental communication efforts, and on how to shift normative narratives focusing on climate catastrophe, environmental disasters, and climate inaction to climate hope, action, and activism. To bring to the forefront the dangers of subscribing to a story of climate chaos, I pose the following questions:

- How can we promote online learning communities where current and future teachers can engage successfully with each other and with their students on the complex issue of climate change?
- How can we encourage current and future teachers to promote sustainable solutions to existing environmental problems?
- What are educators' responsibilities to inspire inquiry-based and critical self-reflection that leads to constructive climate activism in students and teachers?

In my review of current literature, I discuss current research on efforts to address the multifaceted approaches to climate change discussions as well as the negative and long-lasting impact of eco-anxiety on environmental engagement. I provide background on efforts by teachers to create learning spaces that address environmental engagement and climate action. I then use my experiences as a teacher-researcher working at a U.S. institution to provide a narrative exploration of "Environmental Rhetoric," a graduate-level course for current and future English teachers who are interested in connecting environmental learning and awareness of climate change into the language arts classroom. I argue that we need to redefine, blend, and expand existing and limiting boundaries despite and because of the challenges and despite resistance to an environmentally-focused curriculum, and create an environment where transdisciplinary and intersectional perspectives are foregrounded in order to create a new way of addressing climate change and moving towards climate action. Additionally, I show the importance of applying knowledge frameworks that raise awareness of the impact of race, ethnicity, gender, sexuality, socio-economic status, and culture on environmental communication practices, and that acknowledge that environmental education needs to be based on the diverse needs of the communities in which we teach and live. A detailed analysis of the course outcomes leads to further recommendations on how to integrate discussions of the socio-ecological systems in which we live with expanded discussions on how to teach future and current teachers the complex theoretical foundations and ever-changing practical applications of an environmentally forward curriculum that explores future directions for climate change discussions.

REVIEWING CURRENT LITERATURE: FRAMING AND REFRAMING EDUCATION FOR ENVIRONMENTAL CONSCIOUSNESS

The consensus among the geoscientific community that we are experiencing anthropogenic global warming is almost unanimous (Lynas, Houlton, & Perry, 2021; Myers et al., 2021). As Myers et al. (2021) found in their survey data, 99 percent of geoscientists agree that "the Earth is getting warmer mostly because of human activity such as burning fossil fuels" (Myers et al, 2021). However, the high level of agreement in the geoscience community has not yet reached the general population where, as Leiserowitz et al. (2023) point out, 29 percent of Americans think that global warming "is caused mostly by natural changes in the environment" (p. 3). Certainly, the discrepancy between scientific data and the beliefs of the general population has caused a discordant approach to global warming, ranging from demands by a majority to move towards sustainable development to an insistence by a minority that global warming is not happening or is not caused by human activity (Leiserowitz et al., 2023; Cook, 2019, 2021; Kuthe et al., 2020).

The discrepancy addressed by Leiserowitz et al. (2023) shows the need to provide learning opportunities that address global warming and climate change from a variety of perspectives, including scientific, sociological, psychological, economic, cultural and political viewpoints. The following discussion provides an overview of efforts to address climate-influenced inequalities and increase environmental consciousness, how eco-anxiety impacts environmental engagement, and what teachers have attempted in their efforts to teach sustainability and environmental engagement.

Addressing Climate-Influenced Inequalities and Increasing Environmental Consciousness

In 2015, the United Nations Member States adopted the 2030 Agenda for Sustainable Development, including 17 Sustainable Development Goals (SDGs) in order to highlight the need for continuous collaboration and action on such issues as fighting poverty and hunger, increasing access to education, reducing inequalities based on gender, sex, race, and socio-economic backgrounds, promoting climate action, raising awareness of land and water use, and collaborating on and attaining the 17 of the SDGs (United Nations, 2024). In a call to action, global leaders in the 2015 United Nations meeting in New York pointed out that "the future of humanity and of our planet lies in our hands" and, in their final declaration, affirmed an "unwavering commitment to achieving this Agenda and utilizing it to the full to transform our world for the better by 2030" (United Nations, 2015).

The goals outlined in detail emphasize the interrelated nature of a healthy people and a healthy planet, an emphasis that is reinforced in the United Nations 2023 Report on the Sustainable Development Report aptly subtitled "Towards a rescue plan for people and planet" (United Nations, 2023). Li Junhua, the United Nations Under-Secretary-General for Economic and Social Affairs, emphasizes in the 2023 report that “conflict, climate change, the lingering effects of the COVID-19 pandemic and other global challenges are threatening to derail hard-earned progress towards the SDGs” (p. 3). The need for action is emphasized by Secretary-General of the United Nations António Guterres who points out in the 2023 Sustainable Development Goals Report that “unless we act now, the 2030 Agenda will become an epitaph for a world that might have been” (p. 78).

The efforts of the United Nations to develop an agenda for sustainable development that focuses on people, the planet, prosperity, peace, and partnership (United Nations, 2015) have encountered numerous roadblocks. Instead of moving towards a thriving planet and equitable relationships, the 2023 Sustainable Goals Report shows that “of the 140 targets for which trend data is available, … about half of these targets are moderately or severely off track; and over 30 percent have either seen no movement or regressed below the 2015 baseline” (p. 4). Even though the 17 SDGs are intended to create a strong commitment to working towards improving the human condition and slowing the destruction of the natural world, the expected outcomes have not been achieved. Instead, the report emphasizes that hunger levels are back to the devastating levels of 2005, 575 million people will live in extreme poverty if current trends continue, and 84 million children will not be in school. Furthermore, following the same trend, it will take 286 years to close the gender gap (p. 4).

In conjunction with the increased human suffering, the 2023 Report also reminds us that “the short-sightedness of our prevailing economic and political systems” increases “the war on nature” with increasing global temperatures and increasing carbon dioxide emissions (p. 4). Additionally, the authors of the 2023 Report point out that “it could take another 25 years to halt deforestation, while vast numbers of species worldwide are threatened with extinction” (p. 4). The potential “collective failure” to meet the 2030 goals, they argue, will affect “developing countries and the world’s poorest and most vulnerable people” most severely (p. 4). It will, the 2023 Report points out, “fuel greater political instability and displacement, further erode trust in public institutions, upend economies and lead to irreversible existential changes to our natural environment” (p. 48).

Despite the unsettling and disappointing progress towards achieving the SDGs, the 2023 Report continues to emphasize that the goals are still achievable if we don’t continue on the same path and instead focus on a “renewed sense of common purpose across the international community and a global alliance for Sustainable Development Goals-related action across business, civil society, science, young people, local authorities and more” (p. 4). However, the 2023 Report also found that the establishment of the 17 SDGs have led to partnerships within and across governments and countries and have created innovations and data-sharing opportunities. This, the report concludes, “allows researchers to conduct more in-depth analysis, promotes transparency and accountability, and fosters collaboration” (p. 10).

The five key areas emphasized by the report include:

- An accelerated recommitment to national and international transformative action
- An advancement of concrete policies to “eradicate poverty, reduce poverty, reduce inequality and end the war on nature,” especially focusing on women’s and girls’ rights
- An increase in accountability to ensure progress towards the SDGs
- A recommitment at the SDG to the Action Agenda
- A strengthening of the United Nations’ development system “to tackle emerging challenges and address Sustainable Development Goals-related gaps and weaknesses.” (p. 5)

Thus, the report emphasizes that despite the concerning challenges–for example, only 15% of the SDGs are on track–we need "to redouble our efforts to eliminate poverty and hunger, advance gender equality and overcome the triple planetary crisis of climate change, nature and biodiversity loss and pollution" so that we do not "cause immense suffering for current and future generations" (p. 48). The SDGs, they continue, "remain a truly inspiring and unifying compass, and transformative progress can be made even in the face of adversity" by providing support to United Nations Member States in the following three areas:

- Equipping governance and institutions for sustainable and inclusive transformation
- Prioritizing policies and investments that have multiplier effects across the goals
- Securing a surge in Sustainable Development Goals financing and an enabling global environment for developing countries. (p. 48)

As the 2023 report concludes, "there are no excuses not to be ambitious. Never before have we had such an abundance of knowledge, technology and resources to succeed in ending poverty and saving the planet. Never before have we carried such a responsibility to pivot to a bold set of actions" (p. 57). The urgency of this message, in conjunction with the opportunities created by new and increased resources, has shifted past understandings of our positionalities in the fight against climate disasters and has created new perceptions about our roles as climate change activists. For example, according to a briefing series published by the Environmental and Energy Study Institute (EESI, 2023), the number of climate change deniers in the United States has decreased to 16 percent, with most Americans in accord that global warming "is happening, human-caused, and a serious problem" (EESI, 2023). Speakers in the briefing series–Anthony Leiserowitz, Director of the Yale Program on Climate Change Communications; Cary Funk, Director of Science and Society Research at Pew Research Center; Jonathon Schuldt, Executive Director of the Roper Center for Public Opinion Research and Associate Professor of Communication at Cornell University; and Neela Banerjee, Deputy Senior Supervising Climate Editor at NPR–emphasized the bipartisan consensus on needed government action, especially in terms of implementing a carbon tax, planting trees, and investing in a clean energy transition (EESI, 2023). Americans who have experienced extreme weather, the briefing shows, strongly "support policies for grid resilience, disaster preparedness, and greenhouse gas emission reductions" (EESI, 2023).

Similarly, Leiserowitz et al. (2023) in "Climate Change in the American Mind: Beliefs & Attitudes," (Leiserowitz, et al., 2023) report that most Americans agree that global warming is indeed a reality and that they are worried about the outcomes of global warming (pp. 4-5). In addition, the report also highlights that more than 60 percent of Americans disagree with the statement that "it's already too late to do anything about global warming," and more than half disagree with the statement that "the actions of a single individual won't make any difference in global warming" (p. 4).

Leiserowitz et al. (2023) also address the questions that Americans would ask if they had an opportunity to speak with an expert on global warming. They point out that the following questions were most often on the minds of the survey participants:

- What can the nations of the world do to reduce global warming? (74%);
- What can the United States do to reduce global warming? (72%);
- How do you know that global warming is caused mostly by human activities, not natural changes in the environment? (70%);

- Is there still time to reduce global warming, or is it too late? (69%) (p. 4)

These questions show how important it is to address what Fessmann, (2019) calls an "information-deficit model" (p. 1) or even more importantly, counter a misinformation model (Anspach & Carlson, 2020; Cook, 2019, 2022; Enders, et al., 2023; Falkenberg et al., 2022). As Falkenberg et al. (2022) point out, "rapid and effective climate action depends on broad international consensus and collaboration," and misinformation and the accompanying polarization "may risk political deadlock if it fuels antagonism to climate action" (p. 1120). Additionally, misinformation and conspiracy theories surrounding environmental concerns – often distributed through social media – have contributed to conditions that often lead to eco-anxiety, climate skepticism, and climate inaction.

Understanding the Impact of Eco-Anxiety on Environmental Engagement and Education

Many of the students at our university are interested in contributing to climate change action. However, many also are worried that their contributions are not enough, that the university could do more to be green-focused, that the state is behind in addressing climate change, that the country is not keeping to the promises of a climate first nation, and that global warming is indeed global and cannot be curbed by individual countries and much less by the efforts of individuals. These sentiments are wide-spread, and media reports on Greta Thunberg's engagement as a climate activist have led to young people's global climate change protests around the globe (e.g., Fridays for Future, 2024; Extinction Rebellion, 2024). Surveys on young people's climate change engagement conducted over the last years have shown that young people consider climate change one of the most important societal issues, and that they are more engaged in climate change discussions than previous generations (Parry, McCarthy, & Clark, 2022; Neas, Ward, & Bowman, 2022; Sloam, Pickard, & Henn, 2022). Maria Ojala (2018-19), however, points out that despite concern about climate change, young people are not necessarily more engaged as climate activists. Instead, the perceived vastness of the problem prevents young people from participating on a local scale. As Ojala (2018-19) argues, "many young people have a rather pessimistic view of climate change and the future on a global scale. Consequently, many lack a sense of empowerment and agency regarding this threat" (p. 11). This, she continues, creates a situation where "the younger generation, and especially children, are more vulnerable to having their personal wellbeing negatively impacted because of this problem" (p. 11).

The resulting anxiety about how to address climate change, and how to participate as individuals in finding solutions to a problem that is much larger than anything we have experienced in the past, can lead to a number of coping strategies that cause further worry and guilt for young adults (Baudon, 2021; Benoit, 2022; Clayton, 2020; Diffey et al., 2022; Ojala, 2018-19; Pikhala, 2022). For example, Ojala (2018-19) shows that de-emphasizing the problem and not acknowledging the severity of the situation, distancing themselves from the problem and not talking about it, or focusing on individual action for societal problems "can create more distress" for young people and can place "a heavy burden on a young person's shoulders" (p. 13). As Diffey et al. (2022) describe in more detail, young adults' climate related feelings include "worry, fear, helplessness, anger, frustration, grief and betrayal (p. 500). These feelings are especially pronounced when "we believe people are responding without sufficient urgency, we feel despair about the unnecessary harm being caused, angry at the injustice being inflicted by the world's richest people" (p. 501). In addition, "invalidating responses" that lead us to believe that "our feelings are invalid, unimportant, and not based on legitimate concerns about the world" (p. 502) create additional negative climate-related feelings and distress. This "wide range of uncomfortable feelings in the climate crisis," according to Diffey, et al. (2022, p. 506), can be found among young people from diverse cultural, racial, gender, socio-economic, and national backgrounds. They suggest a number of ways in which young people can be supported in coping with their climate-related feelings, including:

- Showing compassion and care for their wellbeing
- Offering spaces to share feelings without being judged
- Taking the climate crisis seriously

If we don't engage in "climate action, climate-related mental health support, intergenerational collaboration and a radically more compassionate world," Diffey et al. (2018-19) argue that it will be difficult to "imagine the better future we could be working towards together" (p. 506).

However, instead of moving towards positive action, high anxiety about climate change, Gorman and Gorman (2019) argue, can lead individuals to ignore and deny the evidence that shows that climate change is occurring. As they point out, "many people either don't believe it or shrug it off" (Gorman & Gorman, 2019). Furthermore, climate change skeptics, according to Gorman and Gorman (2019), resist believing in climate disasters when "that calamity is real, will be spread out over decades, and is of catastrophic proportions that can only be averted if we change almost everything about the way we live" (Gorman & Gorman, 2019). And because denial, "the refusal to accept facts in order to protect us from uncomfortable truths" is considered to be "a primitive defense mechanism," motivating individuals to engage in climate activism runs into problems because "it is very difficult to accept as real a problem that requires this magnitude of solution" (Gorman and Gorman, 2019). In other words, even though we might understand the impact of global warming, "we do not accept the responsibility" and, as Young and Jones (2023) point out, "we can find it surprisingly easy to detach our own daily actions … from the broader goals of getting emissions down to zero as soon as possible" (Young & Jones, 2023).

Even though climate change and global warming is less contested and more accepted, Watson (2021) points out, climate change skepticism has shifted from complete denial to a focus on the results of listening to climate activists. For example, the "altered arguments" used by climate skeptics include that:

- Clean energy will hurt working-class people because of job loss
- Scientists and climate activists are extremists, are overreacting, and are alarmists
- Corporations are already doing the necessary work and are dealing with the problem
- The apocalypse is inevitable and we're doomed. (Watson, 2021)

Efforts to cast doubt on the reality of climate change and to create climate misinformation, Braun (2023) points out, have flourished on Facebook and TikTok, and ads on Google have continued to broadcast climate deniers' messages (Braun, 2023). Braun quotes from an interview with Kathie Treen who points out that, "there is no silver bullet when it comes to the fight against climate misinformation, and a multi-faceted approach is needed, including education, inoculation, correction, and actions by the platforms" (Braun, 2023). Instead, Treen et al. (2020) point out in their discussion of research on online misinformation concerning climate change, that it "has confused the public, led to political inaction, and stalled support for or led to rejection of mitigation policies" (Treen et al., 2020). As they found, "there has been long-standing academic discussion of systematic attempts by various actors […] to discredit climate science and confuse political debate on climate change, by casting doubt on well-supported theories and providing alternative and often fallacious interpretations of observations" (Treen et al., 2020). For example, Treen et al. (2020) identify four variations of climate change misinformation and with it climate change denial:

- trend denial (no significant warming is taking place)
- attribution denial (it is not anthropogenic)
- impact denial (it will not have significant negative impact on humans or the environment
- consensus denial (there is no consensus among climate scientists about anthropogenic climate change). (Treen et al., 2020).

As their research review found, climate change misinformation is "repeated and amplified" through social media and the "echo chambers" created by influencers (Treen et al., 2020). Treen et al. (2020) emphasize the misinformation cycle is able to spread because humans "preferentially consume information that has coherence with their belief systems," and because people trust those within their social network (Treen et al., 2020).

The Role of Teachers and Students in Environmental Education Environments

Addressing eco-anxiety, climate skepticism, and climate inaction has become an essential first step when working towards providing learning opportunities for constructive climate change communication. This means, as Treen et al. (2020) argue, "to educate people in critical thinking techniques to be able to identify misinformation" (Treen et al., 2020). Additionally, Treen et al. (2020) concur that we need to "increase the coverage of climate change in the general curriculum" and to also support "an appropriate level of climate literacy in educators" (Treen et al., 2020). In a pilot case study on climate education in the humanities, Siegner and Stapert (2019) argue that "teaching young people about the causes, consequences, and solutions to climate change" has become a "national imperative for public and private education" (p. 1). The humanities, they argue, "are central to creating an educated citizenry prepared for leadership in mitigating and reversing climate change" because the humanities "inherently incorporates real world contexts, social and political histories, and opportunities for human engagement between students and their surrounding community" (p. 4). For service learning projects on climate change to have meaningful impact, Siegner and Stapert (2019) argue, teachers need to focus on action pathways and need to work with the community and with local officials (p. 14). Preparing teachers to be effective climate change educators, they conclude, "is a crucial step in humanity's response to climate change" (p. 13).

With a focus on students' climate literacy, Dzambo et al. (2021) argue that "an individual's climate literacy is one indicator of a person's likelihood of engaging in responsible environmental stewardship" (p. 501). However, a mindset of climate stewardship, they point out, "can be challenging to instill due to climate change disinformation campaigns, personal political leanings, or biased media coverage" (p. 501). With a more focused discussion on the young individuals who are part of a climate literacy curriculum, Kuthe et al. (2019) point out the importance of paying attention to young individuals' previous experiences, with a need to adapt climate change education "with regards to its content, frame and method, to the particular preconditions of the target group" (p. 376). They emphasize that successful climate change education needs to take into account the following points:

- Attitude: a positive attitude, for example, can increase self-efficacy and a willingness to act
- Personal concern: concern can lead to engagement in climate-friendly activities
- Climate-friendly behavior: ability to change to a climate-friendly attitude
- Multiplicative actions: young individuals can influence climate change literacy in their family and friends
- Knowledge: increased knowledge about the causes and effects of climate change is the basis of increased climate awareness. (p. 376)

Along similar lines, Damico et al. (2020) show that young people's interest in climate change has to be grounded in their understanding of "their own backgrounds, beliefs, biases, and values" (p. 684). As they point out, many of us live by two interrelated stories:

- Humans are the center of existence, separate from nature
- Consumerism is a primary pathway to happiness and fulfillment

To locate and examine these stories, Damico et al. (2020) suggest "inquiry-based instructional approaches" (p. 685) that promote critical thinking about climate change. Such "investigative work" can highlight the destructive nature of subscribing to humans as the center of existence and of seeing consumerism as a main way to achieve happiness (p. 686). Damico et al. (2020) also suggest to focus climate change literacy on "beneficial stories-we-live-by" (p. 686) to strengthen our relationship with each other and with nature. These beneficial stories are guided by indigenous narratives, by a focus on gender equality and climate justice, and by spotlighting youth activism and civic engagement (pp. 686-687). This is especially necessary because being a climate justice literacy educator, they point out, is a challenging task.

To address climate change successfully, Damico et al. (2020) found it important to see climate change as a socioscientific topic connected to science, economics, society, politics, morality, and ethics" (p. 688). Additionally, it is also necessary to understand how texts and information influence us, and to be aware of the polarizing effects of texts. And finally, Damico et al. (2020) also point out that climate change has to be seen in connection with climate justice and climate injustice (p. 688). They propose to follow Greta Thunberg in her efforts to "stop saying 'climate change' and instead call it what it is: climate breakdown, climate crisis, climate emergency, ecological breakdown, ecological crisis and ecological emergency" (para. 1). Awareness of the language we use, Damico et al. (2020) point out, can help teachers and students "identify and examine ecologically destructive and beneficial stories-we-live-by that circulate in an escalating array of complicated, motivated texts about climate change" (p. 689).

In a quest for taking the next step, Beach, Share, and Webb (2017) argue that teachers "have the opportunity to educate our students about climate change" because, as they point out, "we and our students can and must make a difference" (p.2). Climate change, they continue to argue, is not purely a scientific issue, but "raises social, moral, and ethical questions" and requires "understanding experiences of others, rethinking values, critically analyzing arguments and media presentations, and reflecting on the past to better imagine alternative futures" (p. 17). Similar to Damico et al. (2020), Beach, Share, and Webb (2017) propose a critical inquiry approach "where students pose questions that matter and undertake meaningful research" that can lead to "envisioning, advocating, and enacting change" (p. 17). This approach then, provides students with agency because, as Beach, Share, and Webb (2017) point out, "instead of telling students what climate change means or what to do about it, we want to empower them to investigate, collaborate, develop their own conclusions, and devise ways to make a difference, drawing on their knowledge, their concerns, and their emerging literacies. To motivate students, they propose that:

- Students need to learn about the impacts of climate change in personally meaningful ways
- Students need to learn about effective personal and community action
- Students need to be motivated by working collaboratively. (p. 133)

The research explored in this section shows the need to address the interrelated challenges and promises of teaching an environmentally-focused curriculum. As the discussion of the 17 Sustainable Development Goals has shown, climate-influenced inequalities are still–and often more extensively–a part of climate change discussions. The resulting eco-anxiety, climate skepticism, and climate denial create many challenges for enacting positive change for environmental sustainability, and teachers' roles as climate-literacy advocates who teach a climate-focused curriculum are ever more critical.

TEACHING FOR ENVIRONMENTAL CONSCIOUSNESS: THEORETICAL, METHODOLOGICAL AND CONCEPTUAL FRAMEWORKS

To explore successful teaching and learning practices geared towards creating a productive environment for reframing discussions on sustainability and the environment, this hermeneutical case study situated within a U.S. teaching context reflects on an online graduate-level course titled "Environmental Rhetoric in Public Spaces" which I have been teaching as part of a requirement in an online master's program in rhetoric and writing studies. My theoretical, methodological, and conceptual frameworks adapt a teacher-researcher approach combined with social and critical action research and narrative inquiry. Specifically, I explore how teachers and students can gain and increase their critical and analytical skills in order to analyze, reframe, and participate in environmental communication practices at a local, national, and global level. As a teacher researcher, I am especially interested in gaining more knowledge about creating positive learning environments that encourage students to become active and engaged members of their local, national, and global communities. The research process used by teachers to gain in-depth knowledge about best teaching and learning practices encourages teachers to not only ask questions about their own teaching practices but to look beyond their classrooms and, as Burnaford, Fisher, and Hobson (2001) point out, "read, react, and build on what they encounter" in previous studies conducted on similar research questions (p. 51). This approach, they argue, encourages a "community of learners" (p. 52) who can, through analysis of studies in their field of interest, contribute to a body of knowledge that benefits not only their own students but a network of teachers who can participate in a productive and far-reaching knowledge exchange. Furthermore, Kincheloe maintains that critical action research "allows teachers to organize themselves as communities of researchers dedicated to the achievement of their own and ultimately their students' enlightenment" (p. 23).

Additionally, my interest in social and critical action research is especially pertinent when addressing issues related to climate change and sustainability. The urgency of creating an environment where students can learn about diverse global perspectives on climate change discussions and are able to participate as agents of change has become even more pronounced in current reports on global warming (Panda & Maity, 2021; Zandalinas Fritschi, & Mittler, 2021). Foregrounding social and critical action research asks the researcher to pay attention to how, for example, our positionalities in specific environments influence our understanding of political, environmental, gender, racial, and social situations, and how they limit or promote access to positions of influence, power, and privilege. As Canella and Lincoln (2009) argue, social action research seeks to "illuminate the hidden structures of power deployed in the construction and maintenance of its own power, and the disempowerment of others" (p. 55). In this study, I am especially interested in what Canella and Lincoln (2009) call "the interactive nature, and institutionalization, of power, oppression, and injustice" (p. 56) and how, as environmentally-conscious educators, we can encourage students to critically examine their own positionalities as members of the communities to which they belong.

Since many of the students in the program are current and future teachers, the study also explores how online students can use the critical knowledge and skills from the course discussed here, adapt them to meet their communities' needs, and participate in promoting the need for environmentally-conscious teaching practices in their school and in schools across the U.S. and abroad. It examines whether students' use of social action research and critical narrative inquiry in their research can reach an audience interested in and willing to learn more about an environmentally-focused curriculum. As Kincheloe (2003) argues, critical inquiry also encourages teachers to become what he calls "transformative intellectuals" who "treat students as active agents, render knowledge problematic, utilize dialogical methods of teaching, and seek to make learning a process where self-understanding and emancipation is possible" (p. 24).

The course I discuss in the following pages provides some initial answers to the questions that guide my critical narrative inquiry:

- How can we promote online learning communities where current and future teachers can engage successfully with each other and with their students on the complex issue of climate change?
- How can we encourage current and future teachers to promote sustainable solutions to existing environmental problems?
- What are educators' responsibilities to inspire inquiry-based and critical self-reflection that leads to constructive climate activism in students and teachers?

Kincheloe (2003) mentions that "critical teachers as researchers cannot avoid the political role of promoting critical self-reflection" (p. 23) in the communities they influence through their teaching, thus advocating for educators' responsibility to transform current heteronormative standards and to work against discriminatory, unjust, and oppressive systems. However, despite our responsibilities to engage in social action work with our students and in our communities, we often encounter communication barriers based on how we understand and interpret reality. Our constructions of reality – the filters we use to understand the world around us – limit what we see and how we see it. These terministic screens can, as Kenneth Burke (1965, 1966) argues, reflect, select or deflect reality, thus using our understanding of reality to construct knowledge and construct our social realities by directing attention away from or towards specific directions.

Certainly, leaving behind the filters that have helped us shape our realities is not an easy task and requires that we expand our perceptions of reality. Javier Echeverría (2011) proposes an opportunity for such expansion by forming knowledge communities that then lead to knowledge cities followed by knowledge societies. He sees "agoras" as spaces where "knowledge can be expressed freely and accessed by any citizen" (Echeverría, 2011). In essence, we can shift our current knowledge paradigm by establishing spaces – what Echeverría (2011) calls "agoras" – where diverse individuals and groups can meet and where intersectional and heterogeneous knowledge creation is supported and encouraged.

Participants in such knowledge communities can share common concerns, interests, and commitments, and can contribute to shared problem-solving and collective learning. Participants can gather and explore knowledge from diverse communities and transcend the limits of terministic screens. In terms of the environmental crisis and global warming, for example, participants can also transcend the limitations of preconceived notions about how to address climate anxiety, the complex nature of the climate crisis, and possible climate-related social action opportunities.

For my online course on "Environmental Rhetoric," the interdisciplinary and transdisciplinary nature of knowledge communities is especially important for creating action-forward online spaces to address the climate crisis. Instead of relying on preconceived notions of reality, we can develop and explore new knowledge as part of a community, and we can redefine and expand disciplinary boundaries that often limit our approaches to solutions of complex environmental, societal, cultural, and political problems. Addressing a societal, and with it the environmental crisis, from a transdisciplinary perspective, where academics from various disciplines work with non-academic participants from diverse backgrounds, creates spaces that are open to new ways of thinking and seeing. It enables us to acknowledge and challenge the reality of systemic forms of oppression and allows us to emphasize diversity, equity, inclusion, and justice when we address the climate crisis. In other words, I am especially interested in creating and promoting knowledge communities in my course to encourage new and inclusionary ways of learning and problem-solving.

Such inclusionary approaches encourage us to build bridges so that we can successfully engage in social action. Much in line with these inclusionary practices and with proponents of environmental justice, postcolonial ecocriticism, and critical ecofeminism (Ferris, 2008; Gaard, 2017; Sturgeon, 1997), Sturgeon (1997), for example, explains the importance to see "at all times an interactive relationship among inequalities of gender, race, sexuality, class, and nation and revealing the connections between social inequalities and environmental problems to uncover the systems of power that continue to generate the complex problems that we face" (p. 24). Additionally, Ferris (2008) pays close attention to people in poverty and marginalized individuals and groups and highlights the problems encountered by poor people affected by natural disasters such as "unequal access to assistance; discrimination in aid provision; enforced relocation; sexual and gender-based violence; loss of documentation; recruitment of children into fighting forces; unsafe or involuntary return or resettlement; and issues of property restitution" (Ferris, 2008)

In "Environmental Rhetoric," I want my students to acknowledge and discuss the effects of Western capitalism on environmental challenges, and I want them to challenge and avoid easy dualisms, in this case the dualism created and encouraged by anthropocentrism that is reinforced by traditional western heteronormative thinking. Instead, the conceptual framework of the course addresses local, national, and global challenges in climate crisis discussions. Overall, I want students to explore the complex nature of climate change, including race, ethnicity, gender, sexuality, age, socio-economic status, and nationality in discussions of the effects of global warming and the climate crisis.

The contextual framework for the course and the critical narrative analysis and reframing of opportunities for action provided in this section provides the foundation for developing "Environmental Rhetoric." The following section includes detailed information on the course goals, the structural components of the course, opportunities for student engagement and transformation, and opportunities for moving from individual student success to knowledge communities of engaged and critical learners.

DEVELOPING, ANALYZING, AND REFRAMING OPPORTUNITIES FOR ACTION

The overarching goal of "Environmental Rhetoric" is to raise awareness of communication practices in sustainability and climate change discussions, and to encourage social action for positive change at an individual and collaborative level. When I conceptualized this online course, the divisiveness in discussions on global warming could only be described as toxic. Instead of trusting geoscientists who agree that global warming is human-caused (Nuccitelli, 2020; Eyring, 2021), climate change deniers joined forces and created a staggering social media presence that redirected the focus from the devastating global effects of the climate crisis and instead discredited climate-forward scientists, politicians, and researchers as well as individuals and groups focused on finding solutions to the current climate crisis (Leiserowitz, et al., 2023). The importance of developing a course based on critical inquiry became especially apparent when some students in my classes uncritically repeated climate misinformation posted by climate change deniers on Twitter, Facebook, Youtube posts, and Instagram, and when others related their worries about and inability to participate significantly as climate change influencers and activists.

Course Development: Foundations, Objectives, and Practical Approaches

The online course I developed and that I address in this chapter is geared towards online graduate students in a rhetoric, writing, and digital media studies program. Most of the students who attend classes are current and future middle and high school teachers who are interested in learning more about best practices in language arts teaching. In addition to providing theories and methodologies for language arts teaching, the course focuses on the topic of "Environmental Rhetoric" to raise awareness of communication practices in sustainability and climate change discussions in and out of the classroom. Furthermore, the theoretical framework of social action teaching and research encourages future and current teachers to engage in social action teaching for positive change. To achieve this overall goal, we explore how educational and professional communities use communication tools to address sustainability and climate change. We also examine what our own roles and responsibilities are for contributing to successful communication practices with and among diverse individuals and communities.

In the course, then, students are encouraged to apply theoretical and rhetorical frameworks to practical problems that include diverse and multifaceted voices. Specifically, students are asked to explore post-colonial and critical ecofeminist practices to address intersectional opportunities for climate advocacy. They learn to apply rhetorical principles to create successful inclusionary narratives that they can share with their peers, their students, and their communities. They then apply transdisciplinary practices to address the climate crisis and the complexities of sustainable practices. The overall result is the production of multimedia pieces and social media work that contributes to innovative inclusionary climate change approaches. In general, the course goals for "Environmental Rhetoric" include a focus on:

- Theoretical, rhetorical, and methodological frameworks
- Analytical and critical thinking
- Application of learned skills to hands-on projects for eco-social action

After students are introduced to the theoretical, rhetorical, and methodological concepts surrounding the climate crisis and climate-focused communication, they are expected to show that they understand and are able to participate in advancing current knowledge on climate change communication practices and the relationship between climate change communication and the effects of the climate crisis on individuals and groups from diverse social, cultural, racial, gendered, and socio-economic groups and LGBTQIA individuals and groups. Specifically, in terms of theoretical, rhetorical, and methodological skills, students are encouraged to:

- Research, evaluate, and apply rhetorical principles for participating po-actively in climate-crisis communication.
- Recognize that research is a recursive process and apply research data to formulate or evaluate new research questions, using reason and persuasion in a logical argument.
- Demonstrate competent, ethical, and responsible use of information in academic and workplace writing.

To hone their analytical and critical thinking skills, students analyze how cultural artifacts, such as written, digital, visual, and spoken texts, are used to shape identity and with it the reality of individuals and groups in diverse local, national, historical, and global communities. Specifically, students focus on:

- Researching, evaluating, and applying knowledge of significant issues related to global warming, diversity, and sustainable educational and environmental practices.
- Researching, evaluating, and applying historical knowledge of rhetorical terms and concepts to show how contemporary language use about the climate crisis is influenced by culture, race gender, sexuality, and socio-economic status and by knowledge about past and current treatment based on systemic intersectional injustices.
- Tailoring academic and professional work to a specific audience and a specific purpose intended to increase awareness of current eco-social injustices.

Finally, to apply the theoretical, rhetorical, and methodological skills as well as their analytical and critical thinking skills, the course focuses heavily on the application of those skills to hands-on projects that address eco-social action. Students are encouraged to create advanced-level projects that question current norms and dominant cultural assumptions and that can be used in local, national, or global contexts to raise awareness of current environmental injustices and propose opportunities for climate-focused action. Specifically, students work on:

- Designing text and digital projects that show a clear understanding of purpose, audience, and cultural, social, historical, and political contexts, and that show the ethical use of language in diverse academic, public, and professional communities.
- Applying their understanding of how cultural, racial, ethnic, gender, and economic factors influence communication practices in text and digital environments.
- Presenting their original research to professional and non-professional audiences, and showing how the research they conducted led to the action-focused project on sustainability and climate crisis intervention.

To provide students with multiple opportunities to practice their analytical and critical thinking skills and to show their skills as researchers and practitioners in an online space, the course components included text and multimodal responses to readings on global warming, theoretical and applied research opportunities, and action-focused projects that combined their knowledge of theory as well as historical and currents needs for social action. The multiple opportunities for participation in course activities provided students from diverse backgrounds to be successful members of the course community and engage successfully in learning activities as well as with their course colleagues. The course activities included:

- Active participation in online discussions focused around topics of environmental rhetoric, sustainability, climate change, and eco-social action. Topics focused around:
    - Rhetorical principles for climate change communication and environmental rhetoric in global and intercultural settings
    - The importance of knowledge communities in climate change communication and environmental rhetoric
    - Subject matter and genre in climate change communication and environmental rhetoric
    - Social media and global communication on climate change and environmental rhetoric
- Peer-reviewed writing tasks focused on familiarizing students with the conceptual frameworks and interconnected discussions surrounding the climate crisis. Tasks included:
    - Multi-entry blog portfolio to enable students to become positive climate change influencers by writing well-constructed blog entries on climate change communication and environmental rhetoric for a specific public readership.
    - Final essay which provides a critical, descriptive, interpretative, and analytical commentary on the students' growth as writers and their changing perspectives on their roles as influencers and activists in environmental rhetoric and climate change communicators.
    - Multimodal presentation is based on the students' topics for their blog posts, showing their audience the theoretical and rhetorical foundation for their blog posts. The presentation is intended to show that students conducted research and that they understand the rhetorical situation for their blog posts.

The narrative analysis included in the following pages provides details on the immediate and possible long-term impact of an environmentally-conscious curriculum on future and current teachers. It draws on data from three years of teaching "Environmental Rhetoric" to online graduate students who teach in middle and high schools in a largely conservative state that struggles with limited water availability and high danger of wildfires. My narrative analysis allows for integrating data and observational information and provides insights into how a climate-forward curriculum can impact and change current teaching practices at the middle and high school as well as the college level.

Environmental Rhetoric: Analyzing and Reframing Opportunities for Climate Action

Teaching in an online environment makes it especially pertinent to create knowledge communities (Echeverría, 2011) and spaces where students can explore new areas of knowledge, where they can learn from each other, and where they can collaborate and communicate. To make sure that the community of learners in "Environmental Rhetoric" had an opportunity to learn about each other, they were asked to create an introductory video in which they could share information based on the following questions:

- How would you describe yourself in terms of current discussions on the environment, climate change, and sustainability efforts?
- How did you become interested in discussions on climate change?
- What do you consider part of climate change discussions, discussions about our impact on the environment, and sustainability?
- What are your future goals in terms of integrating environmental rhetoric into your current work situations?
- What do you see as the role of this class in helping you achieve your future goals?

Students' video introductions, and the written responses by their classmates, showed how important it is to include opportunities for open dialogue, which, according to Gregory Clark (1990), reinforces the social construction of knowledge and enables us to "continue the cooperative process of working with others to meet our common needs" (p. 1). In such an environment, despite different backgrounds, students are able to come together and explore moments of intersecting experiences. Students addressed their attempts at recycling, their frustrations with wasteful living, composting adventures, the need for more policies that would encourage individuals and corporations to be more sustainable, participating in beach cleanups, creating walkable and bike-friendly cities, and feeling hopeless because of the many failed attempts at creating a more sustainable environment. One of the students discussed her undergraduate coursework in green architecture, and others addressed how important it is to create spaces where participants can discuss solutions to environmental problems. Some wanted to learn more about how to talk successfully with specific audiences and encourage them to participate in climate action. The droughts, wildfires, and water shortages in some parts of the U.S and globally, and the hurricanes and storms in other parts were also a high priority in the video posts and created a feeling of urgency for learning more about climate change and global warming.

The responses to the initial video posts showed an overall agreement that climate change literacy is an area that needs to be addressed if we want to be successful as climate change advocates and talk to family, friends, colleagues, or employers. Additionally, students affirmed comments about the roles of individuals and groups in figuring out environmental measures, what the impact of our spending can be on the environment, how emotionally charged climate change discussions can be, and how important it is to understand our responsibilities as well as our ability to take steps in the right direction. Others mentioned that they had not had any experiences with flooding or droughts, but that the experiences of others in class made them aware how immediate the climate crisis is.

In effect, the introductory conversations were a first step in bringing together individuals from diverse social, economic, cultural, racial, gender, sexuality, and economic backgrounds, providing an initial opportunity to reconsider current filters we use to see reality (Burke, 1965, 1966) and allowing for shifts in perspective based on the experiences of those around us. It created a community of learners who were then able to move to the next step and read and analyze texts (text-based, videos, audio, multimodal) that focused on four main areas, each of which provided theoretical readings as well as application readings, including multiple United Nations websites focused on sustainability goals, social media sites providing hands-on discussions of sustainability and the environment, documentaries, and TedTalks.

The first area included readings on rhetorical principles for climate change communication and environmental rhetoric in global and intercultural settings where students learned about the need to determine the purpose of a specific text and the needs of the text's specific readers. For example, students, employees, managers, and bosses have to adapt their writing to the criteria and values of individuals, groups, or organizations. Therefore, writers have to know which arguments or means of analysis will encourage the reader to further engage with the topic. Furthermore, writers have to use rhetorical principles to understand acceptable lines of argument, acceptable evidence, and structural or linguistic features that lend validity to the text in a particular rhetorical context, all of which will contribute to the text's persuasiveness. The readings in this section (e.g., Armstrong, Krasny, & Schuldt, 2018; Constantino & Weber, 2021; DellaSala, 2021; Newburger, 2021) focused on how the study of rhetoric can increase our understanding of environmental rhetoric and climate change communication in intercultural and global settings.

A second area of study included discussions of the importance of understanding our discourse communities and knowledge communities in climate change communication and environmental rhetoric. Students learned that to address a specific discourse community means to understand communicative practices, including communication channels, norms for genres of writing, and writers' roles, which are often steeped in long-term goals and values. Unlike rhetorical knowledge--which focuses on the immediate task and the specific audience--discourse community knowledge highlights the larger institutional issues at stake in the communicative event. For example, writers need to learn to respect cultural traditions, understand business relationships, or understand the relationship between oral and written communication. The readings (e.g., Bladow & Ladino, 2018; McArdle, 2021; OECD, 2021) focused on how different discourse communities function in relation to climate change communication.

For the third area of study, the course addressed subject matter and genre in climate change communication and environmental rhetoric. The reason for highlighting subject matter and genre is based on the need of successful communicators to possess excellent subject matter knowledge in order to write for a specific purpose. This means that every communicator needs to be able to research information through a variety of channels, understand detailed and often technical materials, learn about new topics, and synthesize quantities of information. Such subject matter knowledge coupled with critical thinking knowledge is necessary to move from knowledge telling to knowledge-transforming communication practices. Genre knowledge, in conjunction with subject matter and discourse community knowledge, requires a communicator to understand a genre's pragmatic purpose, content, structure, and linguistic features as well as how these elements combine in a single text. Communicators also need to understand how the genre's formal features are determined by its role in a given discourse community. Decontextualized, formal rules, students learn in this section, will not lead to effective texts. The readings for this section (DellaSala, 2021; Tinnel, 2011; Zapf, 2016) focused on the importance of subject matter knowledge and genre knowledge to achieve successful communication on sustainability and climate change issues across different platforms.

The final area of study focused specifically on the hopeful aspects of social media in enhancing global communication on climate change and environmental rhetoric. Students were introduced to the role of social media, and how social media has become ubiquitous in promoting the goals of environmental and sustainability organizations, with many of them employing staff to oversee social media efforts. Environmental and sustainability organizations use social media to announce events and activities, share information, and provide information about the main issues with which the organization is involved. In this section, they also learn that even though environmental organizations have done much to engage a social media audience, many are still working towards creating a social media presence that engages a diverse global audience, promotes a positive self-image, and encourages readers to advocate for them. The texts in this section (Rawlinson, 2015; Harper, 2009; Resurreción, 2013) discussed aspects of how we can use social media responsibly and successfully in climate change discussions.

Students' initial responses were guided by questions related to the readings and to their own engagement with the topics addressed in the readings. Additionally, the follow-up comments by their classmates encouraged in-depth discussions on our roles as environmentally engaged communicators and activists whose responsibility it is to educate young individuals. Class participants were encouraged to respond carefully to each other's posts in order to promote critical thinking within a non-hierarchical and non-judgmental environment (Brydon-Miller, 2018; Wlodkowski & Galbraith, 2004) where idea development was encouraged within a largely closed discourse community – the classroom – before the ideas were then strengthened through review and additional reading and discussions and applied in a public space. Thus, the reading responses and the conversations with their classmates were especially important so that students had a solid foundation for developing, revising, and publishing their publicly accessible blog entries.

Students' blog posts focused on the topics addressed in class, with the intent of convincing their audiences to engage in climate-forward actions. Students were asked to connect their action-focused writing to discussions on such topics as climate change activism, food deserts, the global impact of climate change, climate industry and climate change, health and climate change, changing ecosystems, socio-economic inequities, and water resources. To be successful as blog writers, they needed to consider the knowledge that their intended audience already had, and what the assumptions, myths, and narratives were that could make them reluctant to listen to their argument. Doing initial research, and talking to individuals with knowledge on the topic they wanted to explore in more depth were essential parts of the blog writing process. Including a scaffolded approach to public writing – starting with writing that was not publicly accessible and that was reviewed by classmates and the teacher – also increased students' self-efficacy as writers (Bandura, 1982; Chung, Chen, & Olson, 2021) and encouraged them to situate their writing within the larger framework of previously discussed topics.

Much of the research conducted by students focused on educating their audiences on how we, as future and current teachers, can educate students, parents, other teachers, and members of the community on why it is so important to participate as climate activists at the local, national, or global level. Topics discussed included best ways to talk to educators about incorporating environmental education in the classroom, the importance of intersectionality in climate justice discussions, using social media as a tool towards change, listening to and learning from young voices talking about climate change, the role of national leaders, corporation board members and executives, and elected officials in keeping global warming under 1.5 degrees centigrade, a guide to persuading friends, coworkers, parents, teachers, and students to become a climate activist, sustainable consumerism, making the science of climate change approachable to young individuals, and tailoring our environmental lessons to local communities. An indepth exploration of these varied topics was only feasible because students participated as members of a knowledge community whose diverse and transdisciplinary interests (Klen & Meehan, 2015; Kubisch, et al., 2021) enabled them to learn from each other and provide additional resources and feedback.

At the end of the class, students were excited about how much they learned about climate-forward education that could benefit their localized communities. One of the Navajo students in class, for example, mentioned how the course reminded her of her grandfather's and other tribal members' travels to London during the early 1990s to protest coal mining operations on the Navajo Nation. Their reception in London, she pointed out, show the ongoing fight against environmental justice, with Nick Fletcher (1991) from *The Times* reporting that "A touch of the wild West came to town yesterday when four Navajo red Indians went on a warpath against millionaire industrialist Lord Hanson in a battle to protect their land" (Fletcher, 1991). Similarly, a student living close to the Salton Sea in California focused on the importance of highlighting how the chemicals and pesticides create toxic dust which has been long ignored by local and federal authorities. Students were also encouraged to continue work on how, for example, narratives can be used to persuade people to think differently about climate change and to act on their convictions. Students were especially encouraged by the interview assignment which created connections between academic assignments and localized impact of climate change. It provided them with continued excitement to remain active as climate change activists who encourage self- and communal efficacy in their students and communities.

The following section provides more detailed information on how teachers and students can continue their contributions to climate change discussions and climate-forward actions.

ENCOURAGING CLIMATE ACTIVISM FOR TEACHERS AND STUDENTS: RECOMMENDATIONS

Environmental education is not a stand-alone teaching endeavor. Instead, if we want to participate in a world where change and of hope drive action, teachers and students can become essential contributors to an ecological paradigm shift where the complexity of the climate crisis is not a hindrance to action but the impetus for climate action movements.

Stay Informed About Current Efforts Relating to Climate Activism

Current efforts show the engaged and committed young people and the global efforts of individuals and groups to tackle climate change on multiple levels (e.g., Connect4Climate, 2024; Earth.org, 2024; Ellen MacArthur Foundation, 2024; Children's Environmental Literacy Foundation, 2024; Plant for the Planet, 2024; Sunrise Movement, 2024; Extinction Rebellion, 2024). Furthermore, talks and videos (e.g., Battle, 2019; Burchardt, 2021; Collyer, 2023; Mathai, 2023) on the climate crisis and sites such as those maintained by the United Nations (2016, 2023, 2024) can be excellent ways of learning more in-depth information about the multi-dimensionality of the climate crisis. Teachers and students can contribute to ongoing efforts and also create new movements in order to encourage their peers, friends, parents, and community members to participate in climate action in any way they can.

Emphasize the Transformative Nature of First-Hand Stories

The climate crisis and its economic, sociological, and psychological impact has created chasms locally, nationally, and globally. Climate injustices, for example, impact poorer communities more severely than socio-economically advantaged groups. Droughts, wildfires, hurricanes, and heatwaves, and the resulting need for leaving newly uninhabitable areas are becoming more common and more devastating (Battle, 2019; Ferris, 2008). Climate refugees have left their countries and have become part of the community that is looking for a new home in less-affected countries. Learning their stories, and thinking critically about ways to incorporate their experiences into exploring new solutions to the climate crisis is a first step in creating engaged community involvement that reaches beyond known boundaries.

Encourage School and Community Collaborations

Teacher training focuses to a large extent on working with diverse students in a classroom setting. Learning is measured by how well students perform in the classroom and on tests, and little attention is given to collaborative efforts that reach outside the classroom and involve students' communities. However, the climate crisis involves multiple stakeholders and requires new ways of collaborating on multiple levels (Brydon-Miller, 2018; Damico, Baildon, & Panos, 2020; Kubisch, 2020). Such collaborative efforts can lead to new solutions and can encourage participants to become active members of a climate-forward movement.

Connect to Curriculum Resources Available for Adaptation to Local and Regional Needs

Curriculum resources focused on climate action are an important resource for educators who want to make sure that they can participate as agents of change in their schools and communities. Sites such as Stanford's School of Sustainability provide an excellent starting point for adapting climate education goals to the specific needs of the community. Including a middle and high school curriculum on Climate Change Education (2024), Stanford provides detailed information, videos, and websites for engaging students in discussions of climate change. Furthermore, organizations such as the EPA (2024) and NOAA (2024) provide lesson plans and teacher guides for teaching climate change topics, and UNESCO (2023) encourages teachers to green every curriculum and provides guidance for developing climate-action curricula online as of June 2024. Such curriculum resources provide an excellent starting point for ensuring that environmental education becomes part of teachers' and students' educational training.

Connect to Climate Activism Online

Teachers are always acquiring new knowledge in their subject areas and in their pedagogical approaches to make sure that students receive the best education. Similarly, teachers need to continuously update their knowledge on environmental education so that students can be exposed to the newest information on climate change education. Resources such as *Climate Literacy: The Essential Principles of Climate Science* (U.S. Global Change Research Program, 2009) and OECD's (2021) discussion of climate education provide excellent information that can encourage further study of including environmental education into the curriculum. Additionally, a resource site that includes access to current efforts by young individuals and by environmentally-forward organizations, and that provides access to videos or talks (e.g., Connect4Climate, 2024; Ellen MacArthur Foundation, 2024; Plant for the Planet, 2024; Sunrise Movement, 2024; Extinction Rebellion, 2024) can be used to increase students' understanding of their own role in climate change activism.

A CONTINUOUS ECOLOGICAL CHALLENGE: FUTURE RESEARCH DIRECTIONS

Environmental education provides opportunities for participating in climate change conversations and providing the necessary foundation for understanding the systemic nature of our changing climate. Future research into how we can highlight the undeniable effects of climate change on all aspects of our existence is especially important if we want to create increased awareness of our own roles, the roles of our communities, politicians, corporations, and world leaders in slowing down the climate crisis and creating opportunities for positive climate action.

To create sustainable solutions to the continuing environmental challenge also relies on continuous research into how interdisciplinary and transdisciplinary collaborations that include researchers from different academic disciplines as well as community members, city council members, politicians, CEOs, and world leaders who can create knowledge communities and knowledge societies (Echeverría, 2011) where collaborative work can become the starting point for continued explorations of climate solutions. Such interconnected research and scholarship is essential in our search for sustainable solutions to the climate crisis, a new and unprecedented crisis that cannot be solved by old ways of doing scholarship.

Similar collaborations among teachers and students in K-12 education provides an additional area for research partnerships to show what is necessary to create an environment where climate change can be addressed through a number of different but related lenses. For example, the effects of eco-anxiety on climate action, the economic impact of a green car or housing industry, the literary examples provided by environmental literature, the science behind green hydrogen, and the results of strengthening students' eco-literacy all need further research explorations. Encouraging a school environment where students can move from class to class and explore the same issue in their science, English Language Arts, mathematics, or social science classes.

CONCLUSION

The current climate crisis has affected all areas of life in all parts of the globe and has prompted UNESCO (2021) to call for including environmental education as a core curriculum component in all countries by 2025. If we can engage students in environmental communication efforts, and if we can shift the normative narrative from climate catastrophe, environmental disasters, and climate inaction to climate action and activism, we can encourage students and teachers to become climate advocates. It is important to promote learning communities where current and future teachers can engage successfully with each other and with their students on the complex issue of climate change, and where they can promote sustainable solutions to existing environmental problems. If teachers can inspire inquiry-based and critical self-reflection that leads to constructive climate activism in students and teachers, we can create learning spaces that address environmental engagement and climate action.

As my chapter shows, it is important to create spaces where we can redefine, blend, and expand existing and limiting boundaries and create an environment where transdisciplinary and intersectional perspectives are foregrounded in order to create a new way of addressing climate change. If we are serious about stopping the ever more dangerous environmental crisis, we need to continue our education efforts and discussions of the socio-ecological systems in which we live, and we need to continue discussions on how to teach future and current teachers the complex theoretical foundations and ever-changing practical applications of an environmentally forward curriculum that explores future directions for climate change discussions.

REFERENCES

Anspach, N. M., & Carlson, T. N. (2020). What to believe? Social media commentary and belief in misinformation. *Political Behavior*, *42*(3), 697–718. doi:10.1007/s11109-018-9515-z

Armstrong, A. K., Krasny, M. E., & Schuldt, J. P. (2018). *Communicating climate change: A guide for educators*. Cornell University Press.

Bandura, A. (1982). Self-efficacy mechanism in human agency. *The American Psychologist*, *37*(2), 122–147. doi:10.1037/0003-066X.37.2.122

Bateman, J. (2023). July 2023 brought record-high temperatures, devastating floods across the U.S. *NOAA*. https://www.noaa.gov/stories/july-2023-brought-record-high-temperatures-devastating-floods-across-us

Bateman, J. (2023). Record shattering: Earth had its hottest July in 174 years. *NOAA*. https://www.noaa.gov/news/record-shattering-earth-had-its-hottest-july-in-174-years

Battle, C. P. (2019). *Climate change will displace millions. Here's how we prepare*. TEDTalk. https://www.ted.com/talks/colette_pichon_battle_climate_change_will_displace_millions_here_s_how_we_prepare

Baudon, P., & Jachens, L. (2021). A scoping review of interventions for the treatment of eco-anxiety. *Environmental Research and Public Health*, *18*, 18. PMID:34574564

Beach, R., Share, J., & Webb, A. (2017). Teaching climate change to adolescents: Reading, writing, and making a difference. Taylor & Francis. doi:10.4324/9781315276304

Benoit, L., Thomas, I., & Martin, A. (2022). Ecological awareness, anxiety, and actions among youth and their parents: A qualitative study of newspaper narratives. *Child and Adolescent Mental Health*, *27*(1), 47–58. doi:10.1111/camh.12514 PMID:34687125

Bladow, K., & Ladino, J. (Eds.). (2018). *Affective Ecocriticism: Emotion, Embodiment, Environment*. University of Nebraska Press. doi:10.2307/j.ctv75d0g8

Braun, S. (2023, Aug. 11). *Why is climate denial flourishing online?* Deutsche Welle. https://www.dw.com/en/climate-change-denial-still-thriving-online-but-why/a-66159857

Brydon-Miller, M. (2018). Critical learning, community, and engagement: Elements for creating positive learning environments and opportunities for positive change. *Educational Action Research*, *26*(3), 351–353. doi:10.1080/09650792.2018.1469651

Burchardt, J. (2021). How we can curb climate change by spending two percent more on everything. *TEDTalk*. https://www.ted.com/talks/jens_burchardt_how_we_can_curb_climate_change_by_spending_two_percent_more_on_everything

Burke, K. (1965). Terministic screens. *Philosophy and the Arts*, *39*, 87–102.

Burke, K. (1966). *Language as symbolic action: Essays on life, literature and method*. University of California Press. doi:10.1525/9780520340664

Burnaford, G. E., Fischer, J., & Hobson, D. (Eds.). (2001). *Teachers doing research: The power of action through inquiry*. Routledge.

Canella, G. S., & Lincoln, Y. S. (2009). Deploying critical methods for critical social purposes. In N. K. Denzin & M. D. Giardina (Eds.), *Qualitative inquiry and social justice: Toward a politics of hope* (pp. 53–72). Left Coast Press.

Cheng, L., Abraham, J., Trenberth, K. E., Boyer, T., Mann, M. E., Zhu, J., Wang, F., Yu, F., Locarnini, R., Fasullo, J., Zheng, F., Li, Y., Zhang, B., Wan, L., Chen, X., Wang, D., Feng, L., Song, X., & Liu, Y. (2024). New record ocean temperatures and related climate indicators in 2023. *Advances in Atmospheric Sciences*. https://link.springer.com/content/pdf/10.1007/s00376-024-3378-5.pdf?pdf=preview

Chung, H. Q., Chen, V., & Olson, C. B. (2021). The impact of self-assessment, planning and goal setting, and reflection before and after revision on student self-efficacy and writing performance. *Reading and Writing*, *34*(7), 1885–1913. doi:10.1007/s11145-021-10186-x

Clark, G. (1990). *Dialogue, dialectic and conversation: A social perspective on the function of writing*. siu press.

Clayton, S. (2020). Climate anxiety: Psychological responses to climate change. *Journal of Anxiety Disorders*, *74*, 102263. doi:10.1016/j.janxdis.2020.102263 PMID:32623280

Collyer, R. (2023). How to supercharge renewables and energize the world. *TEDTalk*. https://www.ted.com/talks/rebecca_collyer_how_to_supercharge_renewables_and_energize_the_world

Constantino, S. M., & Weber, E. U. (2021). Decision-making under the deep uncertainty of climate change: The psychological and political agency of narratives. *Current Opinion in Psychology*, *42*, 151–159. doi:10.1016/j.copsyc.2021.11.001 PMID:34861621

Cook, J. (2019). Turning climate misinformation into an educational opportunity. In J. Fessmann (Ed.), *Strategic climate change communications: Effective approaches to fighting climate denial* (pp. 13–33). Vernon Press.

Cook, J. (2022). *Understanding and countering misinformation about climate change. Research anthology on environmental and societal impacts of climate change, 1633-1658.* IGI Global.

Damico, J. S., Baildon, M., & Panos, A. (2020). Climate justice literacy: Stories-we-live-by, ecolinguistics, and classroom practice. *Journal of Adolescent & Adult Literacy*, *63*(6), 683–691. doi:10.1002/jaal.1051

Das, L. (2022). Greenwash: What it is and how not to fall for it. *Greenpeace*. https://www.greenpeace.org.uk/news/what-is-greenwashing/

DellaSala, D. A. (2021). Speaking truth to power for the Earth. In *Conservation Science and Advocacy for a Planet in Peril* (pp. 355–369). Elsevier. doi:10.1016/B978-0-12-812988-3.00010-7

Diffey, J., Wright, S., Uchendu, J. O., Masithi, S., Olude, A., Juma, D. O., Anya, L. H., Salami, T., Mogathala, P. R., Agarwal, H., Aboy, K. V., Cote, J., Saini, A., Mitchell, K., Kleczka, J., Lobner, N. G., Ialamov, L., Borbely, M., Hostetler, T., & Lawrance, E. (2022). "Not about us without us"–the feelings and hopes of climate-concerned young people around the world. *International Review of Psychiatry (Abingdon, England)*, *34*(5), 499–509. doi:10.1080/09540261.2022.2126297 PMID:36165749

Dzambo, A. M., Mooney, M., Handlos, Z. J., Lindstrom, S., Hang, Y., & Ackerman, S. A. (2021). Improving Climate Literacy through an Interactive Online Undergraduate Course. *Bulletin of the American Meteorological Society*, *102*(6), 501–506. doi:10.1175/BAMS-D-19-0271.A

Echeverría, J. (2011). Epistemopolis and knowledge communities. *Plenary Speech, International Conference on Technology, Knowledge and Society*, Bilbao, Spain.

Enders, A. M., Uscinski, J. E., Seelig, M. I., Klofstad, C. A., Wuchty, S., Funchion, J. R., Murthi, M. N., Premaratne, K., & Stoler, J. (2023). The relationship between social media use and beliefs in conspiracy theories and misinformation. *Political Behavior*, *45*(2), 781–804. doi:10.1007/s11109-021-09734-6 PMID:34248238

Environmental and Energy Study Institute. (2023). *Public Polling on Climate Change*. EESI. https://www.eesi.org/briefings/view/020923camp

EPA. (2024). *Lesson plans, teacher guides and online environmental resources for educators*. EPA. https://www.epa.gov/students/lesson-plans-teacher-guides-and-online-environmental-resources-educators

Eyring, V., Gillett, N. P., Achuta Rao, K. M., Barimalala, R., Barreiro Parrillo, M., Bellouin, N., Cassou, C., Durack, P. J., Kosaka, Y., McGregor, S., Min, S., Morgenstern, O., & Sun, Y. (2021). Human influence on the climate system. In Masson-Delmotte, V., P. Zhai, A. Pirani, S.L. Connors, C. Péan, S. Berger, N. Caud, Y. Chen, L. Goldfarb, M.I. Gomis, M. Huang, K. Leitzell, E. Lonnoy, J.B.R. Matthews, T.K. Maycock, T. Waterfield, O. Yelekçi, R. Yu, and B. Zhou (Eds.), *Climate Change 2021: The Physical Science Basis. Contribution of Working Group I to the Sixth Assessment Report of the Intergovernmental Panel on Climate Change*. Cambridge University Press. https://www.ipcc.ch/report/ar6/wg1/chapter/chapter-3/

Falkenberg, M., Galeazzi, A., Torricelli, M., Di Marco, N., Larosa, F., Sas, M., Mekacher, A., Pearce, W., Zollo, F., Quattrociocchi, W., & Baronchelli, A. (2022). Growing polarization around climate change on social media. *Nature Climate Change*, *12*(12), 1114–1121. doi:10.1038/s41558-022-01527-x

Ferris, E. (2008). *Natural disasters, human rights and the role of national human rights institutions*. The Brookings Institution. https://www.brookings.edu/articles/natural-disasters-human-rights-and-the-role-of-national-human-rights-institutions/

Fessmann, J. (2019). An unlevel playing field: A primer on the problems of climate change communications. In J. Fessmann (Ed.), *Strategic climate change communications: Effective approaches to fighting climate denial* (pp. 1–11). Vernon Press.

Fletcher, Nick. (1991, Jan. 31). Indians' plea to big chief Hanson. *The Times.*

Gaard, G. (2017). *Critical ecofeminism*. Lexington Books.

Gorman, S., & Gorman, J. M. (2019, Jan. 12). Climate change denial: Facing a reality too big to believe. *Psychology Today.* https://www.psychologytoday.com/us/blog/denying-the-grave/201901/climate-change-denial

Harper, A. B. (Ed.). (2009). *Sistah vegan: Black female vegans speak on food, identity, health, and society*. Lantern Books.

Klenk, N., & Meehan, K. (2015). Climate change and transdisciplinary science: Problematizing the integration imperative. *Environmental Science & Policy*, *54*, 160–167. doi:10.1016/j.envsci.2015.05.017

Kubisch, S., Parth, S., Deisenrieder, V., Oberauer, K., Stötter, J., & Keller, L. (2020). From transdisciplinary research to transdisciplinary education—The role of schools in contributing to community well-being and sustainable development. *Sustainability (Basel)*, *13*(1), 306. doi:10.3390/su13010306

Kuthe, A., Körfgen, A., Stötter, J., & Keller, L. (2020). Strengthening their climate change literacy: A case study addressing the weaknesses in young people's climate change awareness. *Applied Environmental Education and Communication*, *19*(4), 375–388. doi:10.1080/1533015X.2019.1597661

Leiserowitz, A., Maibach, E., Rosenthal, S., Kotcher, J., Goddard, E., Carman, J., Ballew, M., Verner, M., Marlon, J., Lee, S., Myers, T., Goldberg, M., Badullovich, N., & Thier, K. (2023). *Climate change in the American mind: Beliefs & attitudes. Yale Program on Climate Change Communication*. Yale University and George Mason University. https://climatecommunication.yale.edu/publications/climate-change-in-the-american-mind-beliefs-attitudes-fall-2023/

Lynas, M., Houlton, B. Z., & Perry, S. (2021). Greater than 99% consensus on human caused climate change in the peer-reviewed scientific literature. *Environmental Research Letters*, *16*(11), 114005. doi:10.1088/1748-9326/ac2966

Mathai, W. (2023). The tree-growing movement restoring Africa's vital landscapes. *TEDTalk*. https://www.ted.com/talks/wanjira_mathai_the_tree_growing_movement_restoring_africa_s_vital_landscapes

McArdle, R. (2021). Intersectional climate urbanism: Towards the inclusion of marginalised voices. *Geoforum*, *126*, 302–305. doi:10.1016/j.geoforum.2021.08.005

MovementS. https://www.sunrisemovement.org/

Myers, K. F., Doran, P. T., Cook, J., Kotcher, J. E., & Myers, T. A. (2021). Consensus revisited: Quantifying scientific agreement on climate change and climate expertise among Earth scientists 10 years later. *Environmental Research Letters*, *16*(10), 104030. https://iopscience.iop.org/article/10.1088/1748-9326/ac2774/pdf. doi:10.1088/1748-9326/ac2774

Neas, S., Ward, A., & Bowman, B. (2022). Young people's climate activism: A review of the literature. *Frontiers in Political Science*, *4*, 940876. doi:10.3389/fpos.2022.940876

Newburger, E. (2021, Nov 16). What the COP26 climate conference really accomplished. *CNBC*. https://www.cnbc.com/2021/11/16/un-cop26-climate-summit-what-was-accomplished.html

NOAA. (2024). *Toolbox for teaching climate and energy*. NOAA. https://www.climate.gov/teaching/toolbox

Nuccitelli, D. (2020). How we know the Earth is warming and humans are responsible. *Bulletin of the Atomic Scientists*, *76*(3), 140–144. doi:10.1080/00963402.2020.1751969

OECD. (2021). Think green: Education and climate change. *Trends Shaping Education Spotlights*, *24*, 1–12.

Ojala, M. (2018-19). Eco-anxiety. *RSA Journal*, *164*(4), 10–15.

Panda, R., & Maity, M. (2021). Global warming and climate change on earth: Duties and challenges of human beings. *International Journal of Research in Engineering. Science and Management*, *4*(1), 122–125.

Parry, S., McCarthy, S. R., & Clark, J. (2022). Young people's engagement with climate change issues through digital media–a content analysis. *Child and Adolescent Mental Health*, *27*(1), 30–38. doi:10.1111/camh.12532 PMID:34904349

Pihkala, P. (2022). Towards a taxonomy of climate emotions. *Frontiers in Climate*, *3*, 1–22. doi:10.3389/fclim.2021.738154

Rao, M. (2022). Mental health impacts of the climate crisis: The urgent need for action. *International Review of Psychiatry (Abingdon, England)*, *34*(5), 439–440. doi:10.1080/09540261.2022.2128272 PMID:36165748

Rawlinson, M. C. (2015). Food, Health, and Global Justice. *International Journal of Feminist Approaches to Bioethics*, *8*(2), 1–9. doi:10.3138/ijfab.8.2.1

Rebellion, E. (2024). https://rebellion.global

Resurrección, B. P. (2013, September). Persistent women and environment linkages in climate change and sustainable development agendas. *Women's Studies International Forum*, *40*, 33–43. doi:10.1016/j.wsif.2013.03.011

Robinson, D. (2022). Explainer: What is greenwashing and how to avoid it? *Earth.org*. https://earth.org/what-is-greenwashing/

Siegner, A., & Stapert, N. (2020). Climate change education in the humanities classroom: A case study of the Lowell school curriculum pilot. *Environmental Education Research*, *26*(4), 511–531. doi:10.1080/13504622.2019.1607258

Sloam, J., Pickard, S., & Henn, M. (2022). Young people and environmental activism: The transformation of democratic politics. *Journal of Youth Studies*, *25*(6), 683–691. doi:10.1080/13676261.2022.2056678

Smiley, K. T., Noy, I., Wehner, M. F., Frame, D., Sampson, C. C., & Wing, O. E. (2022). Social inequalities in climate change-attributed impacts of Hurricane Harvey. *Nature Communications*, *13*(1), 3418. doi:10.1038/s41467-022-31056-2 PMID:36008390

Smith, A. B. (2024, Jan 8). 2023: A historic year of U.S. billion-dollar weather and climate disasters. *NOAA and Climate.gov*. https://www.climate.gov/news-features/blogs/beyond-data/2023-historic-year-us-billion-dollar-weather-and-climate-disasters

Smith, G. S., Anjum, E., Francis, C., Deanes, L., & Acey, C. (2022). Climate change, environmental disasters, and health inequities: The underlying role of structural inequalities. *Current Environmental Health Reports*, *9*(1), 80–89. doi:10.1007/s40572-022-00336-w PMID:35338470

Sturgeon, N. (1997). *Ecofeminist natures: Race, gender, feminist theory and political action*. Routledge.

Thunberg, G. (2019, May 4). It's 2019. Can we all now... [Twitter post]. https://twitter.com/gretathunberg/status/1124723891123961856

Tingley, B. (2024, Jan 12). 2023 was the hottest year on record, NASA and NOAA say. *Space Forums.* https://www.space.com/nasa-noaa-2023-hottest-year-on-record

Tinnell, J. (2011). Scripting just sustainability: Through green listing towards eco-blogging. *Environmental Communication, 5*(2), 228–242. doi:10.1080/17524032.2011.562519

Treen, K. M. D. I., Williams, H. T. P., & O'Neill, S. J. (2020). Online misinformation about climate change. *Wiley Interdisciplinary Reviews: Climate Change, 11*(5), e665. https://wires.onlinelibrary.wiley.com/doi/full/10.1002/wcc.665. doi:10.1002/wcc.665

UNESCO. (2021). *UNESCO urges making environmental education a core curriculum component in all countries by 2025*. UNESCO. https://www.unesco.org/en/articles/unesco-urges-making-environmental-education-core-curriculum-component-all-countries-2025

UNESCO. (2023). *Greening every curriculum: Development of the greening curriculum guidance.* UNESCO. https://www.unesco.org/en/education-sustainable-development/greening-future/curriculum

United Nations. (2015). Transforming our world: The 2030 agenda for sustainable development. *UN Department of Economic and Social Affairs: Sustainable Development.* UN. https://sdgs.un.org/2030agenda

United Nations. (2023). *The sustainable development goals report 2023, Special edition: Towards a rescue plan for people and planet*. United Nations. https://unstats.un.org/sdgs/report/2023/The-Sustainable-Development-Goals-Report-2023.pdf

United Nations. (2024). *The 17 Goals*. UN Department of Economic and Social Affairs: Sustainable Development. https://sdgs.un.org/goals

U.S. Global Change Research Program. (2009). *Climate Literacy: The Essential Principles of Climate Science*. US Global Change Research Program. https://www.globalchange.gov/reports/climate-literacy-essential-principles-climate-science

Vahlsing, C., & Liscow, Z. (2023, March 14). *The importance of measuring fiscal and economic costs of climate change*. The White House. https://www.whitehouse.gov/omb/briefing-room/2023/03/14/the-importance-of-measuring-the-fiscal-and-economic-costs-of-climate-change/

Watson, S. K. (2021, Dec. 4). 4 new myths about climate change – and how to debunk them. *Popular Science.* https://www.popsci.com/environment/climate-denier-myths/

Wlodkowski, R. J., & Galbraith, M. W. (2004). Creating motivating learning environments. *Adult learning methods: A guide for effective instruction, 3*, 141-164.

Young, C., & Jones, R. (2023, Nov. 1). *Denial is over: Climate change is happening, but why do we still act like it's not?* Phys.org. https://phys.org/news/2023-11-denial-climate.html

Zandalinas, S. I., Fritschi, F. B., & Mittler, R. (2021). Global warming, climate change, and environmental pollution: Recipe for a multifactorial stress combination disaster. *Trends in Plant Science, 26*(6), 588–599. doi:10.1016/j.tplants.2021.02.011 PMID:33745784

Zapf, H. (Ed.). (2016). *Handbook of ecocriticism and cultural ecology* (Vol. 2). Walter de Gruyter. doi:10.1515/9783110314595

ADDITIONAL READING

Baker, C., Clayton, S., & Bragg, E. (2021). Educating for resilience: Parent and teacher perceptions of children's emotional needs in response to climate change. *Environmental Education Research*, *27*(5), 687–705. doi:10.1080/13504622.2020.1828288

Bourn, D. (2021). Pedagogy of Hope: Global Learning and the Future of Education. *International Journal of Development Education and Global Learning*, *13*(2), 65–78. doi:10.14324/IJDEGL.13.2.01

Dolan, A. M. (Ed.). (2021). *Teaching climate change in primary schools: an interdisciplinary approach.* Routledge. doi:10.4324/9781003112389

Hannah, A. L., & Rhubart, D. C. (2020). Teacher perceptions of state standards and climate change pedagogy: Opportunities and barriers for implementing consensus-informed instruction on climate change. *Climatic Change*, *158*(3-4), 377–392. doi:10.1007/s10584-019-02590-8

Hathaway, J. R. (2020). Climate change, the intersectional imperative, and the opportunity of the Green New Deal. *Environmental Communication*, *14*(1), 13–22. doi:10.1080/17524032.2019.1629977

Höttecke, D., & Allchin, D. (2020). Reconceptualizing nature-of-science education in the age of social media. *Science Education*, *104*(4), 641–666. doi:10.1002/sce.21575

Hung, C. C. (2022). *Climate change education: Knowing, doing and being*. Taylor & Francis. doi:10.4324/9781003093800

Khadka, A., Li, C. J., Stanis, S. W., & Morgan, M. (2021). Unpacking the power of place-based education in climate change communication. *Applied Environmental Education and Communication*, *20*(1), 77–91. doi:10.1080/1533015X.2020.1719238

Li, C. J., & Monroe, M. C. (2019). Exploring the essential psychological factors in fostering hope concerning climate change. *Environmental Education Research*, *25*(6), 936–954. doi:10.1080/13504622.2017.1367916

Mayes, E., & Holdsworth, R. (2020). Learning from contemporary student activism: Towards a curriculum of fervent concern and critical hope. *Curriculum Perspectives*, *40*(1), 99–103. doi:10.1007/s41297-019-00094-0

McArdle, R. (2021). Intersectional climate urbanism: Towards the inclusion of marginalised voices. *Geoforum*, *126*, 302–305. doi:10.1016/j.geoforum.2021.08.005

Nairn, K. (2019). Learning from young people engaged in climate activism: The potential of collectivizing despair and hope. *Young*, *27*(5), 435–450. doi:10.1177/1103308818817603

Nation, M. T., & Feldman, A. (2022). Climate Change and Political Controversy in the Science Classroom: How Teachers' Beliefs Influence Instruction. *Science & Education*, *31*(6), 1567–1583. doi:10.1007/s11191-022-00330-6 PMID:35261483

Ojala, M. (2017). Hope and anticipation in education for a sustainable future. *Futures*, *94*, 76–84. doi:10.1016/j.futures.2016.10.004

Rousell, D., & Cutter-Mackenzie-Knowles, A. (2020). A systematic review of climate change education: Giving children and young people a 'voice' and a 'hand' in redressing climate change. *Children's Geographies*, *18*(2), 191–208. doi:10.1080/14733285.2019.1614532

KEY TERMS AND DEFINITIONS

Climate Crisis: Addresses the threat of global warming and environmental catastrophe brought about by climate change.

Climate Justice Activism: Links human rights with climate justice and acknowledges the need to combat all injustices.

Eco-Anxiety: The fear of environmental disasters incurred by climate change and the associated concern for one's future and that of future generations.

Environmental Consciousness: Becoming aware of environmental problems and supporting and committing to efforts to solve environmental problems.

Environmental Education: Provides individuals with opportunities to discuss environmental issues, engage in problem solving, and take action to address the climate crisis.

Social Media Activism: Promoting awareness of social justice issues and showing solidarity by creating opportunities for grassroots movements.

Teacher Training: Provides teachers and educational personnel with opportunities to increase their knowledge about effectively instructing and assessing students.

Chapter 14
Growing Pains:
Moving Campus Gardens Beyond Administrative Barriers

Amara Szrom
University of New Mexico, Albuquerque, USA

ABSTRACT

Campus gardens are invaluable community engagement and experiential learning spaces serving diverse academic disciplines. They are positioned to become fixtures in educational institutions, given community advocacy and national legislative interest. At the time of writing, The University of New Mexico (UNM) misses a valuable opportunity to leverage its student-initiated Lobo Garden as a living laboratory, applied educational learning space, food production area, and hub for community engagement. The University of New Mexico's Albuquerque campus spans about 600 acres, yet only 600 square feet are allotted for outdoor learning. The garden has faced a decade of administrative barriers preventing its integration into student life and institutionalization as a valued campus resource. To explore the variables contributing to thriving and well-integrated university garden programs, 19 in-depth interviews were conducted with university program coordinators, administrators, and key stakeholders in the United States.

KEY FINDINGS

University campus garden programs are critical community engagement and experiential learning spaces. Administrators celebrate gardens as part of their sustainability initiatives, yet these programs must fight for their existence as a campus resource. Through interviews with 12 university campus garden programs across the country, it was found that obstacles to program success manifest as five commonly experienced barriers:

- Weak administrative support.
- Unpaid labor and informal management.
- Unstable funding.

DOI: 10.4018/979-8-3693-2577-3.ch014

- Inadequate student input process.
- Lack of program oversight.

In support of my case study throughout the chapter, these findings lead to informed recommendations to integrate and institutionalize UNM's Lobo Garden program:

- Expand the food production and outdoor classroom area.
- Clarify roles between stakeholders.
- Increase visibility through partnerships.

INTRODUCTION TO LOBO GARDEN, A CASE STUDY

The Lobo Campus Garden provides a vital, authentic space for diverse community connections on the University of New Mexico (UNM) campus. Here, students of many disciplines work side-by-side, pausing to split a sun-ripened melon or nibble snap peas off the vine while discussing and bonding over environmental issues, intended career paths, and family stories. The joy of being outside within the community, which includes the web of relations with other creatures, such as the roadrunners and pollinator insects who frequent the space, is a throughline that keeps students and neighbors returning to volunteer each week. Participants arrive by foot or bicycle to tend to the modest space bursting with blue corn, beans, and squash (traditionally known as the Three Sisters), colorful medicinal flowers, and a small orchard offering juicy cherries, pomegranates, and apples throughout the growing season. A decade of student academic design projects found throughout the garden gives the space an air of an adored community hub and reinforces its position as an outdoor classroom.

Despite students' best efforts and recent increasing administrative interest, the garden historically lacks the institutional backing it needs to be a well-integrated, institutionalized part of student life. Tucked away behind the Real Estate Development building at the farthest northeast point on campus, the garden's hidden location reflects both the administration's inability to recognize a powerful experiential learning opportunity and its refusal to invest in the garden as a campus sustainability hub, as it denied student proposals for a more central site (Riley, 2010). Though the campus spans about 600 acres, only 600 square feet are allotted to grow food and serve as an outdoor classroom.

Most students are unaware of the campus food production space and pass by the adjacent lawn without noticing. One student remarked on his first day of class at Lobo Garden, "I feel like I'm in someone's backyard and I'm not supposed to be here" (personal conversation, August 2021). This lack of visibility stems from five identified institutional barriers that commonly prevent the integration of university garden initiatives across the United States (US).

Lobo Garden is entering its eleventh year and is at a pivotal moment to expand its impact. In this report, 12 US university garden program managers discuss their operational successes and challenges to better understand how Lobo Garden can position itself to be a valued campus resource.

BACKGROUND: THE RISE IN SCHOOL GARDENS

School gardens have become increasingly popular and are positioned to become fixtures in educational institutions, given community advocacy and national legislative interest (Turner, 2016). Most academic research on gardens covers K-12 grade schools and centers on extensive student academic and health benefits. Little is known about university-level garden programs or the management approaches contributing to their operational success. In this paper, "school garden" will refer to research on K-12 grade school gardens and "campus garden" will refer to research on gardens at universities.

Around the 1990s, school gardens began to replace barren dirt, asphalt parking lots, or lawns, adding many benefits, such as dynamic outdoor classrooms, living laboratories, community hubs, and hands-on experiential learning. Academic, health and social benefits are well-documented and covered in the Benefits of School Gardens section.

In the U.S., federal and state legislation promotes the development of school gardens through several key policies, including the Farm to School program and National School Lunch Program. The Farm to School program connects schools with local farmers and encourages the use of locally sourced foods in school meal programs (Farm, 2023). In New Mexico, where Lobo Garden is located, the Public Education Department has established nutrition and wellness standards for all public schools in the state. These standards include guidelines for nutrition education that encompass support for school gardens.

This initiative often involves supporting and utilizing school gardens. An interagency Outdoor Classroom Task Force convened to coordinate and expand the use of school gardens statewide and provide funding (Recommendations, 2021). These efforts indicate positive legislative trends toward school garden implementation.

UNM'S SUSTAINABILITY COMMITMENTS LEAVE LOBO GARDEN BEHIND

In its 2017 Administrative Policies and Procedures Manual, UNM considers sustainability a priority. It acknowledges its position "to educate and prepare future leaders in sustainable practices critical to the future of society and the environment" (section 3.0). It values a "campus culture of sustainability" achieved through interdisciplinary projects and experiential learning, identifying that "flexibility in curricula should be increased so that students can increase their knowledge about sustainability issues" (section 4.1). It commits that the Office of Sustainability will find unique learning opportunities for sustainability practices and "provide opportunities for student involvement in sustainability projects as part of their academic studies" (section 3.2.1). Additionally, UNM's 2040 Strategic Plan aims to "Transform the educational experience by creating supportive, intellectually challenging, exciting, diverse, joyful learning environments inside and outside the classroom to ensure the lifelong success, upward social mobility, and engagement of all learners" (GOAL 2).

Despite these commitments, Lobo Garden is not included in any of UNM's sustainability initiatives. It is not included in campus tours, orientations, policies, or marked on the official university map. This paper explores practical recommendations for advancing Lobo Garden based on interviews with better-integrated programs across the U.S. The lessons of this case study can be applied to overcome operational hurdles that inhibit campus gardens by avoiding the five common administrative barriers covered in my research.

THE BENEFITS OF SCHOOL GARDENS

Greater Academic Success and Student Engagement

For students who struggle in the standard classroom setting, academic success can improve dramatically when linked with outdoor learning. Garden classrooms are versatile teaching spaces for diverse subjects that promote creativity. Forty-eight studies between 1990 and 2010 reported on 25 unique learning outcome categories, including direct academic improvement in science (reported by 93% of studies), math (80%), and language arts (72%), as well as indirect outcomes, such as social development, improved attitude toward academics, motivation, and school bonding (Williams, 2013). Of 4,184 California public school garden programs surveyed, the subjects most frequently taught outdoors were science (95%), environmental studies (70%), nutrition (66%), language arts (60%), and math (59%) (Ozer, 2007).

Results are consistent across programs, student samples, school types, and disparate research methodologies. As a possible explanation for why, research shows that school garden programs strengthen student engagement with academic work (Marley, 2022). Engagement with academic work is defined as constructive, enthusiastic, willing, emotionally positive, and cognitively focused participation in learning activities (Skinner, 2012).

Campus and school gardens enable educators with hands-on opportunities to transform classroom discussions into practical skills. The GeoScience Garden at the University of Alberta is an outdoor simulated field environment where students develop field observation skills in an accessible campus environment, preparing them for field courses in remote locations. Surveys of students interacting with the GeoScience Garden reported significantly greater satisfaction with their ability to collect field data (Waldron, 2016).

Enhanced Social Skill Sets and Ecological Engagement

School gardens provide unique opportunities for students to form social connections across diverse backgrounds through experiential learning while influencing school policies and culture. Urban agriculture education programs prepare youth for future civic engagement by enhancing their understanding of social justice and community assets and building leadership skills (Russ, 2021). Bringing students of diverse backgrounds together through a memorable educational experience sets outdoor classrooms apart from traditional classrooms. Several studies communicate this benefit, describing it as acquiring an "ecocentric identity" in which students perceive themselves as linked to the web of nature, a "positive environmental communication" (Millstein, 2023), a "social-ecological-transactional perspective of human development" (Ozer, 2007), and a place to "foster leadership for sustainability through a transformational education process" (Burns, 2015).

School gardens are systemic interventions and agents of change that promote health and well-being and contribute to positive school environments (Ozer, 2007). Gardening programs promote connections between children and adults, and between schools and communities (Bice, 2018). Campus gardens influence curricula and policies to be more environmentally focused. Policy improvements include incorporating garden food into cafeterias (Calamidas, 2020) and integrating with farm-to-college programs (Merrigan, 2008).

Improving Student Health

Childhood and adolescent obesity have been rising in the U.S. since the 1970s and are currently estimated to be at a shocking 34% (Monferrer, 2022). In response to increasingly sedentary lifestyles and disconnection of the youth from the natural world, school gardens and nutrition programs act as intervention points that lead to improved dietary intake and reduced blood pressure (Davis, 2013). Growing food alongside a nutrition curriculum encourages students to make healthier nutritional choices. Nutrition-related benefits of school gardens include increased knowledge of nutrition (Calamidas, 2020); increased fruit and vegetable consumption (Kim, 2020); and improved attitude toward, increased preference for, and willingness to try fruits and vegetables (Chan, 2022; Savoie-Roskos et al., 2017). Eight studies between 1990 and 2007 on garden-based nutrition intervention programs reported the participants' ability to identify food groups, recognize the benefits of fruits and vegetables, have an increased likelihood to cook, and have an appreciation for other individuals and cultures (O'Brien, 2009).

Food insecurity is a common barrier to academic success for college students. Reported rates of food insecurity among college students range from 20% to 60%, up to three times higher than those of the general U.S. population (Ullevig, 2021; Nazmi, 2019). This is no exception for UNM students; 32% are food insecure (Cargas, 2020). To address food insecurity on campus, students at a university serving a primarily Hispanic population co-created their campus garden with a campus food pantry. They attributed the success of this collaboration to sustained funding from a nominal yearly student fee and management by a clearly defined organizational structure consisting of students, faculty, and staff from various departments that represent nutrition, sustainability, and environmental health (Ullevig, 2021).

Farm-to-college programs are rising, and universities are implementing the "buy local" movement on campuses. These programs create connections between classrooms, cafeterias, and campus gardens by involving faculty, students, cafeteria workers, administrators, and farmers in activities that support good health, nutrition, agriculture, and the local economy (Merrigan, 2008). Farm-to-college programs dramatically increase the fruit and vegetable servings students consume using a meal plan in the dining hall (Joshi, 2008). Yale University's popular Farm-to-college program gives it a competitive edge over other Ivy League universities because prospective college students seek universities committed to sustainability (Merrigan, 2008).

University Role in Fostering Sustainability

To continue being relevant in leadership, universities must adapt to rapidly shifting global, societal, and systemic changes. They must shape students in their formative years to prepare them for successful sustainably-minded civic engagement roles. Professor Tema Milstein's experience initiating Lobo Garden as a class project leads her to believe that the interactive nature of campus food gardens lends students an opportunity to acquire "positive environmental communication": discourses that hearten, inspire, encourage, and engage sustainable and restorative ways of being (Milstein, 2023). Milstein believes that student participation in grassroots planning and management of food gardens provides a material and symbolic transition space that experientially expresses and manifests sustainability transformations far beyond the garden's boundaries. Sustainability must be a multi-departmental practice, not just a degree focus for students.

Campus gardens are interdisciplinary zones that facilitate growth relevant to all students, as the skill sets acquired reach beyond simply growing food. The typical university educational experience relies almost exclusively on classroom-based reading and writing. Academic and campus garden advocate Cathy Sherry explains the collegiate rejection of outdoor learning, stating, "research metrics, hiring and promotion policies, as well as deeply held assumptions about what it means to be intellectual or educated" (Sherry, 2022).

The History of Lobo Garden

The history of UNM's Lobo Garden highlights the critical role of partnerships between students, faculty, and administrators. It was compiled using interviews with previous faculty and administrative stakeholders instrumental in founding the garden. Mary Clark, the Sustainability Manager on the Facilities Management Team, oversees groundskeeping and maintenance; Andrew Marcum a graduate student instructor, and Tema Milstein an associate professor. Teresa Costantinidis, UNM Executive Vice President of Finance and Administration, and Travis Broadhurst, Graduate and Professional Student Association (GPSA) Sustainability Director, were interviewed for their perspective on current UNM sustainability initiatives.

Lobo Garden began as an experiment, movement, and educational process through which students learned how to change the status quo. Only when student initiative and faculty support aligned with administrative acknowledgment of value did Lobo Garden inch forward as a campus resource.

The garden was initiated in 2010 after earlier attempts to grow food on campus met administrative resistance. A previous student, Travis Mackenzie, had tried unsuccessfully to organize a campus community garden four years earlier (Riley, 2010). Before Lobo Garden, the last time food had been grown on campus was seventy years earlier in a World War II victory garden (Milstein, personal interview, 2023). After decades absent of campus tending, the baseline of environmental and cultural norms shifted. University of New Mexico administrators now viewed the idea of campus food production as a liability risk and outlier unnecessary for campus life. For example, they were concerned that garden-grown food might be poisoned and make students sick (Milstein, personal interview, 2023). Therefore, the original campaign was challenged to shift administrators' perceptions of gardens toward viewing them as legitimate educational environments where academic benefits outweigh perceived risks.

In 2010, several formal and informal attempts to establish a garden converged in a perfect storm of social change. At first, students guerrilla-planted, meaning with good intentions but without administrative permission, rogue gardens throughout the highly manicured campus grounds. The initial spark of student and administrative interfacing ignited when UNM's Facilities Management team ripped out an unauthorized vegetable garden planted by a student who lived in a campus residence hall. Key stakeholders, Andrew Marcum and Tema Milstein, recall that the student was inspired to start the garden by their class content in the Sustainability Studies Program. Students gathered to dig up and transfer the vegetable plants on the removal date. The Daily Lobo newspaper arrived to amplify the story, increasing pressure on the administration to honor students' desire for a campus garden space.

Meanwhile, the student group, Seeds of Rebellion, was busy searching for land for their separate community garden idea. Faculty at UNM's Research and Service-Learning Program (RSLP) were also searching for land and faculty to initiate a campus garden, though neither group was connected.

Following the story's publication, Facilities Management offered the student a formal opportunity to find a location for an official campus community garden space, but the student declined. Andrew Marcum, a graduate student and TA instructor, was the first to engage with the Facilities Management team through a formal process to create a garden. Marcum saw the project as an opportunity for his students to learn about land use and social movements through his course, Social Movements in America, under the

Department of American Studies. The event became a class discussion. He wondered about the implicit values that had influenced the administration's decision to eliminate the unsanctioned garden. "Why is it that the university resists students engaging in experiential learning? [The project] can benefit the community, contribute to knowledge about how we use land and resources to support basic human needs and promote equality, access, and opportunity." (personal interview, March 22, 2023).

Marcum and his class leveraged the campus garden project as an experiential learning opportunity, relating, "We used the garden space and the advocacy involved in creating it to understand community organizing and social change." Students presented a garden proposal to the Facilities Management Team as their semester final project. Their presentation included an educational mission and vision statement, addressed previous food safety concerns, and assessed several potential sites to determine an official campus garden site.

In 2011, the site behind the Real Estate Department building was established as Lobo Garden. It was bare, eroding dirt and sparse wild grasses at the time. Assistant professor Tema Milstein's class, Language, Thought and Behavior, was the first to break ground, build terraces, and prepare the land for planting. The timeline moved quickly – her students had planted seeds by the end of the first semester.

Between 2011 and 2014, Lobo Garden was maintained every fall and spring by two classes: Social Movements in America and Sustainability Studies. Marcum, who now serves as an academic director at City University of New York, recalls the academic benefits of his class's engagement and unique opportunity for applied learning beyond the traditional classroom setting:

"It was an interesting and challenging time. I was teaching something in a completely different way and learning new skills at the same time as my students. But it was also fun – it was exciting to go outside and not just sit around in the classroom. We were solving the logistical challenges of building something meaningful on a college campus and within an institution that does not focus on experiential learning."

Milstein remembers that the first class had "tremendous academic benefit," and her students said it was their best class. Students learned theory and discourse in the classroom and then applied it in the garden, becoming agents of change.

Mary Clark details her decade of involvement as a liaison between Lobo Garden and the Facilities Management team. She explains that Lobo Garden has always been a student-driven initiative needing administrative support to thrive: "Facilities Management empowered students to implement the garden - there is no better way to learn than hands-on."

Historically, administrative support has included expertise and oversight from the Facilities Management team when installing permanent infrastructure. Clark identifies the professional learning applications available for students involved throughout the process as essential elements of education offered in university settings; backing these student-led initiatives is the most valuable thing administrators can do.

Clark's insights provide context and enable a better understanding of the diverse stakeholder roles needed to integrate Lobo Garden as a well-utilized campus resource. She understands the garden's value, saying, "There's nothing else like it on campus. It's the perfect place to hold events."

The origin of the campus garden was a unique and powerful educational event that created experiential academic space and left a legacy. The University of New Mexico's Sustainability Studies program has been the academic home of Lobo Garden since 2022. Lobo Garden's transfer to the Sustainability Studies program helped it receive administrative support that was not available for many years.

RESEARCH METHODOLOGY

Only three studies currently evaluate criteria for the long-term success, integration, and institutionalization of school gardens. The first criteria for program evaluation are physical maintenance, academic connection, and administrative value through school policies and events (Ozer, 2007). Developed by researchers at Colombia University and derived from empirical data, the GREEN tool captures how school gardens are effectively established, integrated, and sustained (Burt, 2017). This study shows that, at the deepest level of integration, programs must create school-wide food and environmental policies, utilize robust organizational structures, and have an evaluation process of their goals.

The GREEN Tool Scale was used to rank and analyze the current integration level of Lobo Garden. While overall integration scored low, when categorizing the questions by stakeholder (administration/faculty/student) responsibility, it was clear that the low score came from a lack of administrative participation.

Each interview lasted 45 minutes over telecommunications with affiliates ranging from hired staff, professors, and retired coordinators. A program manager was defined as directly involved with program operations and knows its history. Representation covered both private and public universities. When the interviews were transcribed, the commonality of administrative barriers strongly emerged after the first few interviews.

During interviews, each program manager was asked about their year established, acreage, location on campus, groundskeeping involvement, summertime maintenance, food distribution, academic connection, operating budget, funding source, leadership structure, and department housed.

A "well-integrated" campus garden program met the following criteria:

- Established for more than five years.
- An academic connection: integrated with more than one aspect of learning (i.e., not just a club).
- Diverse involvement of stakeholders.
- Stratified budget categories.

FINDINGS

Four Commonly Reported Program Strengths

High Student Retention and Engagement

Students initially introduced through classes frequently return for volunteering. When students become alumni, they donate both money and time. All six student-initiated programs had previous students holding a staff position, creating an essential throughline for institutional knowledge of program stewardship. One manager said, "The garden is the vehicle, not the end. People come to the site, and the feeling captures them."

Engagement blossoms beyond what traditional classrooms can offer. Regarding student leadership, one coordinator said, "Gardening is only half of it. Students also gain valuable experience that prepares them for the workforce, such as in communication, group problem solving, and independent motivation." When student engagement had room to express itself creatively, managers reported unexpected positive outcomes, such as significant donations, unique academic connections, and keynote visits from international leaders in thought and culture.

Student enthusiasm was evident in sentiments such as, "Even when it's cold or raining, the students want to be outside!" When students are enjoying themselves, they are more open to discussing pertinent subjects that are emotionally taxing, such as climate change or racism. "There's something about sitting in a garden that allows for harder conversations, and it feels natural."

Diverse Academic Connection

Campus gardens improve courses like horticulture, plant pathology, and agroecology; however, universities also use campus gardens to enrich courses in religion, theater, meteorology, psychology, art history, and music.

Multiple Funding Streams

Universities that support their garden programs with internal funding do so through multiple funding streams. They use food contracts, revenue from produce sales, student fees, and the Vice Chancellor's Office.

Median Program Budget of $12,500

Nine of 12 programs operate under $15,000 for operational costs, with a median program budget of $12,500.

Key Findings

Despite these benefits, very quickly into the interview process, a striking narrative appeared. Even when campus gardens are institutionalized as a resource, they still experience administrative barriers to succeeding. Eleven programs reported at least two of the following administrative barriers:

- Weak administrative support.
- Unpaid labor and informal management.
- Unstable funding.
- Inadequate student input process.
- Lack of program oversight.

The following section details each finding with examples from first-hand interviews. To reference the findings in a table, please see the School Data table, included at the end of this section.

Weak Administrative Support

Programs experience adversity when administrators fail to support their gardens as learning spaces. A commonly stated example is that university administrators showcase campus gardens in marketing campaigns to prospective students as proof of their commitment to sustainability. Yet, eight of 12 program managers expressed that their programs are not adequately supported. One manager explained, "We are the crown jewel of [our school]. We draw people to the college. But we also must fight for our existence from year to year."

Five programs receive no internal financial support from the school and rely on grants, unpaid labor, or revenue from produce sales. However, for minimal financial investment, campus gardens are a generative space for experiential learning and community engagement utilized by diverse departments. Nine programs operate below an annual budget of $15,000 for events, workshops, and materials for acres of food production and distribution (Table 4).

A minority of four programs reported that they partner with Facilities Management. Involving Facilities Management to provide oversight and expertise increases the program's institutionalization ability because it creates cohesion across the campus landscape. However, garden programs set themselves apart from Facilities Management by agreeing on sovereign areas of maintenance and responsibility.

Three managers reported that administrators are concerned that growing spaces are visually divergent from the "desirable" highly manicured campus aesthetic. Messy gardens represent the student experiential learning process, which contrasts with the concept of a product-oriented, controlled space, like a golf course. Three managers solved this by posting signage indicating that the spaces are student-led and experimental. Their garden perimeter is maintained to have a manicured look.

Nine program managers reported that their administrators initially viewed campus gardens as a liability before a lengthy process of creating trust. Administrative concerns center on food safety risks and injuries from improper or unsanctioned use of the space, such as homelessness in urban areas. Instead of disinvesting in campus garden programs to avoid risk, risk is best avoided through proactive management and planning. The two programs interviewed which seemed to have the most minor administrative barriers (well-funded, diverse department use) were Dean-initiated.

Programs that sell or distribute produce avoid liability by securing food safety certifications and volunteer safety training. One program located in an urban area mitigates risk through a separate $500/year funding source in case unforeseen circumstances, such as vandalism, occur. No programs reported death or significant damage to the site in their combined 219 years of operation; therefore, liability concerns should inform program development rather than prevent program initiation.

What can administrators do to support campus gardens? Program managers weigh in with examples:

- A Marketing and Advancement Office that fundraises for the garden by selling student-created value-added products to donors.
- Professional development training for faculty wanting to utilize the campus garden as an outdoor classroom.
- Funding infrastructure for amenities, such as shade and seating. An outdoor pizza oven increased student participation at one university.
- A nominal student fee feeds into a Green Initiatives Fund that student groups use for the campus garden.
- The campus garden (i.e., outdoor classroom) is listed online as available to be booked by interested faculty.
- A campus garden tour is included in school orientations that invites people to work for a short period of time in the garden as part of the tour.

Administrative Support at UNM

Lobo Garden only moves forward as a campus resource when administrators and Facilities Management support it. The garden is experiencing increasing administrative awareness, leading to support and institutionalization within UNM. The University of New Mexico's 2022 Impact Report featured the Dean of Arts and Sciences planting at a student-organized, faculty-backed event. This show of goodwill is a step in the right direction. Administrators gain a working knowledge of the program's benefits and shortcomings through engagement with their power to change. Attending garden events is also an excellent opportunity to connect with students. The Dean's visit puts Lobo Garden in an excellent position to expand its reach and ahead of two other programs interviewed, who reported that their administrators have never attended a campus garden event.

Lobo Garden benefits from strong working relationships with the Facilities Management team, which provides valuable land management oversight and expertise. In previous years, the Facilities Team interacted with students through workshops and installations, providing a caliber of professionalism to the garden program. A detailed list of potential administrative roles and action items on page 25.

For minimal financial investment, campus gardens are generative spaces for experiential learning and community engagement that are utilized by diverse departments. The operational costs of Lobo Garden could remain low compared to those of other university initiatives. The University of New Mexico's solar panel initiative will cost at least $350,000 to initiate, and its brand-name, zero-waste, reusable container kiosks cost $15,000 per kiosk (GPSA Sustainability, personal interview, 18 September 2023). Nine of 12 gardens operate on a budget below $15,000 per year.

The garden infrastructure required to facilitate learning is low-tech. One program that has existed for over 40 years uses simple straw bales for student seating. When the bales begin to decompose, they are utilized as mulch and replaced with fresh bales. Staff salaries are excluded from analysis because program structures varied significantly, and it is difficult to conclude this data set.

What does the UNM campus communicate to students about appropriate relationships with nature? What kind of ecocentric identity do students develop on the UNM campus? As designed spaces, the intentional placement of infrastructure communicates a value system. The manicured lawns and oversized cement sidewalks reinforce a separation from, or dominion over, the natural world. Students may appreciate the UNM campus as an arboretum or lush refuge from the sun, yet it does not provide the immersive and rich participatory experiences that one discovers in garden ecosystems. Administrators can make a clear commitment to Lobo Garden by supporting it as an outdoor classroom, or risk being left behind in the campus garden movement.

The University of New Mexico has the smallest acreage allotted to outdoor learning of any university interviewed. For UNM to maintain a relevant leadership position and hold its competitive edge for prospective students concerned about sustainability, it must expand its experiential outdoor learning program.

Unpaid Labor and Informal Management

If campus gardens are without administrative support, they survive through unpaid labor and informal management. Programs started by faculty reported that their highest dissatisfaction was with exploitative, unpaid labor. Seven of 12 programs reported that faculty or managers are not paid for all of their working hours, and five of those seven programs operate without a coordinator. Non-instructional faculty labor maintains (at times, extensively) growing spaces, such as greenhouses, hoop houses, or more than one acre of productive land (Table 4).

Faculty managers reported that alongside their official research and teaching responsibilities, stewardship of campus gardens was considered a "passion project" or "labor of love." One associate professor and informal garden manager explained that all coordination efforts fell to him: "I do it as a part of my departmental service, not a part of my contract. I enjoy my engagement there, but it takes away from my other compensated responsibilities of research and publishing papers."

Another faculty professor, who initiated her garden through grants and received no university funding, explained: "The university doesn't count my maintenance hours, but if I don't maintain the garden, I have nowhere to teach my classes. Even during the summer when I'm not on contract, I still manage the student workers."

Even one manager of a decades-old program reported unpaid labor: "Everybody loves the farm, but nobody wants to give us a staff member."

These programs also experience the additional administrative barrier of unstable funding, leaving professors to spend unpaid hours searching and applying for grants. Programs with formal staff are better integrated with multiple departments, informal management results from a need for more administrative investment in governance and funding for staff members.

Informal management hinders the passage of institutional knowledge. One faculty professor and uncompensated manager commented on the possession of all program responsibility and institutional knowledge: "I don't know what [the administration is] going to do when I win the lottery, die, or retire."

Program managers reported that involving community partners contributes to significant expertise, volunteer labor, and funding. In addition to student volunteers and department faculty, programs partner with local botanical gardens, master gardener programs, agriculture extension offices, and non-profits to bolster their programming and available resources.

Figure 1. Internal funding sources

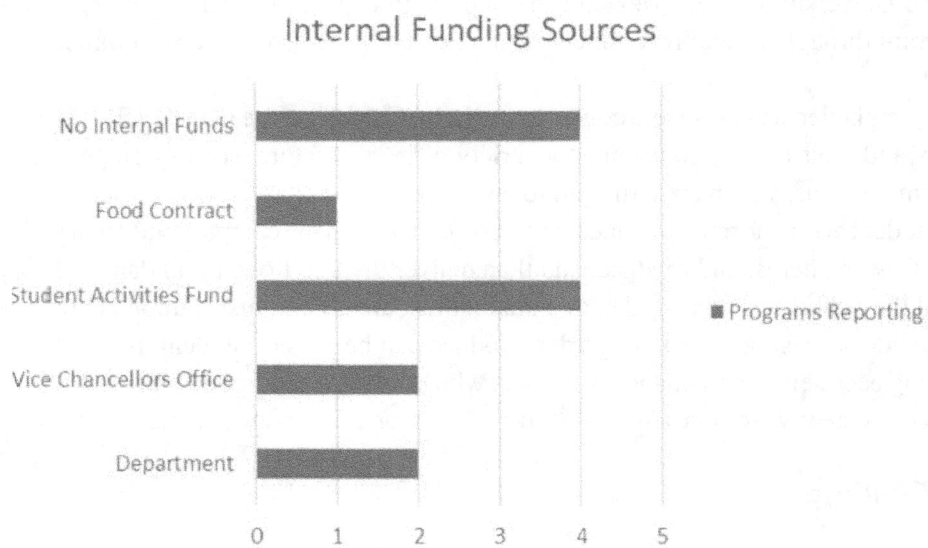

Figure 2. External funding sources

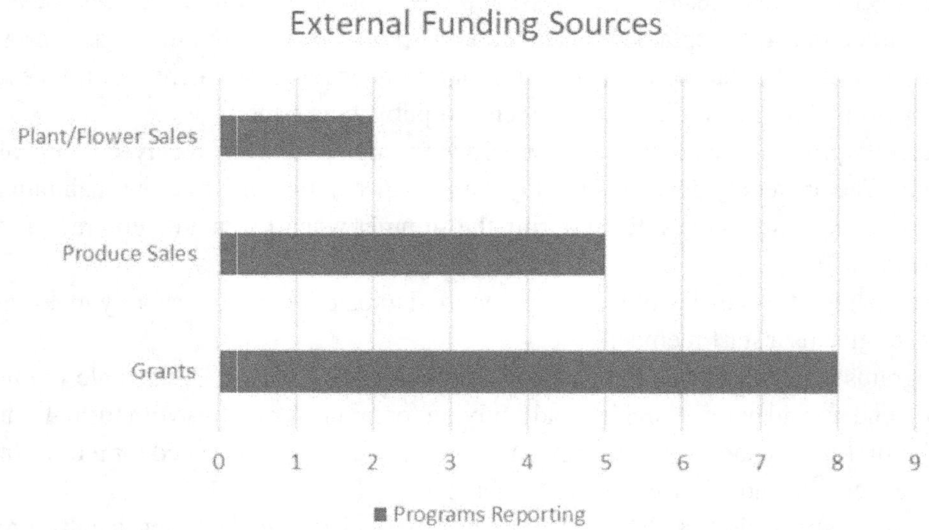

Program Management at UNM

Lobo Garden's management previously employed one graduate student as the course instructor and garden coordinator. Given that students graduate, this leadership design does not provide a throughline of institutional knowledge. In 2023, a faculty liaison was appointed; the first long-term manager to provide an institutional throughline for advocacy and multiyear strategic planning. The faculty liaison is compensated for their efforts to bring visibility and student enrollment to Lobo Garden's courses and activities. The University of New Mexico can continue to expand the stability and reach of the Lobo Garden program through several recommendations that are provided in the Recommendations section of this report.

Previously, eight departments were connected to Lobo Garden before the COVID-19 2020 shutdown. These faculty and their classes contributed a variety of infrastructure, such as grapevine trellises, terraces, and compost bins, which are still used today.

In 2022, student efforts formally connected Lobo Garden with the campus food pantry. Garden volunteers harvest flowers, herbs, and produce, and then deliver them to Lobo Food Pantry. This partnership has increased the visibility of the garden as a student resource. Food distribution is possible through a grant that provides refrigerators where garden produce can be stored. Student use of the campus food bank is growing each semester. Amanda Martinez, who runs the Lobo Food Pantry, detailed: "Our highest request is consistently fresh produce/fresh fruit." (personal interview, August 7, 2023).

Unstable Funding

Relying on external grants that shift from year to year is a significant challenge for eight programs. Programs without stable funding focus on selling produce more than hosting academic research. Six of the eight programs without stable funding sell produce to university dining halls and local restaurants, and run a weekly farm stand to meet costs. The remaining four programs, consistently funded by their university, utilize multiple funding streams that are not their department's responsibility to raise. Student affairs, residence halls, student community engagement centers, dining halls, and agricultural extensions are all examples of financial contributors (Table 4). The program with the most academic connections (more than 24 classes per semester) emphasized the importance of multi-departmental leadership and funding. Their campus garden is a multipurpose space for events, research, and applied learning. This university prioritizes involving as many departments as possible and believes that this policy has contributed to their recent approval of an additional 20-acre, 40-million-dollar expansion project.

For all 12 programs, the highest financial burden is staffing. Yet, well-integrated programs consider a permanent staff member one of the most essential strategies for success. Programs found funding for staff salaries through a variety of departments, supported by a myriad of leadership structures (Table 1).

Funding at UNM

Lobo Garden's funding for materials is from the Facilities Management Department, estimated at about $500 annually. The faculty salary and student teaching assistant stipend are currently Dean-supported and previously came from an external grant. During my two years as garden coordinator, the available budget was unknown for purchases or events; this created confusion around the potential scopes of projects or events that I wanted to create.

Moving forward, UNM's priority investments are infrastructure, such as seating and shade, and a paid coordinator position not occupied by a student. These additions bring visibility and class enrollment because they legitimize the campus garden as a functional outdoor classroom. More visibility and enrollment will bring more funding.

Figure 3. Program operating budget

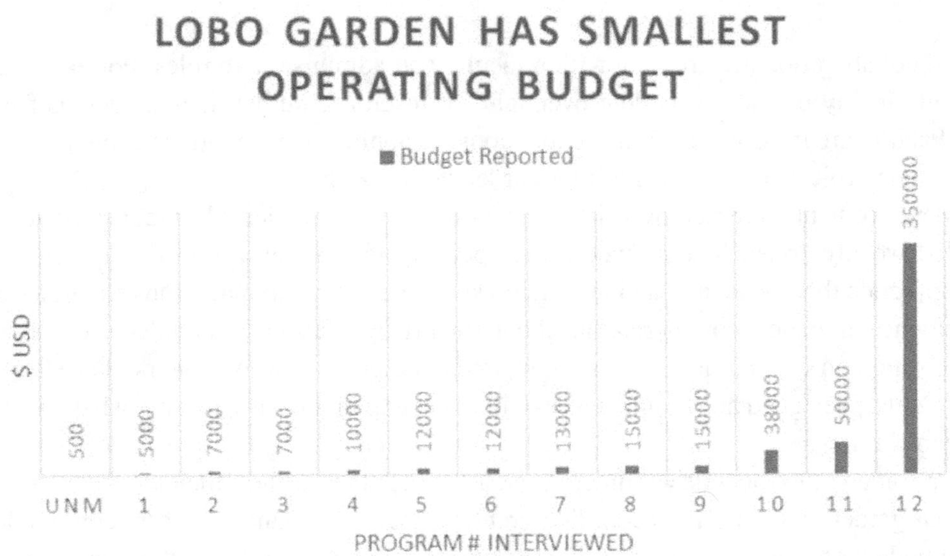

Inadequate Student Input Process

The working relationship between students, faculty, and administrators is critical for developing projects that foster academic connection and community engagement. While student enthusiasm rings clear, eight managers believe their administrators are dragging their feet in taking a supportive role for student experiential learning:

"Campus administrators want to be viewed as if they are doing something about sustainability. But when it comes to students proposing ideas, there is certainly a feeling of distrust that ideas are not coming from a professional. In my work, I noticed a student group pushing a sustainability initiative for a few years. Still, it wasn't until [my department] backed it and created a potential budget that it became credible to the administration."

Administrators may view students as unprofessional and need more trust in their organization. One student organizer turned part-time staff manager said: "We were actively trying to get squashed for many years, but now there's a good number of faculty and administrators who look out for us."

Trust is gained when students organize themselves into official groups that are backed by faculty, create roles of responsibility, and follow through. An adjunct professor and campus garden manager wondered if mistrust explained the slow pace of administrative engagement with student ideas to the point of delaying action until after student graduation. They wondered if this delay was intentionally meant to fragment and create a lack of activism on campus.

Only one program used an official student input process: a project proposal form for students to submit to their coordinator. Their process was guided by a set of values determined by an advisory board that included diverse, cultural, and student-led groups.

Student Input Process at UNM

Support and collaboration are growing within faculty and administrative roles. For example, Facilities Management already provides excellent oversight for installing infrastructure, such as fruit trees and irrigation. Faculty are making academic connections for applied learning in the campus garden through student capstone projects and class credit for garden volunteer hours.

Student engagement increased over the past two years. Some students became consistent volunteers and attended weekly "open hours", where they participated in stewarding the garden and used the grounds to practice their academic and leadership skill sets. There are numerous examples of this. One student studying lm made a commercial for the garden (Boyd, 2021); an architecture student created a professional site plan for an outdoor classroom, and another student, who worked at UNM's Outdoor Adventure Center, designed and taught a weekend workshop in the garden using what he learned in the Lobo Garden class.

However, students also operate within a gray area of understanding how their ideas can be implemented in the campus garden. As the garden coordinator, no protocols or boundaries were communicated to me regarding a budget or guidelines for actualizing student ideas. A student input process is advantageous because students, who spend the most time in campus gardens, are the first to identify essential projects. Students identified infrastructure needs, such as American Disability Act (ADA) accommodations, shade, and bike racks. They imagined creative projects that would contribute to Lobo Garden's biodiversity and community engagement; however, these projects are not always installed without a student input process.

Lack of Program Oversight

Lack of program oversight is defined as a decision maker's failure to solve challenges inherent in their programs. Five of six student-initiated programs need more oversight. Two Dean-initiated programs reported they were well supported and integrated, shifting priorities with the seasons to meet the needs of the garden's plants and student school schedule.

Student-led garden campaigns reported one barrier to oversight: grounds team workers were concerned that the campus garden would become their responsibility after students graduated or if the program fell through. Deciding who is responsible for what amongst stakeholders and creating accountability resolves this concern and supports worker solidarity.

As one program coordinator emphasized: "If we're going to work in food justice, we need to think about the [grounds] workers and not just see them as grumpy people preventing our visionary ideas. They're going to be there after we've graduated and gone."

Seven programs operated in a gray area for years; there was little or no awareness from the school of the program's activities, or the administration was actively trying to prevent the establishment of a campus garden. Programs reported that oversight stabilizes by developing advisory boards, working groups, feasibility studies, strategic plans, and handbooks. Advisory boards ensure that departments are connected and engaged around campus. They are most effective when they include diverse stakeholders outside of the university, such as city planners and indigenous land stewards. Advisory boards can create handoff documents and strategic multi-year plans. It is advantageous for boards to establish early on a program's mission and vision statements, as several program managers reported friction between departments over their garden's purposes.

Working groups more closely relate to daily operations and involve student groups. One program with high student engagement (up to 500 volunteers per semester) detailed that they use a "do-ocracy", where students with greater participation hours have greater decision-making privileges.

Programs must solve the inevitable transience of students. Programs reported employing students in positions of two to four years for less training time. All programs had at least one person in charge who was not a current student. Summertime campus garden maintenance was non-existent for three programs; the remaining nine hired one to two students to maintain the gardens via grants, internships, and federal work-study or community partnerships.

Program Oversight at UNM

Administrative investment clarifies Lobo Garden's purpose within its newly appointed home in the Sustainability Studies minor. This rehoming brings more academic connections and volunteer recruitment. Garden management is now included in regular team meetings between faculty and student coordinators. An advisory board is forming and plans to meet in Spring 2024. It will include a diverse group of decision-makers (who are not part of the university) and the broader community. The advisory board will update the campus garden's mission and vision statements, and establish annual goals. The advisory board will need to aid in long-term planning. It must collect data on engagement and paid opportunities for students, and amplify both academic connection and multi-departmental use.

The board must identify and resolve issues inherent in running a campus garden program, such as summertime maintenance, risk mitigation, student transience, land use and planning, governance, student input processes, and food distribution.

Lobo Garden scored as "Moderately Integrated" on the GREEN Integration Scale. When arranging the questions by responsibility, administrator participation scored low for six of six questions assigned to them. Faculty scored low for four of six and moderate for two of six questions assigned to them. The student role scored high for four of six, moderate for one of six, and low for one of six questions assigned to them.

These rankings illuminate areas for growth and can guide Lobo Garden's newly established governance to identify action steps and set goals to achieve institutionalization. Specifically, these questions, when arranged by stakeholder role, show where more responsibility of stakeholders can benefit the program. The questions can also be revisited for future evaluation and feedback.

SUMMARY

With minimal financial investment, campus gardens are generative experiential learning and community engagement spaces that diverse departments must utilize. Instead of disinvesting in campus garden programs to avoid risk, risk is best avoided through proactive management and planning. Broad governance, including a minimum of one paid staff position, increases a university garden program's capacity to thrive long-term. Providing students with guidelines and oversight for project implementation, as well as clarifying stakeholder roles of responsibility and accountability to solve administrative concerns, are of utmost importance.

Summary: Lobo Garden Program Strengths

- Lobo Garden is experiencing increasing administrative awareness, leading to support and institutionalization within UNM.
- Strong relationships are developing between garden stewards and the Facilities Management team, who provides oversight and expertise.
- Previous relationships with UNM Food, La Posada Dining Hall, eight departments, and the broader community, such as non-profits and neighbors, can be reactivated for funding, volunteers, events, and supplies.
- A newly appointed faculty liaison is the first long-term manager to provide an institutional throughline for advocacy and strategic planning.

Recommendations

At a Glance

- Expand the campus garden to develop food production and outdoor classroom areas.
- Clarify roles between stakeholders for good governance.
- Increase visibility through partnerships.

Figure 4. Campus proposed expansion

Expand the Campus Garden to Develop the Food Production and Outdoor Classroom Area

The lawn bordering the garden could expand the size of the food production area nearly three-fold, increasing visibility and student engagement and providing space for additional garden-classroom benefits, such as research plots, seating, community gathering spaces (e.g., pizza oven), indigenous gardens, pollinator gardens, and more.

Clarify Stakeholder Roles

Clarifying stakeholder roles alleviates concerns of risk or liability by enforcing accountability and contributing to better collaboration. These roles are based on what other programs have accomplished.

Administration: University Leadership

Role:

- Oversight of marketing and advertising.
- Value long-term financial support of the campus garden program.
- Planning for security and safety.

Action Items:

- Add Lobo Garden to the official university map.
- Create a permanent coordinator position.
- Include Lobo Garden in official university policies, such as the UNM Food contract.

- Approve infrastructure for an accessible outdoor classroom (i.e., ADA considerations, bike racks, shade, and seating).
- Support faculty to find consistent funding (i.e., nominal student fees that pool in a Green Initiative Fund for the campus garden).

Facilities Management
Role:

- Provide expertise and oversight on permanent infrastructure installations.

Action Items:

- Identify maintenance standards.

Faculty/Advisory Board
Role:

- Amplify and guide student-led initiatives to legitimize student ideas.
- Cultivate research opportunities.
- Adapt curricula to have experiential components taught in the campus garden.

Action Items:

- Diversify academic connections to Lobo Garden by involving additional departments.
- Create diverse avenues of student access (i.e., work-study positions, grant-funded internships, and community service).
- Create a formal student input process.

Students/Coordinator
Role:

- Organize events and workshops.
- Increase and maintain partnerships.
- Build visibility.
- Enhance food production.

Action Items:

- Work with GPSA Sustainability to achieve State Capital Outlay Request funding.
- Create an official Lobo Garden club group for more access between students and administrators.

Increase Visibility Through Partnerships

The Campus Food Pantry

There is a targeted need for the UNM administration to bolster this partnership through policy and expansion of the growing area, which is the smallest of any university growing space reported in this paper. Collaboration between the campus garden and food pantry can increase capacity building, visibility, and numbers of student volunteers.

Establish an Advisory Board for Governance

The committee should span multiple departments and include student representatives, community partners, and allied administrators. Use this board to craft mission and vision statements and identify specific stakeholder responsibilities. Well-integrated programs have boards that create guiding documents. For example, the University of Portland's three-year strategic plan details reporting, feedback, student involvement, quarterly events, partnerships, strategic goals, and work plans (Student, 2020).

Advisory boards must pick a primary focus for land use. Will the campus garden program function mainly as an outdoor classroom, market farm, public event space, or research area? Two program managers warned that disputes can occur when land use and planning visions clash amongst departments. Many departments utilize Lobo Garden (Figure 1), revealing potential for continued involvement and stewardship.

Community Partnerships

Routine public events and "open hours" for volunteers maximize community engagement. Previously, Lobo Garden benefitted from partnerships with UNM Food, La Posada Dining Hall, eight departments, and the broader community, such as non-profits and neighbors. These partnerships can be reactivated for funding, volunteers, events, and supplies. They can be revitalized through investing in clear leadership structures.

CONCLUSION

Campus gardens provide excellent applied learning opportunities that are not possible to manifest in traditional classrooms. As powerful community engagement generators, campus gardens provide unique opportunities for students to build positive ecocentric identities. All 12 campus garden program managers interviewed emphasized the importance of enthusiastic student participation, which correlates with earlier academic studies of K-12 student benefits. With administrative barriers removed, students, faculty, and administrators co-create thriving, dynamic spaces full of surprising academic connections and community support.

Campus gardens are incredibly productive spaces with almost limitless applications. Twelve universities across the U.S. make diverse academic connections to their outdoor classrooms, not only enriching the expected classes like horticulture, plant pathology, and agroecology, but also branching out to con-

Table 1. School data table

Interview #	Title Student# Programs	Acreage Amenities	Year est. Initiated by	Department Housed
UNM	Public R1 27,353 215	.013 acres	2011 Student- campaign	Geography & Environmental Studies
1	Public R2 27,229 200	12 acres Food Pantry Garden Dialogue Dome Permaculture Demo	2011 Student- campaign	Student Affairs Division Student Community-Engagement Center
2	Public Liberal Arts 8,668 26	.5 acre	1971 Faculty	Environmental Studies & Planning
3	Private Liberal Arts 2,526 50	1 acre	2003 Student- campaign	Sustainability Office
4	Public Land Grant R1 38,808 205	2 acres Greenhouse	2001 Faculty	Horticulture
5	Public Land Grant R1 29,217 250	1.7 acres High Tunnel	2013 Dean	Environmental & Plant Biology
6	Public Liberal Arts & Sciences 2,116 45	3 acres Learning Garden Organic Farm Off-campus	1995 Faculty	Science & Operations
7	Public Land Grant – Ag School 72,982 398	7 acres w/ 20 acre expansion approved. Children's area Visitor's center Food Service Outdoor Gathering Space	1993 Dean	Not affiliated with a single department as a success strategy
8	Private 10,000 150	1/8th acre Seed Library	2006 Student-campaign	Environmental Studies
9	Public Land Grant R1 45,000 350	7 acres (1.5 cultivated) Greenhouse Off-campus	2012 Student-campaign	College of Natural Resources
10	R1 29,500 243	¼ acre Greenhouse 10 satellite gardens	2016 Student-campaign	Botanical Garden
11	Public Land Grant R1 20,000 500	4 acres (1.5) cultivated Off-campus	2015 Faculty	College of Agriculture
12	Public Land Grant R1 49,500 600	10 acre Research Plots Hoop Houses	2020 Student Campaign	Grounds Department
Interview #	**Governance**	**Operating Budget (staffing not included)**	**Funding Source**	**Academic Connection**
UNM	GA/TA Coordinator Faculty Coordinator	$500	Grant	1 class – Lobo Gardens
1	1 staff full-time 1 student coordinator 6 students 10/hrs/wk. 2 interns	$12K $500 – risk mitigation	Student Activities Fund	Food security senior capstone International Students Leadership for Sustainability Graduate students Sustainable Food Systems certificate course
2	2 paid interns Adjunct Faculty	$15K	Instructionally Related Activity Fund	Agroecology Biology Psychology Multicultural Studies

3	2 staff full-time	$7K	Private Donors Grants	10 courses/year: Body and Earth Geography Geology English Religion Art History Architecture Research Projects
4	30 student workers (work study community service)	$12K	Plant Sales Grants Donations Flower Shop	Plant Identification Landscape Mgmt Special Topic
5	Faculty-run Volunteers via Office of Sustainability Summer interns –paid Workstudy Highschool interns Non-profit Collaboration	$7K	Grants 6K Revenue	Sustainable Agriculture
6	Farm Manager –paid Farm Aides 10 – 30 Student Internships AmeriCorps	$50K	62K Revenue	Agroecology Sustainable Agriculture Psychology
7	Faculty Run Garden Manager Vol Coordinator AgriLife events group Unpaid Interns	$350K	Extension Station Vice Chancellor's Office Private Donors Department Fees	Crop Science Landscape Architecture Plant Materials Lab Plant Pathology English Elementary Education Music Weavers and Spinners Entomology Floral Design Theater Meteorology Master Gardeners 4H
8	Adjunct Professor 1 Student Position	$5K	Vice Chancellor's Office	Urban Ag minor Community Engagement Anthropology
9	1 Staff part-time Student Working Groups Intern Coordinators Dean's Advisory Committee Indigenous Land Stewards	$10K	Private Donors	Student Groups Food Systems Minor Research plots
10	Paid Coordinator Student Leadership Team Advisory Board Volunteers Work Study Class Interns for credit	$13K	Dean Dept. Fees	3 classes/semester: Chemistry Food and Ag courses Service-Learning orgs
11	1 Staff – full time 3 students – part time Operations Manager Dean of Ag. Advisory Board	$38K	Dept. Fees	16 for-credit internships 2 classes/semester 2 Americorps 10 work-trade Garden Club Special Events
12	Culinary Services Ag and Life Sciences Student Orgs Paid Interns	$15K	Grants	Ag & Life Sciences Highschool groups

Interview #	**Food Distribution**	**Summer Maintenance**	**Groundskeeping Involved**	
UNM	Campus Food Pantry	N	Y – provide oversight	
1	Local Food Bank	N	Y – maintain safety	
2	Campus Food Pantry Local Food Bank	Y	N	

3	Sell to Dining Hall	Y	N	
4	Local Food Bank	N	N	
5	Sell to Dining Hall, Restaurants Local Produce Auction	Y	N	
6	Campus Farm Stand Local Food Bank	Y	N	
7	Plant Sales	Y	N – use 3rd party company	
8	Campus Food Bank	N	N	
9	Campus Food Bank	Y	Y – provide oversight	
10	Students Forage	Y	N – Botanical Garden provides expertise	
11	Sell to Restaurants	Y	N	
12	Sell to Dining Hall	Y	Y – collaborate on perennial areas	

nect courses in religion, theater, meteorology, psychology, art history, and music. Increasing academic connections leads to more support for physical maintenance, student engagement, and funding.

Campus gardens are excellent opportunities for students to apply the learning of their coursework and design in the field with academic guidance and faculty oversight. The garden is a springboard for limitless academic applications, including but not limited to data visualization, facilitation, community-led engagement processes, collaborative problem-solving, and independent motivation, all steeped in ecological values and in a highly accessible on-campus setting.

Continued research and deeper analysis are required to understand why the majority of program managers interviewed believe administrators do not adequately support their campus gardens. The scope of this study did not cover the motivations of administrators because they were not interviewed alongside their campus garden program managers. What is clear is that university administrations play key roles in legitimizing their campus gardens. Through the implementation of administrative support, including formal policies, paid labor, stable funding, a student input process, and program oversight, universities can reach countless students in their formative years and give them positive and ecologically centered experiences.

REFERENCES

Bice, M. R., Ball, J. W., Bickford, N., Bickford, S., Hollman, A., Coughlin, A., Dinkel, D., Meyer, R., & Ranglack, D. H. (2018). Community Gardens: Interactions between Communities, Schools, and Impact on Students. *Health Educator : Journal of Eta Sigma Gamma*, *50*(1), 2–10. https://files.eric.ed.gov/fulltext/EJ1196092.pdf

Block, K., Gibbs, L., Staiger, P. K., Gold, L., Johnson, B., Macfarlane, S., Long, C., & Townsend, M. (2011). Growing community. *Health Education & Behavior*, *39*(4), 419–432. doi:10.1177/1090198111422937 PMID:22167317

Burns, H., Diamond-Vaught, H., & Bauman, C. (2015). Leadership for Sustainability: Theoretical Foundations and Pedagogical Practices that Foster Change. *International Journal of Leadership Studies*. https://www.regent.edu/acad/global/publications/ijls/new/vol9iss1/6-IJLS.pdf

Burt, K. G., Koch, P., & Contento, I. R. (2017). Development of the GREEN (Garden Resources, Education, and Environment Nexus) Tool: An Evidence-Based Model for School Garden Integration. *Journal of the Academy of Nutrition and Dietetics*, *117*(10), 1517–1527.e4. doi:10.1016/j.jand.2017.02.008 PMID:28389106

Calamidas, E. G., Crowell, T. L., Engelmann, L., & Watkins-Jones, H. (2020). AtlantiCare healthy school edible garden startup grants: A content analysis of post-grant follow-up reports. *Health Education Journal*, *79*(6), 671–685. doi:10.1177/0017896920905622

Cargas, S., & Walsh Dilley, M. (2020). *Basic Needs Insecurity at UNM: 2020 Research Report*. https://basicneeds.unm.edu/basic-needs-at-unm-2020-research-report.pdf

Daniel Boyd. (2021, December 18). *Lobo Gardens* [Video]. YouTube. https://www.youtube.com/watch?v=SXwTp0JqIrE

Davis, J. (2015). Retraction statement: LA sprouts randomized controlled nutrition, cooking and gardening program reduces obesity and metabolic risk in Latino youth. (2015). *Obesity (Silver Spring, Md.)*, *23*(12), 2522. doi:10.1002/oby.21390 PMID:26524103

Duram, L. A., & Klein, S. K. (2015). University food gardens: A unifying place for higher education sustainability. *International Journal of Innovation and Sustainable Development*, *9*(3/4), 282. doi:10.1504/IJISD.2015.071853

Farm to School/New Mexico Grown. (2023). New Mexico Public Education Department. https://webnew.ped.state.nm.us/bureaus/student-success-wellness/nutrition/farm-to-%20school/#:~:text=The%20purpose%20of%20the%20New,the%20state%20for%20school%20meals

Hazzard, E. L., Moreno, E., Beall, D. L., & Zidenberg-Cherr, S. (2011). Best Practices Models for implementing, sustaining, and using instructional school gardens in California. *Journal of Nutrition Education and Behavior*, *43*(5), 409–413. doi:10.1016/j.jneb.2011.05.005 PMID:21906553

Joshi, A., Azuma, A. M., & Feenstra, G. (2008). Do Farm-to-School programs make a difference? Findings and future research needs. *Journal of Hunger & Environmental Nutrition*, *3*(2–3), 229–246. doi:10.1080/19320240802244025

Kim, S., & Park, S. (2020). Garden-Based Integrated intervention for improving children's eating behavior for vegetables. *International Journal of Environmental Research and Public Health*, *17*(4), 1257. doi:10.3390/ijerph17041257 PMID:32075303

Marley, S. A., Siani, A., & Sims, S. (2022). Real-life research projects improve student engagement and provide reliable data for academics. *Ecology and Evolution*, *12*(12), e9593. doi:10.1002/ece3.9593 PMID:36514554

Merrigan, K. A., & Bailey, M. R. (2008). The potential of Farm-to-College programs. *Nutrition Today*, *43*(4), 160–165. doi:10.1097/01.NT.0000303339.75605.9e

Milstein, T., Sherry, C. S., Carr, J. E., & Siebert, M. (2023). "Got to get ourselves back to the garden": Sustainability transformations and the power of positive environmental communication. *Journal of Environmental Planning and Management*, 1–19. doi:10.1080/09640568.2023.2197140

Monferrer, L., Valentín, G. L., & Villanueva, M. S. (2022). Mathematical and Experimental Science Education from the School Garden: A Review of the Literature and Recommendations for Practice. *Education Sciences*, *12*(1), 47. doi:10.3390/educsci12010047

Nazmi, A., Martinez, S. M., Byrd, A., Robinson, D., Bianco, S., Maguire, J., Crutchfield, R. M., Condron, K., & Ritchie, L. D. (2018). A systematic review of food insecurity among US students in higher education. *Journal of Hunger & Environmental Nutrition*, *14*(5), 725–740. doi:10.1080/19320248.2018.1484316

Ohly, H., Gentry, S., Wigglesworth, R., Bethel, A., Lovell, R., & Garside, R. (2016). A systematic review of the health and well-being impacts of school gardening: Synthesis of quantitative and qualitative evidence. *BMC Public Health*, *16*(1), 286. doi:10.1186/s12889-016-2941-0 PMID:27716134

Ozer, E. J. (2006). The Effects of school gardens on students and Schools: Conceptualization and Considerations for Maximizing Healthy Development. *Health Education & Behavior*, *34*(6), 846–863. doi:10.1177/1090198106289002 PMID:16861584

Policy 2100: Sustainability : University Policy. (2023, June 22). The University of New Mexico. https://policy.unm.edu/university-policies/2000/2100.html

Recommendations for Outdoor Learning in New Mexico. (2021). NMLegis.gov. https://www.nmlegis.gov/handouts/ALESC%20111721%20Item%2010%20.1%20-%20Outdoor%20Classrooms%20Task%20Force%20Slides.pdf

Riley, H. (2010, February 8). Students ask for sustainable garden. *Students Ask for Sustainable Garden - the Daily Lobo*. https://www.dailylobo.com/article/2010/02/students_ask_for_sustainable_garden

Robinson-O'Brien, R., Story, M., & Heim, S. (2009). Impact of Garden-Based Youth Nutrition Intervention Programs: A review. *Journal of the American Dietetic Association*, *109*(2), 273–280. doi:10.1016/j.jada.2008.10.051 PMID:19167954

Savoie-Roskos, M. R., Wengreen, H. J., & Durward, C. (2017). Increasing Fruit and Vegetable Intake among Children and Youth through Gardening-Based Interventions: A Systematic Review. *Journal of the Academy of Nutrition and Dietetics*, *117*(2), 240–250. doi:10.1016/j.jand.2016.10.014 PMID:27964852

Sherry, C. S. (2022). Learning from the Dirt: Initiating university food gardens as a cross-disciplinary tertiary teaching tool. *Journal of Outdoor and Environmental Education*, *25*(2), 199–217. doi:10.1007/s42322-022-00100-6

Skinner, E. A., & Chi, U. (2012). Intrinsic Motivation and Engagement as "Active Ingredients" in Garden-Based Education: Examining models and measures derived from Self-Determination Theory. *The Journal of Environmental Education*, *43*(1), 16–36. doi:10.1080/00958964.2011.596856

Student Sustainability Center Annual Impact Report. (2020). PDX.edu. https://www.pdx.edu/student-sustainability-center/sites/studentsustainabilitycenter.web.wdt.pdx.edu/files/2022-07/2020%20-%20 2021%20SSC%20Impact%20Report.pdf

Turner, L., Eliason, M., Sandoval, A., & Chaloupka, F. J. (2016). Increasing prevalence of US elementary school gardens, but disparities reduce opportunities for disadvantaged students. *The Journal of School Health*, *86*(12), 906–912. doi:10.1111/josh.12460 PMID:27866388

Ullevig, S. L., Vasquez, L., Ratcliffe, L. G., Oswalt, S. B., Lee, N., & Lobitz, C. A. (2020). Establishing a campus garden and food pantry to address food insecurity: Lessons learned. *Journal of American College Health*, *69*(6), 684–688. doi:10.1080/07448481.2019.1705830 PMID:31916927

Waldron, J. W., Locock, A. J., & Botey, A. P. (2016). Building an outdoor classroom for field geology: The Geoscience Garden. *Journal of Geoscience Education*, *64*(3), 215–230. doi:10.5408/15-133.1

Williams, D. R., & Dixon, P. (2013). Impact of Garden-Based Learning on academic outcomes in schools. *Review of Educational Research*, *83*(2), 211–235. doi:10.3102/0034654313475824

Chapter 15
Histories, Theories, and Contemporary Considerations of Nature-Based Early Childhood Education:
Roots to Branches

Stacey Alfonso
https://orcid.org/0000-0003-4344-5897
University of Washington, USA

ABSTRACT

This chapter provides a holistic examination of nature-based early childhood education (NBECE), spotlighting forest schools. Extensive research underscores the benefits of nature experiences for children's growth and development as well as nature-connectedness and ecological identity. As such, NBECE in North America has witnessed over a twofold growth from 2015 to 2020, with continued expansion. A similar trend is observed in the UK, where demand for nature-based education, particularly forest schools, has grown since the 1990s, with a significant surge over the last three years. Amid this recent growth, it is crucial to recognize the enduring existence of nature-centric pedagogies in Indigenous and local communities which begins this chapter's exploration. The discussion covers historical, philosophical, and theoretical underpinnings that have profoundly influenced contemporary forest school curricula, instructional practices, and overall approaches within the NBECE domain.

INTRODUCTION

An extensive body of research has identified nature exposure as providing abundant benefits to young children's healthy growth and development (Johnstone et al., 2022; Merrick, 2019) such as enhanced communication skills (Cameron-Faulkner et al., 2018; O'Brien & Murray, 2007), reduced stress and stronger resilience (Lee et al., 2018; Wells & Evans, 2003), social-emotional skills (O'Brien & Murray,

DOI: 10.4018/979-8-3693-2577-3.ch015

2007; Scott et al., 2018), improved academic performance and executive functioning skills (Carr et al., 2017; Tuen Veronica Leung et al., 2019), and increased opportunity for physical development and health (Meyer et al., 2017; O'Brien & Murray, 2007).

With these recognized developmental benefits and the substantial potential to impact a child's social-ecological identity (Barrable, 2019; Basil & White, 2005), it is no surprise that nature-based early childhood education (NBECE) in North America has more than doubled between 2015-2020 (NAAEE et al., 2020) with a continued interest and increase (NAAEE, 2023). Similarly, in the UK, nature-based education, more specifically, forest schools have seen an increased demand and growth over the past decade (Harding, 2021; Obrien, 2019; Smith et al., 2018). While nature-based education can be found globally, this chapter concentrates on the recent growth in North American (US and Canada) and the UK. The recent growth has spurred documentation, research, and a variety of practices to analyze.

Documenting and organizing this growth are crucial for educators, researchers, and policymakers to understand the evolving landscape of educational practices. As such, this chapter presents a comprehensive overview of nature-based early childhood education (NBECE), with a specific focus on the growing practice of forest schools in early childhood. It delves into the historical, philosophical, and theoretical foundations that have profoundly shaped contemporary forest school curricula, instructional practices, and overall approaches within NBECE and forest school. The discussion commences with Indigenous practices and reflection on land-based learning. It transitions to the early childhood and nature-based movements in Europe in the early and middle 20th century, then to the later 20th century expansion of nature-based education in the US, Canada, and the England. This discussion incorporates the significance of political and cultural contexts influencing pedagogical practice and theoretical orientation. The focus on the chapter narrows in on early childhood and more specifically forest school. The context proceeds to curriculum, instruction, and practice while providing examples of various curricula and theoretical considerations, including a critical examination of forest pedagogy arising from the "Common World" perspective. Then finally, the chapter brings together considerations for the future trajectory of research and practice.

NBECE: NURTURING CHILD-NATURE-CONNECTEDNESS AND ECOLOGICAL IDENTITY

As detailed above, a wealth of research clearly supports the benefits of exposing young children to nature. The growth of nature-based preschools has been instrumental in producing this research. Beyond affirming well-explored facets of child development, existing research introduces a less-explored yet crucial realm, child-nature-connectedness. This emerging domain carries profound implications for holistic child development, potentially fostering a transformative shift in ecological identity (Barrable & Booth, 2020; Finch & Bailie, 2015). Delving into this connection entails exploring not only cognitive, emotional, and physical aspects but also delving into the nuanced dimensions of a child's relationship with the natural world, offering a holistic understanding of their interconnectedness with the environment.

A related concept found in the human-nature relationship literature is attachment to place/place attachment. Summarized by Beery and Wolf-Watz (2014), "place-bonding or place-attachment...broadly refers to the emotional bond that may develop between an individual and a particular setting" (p203). As discovered in a recent literature review on children's place attachment (Koller & Farley, 2019), "place attachment" is used to capture a variety of concepts and developmental aspects throughout literature but always addresses the "profound and enduring connections between people and their surroundings" (p492). While attachment theory was developed to encompass personal, community, and environmental domains, it is of particular relevance to human-nature relationships due to the incorporation of natural elements within the environmental aspect. This inclusion has implications for individuals' ecological identity, rendering place attachment an essential piece of human-nature relationship. Additionally, place attachment has been found to be significantly correlated to environmentally responsible behavior (Vaske & Kobrin, 2001) which Ernst & Theimer (2011) interpret as indicative of a positive impact on one's ecological identity. These considerations are important throughout life and seem to be of significant influence during childhood (Beery & Lekies, 2018; Boyd, 2019).

Examining nature-based educational practices in early childhood is imperative since this time is a formative period in a child's life (Crain, 2005; Orman, 2011). Children are learning about themselves, about other humans, and about their surrounding environment. Cognitive and social capacities develop, and relationships become more complex, as children build an understanding of the meaning of their world in the context of their families and society (Crain, 2005; Rogoff, 2003). These budding relationships include a relationship with the living beings, systems, and non-living elements of our natural environment or "nature" herself (Sobel et al., 2015; Warden, 2015). In fact, early childhood has been known to be a pivotal stage for fostering a deep affinity with nature (Beery et al., 2020; DeVille et al., 2021; Kellert & Wilson, 1993; Orr, 1993; Sobel, 2008). Retrospective studies with adults (Alfonso, 2024; Barrows et al., 2022; Chawla, 1999, 2007; Tam, 2013; Wells & Lekies, 2006; Windhorst & Williams, 2015), confirm the importance of early childhood in their pro-environmental behaviors, attitudes, and life-decisions. For example, Chawla (1999) interviewed various environmentalist on a range of issues related to nature and the environment. One of her findings revealed that respondents consistently noted their environmental interests and action at least partially due to quality time in nature, typically in childhood.

Later work from Chawla (2007) points to the importance of nature role models in childhood. This influence of a role model, mentor, or adult guide in nature has been supported through other empirical studies (Alfonso, 2024; Beery et al., 2020; Cudworth, 2020; Linzmayer & Halpenny, 2014; Passmore et al., 2021). This finding is significant when we consider the potential impact teachers in a forest school setting may have on the children's ecological identity and nature-connectedness.

There has been a recent surge in the growth of nature-based education within formalized school settings, specifically in the UK and North America. However, it is essential to acknowledge the influence of pedagogies that prioritize nature and nature-based learning which have existed in various forms throughout history. That is why this chapter next employs a wide global lens to examine the foundational ideas which have become monumentally significant to the recent growth described in this chapter.

LEARNING WITH THE LAND

While the term "environmental education" was first used in North America in the mid-1960s (Gough, 2006), teaching and learning with the land and nature-based knowledges have been practiced by Indigenous peoples long before the settlers created what is now known as the United States or Canada (Merrick, 2019). Many concepts within the field such as reciprocity and respect for all beings, taking care of the land/stewardship, acknowledging and honoring place, and health and wellness have been part of the lives of Indigenous communities since time immemorial (Kimmerer, 2013; OSPI, n.d.; Salmón, 2000).

Throughout histories and presently, Indigenous communities have lived and continue to live in relationship with the land, practicing Traditional Ecological Knowledges, and developing intricate working systems that incorporate nature-cultural relations (Bang & Marin, 2015; Kimmerer, 2013; Pierotti & Wildcat, 2000; Salmón, 2000). Enrique Salmón (2000), an Indigenous scholar from the Rarámuri (Tarahumara) community, shares the word "iwigara." Since his Indigenous language expresses abstract concept of nature differently than the English language, a direct explanation is difficult to articulate, but he explains it as the "total interconnectedness and integration of all life" and equates it loosely to a "kincentric ecology" (p1328). In this way of thinking and being, humans and nature are mutually part of an extended ecological family.

Relating to nature as kin, family, or relatives in a mutually respectful and reciprocal way is widely discussed in literature from Indigenous scholars (Bang et al., 2015, 2020; Pierotti & Wildcat, 2000; Whyte, 2018, 2020). Pierptto & Wildcat (2000) describe this type of mutual relationship via the use of Traditional Ecological Knowledges, in which "human and nature have been and always will be in relation with each other [and] that there is no such thing as nature that exists independent of humans and their activities" (p1334). Indigenous scholars have contributed greatly and continue to significantly provide deeper meaning to the understanding of nature-based education by discussing the concept of relating to nature as kin, family, or relatives in in a way that is core to the human species.

Moreover, cultures worldwide have placed value on nature-based learning and learning with the land centuries before the emergence of the terms "environmental education" or "nature-based learning" (Merrick, 2019; Posey, 1998). In analyzing and advancing "nature-based" learning, it is crucial to recognize the centuries of work done and continuing to be done by Indigenous and local communities. According to Hill et al. (2020), collaborating with Indigenous and local knowledges is critical to inclusively assessing the linkages between nature and people, which is also vital for developing a better understanding of nature-based early childhood education. While these deeply rooted nature concepts have been part of many cultures' traditions and practices for centuries or longer, it was not until the more recent past that the idea of centering "mainstream" education around nature came into the conversation of policy and practice in school or educational settings in the US, Canada and the UK.

ENVIRONMENTAL EDUCATION: A BRIEF FOUNDING HISTORY

Environmental education serves as the encompassing framework within which nature-based early childhood education takes root, subsequently converging into the specialized domain of forest schools. The following section gives a very brief history of converging factors that contributed to the development of environmental education before diving more deeply into nature-based early childhood education and forest schools.

Many scholars writing about the origins of early childhood environmental education credit Friedrich Fröebel, a German educator, as the original "inventor" of kindergartens (Dabaja, 2022b; R. A. Larimore, 2019; MacQuarrie et al., 2015; Sobel et al., 2015) with his design to provide children with opportunities to learn and play, particularly in outdoor settings, hence the name "kindergarten" or "child's garden." This German concept of the mid-19th century was later taken up internationally in various forms. First attributed to nature educator, Gösta Frohm from Sweden, Skogsmulle outdoor education took a stronghold founded on two key principles; children need knowledge and togetherness in nature and children need to learn how to be in and protect nature (Sobel, 2016). With Sweden holding on tightly to the nature roots

Figure 1. Visual representation of environmental education in early childhood encompassing NBECE and forest school

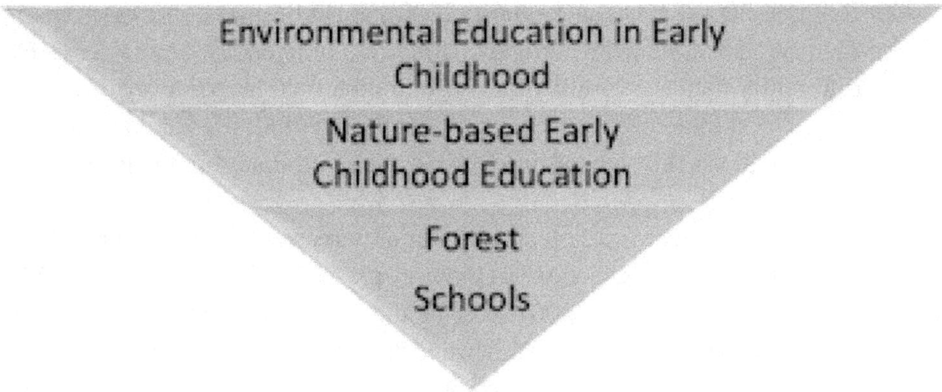

of early education, this concept quickly spread to other Scandinavian countries. By the mid twentieth century forest kindergartens and nature nurseries or "Waldkindergarten" with a strong emphasis on being in and learning with nature, became common in Scandinavian region (Dabaja, 2022a; Mackinder, 2022). In the late twentieth century, this type of nature-based education began to spread elsewhere, notably England, Canada, and the United States.

Concurrently, the environmental education movement in the United Sates was said to have been influenced by the nature studies movement beginning in the 1960s. Nature studies grew from the response to the publicized "environmental crisis" of the time (Gough, 2006). Influenced by the United Nations Conference on the Human Environment in 1972, the original objectives of the environmental education movement were to infuse curriculum at all levels with ecological content "promoting technical training and stimulating general awareness of environmental problems" (p72). Rooted in a cognitive awareness and academic understanding of the environment, environmental education (EE) or later called education for sustainability (EfS) (Barrable, 2019; Gough, 2006) grew across the US, Australia, and Europe to promote sustainable development through the challenges of globalization (Ärlemalm-Hagsér & Elliott, 2017). EfS is often seen as a response to the urgent need for societies to address environmental and social challenges and promote sustainable development, to navigate a rapidly changing eco-political social world. It is applied at various educational levels, from primary schools to higher education, and can be integrated into formal and informal learning contexts. While environmental education and education for sustainability are sometimes used interchangeably in the literature, Barrable (2019) makes an important distinction in contemporary terms: environmental education (EE) aims to educate in, about, and with the environment whereas, education for sustainability (EfS) grew from a focus on sustainable development, including both social and economic dimensions of sustainability. To Barrable (2019) and other scholars of environmental education in early childhood (Boileau, 2013; Monroe et al., 2008; Wilson, 1993; 2019) it is important to note that EE's central goal is to promote environmental literacy and enhance individuals' understanding of the environment and their role in it in order to make informed decisions environmentally sound decisions.

Ruth Wilson (1993) was one of the first environmental educators in the United States to call out the need for a focus on environmental education specifically in early childhood. She also made it a point to emphasize that "environmental education involves much more than learning scientific facts about the natural environment. It's also designed to foster positive attitudes and values about the world of nature and a sense of responsibility towards the natural environment" (p23). She made it clear that environmental education should involve both a cognitive component and affective appreciation. Later scholarly work makes a further distinction and specifies environmental education to support the development of a cognitive component, an affective component, *and* a behavioral component (Beery et al., 2020; Restall & Conrad, 2015; Schultz, 2002). The behavioral thread is said tend to an individual's environmentally responsible behaviors (Restall & Conrad, 2015).

THE GROWTH OF NATURE-BASED EARLY CHILDHOOD EDUCATION (NBECE)

With the previously mentioned influence of Scandinavian Waldkindergartens and the growing momentum of the environmental education movement in North America, Europe, and Australia, nature-based early childhood education was well-positioned to gain traction throughout the US, Canada, and the UK. In the US the first nature-based preschool is said to have opened in 1966 at the New Canaan Nature Center in Connecticut (Larimore, 2016; Merrick, 2019; Sobel et al., 2015). Throughout the 70's and 80's similar NBECE programs began slowly dotting the landscape of the early childhood field in the US, mostly growing from pre-established nature centers (Bailie, 2015). These nature-based preschools typically had a nature-centered indoor space and hold a strong commitment to bringing children outdoors and learning with nature. The influence of nature-based preschool branched out into various forms, including forest school and forest kindergarten which have no indoor space except for perhaps an emergency shelter. The first documented forest school in the US was established in 2007 in Vashon Island, WA by Erin Kenny and Robin Rogers (Sobel, 2016). The founders studied the forest schools extensively, heavily guided by the Waldkindergarten model.

In the UK, momentum of nature-based education budded in England in the early 1990s and continued to spread throughout the UK to Scotland, Wales, and Ireland. The forest school model became prominent with support from the UK's Forest School Association, a professional body and UK-wide voice for Forest School movement. The initial growth was documented to be a direct result of early years staff of Bridgewater College participation in an exchange program with Denmark in 1993 (Dabaja, 2022a; Maynard, 2007; McCree & Cree, 2017; Smith et al., 2018). The growth for forest schools in the UK is often acknowledged as a counterbalance to the increase in indoor time and general decline in and appreciation for outdoor play (Bilton & Waters, 2017; Maynard, 2007).

In the wake of the expansion of nature-based education in the US and the UK, Canada similarly embraced the trend, fostering the growth of its nature-based early childhood programs. According to the Forest Stewardship Council (2012), Carp Ridge Forest School in Ottawa, affiliated with the Child and Nature Alliance of Canada, is recognized as Canada's first forest school, established in 2007. Established by Marlene Powers, Carp Ridge Forest School grew realization that very little environmental education was being offered to children at the early childhood level. The foundation of Carp Ridge quickly sparked the establishment of Forest School Canada, an organization that supports forest school throughout Canada which aided in the quick growth of such programs nation-wide (MacEachren, 2013).

By the late 2000s, nature-based education and forest schools had become a "growth industry" in Canada, the US and the UK (Nelson et al., 2018). As the field of NB-ECE expanded, different program types emerged, leading to the establishment of definitions and guidelines by several professional organizations and academic writers. The subsequent section will provide an overview of these definitions and specifically categorize forest school as part of nature-based education.

Many practitioners and professionals in the field of nature-based early childhood education (NBECE) describe their work to be at the intersection of early childhood education and environmental education (Bailie, 2015; Larimore, 2016). This characterization arises from NBECE addressing both the developmental needs of children and the cultivation of an ecological identity or environmental ethic (Finch & Bailie, 2016). In this educational approach, the focus brings together conventional understanding of early childhood development to include fostering a deep connection with nature and environmental values. While all true NBECE settings would hold this set of double pronged values, there is still much variety in how it is implemented in the field (Larimore, 2018). Nature-based early childhood education can take place in a variety of settings and use different styles or approaches. This can include preschool at a park or nature center, programs in urban areas, farms, zoos, publics schools, or family childcare centers (Merrick, 2019). Nature-based programming can be all outdoors, outdoors with a small indoor center shelter, or an indoor program that spends a lot of time in nature and immerses its curriculum with natural elements and learning. Consider the following three vignettes, each very different but all nature-based programs:

Vignette A: Children romp around a large, forested area known as their "outdoor classroom." A small group is working to gather sticks and logs for a fort which they are constructing against a tree. The heaviness of the large logs requires communication and cooperation. One log is particularly difficult to move but the group manages to scoot it and discover a salamander living under the log. They watch quietly, observing then talk about the salamander's colors, and movement. One child gently touches it with a finger and squeals with delight as it skitters away.

Vignette B: After a morning of exploring a nearby park and walk through a food forest collecting herbs, a preschool group heads to their indoor nature-based classroom. The teacher heats up water and the children pull apart the herbs to make tea. The children observe how the herbs changes the color and taste of the water, and they discuss the different plants they saw on their morning excursion. The teacher talks about the history of the herbs and how they have been used in different cultural practices around the world then reads the book Dr. Wangari Maathai Plants a Forest, about the life of Dr. Wangari Maathai, the Nobel Peace Prize-winning activist and environmentalist from Kenya. The teacher offers a variety of books about herbs in local and global cultures.

Vignette C: Children gather at the front of an urban community center where they check-in with their teacher and make a plan for the day. They head off for their daily exploration of nearby urban green spaces and parks. They carefully cross city streets, stopping to observe a row of ants on the sidewalk and later, fungus growing on a fence. They arrive at a park and gather under a tree for a snack. They share stories about the squirrels they saw and the small rocks they found. One child talks about a dead baby bird she saw on the grown. The children begin to talk about why it may have died and ponder if it was attacked by another animal, pushed out its nest, and consider the impact of nearby branches that were cut to maintain powerlines.

Figure 2. Summary of the growth of nature-based and forest school practices

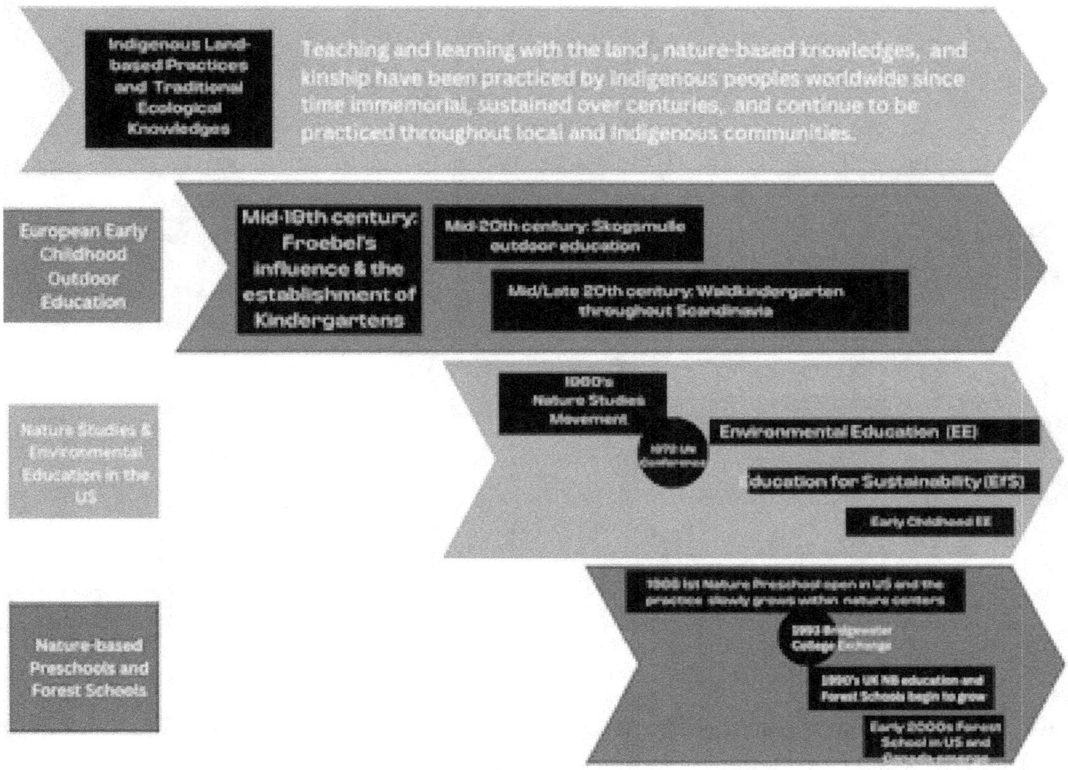

At the core of its existence, nature-based early childhood education is one in which nature is infused into the daily educational environment and is the driving force of the curriculum (Finch & Bailie, 2016; Sobel et al., 2015).

FOREST SCHOOL

A type of nature-based early childhood education mentioned above and growing in popularity is *forest school*. Similar programs are also sometimes called forest kindergartens or outdoor preschools. Forest schools share many of the same goals and principles of general NBECE, but they typically do not have an indoor space, so the teachers and children are outdoors for the entirety of their program. Like nature-based education in general, there are a variety of ways that forest school has been interpreted and implemented in practice (Sobel et al., 2015). Forest school is generally understood to prioritize holistic development (Cree & Robb 2021; Cudworth & Lumber, 2021) and place emphasis on learning that occurs in outdoor environments, and is integrated with the natural world (Cudworth, 2020; Smith et al., 2018; Weiss et al., 2023).[1] The term is often associated with early childhood education, particularly in the United States and Canada but it can span various age groups. For example, the Forest School Association in the UK supports the development of forest schools for preschool-aged children but also elementary/primary age expanding to high school groups as well. While these are significant opportunities for older children

Figure 3. Organizational chart locating forest school within nature-based early childhood education

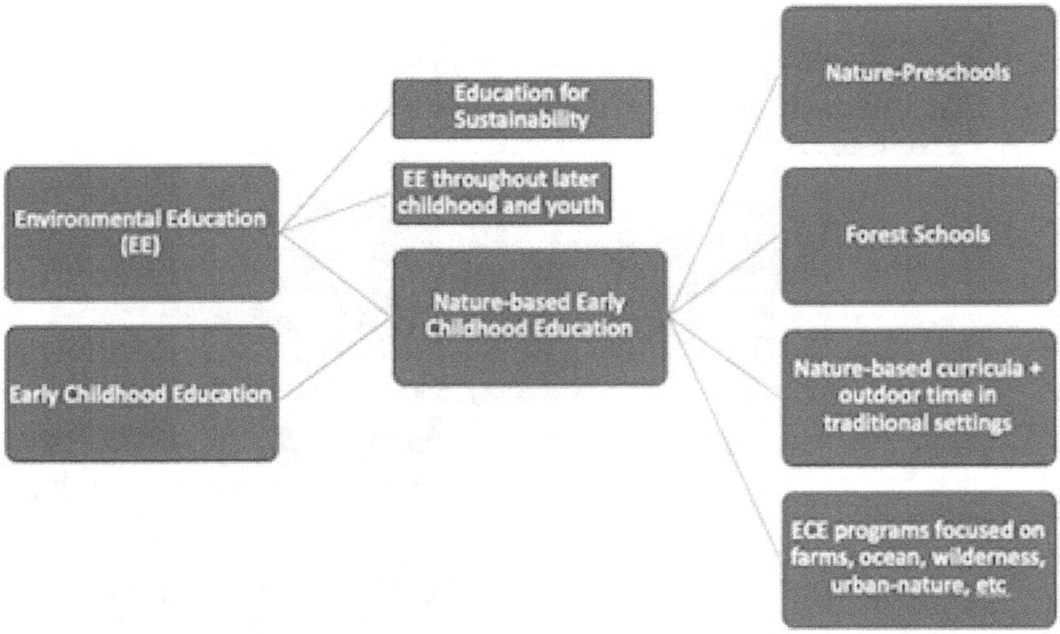

and youth and should not be discounted, this chapter focuses more specifically on the develop of *early childhood* practices and considerations in forest schools.

A core value of forest school for early childhood is placing nature as an additional "teacher" in which nature is a true member of the classroom community and learning experience (Cree & Robb, 2021). Additionally, the forest school setting is one in which risky place is encouraged within a safe learning environment to cultivate curiosity and learning through experience (Cree & Robb, 2021; Finch, 2016). Forest schools typically operate year round allowing children to have the opportunity to experience and learn from the changing seasons, weather patterns, and natural phenomena (Sobel et al., 2016). Importantly, forest school in the preschool years provides opportunity to unpack firsthand issues within environmental justice, sustainability, and ecological literacy.

With the increased popularity of established forest schools, comes the increase in professional organizations that help to support the development of these programs, create guidance and standards, and expand access and opportunity to participate. In United States, The Natural Start Alliance, a subset of the North American Association of Environmental Education, holds the mission to bring together organizations and individuals within the NBECE field. They support practitioners, provide information to families, and advance research in the field(https://naturalstart.org/). The Natural Start Alliance developed Guidelines for Excellence in Early Childhood Environmental Education Programs (2019), created to represent a comprehensive perspectives on environmental education in early childhood, with input from diverse stakeholders such as teachers, childcare professionals, administrators, scientists, curriculum developers, and government agency. The guidelines emphasizing the importance of fostering an appreciation for the natural world in early childhood to build long-lasting positive attitudes and perceptions. These types of policy implementations are happening at a more local level as well. For example, Washington State because the first start in the US to license outdoor preschool programs. As such, the Department of Children, Youth and Families (2022) developed the Outdoor Nature-Based Learning competencies for early childhood educators intended to "highlights the unique aspects of the nature-based learning experience and establishes expectations for educators and administrators who offer this type of learning environment" (p4).

Canada is also actively establishing standards for forest schools to follow via established nature-based networks and professional organizations. The Child and Nature Alliance of Canada centers its mission around "connect[ing] children and youth with nature through policy, research, and practice" (https://childnature.ca). This Child and Nature Alliance, in collaboration with Forest School Canada (2019), has produced a document titled "Principles of Forest School," delineating the definition of Forest/Nature School as:

an educational ethos and practice that centres the Land and the child-at-play. Children and educators build a relationship with the Land through regular and repeated access to the same outdoor space over an extended period of time. Educators support learning through a pedagogical framework that is rooted in place and play, directed and inspired by the child (emergent curriculum), and driven by a process of inquiry. (p1)

Similarly, in the UK, the Forest School Association was established to be a "professional body and UK wide voice for Forest School, promoting and supporting best practice, cohesion and quality forest school for all" (FSA, n.d.). They define 'core ethos' based on six principles. These six principles revolve around best practices for environmental education and holistic development. In summary the principles state that forest school is a long-term process of regular sessions which takes place in a woodland or natural environment promoting holistic development through learner-centered processes and reasonable risk-taking. The following table summarizes the main principles of the above organizations.

These collection of principles from the three organizations above, underscores a robust commitment to intertwining child experiences with nature, in other words, there is a steadfast dedication to learning in, about, and with nature (Barrable, 2019; Warden, 2015). These three terms—"in nature," "about nature," and "with nature"—sound similar however, bear unique nuances that distinguish them from each other.

Learning *in nature* is simply bringing learning outdoors. It does not necessarily include a nature-based component but often does. This can be an art class bringing art material outdoors, a first-grade teacher moving her group to the field for their science lesson to observe and collect data on plants or a preschool teacher setting up an obstacle course using logs, rocks, sticks, and hills for the children to move and navigate their bodies. Learning *about nature* occurs when children are encouraged to learn information from hands on experience but can also be through books and other resources. Learning about nature centers on the study, discourse, knowledge, and conveying of information about the features, processes, and significance of nature. This can include learning about topics such as life cycles, weather, or animals-plant relationships. Learning *with nature* involves teachers or children themselves engaging in learning moments found when exploring nature through a collaborative and interactive relationship between humans and the natural world (Bailie et al., 2023; Sobel, 2016). The natural world becomes a peer, a mediator, or a "third teacher" as children experience the natural world around them (Larimore, 2018; Warden, 2015). It should be noted that *learning with nature* is highly contextualized in different countries, regions, and communities and as case studies within the research from MacQuarrie et al. (2015) illuminate, should account for cultural and situational factors. This highly contextualized nature of forest schools and nature-based learning can feel like a challenge for initial implementation. This

Table 1. National organization and principles

National Organization	Principles
UK: The Forest School Association	1. Forest school takes place over an extended period of time allowing for frequent and regular sessions in a woodland or other natural environment; it is based upon observations, physical and behavioural boundaries and collaborative work between learners and practitioners. 2. Forest School aims to promote the holistic development of all participants; to develop physical, social, cognitive, linguistic, emotional, social and spiritual abilities of the learner and where appropriate to link experiences to home, work and/or school education. 3. Forest School offers learners the opportunity to take supported risks appropriate to themselves and the environment. 4. Forest School is delivered by qualified practitioners who continuously maintain and update their practice and there is a high ratio of adults/practitioners to learners. 5. Forest School uses a range of learner centred processes to create a community for learning and development that integrates play and choice and is responsive to the needs and interests of the learners. 6. Forest School takes place in a woodland or natural setting and aims to foster a relationship with nature and the development of long term pro-environment attitudes through regular personal experiences in nature.
US: The Natural Start Alliance (Affiliate of NAAEE)	• Working toward dual goals of promoting child development and developing children's environmental literacy • Using nature as the central organizing theme for the program • Spending a significant portion of the instructional time outdoors (in some cases, all of the instructional time is spent outdoors) • Employing a child-led, play-based approach to teaching that adapts to children's interests, abilities, cultures, and environments • Recognizing the benefits of children's healthy, appropriate risk-taking as the program manages risks in children's play and learning, particularly in the outdoors
Canada: Child and Nature Alliance and Forest School Canada	1,Takes place in any outdoor space, including urban greenspace, playgrounds, forests, creeks, prairies, mountains, shoreline, and tundra. 2. Is a sustained process of regular and repeated sessions in the same outdoor space, supporting children to develop a reciprocal relationship with the Land, and an understanding of themselves as a part of the natural world. 3. Views children and youth as innately competent, curious, and capable learners. 4. Is led by educators who share power with learners through play-based, emergent, and inquiry-driven teaching and learning methods. 5. Values children's play - self-directed, freely chosen, intrinsically motivated - in and of itself. 6. Views risky play as an integral part of children's learning and healthy development, and is facilitated by knowledgeable, qualified educators who support children and youth to co-manage risk. 7. Relies on loose, natural materials to support open-ended, creative play and learning. 8. Values the process as much as the outcome. 9. Prioritizes building reciprocal relationships with First Nations, Métis, and Inuit, who have been learning from this Land since time immemorial. 10. Practices, policies, and programming reflect and prioritize the building of engaged, healthy, vibrant, and diverse communities through consideration of access and equity in our decisions and actions.

challenge of forest school implementation, along with others, will be addressed in the next section on barriers to forest school implementation.

BARRIERS TO FOREST SCHOOL IMPLEMENTATION

Factors such as community values, cultural traditions, and historical relationships with the land all play a crucial role in shaping the implementation of forest schools within a specific cultural context. Additionally, collaboration with local stakeholders, including Indigenous communities and land stewards, is essential for ensuring culturally sensitive and respectful practices that honor the land and its people. The barrier here is derived from practitioners feeling intimidated or inexperienced but can be addressed through institutional support, ongoing program evaluation, and community relationship building.

If institutional support is not available, it provides a second barrier type, structural barriers, which encompass institutional limitations and societal norms that may hinder the implementation of forest schools. These barriers could include educational policies that prioritize indoor learning environments over outdoor experiences, as well as school or cultural attitudes that undervalue the benefits of nature-based education.

The limitations at institution level, can also be manifested at an educator level as well. Some educators' attitudes and fears about the outdoors can pose significant challenges to forest school. Some educators may be hesitant to embrace outdoor learning due to concerns about their own comfort and expertise in outdoor environments, fear of losing control over classroom dynamics, or uncertainty about how to integrate nature-based activities into their curriculum effectively. Hesitations about forest school can extend to concern for children's safety. Worries about potential risks such as encounters with wildlife, uneven terrain, or adverse weather conditions may lead to reluctance to engage in outdoor activities, despite research indicating the overall safety and benefits of outdoor play (Handler & Epstein, 2010, McClintic & Perry, 2015; Yalcin & Erden, 2021).

Additionally, educators, parents, and community may lack conviction that 'enough' learning and development occur through outdoor play, representing a fundamental barrier to forest school implementation (Debaja, 2023). Addressing misconceptions and raising awareness about the cognitive, social, and emotional benefits of nature-based education is essential for fostering support and engagement in outdoor learning initiatives.

This unique setting also comes with logistical barriers. The first such barrier is location and feasibility issues (Finch, 2016) which may include zoning regulations, environmental audits, and park or land permitting. Developing messaging about the positive impact and engaging influential stakeholders (administration, families, policy-makers, advocacy and community organizations) can also be a logistical and relational hurdle (Sobel, 2016).

Addressing these barriers requires a multifaceted approach that involves collaboration between educators, policymakers, community organizations, and families. Strategies may include advocating for increased investment in green spaces and outdoor learning environments, providing professional development and support for educators to integrate nature-based practices into their teaching, and implementing policies that prioritize equitable access to nature-based education for all children. Additionally, understanding the curriculum and instructional practice of forest schools can be a tool for advocating forest school's implementation and more nature-based learning into traditional teaching practices. As such, the next section details curriculum, instruction, and practice which can often be found in forest schools.

CURRICULUM, INSTRUCTION, AND PRACTICE OF FOREST SCHOOLS

Forest school curriculum, instruction, and practice are unified by some common resources such as principles of Scandinavian Waldkindergartens, grounding in environmental education, and guidance from professional organizations. However, despite these shared visions, the implementation of forest school principles can vary. There's a wide range in the level of structure, types of activities, and settings. Consider for example, three outdoor preschools located within the same city, Seattle WA. First is Fiddleheads Forest school which is located in a 230 acre arboretum in which they "provide routine and structure, and also welcome inspiration and occasional diversion from these routines" (*About Us | Fiddleheads Forest School*, n.d.). Their website places emphasis on teachers as tuned into the natural world and the wonders of the children. It is stated that they will use what they observe to build curriculum, create opportunities, evolve plans, and organize their activities throughout the day and over the course of the school year. Anna's little farm which utilizes a private yard and small urban farm outdoors provides little to no structure and they state "unstructured, imaginative play is the focus of our days at Anna's Little Farm" (*Annaslittlefarm - About Us*, n.d.). Alternatively, holding a more formulated curricula guide, Tiny Trees Preschool which takes place in various public parks, aligns their curriculum with state standards and uses the high-scope curriculum(*Tiny Trees Preschool | Reading, Math and Science at Tiny Trees Outdoor Preschool*, n.d.).

A recent survey done by Harwood et al. (2020) of Canadian forest schools can be considered as a proxy for the overall state of the variety of forest preschools in North America. They found that the majority of respondents, (66%) reported having "an emergent curriculum." On either ends of the spectrum lay a pre-prescribed curriculum (6.8%) or no curriculum/completely exploratory (6.8%). So as this survey illustrates, there can be a great variety in instructional practices. However, it is the guiding principles that are significant in defining forest school practice. The Harwood et al.'s (2020) survey confirms the stabilization of guiding principles as respondents most frequently mentioned guiding principles were "play and exploration" "child-centered" "connection to nature" and "social-emotional well-being." With the increasing prevalence of forest school practices and the variation in implementation of curriculum, a thorough exploration of pedagogy with a critical eye becomes imperative.

A critical analysis of forest pedagogy stems from the "Common World" perspective, a concept rooted in the notion that humans, non-human entities, and environmental elements are intricately interconnected, holding equal significance. This perspective underscores their collective role within the broader ecological system, emphasizing a shared "common world" (Taylor, 2013). In essence, it advocates for recognizing the interdependence and equality of all components within the ecological framework. Working in a common worlds framework, Veronica Pacini-Ketchabaw (2013) warns of pedagogies that simply try to bring children and nature together. She names forest pedagogies as inherently much more complex than most early childhood literature or environmental literature portraits it to be. Pacini-Ketchabaw (2013) uses the term "friction in the forest" to describe the move beyond simply taking children outdoors or basic knowledge about the species that live there. In this essence of going beyond simple exposure and basic knowledge, Pacini-Ketchabaw (2013), along with other common worlds scholars (Cudworth, 2020; Weldemariam, 2020; Taylor & Pacini-Ketchabaw, 2015; Pacini-Ketchabaw & Nxumalo, 2015, Taylor, 2013), call for early childhood educators to lean into the entanglements, uneven histories, and ethical dilemmas of forest pedagogy. For example, Pacini-Ketchabaw (2013) and Cudworth (2020) both discuss in their writings how some beings will live, and some beings will die with our presence of being outdoors for forest school. They make a point that this should be brought up with the children and those teachable moments captured when present. Additionally, Taylor (2013) and Pacini-Ketchabaw (2013) advocate for a critical examination that goes beyond the surface, urging educators to actively engage in challenging colonial histories, questioning the continued presence of settlers on the land, and addressing the erasure of Indigenous peoples' existence and ways of knowing. This entails a deeper exploration and confrontation of historical injustices and ongoing disparities within environmental and nature-based educational context.

Figure 4. Guiding principles as described by teachers
Infographic from Harwood et al. (2020).
Reprinted with permission.

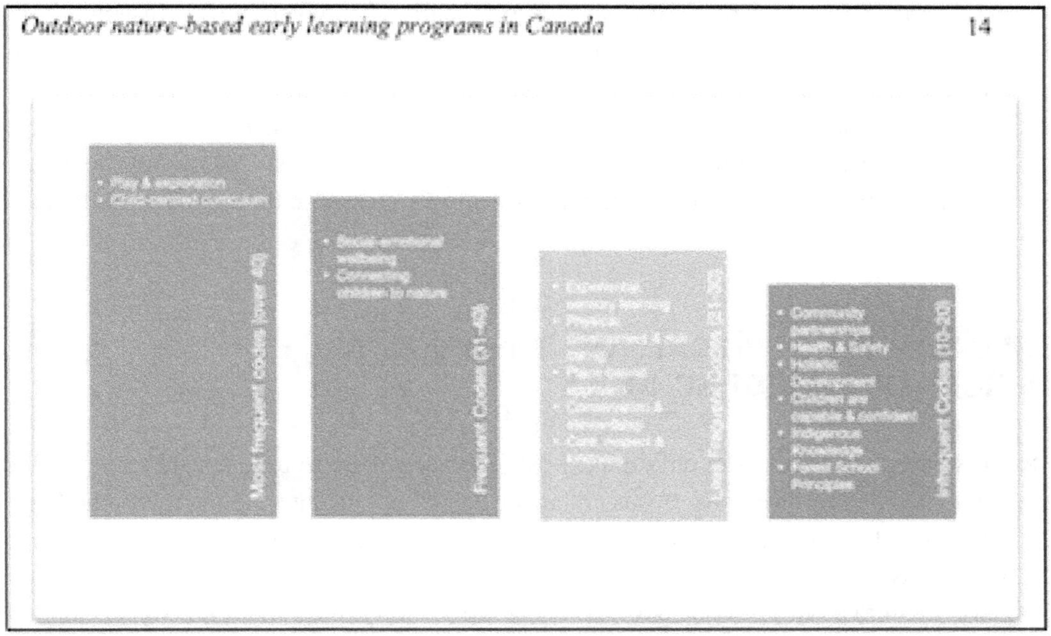

Figure 5. Types of curricular approaches at forest school as described by teachers
Infographic from Harwood et al. (2020).
Reprinted with permission.

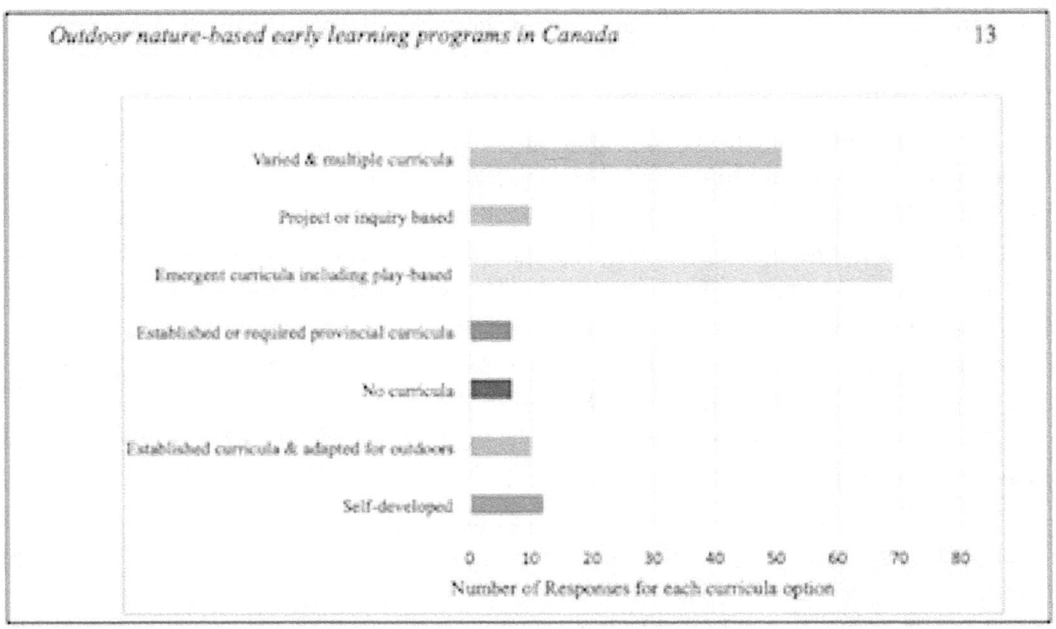

In aligning with common worlds critical perspectives, Nelson et al. (2018) advocate a profound reassessment within environmental education, particularly among forest school practitioners. They emphasize the necessity to break free from the field's dual fixation on replicating an idealized "imagined state of environmental sanctity" and narrowly focusing on only children's development. Instead, they call for a departure from perceiving nature as a limited commodity for the privileged few. The proposed shift involves a transformative vision—a collective reimagining of our connections with nature as living "in relational reciprocity with one another, plants, animals, water, and landscapes" (p6). In order to navigate and set a foundational understanding of the evolving landscape of forest school and the novel critical perspectives, this chapter next outlines the historical, philosophical, and theoretical influences on forest schools modern practices.

HISTORICAL, PHILOSOPHICAL, AND THEORETICAL INFLUENCES OF FOREST SCHOOLS

Forest schools have a desire to offer children exposure to natural environments and foster their connection with the land. However, we know this is not exclusive to our current time. As mentioned previously, Fröebel is known as the founder of kindergarten which at the time of conception, included a strong nature component. He was not alone in thinking about the importance of education for the young child while also considering nature as an essential element. There are several other early philosophers and educators whose theories still resonate today in the field of nature-based early childhood education and therefore are foundational to how forest school practices have evolved.

An Austrian philosopher, Rudolf Steiner, who founded the first Waldorf school in Germany in 1919, build his educational philosophy on the premise of bringing together children, nature, and spirituality (Childs, 1991; Uhrmacher, 1995). Rudolf Steiner's profound influence from the early 20th century endures, as evidenced by the continued esteem for his principles in contemporary Waldorf schools. Present-day Waldorf institutions place a strong emphasis on integrating nature into their educational framework, promoting experiential learning through hands-on experiences (Barnes, 2001; Boland, 2017).

Around the same time that Rudolf Steiner was building the Waldorf approach in Germany, the McMillan sisters of England began implementing a novel approach in early childhood. By bringing children into learning environments called "open-air nursery," they promoted the need for fresh air and exercise for young children. These nurseries aimed to provide a healthier and more holistic environment for young children's development. Unlike traditional indoor classrooms, open-air nurseries were located outdoors, typically in natural settings. They emphasized the importance of fresh air, physical activity, and direct interaction with nature. The curriculum often included activities like gardening, exploring natural objects, and interaction with the environment. This approach sought to create a more natural and stimulating learning environment, recognizing the benefits of outdoor experiences for children's well-being and development. (Liebovich, 2018; Pound, 2015).

Two additional strong influences on the forest school approach are the Montessori philosophy and the Reggio Emilia approach, contributing significantly to nature-based education and more specifically forest school (Sobel et al., 2015). Maria Montessori, a pioneer in education, emphasized the inseparability of the child and nature, advocating for children to draw from natural forces for holistic development of body and spirit (Knight, 2013; Sobel et al., 2015). In the Montessori school approach, characterized by individualized learning and hands-on materials, the curriculum fosters a deep connection between children and their environment. The Reggio Emilia approach, renowned for its project-based curriculum, views children as capable, strong, and inherently curious, aligning with the principles of nature-based education and emphasizing exploration and engagement with the environment. Much like in forest school, the Reggio Emilia approach provides opportunities for children to engage in interactive and experiential learning in order to develop a deep understanding of the world around them (Gandini, 1993, 2003). Both the Montessori and Reggio philosophies contribute valuable insights to the broader framework of forest schools, shaping educational practices that prioritize nature-centric experiences for young learners. A clear indication of the influence of Montessori and Reggio Emilia philosophies on nature-based education is found in David Sobel's (2015) "Nature Preschool and Forest Kindergarten Handbook for Outdoor Learning," where he observes that many practitioners in nature-based early childhood education often seek inspiration from these approaches. Having explored the historical and philosophical landscapes of nature-based education, our journey now turns towards prospective horizons. Transitioning from the established foundations, the upcoming section, "Considerations for Future Trajectory of Research and Practice," navigates the evolving terrain that will negotiate the ongoing narrative of nature-based education.

CONSIDERATIONS FOR FUTURE TRAJECTORY OF RESEARCH AND PRACTICE

As established in this chapter's introduction and supporting evidence, the integration of children with nature presents various benefits. The researched-based evidence of positive outcomes from children and nature coming together has helped spark the growth of forest schools. As forest schools become implemented more widely there is great potential for enhancement of childhood development, and increased opportunities to connect with nature. At the societal level there is potential for increased social cohesion, empathy development, and increased pro-social behaviors. Looking boarding at the environmental/ecological level there are also indicators for positive influence on sustainability, environmental care, and building kinship among multiple species. These various domains of prospective development and improvement generate considerable enthusiasm both within the nature-based education community and in broader contexts.

However, this growth also poses challenges and cautions as we proceed. As evident in the various interpretations enacted (Merrick, 2019; Sobel, 2019) and in the multiple ways teachers perceive their role in a forest school setting (Alfonso, 2024), the field currently lacks standardized practices, making it challenging to ensure consistent quality across different nature-based programs. As the practice grows, there is a need for deeper exploration of forest school pedagogy to ensure that the educational principles align with the desired outcomes for child development and nature-connectedness. Building of standards and expectations is already being done by some of the organizations mentioned above (UK: The Forest School Association; Canada: Child and Nature Alliance and Forest School Canada; US: The Natural Start Alliance) and through research and practitioner publications (see: Evaluating Natureness: Measuring the Quality of Nature-Based Classrooms in Pre-K through 3rd grade, Bailie et al., 2023) and policy (such as the aforementioned Washington State outdoor preschool licensing standards). However, there is more work to be done. As the practice grows, there is a need for deeper exploration of forest school pedagogy to ensure that the educational principles align with the desired outcomes for child development, nature-connectedness, and culturally sustaining practices of Indigenous and local communities.

From a practitioner's role, this could be achieved through seeking out specialized professional development which aligns early childhood and environmental education practices or conducting self-audits using tools such as Bailie et al.'s (2023) Evaluating Natureness. To address culturally sustaining practices, teachers and administrators will need to work with local communities, including Indigenous leaders in their locale to build relationships and authentically engagement with Indigenous perspectives. A strong example of achieving this was exemplified by researchers Julie Ernst and Claire Underwood in partnerships with Thelma Nayquonabe, an Ojibwe Elder and educator in creating a Communities of Practice model to support 15 regional early childhood professionals to co-create empathy practices through a Two Worlds approach, blending Indigenous and Western knowledges (Ernst et al., 2023). From a researcher's perspective this means asking hard questions about what it means to build relationships with nature on stollen land, examining current teaching practices in terms of resisting (or reinforcing) colonialism and Indigenous erasure, evaluating the alignment of practices to theory, and studying interventions which address desired outcomes for child development, nature-connectedness, and culturally sustaining practices of Indigenous and local communities.

Teacher training and preparation is another upcoming challenge. There is a need for specialized training programs for educators to equip practitioners with the skills and knowledge required for effective nature-based teaching in a forest school. Integrating the practice of learning and being "with nature" requires a unique skill set. As an example of the need, we can consider the context of the United States which has seen exponential growth in forest schools and nature-based programs but the higher-education and teacher training lags behind. Currently, in the US only five AAUs offer programs in environmental or sustainable education. 70% of AAUs do not offer any type of course in environmental education, sustainability education, nature-based education, or a related field (Hale, Conerty, & Lindgran, 2023). The Natural Start Alliance (2022) has identified only three university degree program and only five university certificate programs in early childhood nature-based or environmental education, describing the need for additional teacher preparation and training programs as outdoor education becomes licensed in more states. A US survey (Born, 2022) of higher-education institutions revealed a growing interest in nature-based and a desire to support preservice teachers' skills in this area. However, it also identified a clear need for structural supports currently very limited in nature-based higher-education, including finding faculty who are skilled and competent in this area, illuminating he benefits of nature-based education, ability to collaborate or work with agencies that are involved in licensing and assistance in helping pre-service teachers find practical experience and jobs.

There are also the challenges of navigating equity and inclusion within this new field. How to best address the need for nature-based education to be accessible to children from diverse socio-economic backgrounds, ethnicities, and geographic locations is still being investigated (Natural Start Alliance, 2021). As is how to create culturally sustaining and historically relevant context for forest school. The significance of cultural and historical contexts in shaping forest school practices necessitates careful consideration, as this unique approach works directly on the land, often times with a contentious history encased in colonial violence, slavery, and displacement of Indigenous communities (Leather, 2018). Acknowledging and addressing these historical realities is paramount for fostering an inclusive and respectful forest school pedagogy that promotes environmental stewardship while recognizing and rectifying past injustices. The Communities of Practice model (Ernst et al., 2023) mentioned previous is one way to work on this. Other examples enacted by forest schools include implementation of a predetermined antibias curriculum (https://tinytrees.org/equity); incorporating land-based learning ((About

Us I Fiddleheads Forest School, n.d.) and engaging with local and Indigenous community members in teaching practices (Hayden & Wai, 2013).

A massive challenge for researchers and practitioners alike is understanding the presence and persistence of settler-colonial views and how they have penetrated our educational theory and practice. From a forest school perspective, these challenges are alarmingly evident when we acknowledge the violent disruption that settler colonialism has on the relationship between humans and the environment (Tuck & Yang, 2021; Watts, 2013; Whyte, 2018). Indeed, scholars illuminate how current Wester settler-colonial views dominate the US school systems, constructing humans and culture as separate from nature, limiting the epistemology which informs teaching practices (Medin & Bang, 2014; Whyte, 2018). Further research is needed to investigate teaching practices in forest schools and nature-based education as a comparison to traditional US school systems.

It is the task of those engaged in the field of nature-based and environmental education to acknowledge the influence of settler-colonialism and its potential role in shaping an unbalanced perception of humans and nature (Bang & Marin, 2015; Clark, 2019; Gleason, 2021; Hodgins, 2019), fostering unequal experiences in nature (Gleason, 2021; Hatala et al., 2020), and perpetuating hierarchies of power and privilege (Ross et al., 2011; Salleh, 2009).

Examples of previous work addressing this challenge comes in various forms, including studying the implementation of "land education" into early childhood teacher education programs using updated readings, assignments, field experiences, and direct engagements with local Indigenous peoples (Lees et al., 2021), unpacking how teachers in practice resist settle-colonialism (Alfonso, 2024), and using arts-based methodologies to engage with stories about entanglements of settler colonialism early childhood (Clark et al., 2014). This type of work must continue and more must be done and research in the field of education needs to continue to trouble the presence of settler colonialism and build resources for educators and elevate support for policy for culturally sustaining and historically relevant context for forest school.

Ruth Wilson's (2019) philosophical essay "What is Nature" (2019) underscores the dichotomy of kinship versus domination in humans' relationship with nature. Wilson highlights the struggle of kinship vs domination as pivotal in post-humanistic thinking and pedagogy, and suggests that understanding alternative viewpoints of nature and humans as crucially important at the early childhood level, and therefore particularly applicable to NBECE and forest schools which learn directly on the land in outdoor settings.

At the heart of NBECE in general, and particularly within the framework of forest schools, lies the cultivation of caring, reciprocal relationships among children, nature, and their shared environments, their common world. A critical examination of NBECE theories and practices serves as a valuable tool for researchers to address imbalances in human-nature relationships that uphold anthropocentric visions of children and nature (Pacini-Ketchabaw, 2017). This examination gains particular relevance as educators and school programs increasingly embrace outdoor settings or incorporate nature-based learning into their approaches.

To inform responsible and sustainable growth in the field of forest school and other NBECE, Leather (2018) guides practitioners and researchers to be conscientious of not to letting "pragmatic concerns overtake conceptual understanding...[or] emulate practices without necessarily understanding why they are doing things in certain ways and not others" (p. 2). In other words, Leather suggests that forest schools and NBECE initiatives should prioritize a deep understanding of the underlying principles and reasons behind their practices and programs beyond just following established guidelines. Leather (2018) goes on to suggest it is necessary for all forest schools to understand the cultural and theoretical underpinnings of their practice and programs.

One way to achieve this recommendation is by developing training materials that incorporate the philosophical and theoretical background and history (Leather, 2018; Waite et al., 2016). It also involves closely examining the cultural and historical significance of a specific forest school site (Leather, 2018; MacQuarrie et al., 2015; McNair, 2012) and adopting a pedagogy that genuinely considers and respects children's inquiries regarding historical, political, and cultural impacts (Cudworth, 2020; Nxumalo, 2019; Pacini-Ketchabaw, 2013; Taylor, 2013). Leather (2018) contends that this must be done through child-initiated play as well as adult-directed experiences that reflect a socio-cultural perspective of learning, where practitioners play an important role as they respond to children's choices, interests, and activities. Indeed, several researchers who have done field work in forest schools (Cudworth, 2020; Laaksoharju & Rappe, 2017; Linzmayer & Halpenny, 2014) would agree that a socio-cultural framing is important to understanding the child's experience not only in terms of their teacher's role but also the role of peers, the forest, more-than-human species, family, community and wider cultural influences. In corroboration of the sociocultural emphasis of these field studies, Beery et al.'s (2020) study with professionals in the field of nature-based education revealed the majority of these "expert panel" participants placed great significance on the sociocultural context of young children's experiences of nature.

Envisioning the future of forest school within the context of these responsibilities, challenges, and critical perspectives necessitates exploration complex and multifaceted theoretical framework which can challenge conventional notions of human exceptionalism and encourages a reevaluation of the relationships between humans, the environment, and other beings. Posthumanism theory provides a compelling framework for such an exploration. Donna Haraway's groundbreaking essay in 1985, began the conversation of reimagined ways in which humans interact with and are intertwined with their world (Haraway, 1985). Then in her 2016 book "Staying with the Trouble," Haraway challenges our relations to the earth, natural systems, and other beings through her discussion of kinships which considers reciprocal and interdependent relationships (Haraway, 2016).

This transformative perspective fosters novel insights into the coexistence of humans, animals, and nature. Ruth Wilson (2019) offers ways to apply the posthumanism theory to early childhood pedagogy which she believes "is critical to the very survival of the planet while also nurturing the holistic development of children" (p27). Wilson (2019) taps into the significance of kinship building which should be foundational to forest school pedagogy, emphasizing the importance of young children's experiences to build relationships rather than simply defining or learning about nature. Indeed, many other scholars of posthumanism (Cudworth, 2020; Nxumalo, 2018; Nxumalo & Pacini-Ketchabaw, 2017; Pacini-Ketchabaw, 2013; Weldemariam, 2017) agree with the kinship perspective of "thinking with" rather than "thinking about" nature as it leads to meaningful relationships, connections, and emotional ties between children and nature.

Posthumanism offers a rich theoretical framework for exploring nature-based education, particularly in early childhood pedagogy. Future research in this area could delve into the ways in which tenets of posthumanism can inform and enrich nature-based educational practices, fostering deeper connections between young children, multispecies worlds, and the natural ecosystems. This research could examine the role of non-human participants, such as plants, animals, and ecosystems, in shaping educational experiences and outcomes, and explore innovative pedagogical approaches that prioritize relationality, interdependence, and coexistence. By embracing a posthumanism, researchers can push the boundaries of conventional understandings of nature-based education, creating opportunities for more inclusive, holistic, and sustainable educational practices that resonate with the complex entanglements of human and more-than-human worlds.

CONCLUSION

Gaining an in-depth comprehension of the forest school concept within the context of nature-based early childhood education is essential for establishing a robust knowledge foundation for inquiry and effective environmental education practices in early childhood education. This exploration is not only concerned with the historical trajectory of nature-based education but also delves into its profound impact on curriculum, instruction, and pedogeological beliefs, particularly within the developmental context of the forest school setting. By scrutinizing the guiding theories and philosophies underpinning forest schools, this chapter underscores the significance of political and cultural contexts in shaping pedogeological practices and theoretical frameworks.

It is equally important to examine theory which guides the values and philosophies on which forest schools are built. Therefore, this chapter provides a brief overview of forest school's curriculum and instruction, including historical influential figures, while also considering the importance of political, and cultural contexts for pedogeological practice and theoretical grounding. To that extent, this chapter explore posthumanism as a compelling theoretical lens for future research, analysis, and instructional practice in forest school. As we chart the course for the future of forest school amidst responsibilities and challenges, Donna Haraway's (2016) groundbreaking work has initiated a paradigm shift in how we conceive of human relationships with the environment, advocating for reciprocal and interdependent kinships. Building on Haraway's work, scholars like Wilson (2019) Cudworth (2020) highlight the critical application of posthumanism to early childhood pedagogy, specifically in forest school emphasizing the need for fostering relationships and emotional ties with nature. More generally, this chapter serves as a guidepost, providing an overview of forest school's historical evolution, influential figures, and theoretical underpinnings. Beyond a mere exploration of the past, it aims to lay the foundation for future research and practice, opening avenues for nuanced analysis and instructional strategies. In essence, the chapter's comprehensive overview serves as a resourceful guide for various stakeholders in the education sector, fostering a deeper understanding of nature-based early childhood education and its implications for educational theory and practice.

REFERENCES

About Us. (n.d.). University of Washington Botanic Gardens. https://botanicgardens.uw.edu/education/youth-family/fiddleheads-forest-school/about-us/

Alfonso, S. (2024, April 11). *Deconstructing Settler Colonialism in Early Childhood: Nature-Based Preschool Teachers' Nature-Connectedness Journey & Perceptions of Nature. Disruption and Deconstruction with Early Childhood Teachers and Teacher Educators*. 2024 AERA Annual Meeting: Dismantling Racial Injustice and Constructing Educational Possibilities: A Call to Action, Philadelpia, PA.

Annas Little Farm. (n.d.). Seattle Nature Schools. https://www.annaslittlefarm.com

Ärlemalm-Hagsér, E., & Elliott, S. (2017). Special Issue: Contemporary Research on Early Childhood Education for Sustainability. *International Journal of Early Childhood, 49*(3), 267–272. doi:10.1007/s13158-017-0207-3

Bailie, P. (2015). Nature Preschools: The Cross Fertilization of Early Childhood and Environmental Education. In D. Sobel, P. Bailie, K. Finch, E. Kenny, & A. Stires (Eds.), Nature Preschools and Forest Kindergartens: The Handbook for Outdoor Learning (Illustrated edition, pp. 45–58). Redleaf Press.

Bailie, P., Larimore, R., & Pikus, A. (2023). *Evaluating Natureness: Measuring the Quality of Nature-based Classrooms in Pre-k Through 3rd Grade* (1st ed.). Gryphon House.

Bang, M., & Marin, A. (2015). Nature–culture constructs in science learning: Human/non-human agency and intentionality. *Journal of Research in Science Teaching, 52*(4), 530–544. doi:10.1002/tea.21204

Barnes, H. (2001). *Waldorf Education: An Introduction*. AWSNA.

Barrable, A. (2019). Refocusing Environmental Education in the Early Years: A Brief Introduction to a Pedagogy for Connection. *Education Sciences, 9*(1), 61–61. doi:10.3390/educsci9010061

Barrable, A., & Booth, D. (2020). Nature Connection in Early Childhood: A Quantitative Cross-Sectional Study. *Sustainability (Basel), 12*(1), 1. doi:10.3390/su12010375

Basil, C., & White, C. (2005). Environmental Literacy: Providing an Interdisciplinary Context for Young Children. In Readings in Environmental Education an Urban Model (1st ed., pp. 33–40).

Beery, T., Chawla, L., & Levin, P. (2020). Being and Becoming in Nature: Defining and Measuring Connection to Nature in Young Children. *International Journal of Early Childhood Environmental Education, 7*(3), 3–22.

Beery, T. H., & Lekies, K. S. (2019). Childhood collecting in nature: Quality experience in important places. *Children's Geographies, 17*(1), 118–131. doi:10.1080/14733285.2018.1463431

Beery, T. H., & Wolf-Watz, D. (2014). Nature to place: Rethinking the environmental connectedness perspective. *Journal of Environmental Psychology, 40*, 198–205. doi:10.1016/j.jenvp.2014.06.006

Bilton, H., & Waters, J. (2017). Why Take Young Children Outside? A Critical Consideration of the Professed Aims for Outdoor Learning in the Early Years by Teachers from England and Wales. *Social Sciences (Basel, Switzerland), 6*(1), 1. Advance online publication. doi:10.3390/socsci6010001

Boileau, E. Y. S. (2013). Young Voices: The Challenges and Opportunities that Arise in Early Childhood Environmental Education Research. [CJEE]. *Canadian Journal of Environmental Education, 18*, 142–154.

Boland, N. (2017). *Waldorf: An education of its time?* International Forum of Waldorf/Steiner Education.

Born, P. (2022). "It's All over the Map": A Report on Nature-Based Early Childhood Teacher Preparation Programs in the United States. *International Journal of Early Childhood Environmental Education, 9*(2), 69–82.

Boyd, D. (2019). Utilising place-based learning through local contexts to develop agents of change in Early Childhood Education for Sustainability. *Education, 47*(8), 983–997. doi:10.1080/03004279.2018.1551413

Cameron-Faulkner, T., Melville, J., & Gattis, M. (2018). Responding to nature: Natural environments improve parent-child communication. *Journal of Environmental Psychology, 59*, 9–15. doi:10.1016/j.jenvp.2018.08.008

Carr, V., Brown, R. D., Schlembach, S., & Kochanowski, L. (2017). Nature by Design: Playscape Affordances Support the Use of Executive Function in Preschoolers. *Children, Youth and Environments, 27*(2), 25–46. doi:10.1353/cye.2017.0002

Child and Nature Alliance, & Forest School Canada. (2020). *Principles of Forest/Nature School.*

Childs, G. (1991). *Steiner Education in Theory and Practice*. Floris Books.

Clark, V. (2019). Gathering: An A/r/tographic Practice. In B. D. Hodgins (Ed.), *Feminist research for 21st-century childhoods: Common worlds methods* (pp. 121–130). Bloomsbury Publishing.

Clark, V., Pacini-Ketchabaw, V., & Hodgins, B. D. (2014). Thinking with paint: Troubling settler colonialisms through early childhood art pedagogies. *International Journal of Child, Youth and Family Studies, 5*(4.2). *Article, 4*, 2. doi:10.18357/ijcyfs.clarkv.5422014

Crain, W. (2005). *Theories of development: Concepts and applications* (5th ed.). Pearson Education, Inc.

Cree, J., & Robb, M. (2021). *The essential guide to Forest School and nature pedagogy*. Routledge. doi:10.4324/9780367853440

Cudworth, D. (2020). Promoting an emotional connection to nature and other animals via forest school: Disrupting the spaces of neoliberal performativity. *The International Journal of Sociology and Social Policy, 41*(3/4), 506–521. doi:10.1108/IJSSP-09-2019-0188

Cudworth, D., & Lumber, R. (2021). The importance of Forest School and the pathways to nature connection. *Journal of Outdoor and Environmental Education, 24*(1), 71–85. doi:10.1007/s42322-021-00074-x

Dabaja, Z. F. (2022a). Reviewing two decades of research on the Forest School impact on children: The sequel. *Education, 50*(6), 737–750. doi:10.1080/03004279.2021.1905019

Dabaja, Z. F. (2022b). The Forest School impact on children: Reviewing two decades of research. *Education, 50*(5), 640–653. doi:10.1080/03004279.2021.1889013

Dabaja, Z. F. (2023). The strands of the Forest School implementation challenges: A literature review. *Prism.*

Ernst, J., & Theimer, S. (2011). Evaluating the effects of environmental education programming on connectedness to nature. *Environmental Education Research, 17*(5), 577–598. doi:10.1080/13504622.2011.565119

Ernst, J., Underwood, C., & Nayquonabe, T. (2023). Everyone Has a Piece of the Story: A Community of Practice Approach for Supporting Early Childhood Educators' Capacity for Fostering Empathy in Young Children through Nature-Based Early Learning. *International Journal of Early Childhood Environmental Education, 11*(1), 34–62.

Finch, K. (2016). The Dollars and Sense of Business Planning and Budgeting in D. Sobel, P. Bailie, K. Finch, E. Kenny, & A. Stires (Eds.), Nature Preschools and Forest Kindergartens: The Handbook for Outdoor Learning (Illustrated edition, pp. 158-190). Redleaf Press.

Finch, K. (2016). The Risks and Benefits of Nature Play in D. Sobel, P. Bailie, K. Finch, E. Kenny, & A. Stires (Eds.), Nature Preschools and Forest Kindergartens: The Handbook for Outdoor Learning (Illustrated edition, pp. 137-157). Redleaf Press.

Finch, K., & Bailie, P. (2016). Nature Preschools: Putting Nature at the Heart of Early Childhood Education. *Occasional Paper Series, 2015*(33). doi:10.58295/2375-3668.1020

F. S. A. (n.d.). *What is the Forest School Association? | Forest School Association.* https://forestschoolassociation.org/the-forest-school-association/

F. S. C. (2012). *History of Forest and Nature School.* Forest School Canada. http://www.forestschoolcanada.ca/home/about-forest-school/history-of-forest-school

Gandini, L. (1993). Fundamentals of the Reggio Emilia Approach to Early Childhood Education. *Young Children, 49*(1), 4–8.

Gandini, L. (2003). Values and Principles of the Reggio Emilia Approach. In Insights and inspirations from Reggio Emilia: Stories of teachers and children from North America.

Gleason, T. (2021). *A Critical Ethnography of an Outdoor School: Reimagining the Relationship Between Science Education and Climate Change Politics.* Routledge. doi:10.4324/9780429331626

Gough, A. (2006). A Long, Winding (and Rocky) Road to Environmental Education for Sustainability in 2006. *Australian Journal of Environmental Education, 22*(1), 71–76. doi:10.1017/S0814062600001671

Hale, J., Conerty, J., & Lindgren, S. (2023, October 12). *Environmental and Sustainability Education Programs at AAUs: A Review of the Field* [Comboned Panel Paper Presentation]. North American Association of Environmental Education Research Symposium, PheedLoop Online Conference Platform.

Handler, D., & Epstein, A. S. (2010). Nature education in preschool. *Highscope Extensions, 25*(2), 1–7.

Haraway, D. J. (1985). A manifesto for cyborgs: Science, technology, and socialist feminism in the 1980s. *Socialist Review, 5*(2), 65–107.

Haraway, D. J. (2016). *Staying with the Trouble: Making Kin in the Chthulucene.* Duke University Press.

Haring, N. (2021). *Growing a Foest School: From the roots up!* (1st ed.). Forest School Association.

Harwood, D., Boileau, E., Dabaja, Z., & Julien, K. (2020). Exploring the national scope of outdoor nature-based early learning programs in Canada: Findings from a large-scale survey study. *The International Journal of Holistic Early Learning and Development, 6*, 1–24.

Hatala, A. R., Njeze, C., Morton, D., Pearl, T., & Bird-Naytowhow, K. (2020). Land and nature as sources of health and resilience among Indigenous youth in an urban Canadian context: A photovoice exploration. *BMC Public Health, 20*(1), 1–14. doi:10.1186/s12889-020-08647-z PMID:32312240

Hill, R., Adem, Ç., Alangui, W. V., Molnár, Z., Aumeeruddy-Thomas, Y., Bridgewater, P., Tengö, M., Thaman, R., Adou Yao, C. Y., Berkes, F., Carino, J., Carneiro da Cunha, M., Diaw, M. C., Díaz, S., Figueroa, V. E., Fisher, J., Hardison, P., Ichikawa, K., Kariuki, P., & Xue, D. (2020). Working with Indigenous, local and scientific knowledge in assessments of nature and nature's linkages with people. *Current Opinion in Environmental Sustainability*, *43*, 8–20. doi:10.1016/j.cosust.2019.12.006

Johnstone, A., Martin, A., Cordovil, R., Fjørtoft, I., Iivonen, S., Jidovtseff, B., Lopes, F., Reilly, J. J., Thomson, H., Wells, V., & McCrorie, P. (2022). Nature-Based Early Childhood Education and Children's Social, Emotional and Cognitive Development: A Mixed-Methods Systematic Review. *International Journal of Environmental Research and Public Health*, *19*(10), 10. doi:10.3390/ijerph19105967 PMID:35627504

Kimmerer, R. (2013). *Braiding Sweetgrass: Indigenous Wisdom, Scientific Knowledge and the Teachings of Plants*. Milkweed Editions.

Knight, S. (2013). Forest School and Outdoor Learning in the Early Years. *Forest School and Outdoor Learning in the Early Years*, 1–176.

Koller, D., & Farley, M. (2019). Examining elements of children's place attachment. *Children's Geographies*, *17*(4), 491–500. doi:10.1080/14733285.2019.1574336

Larimore, R. (2016). Defining Nature-Based Preschools. *International Journal of Early Childhood Environmental Education*, *4*(1), 32–36.

Larimore, R. A. (2018). Using Principles of Nature-Based Preschools to Transform Your Classroom. *YC Young Children*, *73*(5), 34–41.

Leather, M. (2018). A critique of "Forest School" or something lost in translation. *Journal of Outdoor and Environmental Education*, *21*(1), 5–18. doi:10.1007/s42322-017-0006-1

Lee, M. J., Oh, W., Jang, J. S., & Lee, J. Y. (2018). A pilot study: Horticulture-related activities significantly reduce stress levels and salivary cortisol concentration of maladjusted elementary school children. *Complementary Therapies in Medicine*, *37*, 172–177. doi:10.1016/j.ctim.2018.01.004 PMID:29609930

Leung, V. T. W., Yee Tiffany Tam, T., Pan, W.-C., Wu, C.-D., Candice Lung, S.-C., & Spengler, J. D. (2019). How is environmental greenness related to students' academic performance in English and Mathematics? *Landscape and Urban Planning*, *181*, 118–124. doi:10.1016/j.landurbplan.2018.09.021

Liebovich, B. (2018). The McMillan Sisters, The Roots of the Open-Nursery, and Breaking the Cycle of Poverty. *HSE–Social and Education History*, *7*(1), 78–96. doi:10.17583/hse.2018.2925

MacEachren, Z. (2013). The Canadian Forest School Movement. *LEARNing Landscapes*, *7*(1), 1. doi:10.36510/learnland.v7i1.639

MacQuarrie, S., Nugent, C., & Warden, C. (2015). Learning with nature and learning from others: Nature as setting and resource for early childhood education. *Journal of Adventure Education and Outdoor Learning*, *15*(1), 1–23. doi:10.1080/14729679.2013.841095

Maynard, T. (2007). Forest Schools in Great Britain: An Initial Exploration. *Contemporary Issues in Early Childhood*, *8*(4), 320–331. doi:10.2304/ciec.2007.8.4.320

McClintic, S., & Petty, K. (2015). Exploring Early Childhood Teachers' Beliefs and Practices About Preschool Outdoor Play: A Qualitative Study. *Journal of Early Childhood Teacher Education, 36*(1), 24–43. doi:10.1080/10901027.2014.997844

McCree, M., & Cree, J. (2017). Forest School: Core principles in changing times. *Children Learning Outside the Classroom from Birth to Eleven*, 222–232.

McNair, L. (2012). Offering Children first hand experiences through forest school: Relating to and learning about nature. In T. Bruce (Ed.), *Early Childhood Practice: Froebel Today* (pp. 86–99). SAGE Publications, Inc. doi:10.4135/9781446251287.n6

Merrick, C. (Ed.). (2019). *Nature-Based Preschool Professional Practice Guidebook.* North American Association for Environmental Education.

Meyer, J., Müller, U., & Macoun, S. (2017). Comparing Classroom Context and Physical Activity in Nature and Traditional Kindergartens. *Children, Youth and Environments, 27*(3), 56–77. doi:10.1353/cye.2017.0020

Monroe, M. C., Andrews, E., & Biedenweg, K. (2008). A Framework for Environmental Education Strategies. *Applied Environmental Education and Communication, 6*(3–4), 205–216. doi:10.1080/15330150801944416

NAAEE. North American Association for Environmental Education, & Natural Start Alliance. (2019). Nature-Based Preschool Professional Practice Guidebook. North American Association for Environmental Education.

NAAEE. (2023). *Natural Start.* Natural Start. https://naturalstart.org/

NAAEE, North American Association for Environmental Education's Natural Start Alliance. (2019). *Nature-Based Preschool Professional Practice Guidebook. North American Association for Environmental Education., & Natural Start Alliance.*

Natural Start Forum: Enrolling with Equity in Mind. (2021). Natural Start. https://naturalstart.org/bright-ideas/natural-start-forum-enrolling-equity-mind

Nelson, N., Pacini-Ketchabaw, V., & Nxumalo, F. (2018). Rethinking Nature-Based Approaches in Early Childhood Education: Common Worlding Practices. *Journal of Childhood Studies, 43*(1), 4–14. doi:10.18357/jcs.v43i1.18261

Nxumalo, F., & Pacini-Ketchabaw, V. (2017). "Staying with the trouble" in child-insect-educator common worlds. *Environmental Education Research, 23*(10), 1414–1426. doi:10.1080/13504622.2017.1325447

O'Brien, L. (2019). Learning outdoors: The Forest School approach. In S. Waite (Ed.), *Outdoor Learning Research.* Routledge.

O'Brien, L., & Murray, R. (2007). Forest School and its impacts on young children: Case studies in Britain. *Urban Forestry & Urban Greening, 6*(4), 249–265. doi:10.1016/j.ufug.2007.03.006

Orman, J. E. (2011). *Educational Psychology: Developing Learners* (7th ed.). Pearson Education, Inc.

OSPI. (n.d.). *Since Time Immemorial: Tribal Sovereignty in Washington State.* Office of Native Education/OSPI. https://www.k12.wa.us/student-success/resources-subject-area/time-immemorial-tribal-sovereignty-washington-state

Pacini-Ketchabaw, V. (2013). Frictions in Forest Pedagogies: Common Worlds in Settler Colonial Spaces. *Global Studies of Childhood, 3*(4), 355–365. doi:10.2304/gsch.2013.3.4.355

Posey, D. A. (1998). The "Balance Sheet" and the "Sacred Balance": Valuing the Knowledge of Indigenous and Traditional Peoples. *Worldviews: Global Religions, Culture, and Ecology, 2*(2), 91–106. doi:10.1163/156853598X00109

Pound, L. (2015). Margaret and Rachel McMillan. *Nursery World, 2015*(17), 34–35. doi:10.12968/nuwa.2015.17.34

Ragoff, B. (2003). *The Cultural Nature of Human Development* (1st ed.). Oxford University Press.

Restall, B., & Conrad, E. (2015). A literature review of connectedness to nature and its potential for environmental management. *Journal of Environmental Management, 159*, 264–278. doi:10.1016/j.jenvman.2015.05.022 PMID:26087657

Ross, A., Snodgrass, J. G., & Sherman, K. Delcore, Henry. D., & Sherman, R. (2011). Indigenous peoples and the collaborative stewardship of nature: Knowledge binds and institutional conflicts. Left Coast Press.

Salleh, A. (2009). *Eco-sufficiency & global justice: Women write political ecology*. Pluto Press London.

Salmón, E. (2000). Kincentric Ecology: Indigenous Perceptions of the Human–Nature Relationship. *Ecological Applications, 10*(5), 1327–1332. doi:10.1890/1051-0761(2000)010[1327:KEIPOT]2.0.CO;2

Schultz, P. W. (2002). Inclusion with Nature: The Psychology Of Human-Nature Relations. In P. Schmuck & W. P. Schultz (Eds.), *Psychology of Sustainable Development* (pp. 61–78). Springer US., doi:10.1007/978-1-4615-0995-0_4

Scott, J. T., Kilmer, R. P., Wang, C., Cook, J. R., & Haber, M. G. (2018). Natural Environments Near Schools: Potential Benefits for Socio-Emotional and Behavioral Development in Early Childhood. *American Journal of Community Psychology, 62*(3–4), 419–432. doi:10.1002/ajcp.12272 PMID:30222858

Smith, M. A., Dunhill, A., & Scott, G. W. (2018). Fostering children's relationship with nature: Exploring the potential of Forest School. *Education 3-13, 46*(5), 525–534. doi:10.1080/03004279.2017.1298644

Sobel, D. (2016). Finding the right place to put down roots: Facilities and Infrastructure in D. Sobel, P. Bailie, K. Finch, E. Kenny, & A. Stires (Eds.), Nature Preschools and Forest Kindergartens: The Handbook for Outdoor Learning (Illustrated edition, pp. 191-210). Redleaf Press.

Sobel, D., Bailie, P., Finch, K., Kenny, E., & Stires, A. (2015). Nature Preschools and Forest Kindergartens: The Handbook for Outdoor Learning (Illustrated edition). Redleaf Press.

Taylor, A. (2013). *Reconfiguring the natures of childhood.* Routledge. doi:10.4324/9780203582046

Taylor, A., & Pacini-Ketchabaw, V. (2015). Learning with children, ants, and worms in the Anthropocene: Towards a common world pedagogy of multispecies vulnerability. *Pedagogy, Culture & Society, 23*(4), 507–529. doi:10.1080/14681366.2015.1039050

Tiny Trees Preschool. (n.d.). *Tiny Trees Preschool: Quality, Affordable, Outdoor Preschool in Seattle & Puget Sound*. Tiny Trees Preschool. https://tinytrees.org/

Tuck, E., & Yang, K. W. (2021). Decolonization is not a metaphor. *Tabula Rasa*, *38*(38), 61–111. doi:10.25058/20112742.n38.04

Uhrmacher, P. B. (1995). Uncommon Schooling: A Historical Look at Rudolf Steiner, Anthroposophy, and Waldorf Education. *Curriculum Inquiry*, *25*(4), 381–406. doi:10.1080/03626784.1995.11076190

Vaske, J. J., & Kobrin, K. C. (2001). Place Attachment and Environmentally Responsible Behavior. *The Journal of Environmental Education*, *32*(4), 16–21. doi:10.1080/00958960109598658

WA DCYF. (2022). *Nature-Based Learning Competencies*. DCYF.

Warden, C. H. (2015). Learning with nature: Embedding outdoor practice. *Learning with Nature*, 1–128.

Watts, V. (2013). Indigenous Place-Thought and Agency Amongst Humans and Non Humans (First Woman and Sky Woman Go On a European World Tour!). *Decolonization*, *2*(1), 1. https://jps.library.utoronto.ca/index.php/des/article/view/19145

Weiss, T., Kahn, P. H. Jr, & Lam, L.-W. (2023). Children's interactions with relatively wild nature associated with more relational behavior: A model of child-nature interaction in a forest preschool. *Journal of Environmental Psychology*, *86*, 101941. doi:10.1016/j.jenvp.2022.101941

Weldemariam, K. (2017). Challenging and Expanding the Notion of Sustainability Within Early Childhood Education: Perspectives from Post-humanism and/or New Materialism. In O. Franck & C. Osbeck (Eds.), *Ethical Literacies and Education for Sustainable Development: Young People, Subjectivity and Democratic Participation* (pp. 105–126). Springer International Publishing. doi:10.1007/978-3-319-49010-6_7

Wells, N. M., & Evans, G. W. (2003). Nearby Nature: A Buffer of Life Stress among Rural Children. *Environment and Behavior*, *35*(3), 311–330. doi:10.1177/0013916503035003001

Wells, N. M., & Lekies, K. S. (2006). Nature and the Life Course: Pathways from Childhood Nature Experiences to Adult Environmentalism. *Children, Youth and Environments*, *16*(1), 1–24. doi:10.1353/cye.2006.0031

Whyte, K. (2020). Too late for indigenous climate justice: Ecological and relational tipping points. *Wiley Interdisciplinary Reviews: Climate Change*, *11*(1), e603. doi:10.1002/wcc.603

Wilson, R. (2019). What Is Nature? *International Journal of Early Childhood Environmental Education*, *7*(1), 26–39.

Wilson, R. A. (1993). The Importance of Environmental Education at the Early Childhood Level. *International Journal of Environmental Education and Information*, *12*(1), 15–24.

Yalcin, F., & Erden, F. (2021). A Cross-Cultural Study on Outdoor Play: Teachers' Beliefs and Practices. *Education in Science*, *46*(206), 131–159.

KEY TERMS AND DEFINITIONS

Ecological Identity: An individual's sense of connection and identification with the natural world, contributing to pro-environmental attitudes and behaviors.

Environmental Education: Teaching and learning about the environment and nature with a focus on developing awareness, knowledge, attitudes, skills, and a sense of responsibility towards environmental issues.

Forest Schools: Outdoor education programs which take place 100% outdoors, emphasizing child-led, experiential learning.

Learning "With Nature": An philosophy that acknowledges nature as a co-teacher emphasizing direct experiences in natural environments, recognizing nature as a source of inspiration, curiosity, and knowledge.

Nature-Based Education: An approach to learning that integrates natural elements, outdoor experiences, and environmental themes into educational practices. Nature-based education infuses nature as a driving force or its pedagogical approach.

Place Attachment: The emotional bond formed between individuals and a specific environment.

Posthumanism: A theoretical framework challenging traditional human-centric views, promoting interconnectedness between humans and the environment.

ENDNOTE

1 For further exploration of forest school principles with illustrated examples, see *Nature Preschools and Forest Kindergartens: The Handbook for Outdoor Learning* by Sobel, et al. (2015) and *The essential guide to Forest School and nature pedagogy* by Cree,& Robb (2021).

Chapter 16
Educating the Local Population About the Need of Environmental Protection Through Participatory Educational Theatre (PET) With a Focus on Sundarbans:
Participatory Educational Theatre and Environmental Protection

Raghav Prakash
O.P. Jindal Global University, India

Aliya Naz
O.P. Jindal Global University, India

Abhiroop Chowdhury
https://orcid.org/0000-0001-6985-0722
O.P. Jindal Global University, India

ABSTRACT

Participatory and community theatre are applied from radical theatre traditions. Participatory education theatre (PET) has been aiding natural resource conservation projects around the globe. This book chapter elaborates on the role of PET as an effective tool in implementing socio-environmental projects. The genesis and evolution of PET from traditional theatrical forms are showcased in this chapter. The benefit of PET is turning the passive audiences into active participants, which helps in project implementation. As dramatic pretend play games (DPPG) help enlist behavioural changes in audience, PET also caters to more acceptability of any project's critical take-home message to a broader audience. This concept was tested in a socio-environmental project in the disaster-vulnerable Indian Sundarbans, with positive outcomes and has been instrumental in project compliance with the United Nations Sustainable Development Goal (SDG) 10 (reduce inequalities) targets by challenging the existing class, creed, and religion-based social dogmas.

DOI: 10.4018/979-8-3693-2577-3.ch016

Table 1. List of abbreviations

S.No.	Abbreviation	Full- Form
1	TIE	Theatre in Education
2	PET	Participatory Theatre in Education
3	DIE	Drama in Education
4	TFD	Theatre for Development
5	TCA	Theatre for Community Action
6	PT	Participatory Theatre
7	DPPG	Dramatic Pretend Play Games
8	SDG	United Nations Sustainable Development Goals

INTRODUCTION

Participatory Theatre in Education (PET), theatre in school education and theatre for the development of an individual (Kumrai et. al, 2011). This study focuses primarily on how PET is being used to educate the population about environmental conservation. The authors have also discussed other art forms and how they have been linked to educating the population. The Indian Sundarbans has hosted many PETs for educating the population about mangrove conservation and how important these forests are for the survival of the indigenous population. A few forms of theatre have been listed to understand how each art form is playing a crucial role in educating and raising awareness for the conservation of the environment.

Sloman 2012 highlighted two major forms of theatre: (1) conventional- in which the actors perform for a wider audience and pass on a message. This is seen as a form of spreading awareness. The other form (2) Participatory theatre involves the audience giving a higher sense of participation and voice to the audiences. Slowman (2012) discussed that PET leaders highlighted the art used to rehearse change. Conventional theatre, and shared that the spectator is seen to give power to the dramatic character. However, in Boal's theatre, the spectator was taking action. This had more meaning since the community understood and acted according to their abilities according to the situation. Benefits of PET has are as follows (1) building the capacity of groups, individuals, and communities with the power to strengthen and energize; (2) giving voice to the weak and oppressed communities, challenging power structures and being a tool for advocacy; (3) it supports in raising awareness and educating individuals on a given subject; (4) PET is aiding in positive dialogue; (5) building on skills that people already possess; (6) and it is entertaining and lively which gives an incentive to the communities to get involved (Slowman, 2012).

It was observed that community-based natural resource management had been linked to multiple system dynamics and structural, governance and sociocultural factors. The article by Heras and Tabara, 2015, opined that long-term collective action for sustainable use of resources requires a high level of system awareness by the communities. Conservation theatre is described as a form of participatory or community theatre. In this type of theatre, the local people become a part of their plays as protagonists. The local community members are seen performing at a central stage, showcasing their emotions and feelings and sharing a message about social or political action (Heras and Tabara, 2015). Conservation theatre aims to help address the most concerning and pressing issues the community is facing at that time. A case study from Cheran, Mexico, focuses on the importance of conservation theatre for spreading awareness against logging and organized crime in the forests. This community environmental agitation was led by women in resolving the issues.

Spooner et.al, 2019 observed that theatre has been recognized as a form of education in Zoos. The theatre was described as a scripted performance intending to share a story or convey a message. Even in a non-zoo setting, it was seen how educational theatre has been helping school-going children and wider audiences communicate a message. Moreover, the population is seen to stay longer when there is a theatre performance compared to when an exhibit is presented by itself (Spooner et.al, 2019). The zoo keepers conveyed to local population about the need for undertaking such a venture with focus on conservation in Tanzania and educating people about natural resources, sustainability, planting of trees.

Research by Aghamelu, 2023, quoted, "One cannot effect change without being involved in the larger process of change" This refers to community-based theatre involving the population and collaborating with the host community to understand and resolve pressing issues. It was opined that theatre is seen as one of the most convenient and simple forms of educating people. The author shared that theatre appeals to all classes, is entertaining, and at the same time educative. It has been showcased that community theatre is unique and has its way of highlighting critical issues making the population aware of it and coming up with viable solutions. The author observed that community-based theatre is used to come up with new strategies to tackle problems and drama is used as a source for educating the population. In this type of art, entertainment is not the main aim, rather dialogue and formation of ideas is given more importance. Furthermore, it aids in motivating the population to change and adapt to a given situation (Aghamelu, 2023).

It was highlighted that theatre has been used since the mid-1970s in the Sub-Saharan African region. Numerous universities since then have turned to theatre and performing arts for educating the local communities. Moreover, it was also opined that the use of theatre has opened up channels for debates around nature, social issues, and the problems affecting them as a community. The author highlighted how theatre has been used in education, health, agriculture, governance, etc. across the globe to educate people (Kerr et.al, 2022).

Heras and Tabara shared that majority of theatre experiences for community-based natural resource management (CNRM) were in the Global South. This is due to the direct engagement of the participative audiences to understand the management of the community's natural resources. The study highlighted in the Global North, that theatre is more focused on finding apt solutions to climate change, and modification of environmental perceptions and consumption patterns. The authors shared that Africa is using theatre for development. For instance, the study detailed through a 5-year process, a seven-country alliance was co-ordinated between local NGOs, technical agencies, national governments and a theatre group. The actor-facilitators were working with the local communities in action research and came up with several performances highlighting the personal experiences and challenges associated with CNRM as follows- (1) the performances were showcased in various local settings, and the insights were systematically collected; (2) after 2 years of performances the facilitators shared the experiences gathered in different countries; (3) all these inputs were then used to make a new theatrical performance which was showcased in Africa and also played internationally for audiences of policymakers (Heras and Tabara, 2014). The study also shed light on similar projects in Mali and Zimbabwe. The theatre has helped collect feedback on environmental protection programs, amplify baseline information, and support monitoring, facilitation, and communication between communities and project workers.

HISTORY OF EDUCATIONAL THEATRE

Educating the population through theatre peaked during the 1970s, 1980s, and 1990s. Today, the apt term for drama and theatre used to educate individuals is "applied theatre." Research conducted by Jackson and Vine, 2013, discusses how theatre in education in Britain began in the 1960s with a direct response by schools and theatre institutions. Professional theatre institutions started it and soon got supported by local arts boards and schools in the country. It was recorded for almost three decades that theatre had carved a niche for itself in the education domain in crucial and unusual ways. Theatre in Education (TIE) was seen as a powerful tool in bringing about change and educating local communities in the UK. During the 1960s and 1970s, theatre was shown to encourage children to learn new ideas and understand the world's ways. This gave them a practical approach to life. The authors shed light on how theatre in education (TIE) has been used in schools, museums, prisons, and building-based theatres (Jackson and Vine, 2013). Even during the 90s, it was highlighted how TIE focused on multiple topics- environment, racism, local language learning, history, science, and health.

Martin's study focused on the African region and highlighted how any ritual linked to birth, marriage, and death has its cultural practice. Moreover, drama and theatre are an integral part of these rituals and have become a part of the culture of many African countries and communities (Martin, 2004).

EVOLUTION OF THEATRE

The study quoted, "The critic Bernard Beckerman claimed that theatre happens whenever "one or more human beings, isolated in time and space, present themselves to another or others" (Jha, 2019). Theatre is often seen as a stage performance for audiences, but it is much more than a performance for entertainment purposes. It is used to educate people about religious and social issues and spread awareness. The study shed light on how theatre for development was inspired by the work of Paul Freire, who believed individuals gain knowledge through life experiences but are made to believe they are irrelevant by the oppressor. Augusto Boal, a student of Freire, introduced "theatre of the oppressed." This form of theatre is unique as it is, by and for the people, and helps in resisting the daily oppression of people. Theatrical performances were first showcased in Greece. The word theatre is of Greek origin and comes from the Greek word "Theaomai," which means "to see." Theatre became the primary mode of communication and storytelling in Greece due to the importance of the spoken word. European theatre saw a setback during the 5th to 10th centuries. This period is called the "Dark Age" or Middle Age. Indian theatre had different periods which different rulers marked. It was discussed how Indian theatre originates from Sanskrit theatre from the second century BC. It was an essential form of communication till the 17th century. The research divided Indian theatre into three stages: Ancient, Medieval, and Modern.

Ancient Indian theatre showcased the rich traditions and cultures of the time. It started with the religious rituals of the Vedic Aryans. Medieval theatre highlighted the truth of life with the involvement of dance and music. 'Loknatya' was introduced in the 16th century and gained very high importance. The British rule highly influenced modern theatre. It mainly focused on the political changes during those 200 years. The modern theatre also highlights day-to-day activities. The community's performances helped them understand social, economic, and political problems and how they can be resolved. The study opined how Bengali theatre was used to educate the masses on relevant topics (Jha, 2019).

Research conducted by Jha 2019, highlights the evolution of theatre and its numerous uses in society. Jha shared that arts and education engage with societies so they make healthy choices. Theatre is a tool here as it bridges the gap and aids in creating awareness among the people. Theatre is a powerful asset when discussing taboo/ feared and susceptible topics. The study shed light on 'forum theatre,' which enables the spectators to act directly rather than observe the performance. Forum theatre aims to empower the community not to find an optimum solution but to put many options on the table. This technique is instrumental in making women understand their rights and legal aspects.

PET AND OTHER THEATRICAL ART FORMS

According to Valverde 2003, three major categories relating to educational practice and theatrical arts in his research were Theatre in Education (TIE), Drama in Education (DIE), and Educational Drama. The study shed light on TIE having a unique feature of being a play and a whole program that educates children. It is observed that the staging of a play covers some of the cross-curriculum, which is integral for the children. Work inspiration usually comes from topics covered in school/ general relevance. The children's participation is considered to make it a meaningful and educational experience. Valverde opined how identifying feelings, communication, respect for self and others, conflict resolution, and anger management are best seen through interactive drama activities and developing stories in which children's decisions define the outcome of events. It was observed that these procedures rest on the principles that inform the current drama trend in education (Valverde, 2003). Furthermore, drama is seen as an extension of the play and how it helps children- (1) children learn best through direct experience; (2) drama "allows our students to explore the foundations of surface reality" (Wessels, 1987; Valverde, 2003). Orr opined that applied theatre aimed to give students a chance to understand different situations and develop solutions. The interesting fact here is that the answer does not necessarily have to be correct; the exercise should aid in thinking to come up with answers.

The study conducted by Jackson, 1993, stated that there are multiple forms of TIE. The author defines them and showcases the differences in the art.

1. Applied Theatre is defined as a wide range of participatory, politically inspired, socially engaged, and non-traditional theatre practices. This is seen to be performed in front of audiences that are not considered mainstream. This type of TIE is conducted in spaces not designed for theatre.
2. Theatre for Development (TFD)- This type of theatre is used for development. The research highlighted how third-world countries make the maximum use of TFD. Furthermore, non-governmental organizations are seen to lead the way in using TFD, for example- "Save the Children." Lastly, the author also sheds light on how this technique is being used to educate the population about HIV/ AIDS and promote health safety.
3. Theatre of the Oppressed- This form of theatre was introduced by Augusto Boal and is considered part of participatory theatre methodology. This form of theatre aims to drive a shift from making the spectator into a spectator. Simply changing the audience from mere bystanders into a decision-making community. In Boal's words- "Its rehearsal for revolution."
4. Education Theatre- This term is most popular in the USA. The study opined how this term implied theatre is particularly used for educational purposes and a younger audience. Furthermore, several universities offer programs specializing in theatre and drama for students to learn the art.

5. Museum Theatre- This type of theatre is seen to be performed in museums/ historic sites and is done by professional actors. It mediates knowledge about the site's history and gives historical context to the audience.
6. Process Drama/ Drama in Education- The research showcased how this type of art is performed in schools. There is a bias towards theatrical practices, and it is usually done for a large audience. The primary goal of this type of theatre is to generate awareness and educate the students on various topics. Furthermore, it is seen to aid in the child's development of imagination, improvisation, and becoming a better team player. The study conducted by Jackon has shed light on these forms of theatre, providing an overview of the differences in these art forms (Jackson and Vine, 2013). Table 1 describes a few forms of TIE and how they are being used globally.

Table 2. Forms of theatre in education and their use by researchers

S.No.	Title of Paper	Data Presented	Remarks	References
1	Inter-acting with the Past - the use of participatory theatre at museums and heritage sites	The authors shed light on how theatre is a significant art form and a tool for educating communities. It was recorded how interactive theatre is being used in numerous heritage sites to educate the people on the history of that site and its location. The study highlighted how these dramas occur around/ open sites such as Blists Hill Victorian Village at Ironbridge, UK.	The study showcased the importance of interactive theatre in museums and heritage sites. It highlights the different ideas behind having dramas conducted in these locations. Anthony, 2009	Anthony, 2009
2	Deep Listening in a Feminist Popular Theatre Project: Upsetting the Position of Audience in Participatory Education	Butterwick and Selman conducted a study to understand the importance of coalition and community in women's action movements. The project consisted of 14 women in the age groups of 20-40. The theatrical performances helped me understand the barriers in coalition and how they can be addressed.	The paper focused on how the understanding of women's coalition can be achieved through theatre involvement.	Butterwick and Selman, 2003
3	The effectiveness of participatory theatre with early adolescents in school-based sexuality education	Ponzetti Jr. et al. highlighted how theatre could be used to make a difference in the thinking of youth regarding sex education. The authors also pointed out that participatory theatre is nascent in sex education. It was observed that there was some fundamental change in attitude and thinking with the use of PET.	This study aimed to shed light on how theatre is used to create sexual awareness among the youth.	Ponzetti Jr. et.al, 2009
4	You know the homophobic stuff is not in me, like us, it's out there". Using Participatory Theatre to challenge heterosexism and heteronormativity in a South African school	The study focused on forum theatre, which helps in critical reflection, debating ideas, and learning and unlearning concepts around heterosexism and heteronormativity. The research was conducted on school children between the ages of 15 and 18 in a co-educational free state school in Africa. It was recorded how students bemoan heterosexism and heteronormativity. The group was seen to challenge examples of prejudice and behavior by teachers, though continuing to avoid everyday examples of heterosexist exclusions and privileges.	The research conducted by Francis highlighted the use of forum theatre and how it aided in understanding heterosexism and heteronormativity.	Francis, 2013
5	Helping them to help themselves: case study of an integrated popular theatre approach in Africa	Chinyowa shares how, through the use of theatre for community action, the rural sector in the Bulilima-Mangwe district could develop policies in line with HIV/AIDs. It was observed how the theatre group embarked on a journey to educate the other villages and districts in the area. Furthermore, the research highlighted how TCA addressed the issue of treating development as a process rather than an event.	This research aimed to understand theatre for community action (TCA). The author wanted to educate/ raise policy awareness on HIV/ AIDS in the rural sector of Africa.	Chinyowa, 2007

A study shed light on forum theatre changing the role of passive audiences into active and double-role audiences. The author opined that this is very helpful in understanding situations. For instance, few audience members play some conflictual situation related to the community. The other participants then intervened and offered valuable solutions to help the community tackle that issue. This gives the people a voice and a sense of authority in making important decisions (Kaptani and Yuval Davis, 2017). Freebody and Finneran argued participatory theatre is a form in which the social imagination is embodied in playing 'others.' The authors highlight that the drama is felt for ourselves but also being played from a different individual's perspective. This means we act ourselves through someone else's lens. The research quoted, "We walk, talk, feel, and respond to the argument of the drama through the lens of difference. We are ourselves but also not ourselves" (Freebody and Finneran, 2015). The study showcased that participatory theatre allows the population to view ideas and situations from different perspectives. This aids in broadening their horizon and not restraining their minds.

Middlewick et al. 2012 showcased forum theatre being applied in the healthcare sector. It has been used in healthcare education, midwifery, and medicine to understand issues students find challenging, such as empowerment, valuing diversity, and accepting beliefs (Thistlethwaite and Ewart, 2003; Freshwater and Stickley, 2004; McClimens and Scott, 2007; Kemp, 2009). A session is prepared in forum theatre before the actors are given a script. After the performance, the audience is involved, and changes are made to develop better outcomes for the problems discussed in the play. If the audience wants, they can even step in and participate themselves to showcase their work. Pincock 2004, quoted, "Communication is at the heart of all complaints, and forum theatre can empower professionals to deal more effectively with complaints at earlier stages and thus prevent these situations from escalating" (Middelwick et.al, 2012). Forum theatre is seen to use poor practice scenarios. The study shared an example of relevance, "by showing an impatient, bullying, patronising way of addressing a service user and explores the choices that are available to practitioners via actors assuming the roles of particular healthcare professionals, service users and carers" (Middlewick et.al, 2012). The issues raised by characters can be explored in detail. This is another feature of forum theatre that is widely accepted. This helps students/ audiences to start discussing possible outcomes for issues raised in the theatrical performance. It was recorded that forum theatre engages students at a deeper level of reflective thinking. The facilitator motivates and inspires the students to think and change the outcome of the issue discussed in the play.

In an article by Singhal, a story was shared about women in prisons and how "theatre for the oppressed" was seen to gain power. The author highlighted that in Westville Prison in Durban, South Africa, black women were sentenced to lifelong imprisonment for murdering their partners. A woman was continuously abused by her husband, but due to the lobola (dowry/ bride price), her family did not support her and asked her to work out her issues and make the marriage work. After this, she went to the police, but the police officers knew her husband through a social network and called him. The husband then arrived at the police station and started to beat his wife, but the police did nothing about it. After losing all hope, she went to an assassin and asked him to kill her husband. This crime sent her to prison, where she enacted this scene in front of other women. This was known as the "Theatre of The Oppressed." This theatre form aims to help women gain confidence, discover the value and power of social cohesion, and question the oppressive structures underlying their condition. Furthermore, the study highlighted that the power of prison theatre lies in its participatory, socially engaging, and autobiographical narrative and the ability to connect oppressed and oppressive structures in a non-threatening manner. The research quoted, "Further, Prison Theatre embodies a process of participation that is empowering both as a means (for the oppressed poor Black women inmates) and as an end (in terms of the structural outcomes that are generated)" (Singhal, 2004).

The study further detailed on participatory communication being relatively new, and gaining momentum in the 1970s (Jacobson, 1993; Uphoff, 1985; Singhal, 2004). It was observed that all participation is communication-driven, but not all communication is participatory (Fraser and Restrepo-Estrada, 1998; White and Nair, 1999; Singhal, 2004). Participatory communication is defined as working with and by the people, compared to working on and for the people. In order to highlight the differences, the study opined on two approaches to understanding participatory communication. The first approach is based on the dialogic pedagogy of Brazilian educator- Paul Freire. The second approach, the participatory community media approach/ alternative communication approach, revolves around access, participation, self-management, and self-determination. The Freirean theory of dialogue is found to stress more on interpersonal and group dialogue in a community setting. This theory has been applied in community development, literacy education, participation, and transformation. On the other hand, the participatory media communication approach is centred around public and community access, participation of people in designing messages and media productions, and self-management of communication enterprises. This approach is applied in community radio and television, street theatre, folk media, community informatics, internet, and telecentres (Singhal, 2004).

The study quoted, "There is no act too small or bold. Social change is the history of millions of small and large actions coming together at critical points to create a power that governments cannot suppress."- Howard Zinn (Adhikari and Saha, 2021). The study focused on India and how theatre, specifically street theatre, played a significant role in educating the citizens. For instance, the first play highlighting contemporary rule life themes was Dinabandhu Mitra's Nil Darpan. This drama performance emphasized how vital a tool theatre is in awakening protest and educating the country. Adhikari and Saha opined how theatre played a crucial part in India's independence movement, which led to the Dramatic Performances Act of 1876. The paper highlighted how "mass education and rural extension" refer to educating on health, family planning, sanitation, well-being, and other social concerns. Furthermore, such usage of theatre aided in educating the masses, forming a national identity among them, and trying to form new policies for their development (Adhikari and Saha, 2021).

Research conducted by Sloman 2012, highlighted how two forms of theatre exist: (1) conventional- in which the actors perform for a wider audience and pass on a message. This is seen as a form of spreading awareness. The other form (2) Participatory theatre involves the audience giving a higher sense of participation and voice to the audiences. Slowman opined how the PT leaders highlighted the art used to rehearse change. Furthermore, the paper recorded how Thyagarajan opined that in conventional theatre, the spectator is seen to give power to the dramatic character. However, in Boal's theatre, the spectator was taking action. This had more meaning since the community understood and acted according to their abilities according to the situation. Slowman shed light on how PET has many benefits, such as (1) building the capacity of groups, individuals, and communities with the power to strengthen and energize; (2) giving voice to the weak and oppressed communities, challenging power structures and being a tool for advocacy; (3) it supports in raising awareness and educating individuals on a given subject; (4) PET is aiding in positive dialogue; (5) building on skills that people already possess; (6) and it is entertaining and lively which gives an incentive to the communities to get involved (Slowman, 2011).

In an article by Kerr et.al, 2022, it was highlighted that theatre has been used since the mid-1970s in the Sub-Saharan African region. Since then, numerous universities have turned to theatre and performing arts to educate the local communities. Moreover, it was also opined that the use of theatre has opened up channels for debates around nature, social issues, and the problems affecting them as a community. The author highlighted how theatre has been used in education, health, agriculture, governance, etc. across the globe to educate people (Kerr et.al, 2022).

In a study by Zakopolous et al. 2023, sustainable development and education on sustainability were provided with the use of theatre. For instance, Keles developed a course map in Turkey to educate undergraduate students on sustainable production and life. He did this through the use of a three-faceted drama: (1) warming up, (2) improvisation, and (3) role-playing activities. This process helped the students think about the life-cycle approach and critically assess the situation of sustainability and how relevant sustainable production is today (Zakopolous et.al, 2023). Fernandez claimed that students are not only readers in the classroom but also transform into actors and start making decisions that are of social and ecological relevance. The reason for this is the nature of drama/ theatre being used as a tool in teaching. This makes the students more aware and increases their sensitivity towards socio-environmental problems, urging them to come up with solutions (Moreno Fernandez, 2020). Bakersville and Anderson recorded students seeing themselves as environmental scientists while performing in a theatre activity. The children try to come up with answers to solve the issue of climate change. The issue of climate change is seen as both- an environmental and social concern. This is because it affects the entire society (Bakersville and Anderson, 2021). Sajadi admits that arts can be seen as engaging activities that aid in holistic comprehension of self, others, and the environment. This further empowers students with sensitivity and ethical emotions relating to the environment (Sajadi, 2021).

PSYCHOLOGICAL ASPECT OF THEATRE

Chivandikwa and Makumbirofa discuss that active participation is aiding in the psychological development of children. The study showcased that through participatory education in theatre, students developed confidence, intellectual curiosity, and social skills. The authors shed light on the drama- 'Albatross'. The students were seen to be more aware of their environment and accepting of other children's behaviour. Furthermore, the authors highlighted that there was more learning than communication skills. For instance, the students must have gained some incidental thought processes and ideas while performing. Theatre has helped shape children holistically due to the development of sensuous and intellectual involvement (Chivandikwa and Makumbirofa, 2011).

A study by Goldstein and Lerner shed light on 'Dramatic Pretend Play Games (DPPGs) and the affect they have on children's growth and development. DPPG is simply an activity led by an adult for a child/ group of children in which they follow a prompt about a character, situation, emotion, or idea. This has been proven to help enhance social knowledge and control of emotions. These activities are mainly done for pre-school level children who are given basic DPPG topics such as pretending it's your first day at school/walking like a tiger, etc. The school teacher sometimes prompts the children to stay on topic and keep their focus limited to the given situation. These activities aim to make children aware of how others think, make sense of emotional states, and act according to other individual's behaviour. The research quoted, "By engaging with a variety of emotional states in a false, controlled way, with a definite beginning and end, and a clear causal relationship between activity and emotion within the "world" of the activity, children may gain an understanding of the causes and consequences of emotion" (Goldstein and Lerner, 2017).

IMPACT OF PARTICIPATORY EDUCATION IN THEATRE ON THE ENVIRONMENT

Research conducted by Asante, discusses environmental health problems and shares that they have been persistent in the regions of Ghana, especially the coastal area of Woarabeba. There is no proper means for the disposal of solid waste. Due to not having any reusable containers, the fishing community members are forced to bury the waste in the sea or throw it on the beach due to not having any reuses containers. This has caused the creation of "garbage hills." This is a serious problem as children are seen to scavenge the waste, causing multiple health problems such as cholera, typhoid fever, malaria, etc. Furthermore, the coastal area is subjected to open defecation, which adds to the existing issue of insanitation. The study used PET as a means to address the issue of insanitation in the Woarabeba coastal area. The community performed improvised drama from the data and outcomes of the community activities, which the audience could relate to. This made them spec-actors, which allowed them to understand the severity of the issue much faster. The audience shared their views and ideas on how the problem of open defection can be brought to an end. The study highlighted on how only one public toilet was set up in the area. The community took action to fix it, make it more convenient for both sexes and, most importantly, make it functional. This helped solve the issue of open defecation (Asante, 2021).

Brown et al. (2017) shed light on S.A.F.E. Kenya's production house and how it engaged PET in the coastal regions of Vanga and Gazi. The production was called 'Gangawazi, ' which involved several community members, mainly fishermen, farmers, and their families, working alongside chiefs, government advisors, and teachers. The production aimed to highlight the major economic, social, and ecological challenges associated with a volatile environment. The study quoted, "In the space of the 45-min play, issues such as sustainable fishing, mangrove cutting and planting, economic diversification and local supply chains, fuel management, sustainable farming, and biodiversity enhancement were all touched on, alongside more cross-cutting issues of community governance and planning" (Brown et al., 2017). The most intriguing aspect of this production was how multiple characters lost their love. This shocked most of the audience and made them question what could have been done differently. This inspired a discussion to develop apt solutions to such a grave issue. The authors recorded how the production was performed in different coastal areas of Southeast Kenya and engaged approximately 4700 individuals.

Golden Tree Productions did a single performance named 'Weather the Storm' in the UK in the coastal town of Porthleven. This site-specific theatre took the audience from a pub quiz to Seaview Terrace, a fictional street close to the shoreline. It showed how homes were flooded, and people had to be evacuated from the area due to a violent storm. The production dealt with exploring the emotions around loss and grief. The Porthleven's Fishermen sang in a choir, and the fishermen, local artisans, coastguards, and stakeholders were engaged in the drama performance. The performance highlighted many emotions around coastal defence, for instance, where the line has to be drawn, the numerous strands and stressors, and ultimately, the future of the coastal communities in times of rapid change. Weather the Storm engaged 215 people in a single performance (Brown et al., 2017).

A study conducted by McDonagh et al. 2023, showcased the issues faced by coastal and estuarine communities by climate change. The performance was titled "On the Edge" and collaborated with The National Youth Theatre (NYT) and The University of Hull. It was funded by a UK climate resilience impact award and part of MELT, i.e., a multi-year response to the climate crisis led by the NYT and the University of Hull (McDonagh et al., 2023). The performance aimed to highlight the conditions of the communities living in coastal regions of the UK and the impact climate change has on the environment and socio-economic status. The Performance was conducted on a global policy stage at COP 26 in Glasgow, UK, in 2021. The research quoted, “The co-creative process was characterized by an intensive two-way dialogue between academic researchers and young creatives aged 18–26 facilitated via online development workshops (featuring an initial cast of 200 young people from across the UK, and a final ensemble of twelve performers) and in-person rehearsals, a live preview performance and Q&A session at the NYT’s Production House the week before COP as part of the Islington Green Festival” (McDonagh et al., 2017). The final performance was titled “I Don’t Care” and was directed by Adeola Yemitan. The play was a product of collaboration between NYT and members of the Risky Cities Team. The motive was to make the youth aware of how to tackle flooding, coastal change, high deprivation of socio-economic status, unemployment, etc. The performance ended with the actors and community members having a round table discussion on how to solve these significant issues.

APPLICATION OF PET IN MANGROVE CONSERVATION OF INDIAN SUNDARBANS

Sundarbans is the largest contiguous mangrove forest nestled between the Bay of Bengal in the South, the Hooghly River in the West, the Padma River in the East, and the city of Kolkata, agricultural tracks of West Bengal (a state in India), Bangladesh (a sovereign nation) in the North. This forest is also home to the enigmatic mangrove-dwelling tigers. The Indian part of the Sundarbans (40% of the total area) is conserved as per The Wild Life (Protection) Act, 1972, which limits the construction of industries or similar developmental projects in the region. This results in a crunch in liquid cash, which becomes more acute during any disaster incident (Chowdhury et al. 2020). The liquid cash crunch is the fundamental reason behind natural exploitative activities (Hebinck 2018). Hence, promoting environmentally friendly, alternative livelihood options deters the local population from poaching or other nature-destructive activities. A project was implemented at Satjelia island of Indian Sundarbans between 2012 and 2017 (Fig 1), focusing on mangrove conservation by promoting alternative livelihood initiatives amongst 2100 disaster-vulnerable families (Chowdhury et al. 2016, 2018, 2022, 2023).

Figure 1. From the map of India, (a) the state of West Bengal has been zoomed out; (b) highlighting the region of intervention of the project marked by sign; (c) the PTEs were held in Satjelia island, Chotto-mollakhali, Rangabelia, Bali, Kalidaspur, and Bara mollakhali marketplaces

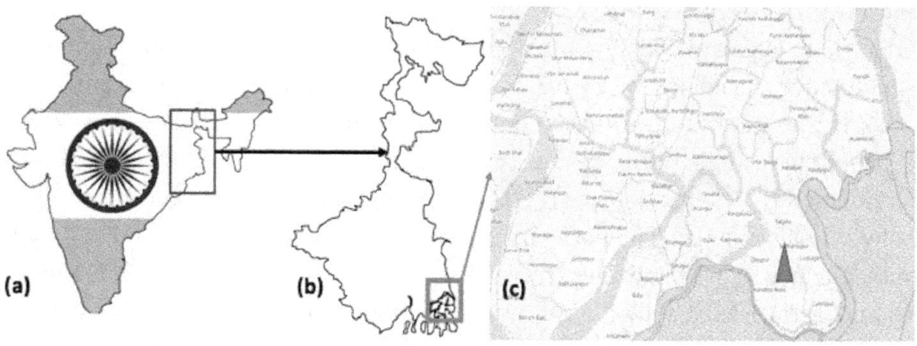

The deliverables of the project were,

- Promote green livelihood options such as poultry, small business, and vermiculture to augment the liquid cash crunch observed in the region due to the paucity of employment options
- Restoration of degraded mudflats through mangrove plantations
- Promote organic agriculture to reduce the chemical footprint in the fragile mangrove ecosystem
- Promoting local pond-based water resource management to combat salinity intrusion in the region's groundwater

PET was used as a tool to communicate the project's success stories. Jatra, or local theatre, is a popular mode of entertainment for the people of Indian Sundarbans, dispersed amongst 54 habitable islands, only connected through ferry or water-borne vessels. Jatra originated in provincial eastern India and are loosely connected song and dance sequences depicting a story of local significance (Sarkar, 1975). Etymologically, the origin of the word 'Jatra' may indicate 'travel' according to the Bengali language. The troupe members generally travel to distant rural areas of Bengal to enact the plays. Sundarbans are also known for their syncretic culture, and the myth of 'Bon bibi,' where ostracised become the saviour and protector (Jalais 2008). The entire play of Bon bibi is a popular entertainment for the people of Sundarbans (Fig 2).

Figure 2. A 'Bon bibi' jatra at Indian Sundarbans: (a) start of the show with the narration of the genesis of 'Bon bibi', the protecting goddess of the forests; (b) the tiger is represented as a villain in the play, attacking the villagers; (c) goddess 'Bon bibi blessing orphan 'Dukhey' the protagonist of the play, with blessings of protection from 'Shah Jangali' or 'Tiger' daemon (d)

Similarly, PET was developed as a song and dance sequence, bearing semblance to the already popular 'Bonbibi Jatra' format but based on the 'importance of mangrove conservation' theme. A 1-hour long performance was arranged during the market day in different islands of the Indian Sundarbans between October and November of the year 2012.

As road connectivity is limited in Indian Sundarbans and water vessel transportation is the only option, a grand market cannot be organized on every island simultaneously. Hence, different islands have markets set on different days, and sometimes, the name of the nearby ferry ghat or local ports is derived from that. For example, one of the famous grand markets locally called "Bazar" is held on Wednesday (Budhwar-local language) on the island of Satjelia, which has two revenue villages, namely Satjelia gram panchayat and Lahripur gram panchayat. People come from different islands during this weekly market, so it is the best age to spread mangrove conservation to a large audience. The parental form of PET, 'the Jatra', requires a monetary investment in purchasing a ticket; this income is required to pay the actors and the entire team of performers while making this form of play sustainable. However, it requires time, resources and energy to master a poetic performance, so it tends to become repetitive if it is based on only one location. In the case of Indian Sundarbans, the performers travel from island to island, repeating the same Jatra performance, which brings diversity to the audience.

Following the same format, PET was organised in such a way that needs to be enacted at different islands to attract the maximum crowd. Generally, in a PET performance 50-100 audience strength is observed during the market days. Unlike Jatra, there is no ticket charge for the audience. Without monetary commitments, the chances increase to reach a wider audience, and the conservation theme can easily compete with any popular Jatra of the region. Sundarban is known to have a majority of the socially ostracised sections of Hindu society works of Jalais 2014, monetary commitments always limit and set up a socio-economic barrier. In this remote, forgotten, less talked about corner of the world's most populated democracy, implementing UN SDG 10 (Reduce inequalities) is difficult. Small interventions such as PET, where any audience, irrespective of caste and creed, gender affiliation, and economic standards were welcomed, played an indirect role in spreading the notion of UNSDG 10 to reduce inequalities to the masses. The choice of participation in such performances is not limited by the 'ability to pay' but rather by interest in the cause. This is the fundamental concept behind using the PET toolset in this conservation project. Overall, PET is a fundamental tool to communicate the conservation project success stories to a larger mass, inspiring the local grassroots-level population to conserve mangroves and align natural resources.

Chowdhury et.al 2016 highlighted the need for PET in educating the population of Sundarbans. In Sundarbans, a training program was being conducted to understand the devastating effects of climate change and sea level rise. the study quoted, "This component was aimed to propagate the ideas of AIGA, MFI, Disaster preparedness' to non- group members, i.e. the whole community residing in and around Satjelia Island." The major reason for this study was to educate the population on resource conservation. The authors discussed that PET was used in order to engage everyone and for a faster learning experience. The trainees and project workers were involved in acting. There was severe competition amongst the audience to be a part of these dramatic performances. This was mainly so they receive some popularity and recognition amongst their community. Multiple shows took place on educating the population on resource conservation and a random survey highlighted that PET was aiding in the learning of these marginalized communities (Chowdhury et.al, 2016).

CONCLUSION

Theatre has played an essential role in communicating social messages since immemorial. The chapter has captured the essence of theatre as an art form and highlights the importance PET as a tool for educating the population. In addition, participatory educational theatre is a theatrical tool to convey implementation project success stories to a broader audience; there is an increase in the adoption of PET model in several socio-environmental projects across the globe with positive results; the case study of a mangrove conservation project of Indian Sundarbans also highlights how PET can circumvent transportation austerities inaccessibility and social differences to successfully communicate, inspire a wider population on broader mangrove conservation themes. It is also observed that PET aid in reducing inequalities and transcends the traditional stereotypes of caste, creed and religion, which can aid projects in complying with the targets of UNSDG 10 (Reduced inequalities).

REFERENCES

Adhikari, A., & Saha, B. (2021). Contextualizing Social Theatre and its History: An Indian Perspective. [IJARESM]. *International Journal of All Research Education and Scientific Methods*, *9*(7), 497–502.

Aghamelu, H. U. (2023). Community theatre and environmental education: the ukwu orji bus park project. *Ama: Journal Of Theatre And Cultural Studies, 16*(1). https://www.journals.ezenwaohaetorc.org/index.php/AMA/article/view/2270

Allsop, J., & Jones, K. (2008). Withering the citizen, managing the consumer: Complaints in healthcare settings. *Social Policy and Society*, *7*(2), 233–243. doi:10.1017/S1474746407004186

Asante, E. (2021). Participatory theatre in environmental health education: the woarabeba experience. *Social Science, 4*(12), 121-128.

Banham, M. (Ed.). (2004). *A history of theatre in Africa*. Cambridge University Press.

Baskerville, D., & Anderson, D. M. (2021). Responding to climate change: Developing primary children's capability to engage with science through drama. *Science and drama: Contemporary and creative approaches to teaching and learning*, 93-105. . doi:10.1007/978-3-030-84401-1_6

Brown, K., Eernstman, N., Huke, A. R., & Reding, N. (2017). The drama of resilience: Learning, doing, and sharing for sustainability. *Ecology and Society*, *22*(2), art8. doi:10.5751/ES-09145-220208

Butterwick, S., & Selman, J. (2003). Deep listening in a feminist popular theatre project: Upsetting the position of audience in participatory education. *Adult Education Quarterly*, *54*(1), 7–22. doi:10.1177/0741713603257094

Cernea, M. M. (1984). The role of sociological knowledge in planned rural development. *Sociologia Ruralis*, *24*(3-4), 185–201. doi:10.1111/j.1467-9523.1984.tb00643.x

Chinyowa, K. C. (2007). Helping them to help themselves: Case study of an integrated popular theatre approach in Africa. *Research in Drama Education*, *12*(2), 133–147. doi:10.1080/13569780701320957

Chivandikwa, N., & Makumbirofa, R. (2008). *Theatre*. Life Skills and Participatory Learning.

Chowdhury, A., Iyer, A. S., Bhattacharyya, S., Khan, S. U., & Naz, A. (2022). Organic farming in the tropical, disaster-prone coasts: Interventions, major challenges and a success story from Sundarban Biosphere Reserve, India. [). IOP Publishing.]. *IOP Conference Series. Earth and Environmental Science, 1084*(1), 012052. doi:10.1088/1755-1315/1084/1/012052

Chowdhury, A., Maiti, S. K., & Bhattacharyya, S. (2016). How to communicate climate change 'impact and solutions' to vulnerable population of Indian Sundarbans? From theory to practice. *SpringerPlus, 5*(1), 1–17. doi:10.1186/s40064-016-2816-y PMID:27516957

Chowdhury, A., Maiti, S. K., & Bhattacharyya, S. (2016). How to communicate climate change 'impact and solutions' to vulnerable population of Indian Sundarbans? From theory to practice. *SpringerPlus, 5*(1), 1–17. doi:10.1186/s40064-016-2816-y PMID:27516957

Chowdhury, A., Nath, R., & Rosencranz, A. (2020). The sunderbans in crisis. *Economic and Political Weekly, 55*(34), 5.

Chowdhury, A., Naz, A., Bhattacharyya, S., & Sanyal, P. (2018). Cost–benefit analysis of 'Blue Carbon' sequestration by plantation of few key mangrove species at Sundarban Biosphere Reserve, India. *Carbon Management, 9*(6), 575–586. doi:10.1080/17583004.2018.1518105

Chowdhury, A., Prakash, R., Bhattacharyya, S., & Naz, A. (2023). Role of ponds as a local practice in Mitigating salinity intrusion threats at coastal aquifer: A case study from Sundarban Biosphere Reserve, India. In *Indigenous and Local Water Knowledge, Values and Practices* (pp. 287–306). Springer Nature Singapore. doi:10.1007/978-981-19-9406-7_17

Francis, D. (2013). You know the homophobic stuff is not in me, like us, it's out there". Using Participatory Theatre to challenge heterosexism and heteronormativity in a South African school. *South African Journal of Education, 33*(4), 1–14. doi:10.15700/201412171338

Fraser, C., & Restrepo-Estrada, S. (1998). *Communicating for development: human change for survival.* IB Tauris and Co Ltd.

Freebody, K., & Finneran, M. (Eds.). (2015). *Drama and social justice: Theory, research and practice in international contexts.* Routledge. doi:10.4324/9781315755977

Freshwater, D., & Stickley, T. (2004). The heart of the art: Emotional intelligence in nurse education. *Nursing Inquiry, 11*(2), 91–98. doi:10.1111/j.1440-1800.2004.00198.x PMID:15154888

Goldstein, T. R., & Lerner, M. D. (2018). Dramatic pretend play games uniquely improve emotional control in young children. *Developmental Science, 21*(4), e12603. doi:10.1111/desc.12603 PMID:28913920

Hebinck, P. (2018). Poaching: Between Conservation from Below, Livelihoods and Resistance. In Nature Conservation in Southern Africa (pp. 257-292). Brill.

Heras, M., & Tàbara, J. D. (2014). Let's play transformations! Performative methods for sustainability. *Sustainability Science, 9*(3), 379–398. doi:10.1007/s11625-014-0245-9

Heras, M., & Tàbara, J. D. (2016). Conservation theatre: Mirroring experiences and performing stories in community management of natural resources. *Society & Natural Resources, 29*(8), 948–964. doi:10.1080/08941920.2015.1095375

Jackson, A. (2000). Inter-acting with the past-the use of participatory theatre at museums and heritage sites. *Research in Drama Education*, *5*(2), 199–215. doi:10.1080/713692887

Jackson, T. (Ed.). (1993). *Learning through theatre: New perspectives on theatre in education*. Psychology Press.

Jacobson, T. L. (1993). A pragmatist account of participatory communication research for national development. *Communication Theory*, *3*(3), 214–230. doi:10.1111/j.1468-2885.1993.tb00071.x

Jalais, A. (2008). Bonbibi: Bridging Worlds. *Indian Folklife*, *28*, 6–8.

Jalais, A. (2014). *Forest of tigers: people, politics and environment in the Sundarbans*. Routledge. doi:10.4324/9780203085516

Jha, R. (2019). Evolution of Theatre as a means of Social Development. SOCIAL SCIENCE GAZETTEER.

Kapstein, A., & Goldstein, T. R. (2019). Developing wonder: Teaching theatre for the very young through collaboration with developmental psychology. *Youth Theatre Journal*, *33*(1), 52–69. doi:10.1080/08929092.2019.1580648

Kaptani, E., & Yuval-Davis, N. (2008). Participatory theatre as a research methodology: Identity, performance and social action among refugees. *Sociological Research Online*, *13*(5), 1–12. doi:10.5153/sro.1789

Keleş, Ö. (2015). Creative drama in the teaching of education for sustainable development: Sample activity about sustainability and sustainable production. *Journal of Environmental Sustainability*, *4*(4), 4.

Kemp, J. (2009). Exploring empowerment issues with student midwives using forum theatre. *British Journal of Midwifery*, *17*(7), 438–439. doi:10.12968/bjom.2009.17.7.43058

Kerr, R. B., Young, S. L., Young, C., Santoso, M. V., Magalasi, M., Entz, M., & Snapp, S. S. (2022). Farming for change: developing a participatory curriculum on agroecology, nutrition, climate change and social equity in Malawi and Tanzania. In *Critical Adult Education in Food Movements* (pp. 29–46). Springer Nature Switzerland. doi:10.1007/978-3-031-19400-9_3

Kumrai, R. R., Chauhan, V., & Hoy, J. (2011). Boundary crossings: Using participatory theatre as a site for deepening learning. *Teaching in Higher Education*, *16*(5), 517–528. doi:10.1080/13562517.2011.570443

McClimens, A., & Scott, R. (2007). Lights, camera, education! The potentials of forum theatre in a learning disability nursing program. *Nurse Education Today*, *27*(3), 203–209. doi:10.1016/j.nedt.2006.04.009 PMID:16790301

McDonagh, B., Brookes, E., Smith, K., Worthen, H., Coulthard, T. J., Hughes, G., Mottram, S., Skinner, A., & Chamberlain, J. (2023). Learning histories, participatory methods and creative engagement for climate resilience. *Journal of Historical Geography*, *82*, 91–97. doi:10.1016/j.jhg.2023.09.002

Middlewick, Y., Kettle, T. J., & Wilson, J. J. (2012). Curtains up! Using forum theatre to rehearse the art of communication in healthcare education. *Nurse Education in Practice*, *12*(3), 139–142. doi:10.1016/j.nepr.2011.10.010 PMID:22094107

Moreno-Fernández, O. (2021). Theatre as a Teaching Resource to Work on Socio-environmental Problems in Primary Education. *International Electronic Journal of Environmental Education, 11*(1), 79–90. doi:10.18497/iejeegreen.817608

Orr, S. H. (2015). Training the peer facilitator: Using participatory theatre to promote engagement in peer education. *Research in Drama Education, 20*(1), 110–116. doi:10.1080/13569783.2014.983466

Pérez Valverde, C. (2003). Theatre in education (TIE) in the context of educational drama. Oxford University Press.

Pincock, S. (2004). Poor communication lies at heart of NHS complaints, says ombudsman. *BMJ (Clinical Research Ed.), 328*(7430), 10. doi:10.1136/bmj.328.7430.10-d

Ponzetti, J. J. Jr, Selman, J., Munro, B., Esmail, S., & Adams, G. (2009). The effectiveness of participatory theatre with early adolescents in school-based sexuality education. *Sex Education, 9*(1), 93–103. doi:10.1080/14681810802639905

Sajadi, N. (2021). Becoming conscious through the arts: A road to social sustainability. *Visions of sustainability for arts education: Value, challenge and potential*, 153-160. . doi:10.1007/978-981-16-6174-7_14

Sarkar, P. (1975). Jatra: The popular traditional theatre of Bengal. *Journal of South Asian Literature, 10*(2/4), 87–107.

Singhal, A. (2001). *Facilitating community participation through communication*. UNICEF.

Singhal, A. (2003). Entertainment-education through participatory theater: Freirean strategies for empowering the oppressed. In *Entertainment-Education and Social Change* (pp. 399–420). Routledge. doi:10.4324/9781410609595-28

Singhal, A. (2004). Empowering the oppressed through participatory theater. *Investigación & Desarrollo, 12*(1), 138–163.

Sloman, A. (2012). Using participatory theatre in international community development. *Community Development Journal: An International Forum, 47*(1), 42–57. doi:10.1093/cdj/bsq059

Spooner, S. L., Jensen, E. A., Tracey, L., & Marshall, A. R. (2019). Evaluating the impacts of theatre-based wildlife and conservation education at the zoo. *Environmental Education Research, 25*(8), 1231–1249. doi:10.1080/13504622.2019.1569201

Thistlethwaite, J. E., & Ewart, B. R. (2003). Valuing diversity: Helping medical students explore their attitudes and beliefs. *Medical Teacher, 25*(3), 277–281. doi:10.1080/0142159031000100346 PMID:12881050

White, S. A. (1999). *The art of facilitating participation: Releasing the power of grassroots communication*. Sage Pub.

Zakopoulos, V., Makri, A., Ntanos, S., & Tampakis, S. (2023). Drama/Theatre Performance in Education through the Use of Digital Technologies for Enhancing Students' Sustainability Awareness: A Literature Review. *Sustainability (Basel), 15*(18), 13387. doi:10.3390/su151813387

Chapter 17
The Power of Nature-Based Learning in Language Arts Curriculum:
The Benefits of Including Nature in English Class

Katrina Broughman
Amherst Education Center, USA

ABSTRACT

Rarely are English language arts (ELA) educators actively encouraging outdoor exploration in the learning process. Yet, ELA skills (writing, reading, researching, and communication) can be used to communicate elements of environmental literacy for the 21st century learner in Virginia (VA) and beyond. This can be done in alignment with VA's standards of learning (SOL) through nature-based learning (NBL), which encourages thinking beyond science, technology, engineering, and math (STEM). Including ELA skills and strategies, historically not the focal point of environmental education (EE) initiatives, encourages outdoor exploration in 6-12 secondary ELA curriculum development. Educators, nature centers, parents, and community leaders can all be stakeholders in communicating elements of environmental literacy through NBL-ELA instruction. Therefore, this chapter promotes NBL as a sound option for improved teaching and learning when creating and implementing interdisciplinary curriculum. It provides research-based tools, strategies, tips, and resources useful for various ELA stakeholders.

INTRODUCTION

Written and verbal communication skills are incorporated not only in secondary English Language Arts (ELA) curriculum (grades 6-12 in the United States) but essentially every classroom. Used daily, writing allows self-expression when learners formulate formative and summative responses. Similarly, reading is a communicative essential to help students comprehend myriads of informative texts. Furthermore,

DOI: 10.4018/979-8-3693-2577-3.ch017

cultivating research skills identify reliable sources, avoid plagiarism, and make discoveries. Presentation, collaboration, and communication skills allow deeper comprehension and stronger connections; learners recognizing their potential yield great power through the written word by persuading, explaining concepts, reflecting, and discussing ideas.

Why not make more effort to understand the planet and each other for a better world through effective communication skills and interdisciplinary instruction by actively exposing issues like climate, pollution, and extinction across disciplines beyond STEM? Teaching English outside and reading or researching about environmental issues in class, some mistook me for a science teacher. I was challenged or met with resistance when paving avenues for environmental advocacy, social justice, and community health, so my pedagogy transformed. I saw Virginia's math and science teachers teaching outside. Why not language arts too? As a result, my questions evolved more deeply: how do we optimize outdoor exploration for secondary ELA curriculum?

Nature-based learning (NBL), a cross-curricular model, can be successful in informal, formal, or non-formal ELA settings through interdisciplinary learning (Jordan & Chawla, 2019). Using outdoor elements of landscapes, NBL connects learners with real-world opportunities (Newman, 2020). When used in education to increase positive attitudes and ensure its permanence, NBL helps learners value elements of nature and have positive attitudes towards it (Sevin, 2020). This significantly benefits environmental attitudes and consciousness (Nkoana, 2020), guiding leaders to think holistically, act sustainably, and care more about the natural world around them (Jin-Hyok et al, 2020).

This is not a new idea, I discovered; some teachers and informal pioneers have taught outside for decades (Watson & Doué, 2007). In fact, progressive educational philosophers like John Dewey have promoted outdoor learning experiences for centuries. Observations reveal most schools consistently limiting perceptions and accessibility of outdoor learning to science, technology, engineering, and math (STEM) but subjects like ELA rarely were considered. These common misconceptions dominate hegemonic views on a systemic level where settings like schoolyards limited to STEM. This issue is deeply rooted in the outdated Prussian educational model still used by most American school systems today (Glavin, 2017).

So, how is going outside related to teaching English? Various studies reveal nature having positive effects on learning and development. Research shows sitting indoors all day is bad for internal systems, or what Bell & Ashwood define as one's *in*-vironment (2016). However, seldom an English teacher considers either their own or their learners' *invironments*. Unfortunately, many keep students indoors and in front of screens rather than encouraging natural, unstructured play and exploration. As virtual learning became a powerful norm worldwide after Covid, the majority more often strayed indoors for hours, expected to sit quietly. Unfortunately, this is often part of daily routines without much encouragement for independent movement or exploration beyond their desks, let alone outside. Nevertheless, most youth will choose to go outside because of an innate desire for the outdoors (Alber, 2016).

Other pathways like environmental education, environment-based education, place-based learning, project-based learning, inquiry-based learning, climate education also suffice due to their nature-related skills, behaviors, and attitude gains (Sevin, 2020). Active research and documentation show NBL is the best option for improved teaching and learning in ELA (Hurley, 2009). Incorporating NBL into ELA is the best method since it assists in stronger communication, listening, speaking, reading, and writing skills by providing learning opportunities to explore the world (Gere, 2009).

Within the scope of language arts, NBL is best defined as utilizing outdoor exploration for literacy-based instruction; it enhances growth. NBL is also useful for interdisciplinary learning when incorporated because it allows opportunities to develop problem-solving and decision-making skills about global topics or challenges. NBL experiences reinforce comprehensive skills like creativity, collaboration, growth mindset, citizenship, and critical thinking. NBL can be incorporated into ELA: it paves the way for multiple instructional methods and engaging experiences. So, where does the English teacher start? Collaboratively or independently initiating NBL into their lesson plans, of course; anyone can use this as an alternative approach to traditional learning models found in most of today's schools.

As our world continues to deal with a burning climate crisis, faulty leadership (Klein, 2019), and gaps in opportunities, Jorgenson et al (2019) encourage educators to adopt strategies and visions like NBL. Therefore, NBL is the best way to connect ELA teachers and students to many different outside settings and perspectives. This is mainly because NBL offers increased rigor and relevance (Jordan & Chawla, 2019). By promoting environmental action in new contexts via energy education and climate change, ELA stakeholders can address multi-actor networks, collective action, and socio-technological innovation in order to guide transitions (Jordan & Chawla, 2019) in their respective communities. This movement towards social and environmental change comes from diverse perspectives, commitments, and motivations (Derr, 2020). Jordan & Chawla (2019) argue that a well-designed nature-based program should be accessible to all regardless of a child's background or forms of identification within society. They advocate for deeper comprehension of various forms of contact with the land to enhance development and learning in order to guide policy decision-making and practice.

Applying NBL within ELA instruction is not limited to the United States. Globally, educators are finding ways to practice NBL despite a historical lack of outdoor exploration in the typical ELA classroom and other public learning environments so students learn to value various elements of the earth (water, soil, etc.), becoming more likely to take care of and have positive attitudes towards it and each other. In Turkey, Sevin (2020) found nature education projects incorporating ELA skills like research and writing increasing positive attitudes towards the planet and ensuring its permanence (Jin-Hyok et al, 2020). These secondary-level programs contributed significantly to both environmental attitudes and consciousness of individuals involved. Similar results were found in Africa (Nkoana, 2020).

Related methods and terms function within ELA-NBL curriculum development and learning experiences. Foundationally, Richard Louv (2005), author of *Last Child in the Woods*, coined the term Nature Deficit Disorder (NDD) almost twenty years ago. Closely related to Attention Deficit Disorder (ADD), Louv's concept of NDD draws parallels between restlessness and lack of natural views in a learner's daily life (2005). This text has become required reading material for educators and parenTs saving our youth from NDD or addressing ADD in alternative ways. Any ELA teacher or related stakeholder not familiar with it should pick up a copy and start reading as soon as they can.

To avoid NDD and develop NBL in ELA curriculum, we should use accessible land, like a school yard or potted plants. Nancy Striniste, EarlySpace founder, says adults are former children, so it is our responsibility to arrange spaces establishing the right to explore (2020). Similarly, Cheryl Charles (2009), co-founder of Louv's Children & Nature Network (CCAN), argues for natural guides for building positive social change. Ultimately, this can be incorporated into language arts curriculum, from news articles to character analysis. Disabled students also suffer from a lack of exposure to nature; Striniste (2020) advocates for considering every learner when planning activities that involve going outside.

Purposeful engagement can develop and reinforce by exploring vital topics like climate change while considering relationships beyond classrooms which helps emphasize global studies and tending to the planet (Stapleton 2019; Nkoana, 2020; Ojala, 2012). Themes specific to climate issues that apply include engagement of deliberative discussions, interaction with real scientists, addressing of misconceptions, and implementation of community projects (Monroe et al, 2019). Furthermore, outdoor learning can engage youth and their families with nature (Stiniste, 2019; Deringer et al, 2020). Advocates for climate education also support programs that incorporate ELA through integration of environmental literacy and global citizenship (Suryawati et al, 2020; Muradova et al, 2020; O'Neill et al, 2020, Husamah et al, 2020). One example of a program incorporating language was the high school Green Team summer program in 2018 in Buena Vista, VA, organized by Elise Sheffield (2020), education director at Boxerwood Nature Center. Interns applied competitively and developed skills related to energy conservation, waste management, and water use in the Chesapeake Bay watershed.

Learning is a process that continues throughout life. The environment, a key component to integrated outdoor discovery, allows historical perspectives to be more succinct amongst youth and teachers alike and supports intergenerational learning approaches (Russ & Kransy, 2017). In this way, ELA stakeholders can plan alongside other disciplines like social studies and science to collaboratively optimize interdisciplinary lessons or engaging cross-curricular research projects while maintaining culturally responsive practices. According to Steele & Scott (2016), this pathway for including language skills for learners also benefits social emotional learning needs. Essentially, unstructured outdoor time alongside loved ones can build confidence. Likely to encourage ownership, control, and problem-solving, NBL can also require teamwork, cooperation, and negotiation. Outdoor learning can be utilized in intergenerational learning (Russ & Kransy, 2017). Boxerwood, for instance, developed family-friendly programs and workshops on topics like poetry, beekeeping, and nature sounds (Sheffield, 2020).

Similarly, placed-based learning (Kudryavtsev et al, 2012), which has commonalities with NBL, can be implemented along a continuum. It starts with understanding the self: one's communal role in the concentric rings of classroom, school, community, nation, and world (Getting Smart, 2020). The concept of place can be interpreted socio-politically, economically, and ecologically, so its scale can also be understood through these lenses, allowing for deeper transfer, understanding, and application of skills and content (Russ & Krasny, 2017). Providing more relevance to teachers and learners through problems, projects, entrepreneurship, service learning, community-centered design, and inquiry can help reimagining possibilities of next-generation learning (Weinstein et al, 2015). It can also play a role in a learner's impact and understanding of place through their social networks, who they follow, and groups they like, which all connect their physical places to their global digital ones (Getting Smart, 2020).

Different people can see the same place differently, so educators should recognize various aspects a place can evoke or how it can influence one's well-being, interactions, descriptions, emotions, values, respect, and perceptions (Steele & Scott, 2016). Connections enrich and revitalize local areas, benefiting the community (Weinstein et al, 2015). When combined with personalized learning models, highly functioning, engaged and informed citizens are manifested, whether rural or city (Russ & Krasny, 2017).

In *Nature Play at Home,* Striniste (2019) says that childhood has changed. Therefore, by inspiring imagination through crucial and foundational play and exploration, we can see more clearly the value of pretending. Like in literature, many characters are not who they seem, so these unlimited pathways to personalization let students experience communal ownership for continued improvement and sustainability (Getting Smart, 2020). Place-based learning sits at the heart of educational ecosystems considering NBL because it is a learner-centered approach that uses next-generation tools and allows integration within community functions (Weinstein, 2015). This is an exceptional foundation for a society not just democratic, but also participatory and informed. Educators, researchers, and other stakeholders are just starting to comprehend how to implement and scale place-based learning with the intention to deliberately boost access, outcomes, and equity (Getting Smart, 2020).

Although some project- and place-based learning strategies do provide access to nature, NBL is unwaveringly the most effective tool for getting learners to adapt to natural area resources through outings, activities, and family participation (Santamaria, 2020). When applied in Indonesia to play a community game with English language learners, Fajaria (2013) found that this enhanced the learning process of topics like cuisine or tourism: 90% of her participants improved general English proficiency and 85% had many ideas related to nature after participating.

Certainly, project-based learning provides the most additional connections to NBL (Sheffield, 2020; Suryawati et al, 2020). An active methodology for learning meant to teach issues in a social context through projects (Santamaria, 2020), it is based on John Dewey's idea that learners need to be involved in the environment around them to be socialized into active citizens. Participation is key, encouraging responsibility and autonomy as roles are centralized in the process. Therefore, both informal and formal stakeholders can incorporate projects into related programs.

According to Russ & Krasny (2017), communities interested in environmental arts can also augment powerful aesthetic experiences and stimulate inventive environmental activism, since creativity generates attention and offers ingenious solutions for environmental challenges. Otherwise called artistic or creative activism, "environmental arts are becoming part of the curriculum" in museums, universities, schools, and community centers (2017). As technology advances, well-designed programs like these should encourage questions and investigation (Santamaria, 2020). This will help awaken a student's scientific spirit that is increasingly latent in schools today, motivating them to participate more than what is currently typical in the classroom (Warner, 2015). Santamaria's (2020) research models positive coexisting relationships, such as when parents, neighborhood gardeners, or local experts become actively involved in youth learning. Santamaria's (2020) research touches deeply on science learning (reinforcing the misconception that STEM is the only avenue for secondary NBL frameworks), but clearly instructors can also utilize project-based lesson plans and incorporate opportunities while reinforcing essential concepts and skills like writing, vocabulary development, the research process, or reading comprehension.

Overall, the goal of incorporating NBL in the ELA curriculum for older learners clearly is to stimulate interest plus understanding the importance of asking questions and thinking critically. For example, Elise Sheffield (2020) at Boxerwood Nature Center in Lexington, VA, partners with community members while using writing, reading, listening, and speaking skills during programs in order to drive inquiry and investigation. Sheffield (2020) gives three reasons to utilize actionable Eco-Project-Based Learning (E-PBL) as "the main course, not dessert!" at Boxerwood: 1) to sustain our world, 2) for our society, and 3) for our schools. Together, these enhance the profile of a High School Graduate or 21st century learner (collaboration, critical thinking, communication, citizenship skills, and creative thinking).

Additionally, Sheffield (2020) advises that stakeholders start with an investigative actionable question simple enough for students can parrot back. Alongside Sheffield, community partners reinforce these basic inquiries as they work together with her and/or her team to incorporate VA's standards of learning (SOL) content, work/skills, and career development in order to execute a successful program collaboratively. Therefore, nature-based educational opportunities, when they incorporate community partners, can greatly enhance interdisciplinary connections regardless of rural or urban setting (Hays & Ledford, 2020). Objectives are centered around constructive inquiry and investigation, which are most effective in the overall learning process regardless of subject area (not just science) when they involve relevant issues, are goal-oriented, remain flexible, and are context-specific (Santamaria, 2020).

According to Sheffield (2020), stakeholders must focus on one community and take care of their Nurturing Environmental Stewardship Together (NEST) to make the action journey real through partners, staff, and volunteers from within the locality who work with youth in classrooms, alongside waterways in programs and camps. This incorporates the community framework and school surroundings (Broda, 2002). Progressivism, driven by student questions and interest arising from their experiences, encourages experimentation and shared decision-making on student-selected topics; books become tools rather than the learner's authority. Through dialogue, inquiry, and multifaceted perspectives on controversial issues, students can participate in community-based learning (Cohen, 1999).

Similarly, Calis & Yildirim (2020) discovered that project-based prediction, observation, and explanation have a significantly positive effect on students' attitudes and behaviors towards nature in contrast to teacher-centered instruction. NBL methodology supports development of positive attitudes and behaviors in youth and can facilitate growth within various curricula utilizing language arts. When woven into the fabric of public spaces, language can also encourage intergenerational pathways for family learning about the earth and each other (Harris, 2016).

Other findings reveal youth at the forefront of outdoor-based research. Recently, Jorgenson et al (2019) discovered that researchers and environmental educators are focusing on promoting and researching energy conservation behavior while emphasizing youth and children. In addition, research has started focusing on the transitions of energy at both national and regional levels (Jorgenson et al, 2019). At the middle and high school level, research skills are an important element of the ELA curriculum, so energy use is a hot topic that students could explore (Keskin et al, 2020). The attitudes that teens themselves have about energy use should be recognized too (Toth et al, 2013).

In the end, many terminologies, strategies, and methods are related to nature-based learning (NBL) and English Language Arts (ELA) instruction. Educators within the professional ELA community and beyond who are interested in moving towards a NBL model should familiarize themselves with these terms as they navigate the waters of connecting the outdoors and secondary instruction.

A BRIEF GUIDE ON HOW TO TEACH WRITING SKILLS WITH NATURE-BASED LEARNING

Many would agree that teaching the writing process is one of the most challenging parts of language arts instruction. Believe it or not, this skill is also one of the best ways to connect students to nature. Therefore, concrete ways are provided in the table below for various kinds of writing skills (see Tables 1, 2, 3, 4, 5, and 6) that teachers, parents, youth leaders, administration, and other related stakeholders can utilize NBL directly in English language arts (ELA) curriculum at the secondary level (Watson & Doué, 2007). Essentially, this next section incorporates written communication skills, narrative writing, expository writing, persuasive writing, reflective writing, nature journaling, research-focused writing, and blogging. There are also suggestions for creative writing (poetry, short stories, prose, etc).

Blogging

Another way to publish writing is through blogging. Students can create public texts for diverse audiences about a variety of environmental topics in online blog platforms. These blog posts can be part of group assignments or for individual learners. Students can blog about different endangered species and discuss issues behind this topic. Specifically, they could write about what events or changes have to occur before recovery projects could begin, how the loss of these species would affect people and the environment,

Table 1. Narrative writing

Writing Skill in Focus	*Essential or Guiding Questions*	*Real Example, Lesson, Activity, or Experience*
Narrative Writing Students can write narrative paragraphs or essays while outside in response to questions.	What happened here 10, 50, or 100 years ago? What does this location remind me of? Does this natural area bring up something from class for you? How could this spot be seen from the perspective of an animal, plant, or another person? What are the meanings of nature-related symbols used by locally or regionally based indigenous people?	**Spotlight: Mythical Creatures** One way to explore the meanings of symbols is through narrative writing. Students can create cartoon characters from a mythical country, then, through captions and analysis, write a narrative story about their characters' experiences. These could be shared through a class book, magazine, or blog. Or, students can record themselves reading these stories aloud, as well as publish them, or share what they wrote with others. Teachers should seriously consider finding ways for combining this exploration of symbols with other courses at the school such as social studies. Civics lessons, for example, can demonstrate the similarities and differences of different cultures and countries that exist around the globe.

whether loss and/or recovery are important and why, which conflicts are involved in recovery, who official players involved are, how public perception influences decisions related to these issues, why it is important that local people understand the species' natural history and its role in the ecosystem, and what is currently being done (or what still needs to be done) to protect a species in danger.

Writing Newspaper Articles

Writing newspaper articles for school projects can allow incorporating the environment into the ELA curriculum through journalism. After analyzing diverse issues related to aquatic creatures and their habitats, students can develop opinions about related issues, then communicate them in writing. Students should be encouraged to look at real models and analyze different sections of a newspaper. They should also be able to research, photograph, and visit local places that offer resources, then gather information and properly credit any sources they have used. Once finished with their articles, students can work together in teams to put their writing together into a template and distribute it. Plus, students can establish a corner for current events or visit a nearby newspaper and ask for their articles to be submitted for public view. Students can also continue beyond one publication and publish weekly, monthly, quarterly, etc. Newspapers can be published online as well. Additionally, students can send their article to real newspaper companies for feedback and suggestions on how to improve their work. These outlets may

Table 2. Reflective writing

Writing Skill in Focus	*Essential or Guiding Questions*	*Real Example, Lesson, Activity, or Experience*
Reflective Writing Reflective essays, journal entries, and even paragraphs can all be used to help students think about different experiences or ideas.	What did you like and/or didn't like about today's lesson? What did you learn during your outdoor adventure? Your connection(s) to the world around you are . . . ? How did a lesson about caring for the earth change your perspective? What is a past experience you have had outdoors? What future experience would you like to have outside of the classroom?	**Spotlight: A Code of Ethics** In this example, students can reflect on their own philosophy in regards to the environment, then create a personal enviro-ethics code. Students can also create a life map in order to share their ideas with a peer, the class, or community. Students can also keep a journal to continue reflection and add to their initial ideas. This demonstrates to students during the learning process that our ideas and beliefs are not meant to be stagnant, but are constantly changing, so we should always be open to new information.

even consider publishing students' work (with permission), allowing for further real-world connections to issues like climate change (Verlie, 2019).

Students can also choose an issue concerning aquatic life or other environmental issues in their community and write an article for a local newspaper that reflects high standards and professional quality. For example, youth from several schools in northern Moldova, a small country in Eastern Europe, participated in a recycling project that incorporated environmental issues such as why open dumping is bad for local aquatic ecosystems. After student leaders were trained in dealing with these ecological issues, they returned to their school communities to educate others and start a recycling program on campus. From that point, participants had the chance to write short explanations of the activities that took place. Obviously, this can be adapted to whatever the needs are of your own school and surrounding community. All it takes is a quick community assessment of what is already working and what needs to be improved.

For a different kind of challenge, students can learn how to write an editorial (an opinion piece). This could be about an advertisement they created that incorporates wildlife that they share with a local newspaper. Similarly, they could create advertisements for protecting wildlife to go along with these editorials. They can also develop opinion pieces about topics like school recycling or composting programs (whether they exist, if they are working, what could be done better, and what is already being done well). This could be in order to encourage a decrease in material and food waste on campus or in the community. These can be used also in order to encourage others to recycle materials rather than throw them away. Editorials can also be shared with local TV and radio stations interested in interviewing students for a morning or evening report. After all, how neat would it be for students to be on television or the radio? Students can also work with recycling organizations to learn more about the process in their community.

Creative Writing

There are lots of ways to incorporate creative writing and nature. Short stories, poetry, and prose can all allow students to express their experiences with the outdoors (Schuster, 2013). A plethora of activities involve creative writing and nature-based activities (see Table 3). Many teens will even choose, when given the option, to write outside for their own purposes (Henderson, 2013).

One way students can creatively incorporate nature into short stories that they write. They could write an ending to the story called "The Twins" by Dr. Clifford Knapp and Suzanne Iudicello. After reading and discussing the story, students can write their own ending for it. This allows an opportunity to describe feelings students have about hunting, compare attitudes to others', and develop judgments about whether hunting is appropriate. Teachers can adapt this lesson to any story that incorporates nature by removing the ending and letting students use their imagination to lead the way.

Another avenue is songwriting and lyric analysis. After listening to a song about a strong environmental statement, students can analyze lyrics and/or write their own about a specific issue, an animal of preference, or how they feel about nature. They can share their song with classmates, friends, family, or any others interested in hearing it. As an extension, students may look for songs that incorporate negative or positive vocabulary about nature, such as words that are harmful or hopeful. Students can even look out for songs specifically calling for action or review music videos with environmental themes.

Here is another option: after discussing how animals adapt, students can write stories about how they see the world and why. Students could create a short story or biography about a local fish. Or, they can write a poem or song about a plant or animal facing extinction. By comparing and contrasting the

Table 3. Descriptive writing

Writing Skill in Focus	*Essential or Guiding Questions*	*Real Example, Lesson, Activity, or Experience*
Descriptive Writing Students can describe in writing several many nature-based experiences	How do simulated field trips of riparian places build new vocabulary? What is a detailed description of a local fish species? What attitudes about the environment exist? Animals we like or don't like, our first impressions of them, and how our impressions of them are changed as we learn more about them through research are . . . ? A great plot for a mythical story about a real aquatic creature would be . . .? Research and describe an animal or water source. What are various reasons and ways water is important to both people and wildlife? Compile a class book of short stories or poetry. Who can we share it with?	**Spotlight: Nature Journaling** Students can connect journaling and nature in order to be engaged with the outdoors. Using scientific-based writing to record weather, time of day, moon phase, direction, etc. allows students as writers to enjoy documenting their time outdoors or appreciate simply recording observations while sitting quietly. Descriptive writing allows nature to be captured while students are observing it. Figurative language, for instance, has been used by countless poets, novelists, and beyond to describe the natural environment around them. Writers can share personal experiences through narrative or reflective writing by telling a story about time outside or reflecting on a past outdoors memory. In a similar activity, students can describe in their journals the differences they see in the environment that are the result of both casual and detailed observations, then give reasons for why looking closely at any environment is important. Originally designed for younger students, this is still applicable for secondary learning. Students do this by writing what they remember seeing in a setting that is familiar, then checking how accurate they were and discussing the results. After that, students apply what they experienced and their new skills in order to think about an outdoor setting that is unfamiliar to them. Further extensions could consider patterns, shapes, categories, living things (plants, animals, etc.), and distinguish between qualitative and quantitative observations. Students interested in fishing or who fish regularly, for example, can create journals to record experiences while fishing. These can include a chart to record dates, times of fishing, weather, the name and location of the water body, what kind of bait and gear they used, the kinds of fish that were caught and their sizes, whether any fish were kept or released, how the fish were cooked, and how they tasted. Students should be able to leave space for recording descriptions of their experiences and any interesting observations. They can hold images and different words, plus artifacts like sand or leaves. Students can also select animal habitats and experience them as closely as possible without disturbing them, then describe that in detail. They can even connect their journals to aquatic creatures. Thinking about early explorers in North America who traveled by waterways such as LaSalle, Lewis and Clark, and others and journaled about what they saw, students can pick a time period and imagine being a brave explorer venturing into territory unknown. Students can write imaginary accounts of what they see, such as wildlife, terrain, vegetation, water quality, and their feelings about hair-raising adventures.

experiences of a human versus an animal, students can even create short stories about how one or the other would experience a storm, tornado, hurricane, earthquake, or other natural disaster.

After learning about different forms of poetry, students can also write their own poem and focus on elements of nature like water throughout their entire writing process. Forms can include haiku, cinquain, diamante, or prose. If desired, final poems can be displayed in a circle that forms around a map of the world. Teachers can connect poems with yarn along with where their bodies of water are located internationally. Additionally, teachers can lead discussions on metaphors found in poetry about nature. One example is talking with students about the meaning of "one body of water" which can emphasize the idea that all waters of the planet connect together (Aquatic Wild, 2016, pg. 163).

Animals are another great topic. Upon going outside and imagining themselves as animals, students can write poetry about their ideas. This imaginative expression in literary form allows students to be affected by the imagery and sound of language. As an extension, poems can be arranged onto a slideshow using digitized presentations along with photographs of animals or clipart. This can also be adapted for other critters such as insects, spiders, fungus, or plants.

Research-Focused Writing

Students can write about a variety of nature-based topics while using information that they found during the research process (see Table 4). Examples include writing an expository essay about the inner workings of something in nature, which students can connect with what they have learned in science class. Or, students could write a persuasive essay about something regarding nature, such as ways to protect a habitat or how to start a new recycling program. During the research process, students can also interview people or local experts, write down those ideas, and incorporate them into their final essays, articles, etc.

Students can write biographies about locals who have advocated for the environment in some way after interviewing them. If desired, they can even create visual displays of finished products around the school or digitally on a website. They could even form a group to address related problems and include these people in the conversation. Students could also view various presentations or films about other conservationists. Another extension would be to share a copy of their biographical reports with the school librarian or public library in order for them to be included in local collections, since these could be a significant contribution to local history. Students could also get in touch with nearby historical societies or community organizations interested in local history for further support and collective action.

A great way to collect information for a research project is through interviews. For example, students interviewed local experts (gardeners and scientists) during the project "Pollinator Posters" at Genesis Alternative School in Staunton, VA. These experts were delighted to interact with the students and share their knowledge, and the students had a great time meeting people from the local community.

Interviewing people about their experiences is a wonderful option students can use to collect information for various activities. From a research paper to a narrative memoir, students can schedule conversations that provide insight to a topic of interest, whether about the history of school-based nature walks, life in rural communities, protection of endangered wildlife, or the health of a local waterway.

RESEARCH WRITING SPOTLIGHT: WE'RE IN THIS TOGETHER

After interviewing at least ten people using a student-generated list of questions about discussed issues in the environment that affect wildlife, learners can rank mentioned concerns from most to least. Then, students can categorize concerns based on whether the problems affect people or wildlife. They can also

Table 4. Persuasive writing

Writing Skill in Focus	*Essential or Guiding Questions*	*Real Example, Lesson, Activity, or Experience*
Persuasive Writing By effectively persuading audiences to agree or disagree with a position is one way students can use writing to incorporate nature.	Why should we protect a local habitat? Why should endangered species found be protected near you? Why include many perspectives when taking care of nature? Have you ever been to visit a specific nature center? What kind of conservation messaging can we use to convince others to act on an environmental issue? What can we do when outdoors to improve one's (or water's) health? How do we install a small pond at school? Which differences matter to you the most in letters to interest groups? Why should we agree with your view? What are you advocating for?	**Spotlight: Legislation and Wildlife** Students can incorporate nature into writing it through position papers about current legislation involving wildlife. After researching the issue, students should write an essay persuading their audience to agree with them about how legislation should be handled. If desired, students can send polished essays to local representatives in a letter as a demand for action for or against the issue. Students can also share publicly through presentations, blogs, podcasts, or other communication to encourage more discussion. John Muir, author of *My First Summer in the Sierra,* is a writer famous for his passionate nature journaling. He wrote detailed notes and observations about his time outdoors; he also drew illustrations to support his writing.These led to changes in legislation about wildlife thanks to a friendship he formed with Teddy Roosevelt while in Yosemite wandering the mountains. Teachers can introduce similar opportunities regarding research or discussion that analyzes different perspectives. The combination of nature journaling and research about naturalists from the past- and present-day, helps students analyze text written by these people. Next, students can start their own journals. They can also compare theirs to those written by naturalists to find differences and similarities, like how nature is exemplified. Students can even mimic a short selection from one a naturalist's journals while describing their own environments instead. Therefore, the value of journaling should also be discussed, since one can log important data and keep record of feelings, observations, and impressions then refer to them later. These can also reflect changes happening in ecosystems, animal populations, and vegatiative types, or attitudes about those topics.

Table 5. Expository writing

Writing Skill in Focus	*Essential or Guiding Questions*	*Real Example, Lesson, Activity, or Experience*
Expository Writing Students can write in order to explain about topics, themes, and real-life experiences or issues.	What connections exist between animals and humans? Why is it important to take care of natural resources? What are some ways to care for habitats of endangered wildlife species? How can we include diverse voices in addressing environmental issues? Which details of a healthy or unhealthy wetland ecosystem do we know or not know? What is the story of a wildlife critter facing extinction? How has your point of view developed over time about an impacted habitat? How have you been affected by changes alongside evidence supporting your perspective based on observations?	**Spotlight: Wildlife Influences and Values** After reading various articles about examples of how wildlife has influenced societal human development, students can possibly create an annotated bibliography of an element based on that. From there, students can write about the influences and values that wildlife has on human societies. This can be combined with civics or social studies class by making connections to how society develops value systems or how those can be influenced and changed. To enrich further, students could record essays in a podcast or do a presentation to share findings. This can also be done in combination with a persuasive call to action by students to protect local wildlife by offering their advice on best practices with the public.

analyze the components of a habitat and decide whether the same issues affect both wildlife and humans. Typically, students discover that wildlife and people share the same needs.

Later, students can summarize their discoveries and emphasize that we are all in this together. Or, they can write a short story about the same topic. Students can also pick a specific country to research how one of its citizens would answer this activity's questions. As a further extension, the class can publish a book or magazine with the results of their interviews as well as the short stories or essays.

Creating a narrative memoir about real-life experiences is another way to connect the outdoors to writing. This is particularly meaningful for both interviewer and interviewee when one is a generation or more older than the one doing the interview. Fresh perspectives can be gained from these conversation(s). Below is an excerpt from "Growing Up in Nature: from Elsie's Memoir" written by her granddaughter, KB, in 2021. This exemplifies how the interviewer (KB) incorporates interwoven experiences with nature that the interviewee shared.

By reminiscing on memories during childhood, Elsie provided a vivid look into growing up as a white red-headed female in rural Appalachian Virginia during the 1930s (Table 6 below):

CONCLUSION: WHAT DO WE DO NEXT?

Clearly, there are many ways that writing can be incorporated into nature-based learning (NBL). Similarly, there are a variety of opportunities for English language arts (ELA) teachers to include the outdoors into writing assignments (Watson & Doué, 2007). Nevertheless, any English teacher knows that English class is more than just writing, and that writing is best taught alongside reading. However, our current educational system, at least in the United States based on my research and experience, is not supporting NBL. Despite typically being aware of most of these concepts, the leaders of our current public school systems are often not supporting NBL or related learning opportunities for students, especially not in secondary ELA learning environments or during active campaigns for new curriculum development initiatives. Such continued assimilation of one-size-fits-all approaches nationwide greatly inhibits any kind of diversity initiatives in America (Lash, 2018) like braille trails (Nature for the Blind, 2020) or provenly effective Montessori methods. This cultural dissonance is also making it difficult to resolve environmental justice and environmental racism across the planet as well (Holifield, 2001). Ultimately, we need NBL and EE to include a critical consciousness of race in place-based learning too (Miller, 2018).

Research amply suggests that sustained time indoors is unhealthy and potentially harmful and that outdoor activities are good for our health (Louv, 2005; Striniste, 2020; Lieberman, 2013; Russ & Krasly, 2017). After all, we are all wired to be outside (Worrall, 2017). Yet, many educators remain simply afraid or still uncomfortable with taking students outside (Grosvenor & Myers, 2020; Glackin, 2018; Connolly & Haughton, 2017; Louv, 2016). So, it is not surprising that today's youth are increasingly detached from nature due to increased screen time (Newman, 2020; Kahr, 2020; Louv, 2005; Schuyler, 2020). Nor is it a shock that many still struggle with the idea of climate change despite the vast research supporting it. Ultimately, it is my hope that NBL will be more greatly incorporated into future ELA curriculum for secondary learners worldwide. Leiringer and Cardellino (2011) are not alone when they advocate for transformations within our school systems so that we are applying more effective designs for all learners in every discipline for years to come.

Table 6. Memoir Writing

. . . Elsie's family had an outside toilet they used in the warmer months. "Course it had a door on it and a wooden latch," Elsie explained. "You kinda just lifted it up." When one hole for an outhouse filled up, someone in Elsie's family, typically one of her brothers or her father, would dig another one. According to Elsie, people who had two holes tended to have more money. Based on observations growing up, she paid attention to her neighbors' toilets too. Elsie soon realized that even though her family was not as poor as other families, they had one big issue with their outdoor toilet. "Them rich families had a two-holer," Elsie exclaimed with a mischievous grin as she continued. "But we only had a one-holer. Ain't but one person could use it at a time." Elsie and her family heated their log cabin with various kinds of firewood. When it was chilly, her older brothers, Clayton and Howard, rigorously chopped up wood daily. They would cut down trees on the property and drag them down to the house before beginning the chore of chopping. Despite splinters, sweat, and dehydration, her brothers and their father, Renard, would return once more to the chopping block each wintery day to prepare wood for the stove. Elsie was certainly a troublemaker, but she had always been a thoughtful gal too. Recognizing this woodchopping chore was a difficult and strenuous task for her father and brothers, Elsie became determined to go out of her way to provide assistance, but she wanted to make sure that her aid was a charming surprise. Ready to make a name for herself, she got the bright idea one day to do just that. Making sure her brothers were not around to notice what she was planning to do (their family didn't have a vehicle), she watched attentively as they caught the Greyhound bus to help their father with some errand-running in the nearby "large" city of Lexington. Once her brothers and father had boarded the bus and were off the property, Elsie and her cousin Polly carefully gravitated over to the wood pile. Since Elsie felt like she accurately remembered how she had seen her father and older brothers perform this task time and time again successfully, she was confident she could perform the same work. Before Polly could say anything to change her mind, Elsie got her hands on the ax her father and brothers used for chopping. Then Elsie lifted the heavy ax up high and swung it down onto the log she had placed a moment ago on the chopping block. Swinging and swinging, she continued to strike the pieces of firewood as she used her memory of her dad and brothers' process as her guide. Elsie was so concentrated on completing her self-elected chore that she didn't notice her cousin Polly slowly moving closer to the chopping block. Before Elsie knew it, Polly was within swinging range. A second later, Elsie nicked Polly in the neck. Barely, thank goodness. Thankfully, there was no severe damage done to the poor girl. However, as the blood began trickling down, dripping off of the ax and onto the freshly cut wood, Elsie realized that she needed to take further action. She immediately took Polly home so that she could be attended to properly by an adult. When Polly's mother saw the injury, she was absolutely furious. However, Polly's older sister said, "At least she had the gumption to bring Polly home!"

REFERENCES

Alber, R. (2016). *New Teachers: Inspire Your Students to Write, Write, Write*. Edutopia. https://www.edutopia.org/blog/new-teachers-inspiring-your-students-write-write-write-rebecca-alber

Bell, M., & Ashwood, L. (2016). *An Invitation to Environmental Sociology. Fifth Addition*. Sage Publications, Inc.

Broda, H. (2002). Learning In and For the Outdoors. *Middle School Journal*, *33*(3), 34–38. www.jstor.org/stable/23047005. doi:10.1080/00940771.2002.11494672

Busch, K. C., Henderson, J. C., & Stevenson, K. T. (2019). Broadening epistemologies and methodologies in climate change education research. *Environmental Education Research*, *25*(6), 955–971. doi:10.1080/13504622.2018.1514588

Calis, D., & Yildirim, H. I. (2020). The Effect of Prediction, Observation, Explanation Supported Project-Based Environmental Education on the Levels of Attitude and Behavior Toward the Environment. *Educational Policy Analysis and Strategic Research*, *15*(1), 22–43. doi:10.29329/epasr.2020.236.2

CCE. (2019). *Finland Education Model*. Council for Creative Education. https://www.ccefinland.org/finland-education-model-c1yvm

Charles, C. (2009, October 2). The Ecology of Hope: Natural Guides to Building a Children and Nature Movement. *Journal of Science Education and Technology*. https://link.springer.com/article/10.1007/s10956-009-9193-z?error=cookies_not_supported&code=e548e227-9255-4ed1-979f-25a6849262ea

Cheeseman, A., & Wright, T. (2020). Perspectives from parents: An investigation of the longer-term benefits of an earth education summer camp program. *Applied Environmental Education and Communication*, *19*(2), 117–128. doi:10.1080/1533015X.2018.1530621

Choe, J.-H., Kim, C.-H., & Ri, G.-H. (2020). An investigation on the environmental knowledge and attitudes of senior middle school students in the Democratic People's Republic of Korea. *International Research in Geographical and Environmental Education*, *29*(2), 146–162. doi:10.1080/10382046.2019.1678276

Cohen, M. (1999). *Section III - Philosophical Perspectives in Education*. Oregon State. https://oregonstate.edu/instruct/ed416/PP3.html

Connolly, M., & Haughton, C. (2017). The perception, management and performance of risk amongst Forest School educators. *British Journal of Sociology of Education*, *38*(2), 105–124. doi:10.1080/01425692.2015.1073098

Deringer, S. A., Hanley, A. W., Hodges, J. S., & Griffen, L. K. (2020). Improving Ecological Behavior in Outdoor Recreation Through Mindfulness Interventions: A Mixed Methods Inquiry. *Journal of Outdoor Recreation, Education, and Leadership*, *12*(2), 2. doi:10.18666/JOREL-2020-V12-I2-9802

Derr, V. (2020). Diverse perspectives on action for positive social and environmental change. *Environmental Education Research*, *26*(2), 219–237. doi:10.1080/13504622.2020.1715925

Fajaria, I. (2013, January). Outdoor Class Project: The Potential Benefits to Foster EFL Students' Language Proficiency. *Advances in Language and Literary Studies*, *4*(1), 140–144. Retrieved March 22, 2020, from https://eric.ed.gov/?id=EJ1129682. doi:10.7575/aiac.alls.v.4n.1p.140

Gere, A. R. (2009). *Writing Outside of School*. NCTE. http://www.ncte.org/library/NCTEFiles/Resources/Journals/CC/0191-sep09/Chron0191BriefWriting.pdf

Glackin, M. (2018). 'Control must be maintained': Exploring teachers' pedagogical practice outside the classroom. *British Journal of Sociology of Education*, *39*(1), 61–76. doi:10.1080/01425692.2017.1304204

Glavin, C. (2017). *Prussian Education System*. K12 Academics. https://www.k12academics.com/Education Worldwide/Education in Germany/History/prussian-education-system

Grosvenor, I., & Myers, K. (2020). "Dirt and the child": A textual and visual exploration of children's physical engagement with the urban and the natural world. *History of Education*, *49*(4), 517–535. https://proxy.marybaldwin.edu:2238/10.1080/0046760X.2019.1701097. doi:10.1080/0046760X.2019.1701097

Harris, K. I. (2016). Let's play at the park! Family pathways promoting spiritual resources to inspire nature, pretend play, storytelling, intergenerational play and celebrations. *International Journal of Children's Spirituality*, *21*(2), 90–103.. doi:10.1080/1364436X.2016.1164669

Hays, D., & Ledford, L. (2020). *Collaboration for Impact: Modeling an Integrated Partnership*. Virginia Association for Environmental Education Conference. Amherst, Virginia.

Henderson, P. J. (2013). *A Multi-Case Study of Teens Who Write Outside of School for Their Own Purposes*. TRACE. trace.tennessee.edu/cgi/viewcontent.cgi?article=3634&context=utk_graddiss

Holifield, R. (2001). Defining Environmental Justice and Environmental Racism. *Urban Geography*, *22*(1), 78–90. https://www.researchgate.net/profile/Ryan_Holifield/publication/250171256_Defining_Environmental_Justice_and_Environmental_Racism/links/5694093108ae425c6896255b/Defining-Environmental-Justice-and-Environmental-Racism.pdf. doi:10.2747/0272-3638.22.1.78

Hurley, S. (2009). *The Benefits of Taking Students Outside to Inspire Writing*. Edutopia. https://www.edutopia.org/student-motivation-writing-outdoors

Husamah, H., Miharja, F. J., & Hidayati, D. A. (2020). *Environmental Literacy of Islamic Boarding School Students: Study in MA Bilingual-Sidoarjo*. Prisma.

Jordan, C., & Chawla, L. (2019). A Coordinated Research Agenda for Nature-Based Learning. *Frontiers in Psychology*, *10*, 766. doi:10.3389/fpsyg.2019.00766 PMID:31065246

Jorgenson, S. N., Stephens, J. C., & White, B. (2019). Environmental education in transition: A critical review of recent research on climate change and energy education. *The Journal of Environmental Education*, *50*(3), 160–171. doi:10.1080/00958964.2019.1604478

Keskin, C., Akcay, H., & Kapici, H. O. (2020). The effects of environmental science projects on middle school students' behaviors and attitudes. [IJTES]. *International Journal of Technology in Education And Science*, *4*(2), 160–167. doi:10.46328/ijtes.v4i2.84

Klein, N. (2019). On Fire: The (Burning) Case for a Green New Deal. New York: Simon & Schuster. Kaufman, C. (September 16, 2013). "Time to Write? Go Outside." New York Times. https://opinionator.blogs.nytimes.com/2013/09/16/time-to-write-go-outside/

Kudryavtsev, A., Stedman, R. C., & Krasny, M. E. (2012). Sense of place in environmental education. *Environmental Education Research, 18*(2), 229 - 250. 10.1080/13504622.2011.609615

Lash, C. L. (2018). Making Americans: Schooling, Diversity, and Assimilation in the Twenty-First Century. *The Russell Sage Foundation Journal of the Social Sciences : RSF*, *4*(5), 99–117. doi:10.7758/rsf.2018.4.5.05

Leiringer, R., & Cardellino, P. (2011). Schools for the twenty-first century: School design and educational transformation. *British Educational Research Journal*, *37*(6), 915–934. www.jstor.org/stable/23077016. doi:10.1080/01411926.2010.508512

Lieberman, G. (2013). *Education and the Environment: Creating Standards-Based Programs in Schools and Districts*. Harvard Education Press.

Louv, R. (2005). *Last Child in the Woods: Saving Our Children from Nature Deficit Disorder*. Algonquian Books of Chapel Hill.

Louv, R. (2016). Nature Smart. [World Edition]. *National Wildlife*, *54*(2), 40–43. proxy.marybaldwin.edu:2060/login.aspx?direct=true&db=a9h&AN=112738502&site=ehost-live

Mechler, H. M. (2016). Fostering Critical Thinking in Young Children. *Texas Child Care Quarterly, 40*(2). https://www.childcarequarterly.com/pdf/fall16_thinking.pdf

Miller, H. K. (2018). Developing a critical consciousness of race in place-based environmental education: Franco's story. *Environmental Education Research, 24*(6), 845–858. doi:10.1080/13504622.2017.1357802

Monroe, M. C., Plate, R. R., Oxarart, A., Bowers, A., & Chaves, W. A. (2019). Identifying effective climate change education strategies: A systematic review of the research. *Environmental Education Research, 25*(6), 791–812. doi:10.1080/13504622.2017.1360842

Muradova, L., Walker, H., & Colli, F. (2020). Climate change communication and public engagement in interpersonal deliberative settings: Evidence from the Irish citizens' assembly. *Climate Policy, 20*(10), 1322–1335. doi:10.1080/14693062.2020.1777928

Nature for the Blind. (2020). *What is a Braille Trail?* Nature for the Blind. <http://www.natureforth-eblind.com/what-is-a-braille-trail>

Nkoana, E. M. (2020). Exploring the effects of an environmental education course on the awareness and perceptions of climate change risks among seventh and eighth grade learners in South Africa. *International Research in Geographical and Environmental Education, 29*(1), 1, 7–22. doi:10.1080/10382046.2019.1661126

O'Neil, J. M., Newton, R. J., Bone, E. K., Birney, L. B., Green, A. E., Merrick, B., Goodwin-Segal, T., Moore, G., & Fraioli, A. (2020). Using urban harbors for experiential, environmental literacy: Case studies of New York and Chesapeake Bay. *Regional Studies in Marine Science, 33*, 100886. doi:10.1016/j.rsma.2019.100886

Ojala, M. (2012). Hope and climate change: The importance of hope for environmental engagement among young people. *Environmental Education Research, 18*(5), 625–642. doi:10.1080/13504622.2011.637157

Panula, E., Jeronen, E., & Rodrigez-Aflecht, G. (2020). Nature Is Something We Can't Replace': Mexican Students' Views of the Landscape They Want to Conserve. *Education Sciences, 10*(13). https://files.eric.ed.gov/fulltext/EJ1241172.pdf>

Pierre, J. (2019). *Racism is a Public Health Crisis*. The Institute for Policy Studies. https://ips-dc.org/racism-is-a-public-health-crisis/

Powell, R. B., Stern, M. J., Frensley, B. T., & Moore, D. (2019). Identifying and developing crosscutting environmental education outcomes for adolescents in the twenty-first century (EE21). *Environmental Education Research, 25*(9), 1281–1299. doi:10.1080/13504622.2019.1607259

Reid, J. L., Kagan, S. L., Hilton, M., & Potter, H. (2015). *A Better Start: Why Classroom Diversity Matters in Early Education.* PRRAC: Poverty Race and Research Council. https://www.prrac.org/pdf/A_Better_Start.pdf

Russ, A., & Krasny, M. (2017). *Urban Environmental Education Review*. Cornell University Press. www.jstor.org/stable/10.7591/j.ctt1qv5qhq

Santamaría, C. N. (2020). Families, experiments, and nature: Learning science through project-based learning. *School Science and Mathematics*, *120*(8), 467–476. https://proxy.marybaldwin.edu:2238/10.1111/ssm.12438. doi:10.1111/ssm.12438

Schuyler, C. (2020). Nature Rx Presentation. *Foxcroft Environmental Education Conference*. Foxcroft School, Virginia.

Sevin, S. (2020). The Change of Secondary School Students' Environmental Consciousness, Attitude and Behaviors with Nature Education Project. *Higher Education Studies, 10*(2), 82-94.

Stapleton, S. R. (2019). A case for climate justice education: American youth connecting to intragenerational climate injustice in Bangladesh. *Environmental Education Research*, *25*(5), 732–750. doi:10.1080/13504622.2018.1472220

Stapleton, S. R. (2020). Toward critical environmental education: A standpoint analysis of race in the American environmental context. *Environmental Education Research*, *26*(2), 155–170. doi:10.1080/13504622.2019.1648768

Steele & Scott. (2016). Emotionality and Learning Stories: Documenting How We Learn How We Feel. *Canadian Journal of Environmental Education*, *21*(6), 1–19. https://cjee.lakeheadu.ca/article/view/1426/862

Striniste, N. (2019). *Nature Play at Home: Creating Outdoor Spaces That Connect Children with the Natural World*. Timber Press.

Striniste, N. (2020). *"Nature Play at Home" presentation*. Virginia Association for Environmental Education Conference. Amherst, Virginia.

Suryawati, E., Suzanti, F., Zulfarina, Z., Putriana, A. R., & Febrianti, L. (2020). The Implementation of Local Environmental Problem-Based Learning Student Worksheets to Strengthen Environmental Literacy. *Jurnal Pendidikan IPA Indonesia*, *9*(2), 169–178. doi:10.15294/jpii.v9i2.22892

Toth, N., Little, L., Read, J. C., Fitton, D., & Horton, M. (2013). Understanding teen attitudes towards energy consumption. *Journal of Environmental Psychology*, *34*, 36–44. https://www.sciencedirect.com/science/article/pii/S0272494412000722. doi:10.1016/j.jenvp.2012.12.001

United States Environmental Protection Agency (EPA). (2018). *What is Environmental Education?* EPA. https://www.epa.gov/education/what-environmental-education>

Verlie, B. (2019). Bearing worlds: Learning to live-with climate change. *Environmental Education Research*, *25*(5), 751–766. doi:10.1080/13504622.2019.1637823

Walden University. (2019). *7 Ways to Teach Critical Thinking in Elementary Education*. Walden University. https://www.waldenu.edu/online-bachelors-programs/bs-in-elementary-education/resource/seven-ways-to-teach-critical-thinking-in-elementary-education

Warner, J. (2015). The High School/College Writing Classroom Disconnect. *Inside Higher Ed*. https://www.insidehighered.com/blogs/just-visiting/high-schoolcollege-writing-classroom-disconnect

Watson, P., & Doué, W. (2007). The Outdoor School Inspires Potentially Highly Qualified Middle Level Teachers. *Middle School Journal, 39*(2), 50–58. www.jstor.org/stable/23048339. doi:10.1080/00940771.2007.11461624

Weinstein, N., Balmford, A., DeHaan, C. R., Gladwell, V., Bradbury, R. B., & Aman, T. (2015). Seeing the Community for the Trees: The Links Among Contact with Natural Environments, Community Cohesion, and Crime. *Bioscience, 65*(12), 1141–1153. doi:10.1093/biosci/biv151

Worrall, S. (2017). *We are Wired to be Outside*. National Geographic. https://news.nationalgeographic.com/2017/02/nature-fix-brain-happy-florence-williams/>

Wu, C. (2020). *Environmental education adapts to virtual learning*. Chesapeake Bay Program. chesapeakebay.net/news/blog/environmental_education_adapts_to_virtual_learning,

Chapter 18
Empowering Special Education:
Fostering Environmental Education, Nature Awareness, and Sustainable Practices

Finita Glory Roy
https://orcid.org/0009-0008-7825-455X
O.P. Jindal Global University, India

ABSTRACT

This chapter explores the pivotal role of special education in promoting environmental education, fostering heightened nature awareness, and integrating sustainable practices. It underscores the importance of inclusive environmental education, tailored to meet the diverse needs of students, particularly those with disabilities. Through sensory experiences in nature and hands-on learning, special education enables a deeper connection between students and the environment. The integration of sustainable practices, such as waste reduction, energy conservation, and sustainable gardening, forms the core of this inclusive education approach. By empowering students to actively engage with and apply sustainable principles in their daily lives, special education not only imparts knowledge but also instills a sense of responsibility towards environmental preservation. Ultimately, special education serves as a catalyst for nurturing environmentally conscious individuals, ensuring no one is excluded from contributing to a greener and more sustainable future.

INTRODUCTION

Amidst a myriad of environmental hurdles, it is now more crucial than ever for everyone, regardless of their abilities, to delve into environmental education, amplify their connection with nature, and embrace sustainable habits. Special education, specifically, holds a pivotal position in guaranteeing inclusivity in the journey towards a more eco-friendly and sustainable tomorrow. This segment delves into the importance of environmental education within special education, the profound influence of heightened nature consciousness, and methods to infuse sustainable practices into special education curricula. Teachers across the world are putting ecological intelligence and sustainability pedagogies into practice to confront

DOI: 10.4018/979-8-3693-2577-3.ch018

environmental calamities (Goleman, Barlow, & Bennet, 2010; Ju & Kim, 2011; Lewis, Mansfield, & Baudain, 2008; Marouli, 2002; Riordan & Klein, 2010; Stone, 2010; Williams, 2008). Given the serious environmental challenges we face, ecological thinking and sustainability education are required at every school level (Bowers, 2009; Capra, 2007; Cortese, 1999; Duhn, 2012; Marouli, 2002; Williams, 2008). To promote ecological intelligence and sustainable living practices, educators are developing curricula and establishing vibrant collaborations globally (Armstrong & Grant, 2004; Davis, 2005; Johnston, 2009; Ju & Kim, 2011; Stone, 2007).

Most of these applications have been primarily used by general education teachers and students. Programs that cater to special populations are scarce, and little study has been done on special educators teachers who work with students who have special needs, and how they might help their students grow up to be responsible, knowledgeable people. While there is little published information specifically addressing how students with disabilities can develop ecological intelligence, there is a wealth of literature discussing the establishment of just social practices and relationships within communities (Artiles, Trent, & Palmer, 2004; Edgar, Patton, & Day-Vines, 2002; Ford, 2004; Irvine, 2012; Ladson-Billings, 1995; Lamar-Dukes, 2010).

In the face of numerous environmental challenges, there is an urgent need for universal engagement in environmental education, regardless of individual abilities. Heightening our connection with nature and adopting sustainable habits are now imperatives. Special education plays a pivotal role in ensuring inclusivity on the path toward a greener, more sustainable future. This exploration delves deeply into the significance of environmental education within special education, the impactful influence of heightened nature consciousness, and strategies to seamlessly integrate sustainable practices into special education programs.

Inclusivity lies at the heart of special education, emphasizing the need for adaptable and tailored environmental education programs. This demands innovative teaching methods, accessible learning materials, and robust support systems to ensure that every student, irrespective of their abilities, can actively engage and benefit from environmental education initiatives.

Heightened nature awareness transcends appreciating the beauty of natural landscapes. It's about comprehending the inherent link between nature and human existence. Within special education, fostering this heightened awareness becomes a cornerstone. It moves beyond surface admiration, aiming to instill a profound understanding of nature's crucial role in our lives. Tailored programs within special education serve as conduits for this deeper comprehension, enabling students to grasp the intricate relationship between the natural world and human existence. By nurturing this profound respect for nature's role, special education equips students to appreciate nature beyond aesthetics, embracing its significance in our existence.

Practical implementation of sustainable practices within special education becomes imperative in cultivating eco-consciousness. This involves embedding sustainability into daily routines and activities. By integrating waste reduction, energy conservation, and sustainable projects into the curriculum, students don't just learn about sustainability; they actively apply it. These hands-on experiences offer insights into how individual actions impact the environment, fostering a sense of responsibility and empowerment among students.

The infusion of sustainable practices into special education programs goes beyond theoretical teachings. It's about nurturing a mindset that transcends the classroom—a mindset ingrained in daily habits and choices. Students, through these initiatives, not only develop academic knowledge but also cultivate a deeper connection with the environment. This connection becomes foundational, empowering them to become proactive and conscientious contributors to a sustainable future.

In essence, this chapter emphasizes that the amalgamation of environmental education, heightened nature awareness, and sustainable practices within special education is not just about addressing disabilities; it's about empowering individuals of all abilities to become responsible stewards of our planet. It's about fostering a collective commitment to sustainability, ensuring that no one is left behind in the collective endeavor toward an eco-friendlier and more sustainable world.

The Importance of Environmental Education

Environmental education stands as a comprehensive approach, transcending mere comprehension of ecosystems and conservation. It embodies a holistic ethos, aiming to instill a profound sense of responsibility towards our environment while advocating for sustainable lifestyles. Within the realm of special education, the imperative for environmental education remains undiminished. Infusing environmental topics into the curriculum not only enriches the educational journey but also equips students with indispensable knowledge and skills crucial for their holistic development and future well-being. Researchers have discovered that including outdoor education can successfully support student inclusion and more favorable attitudes toward attending school for adolescents with emotional and behavioral issues (EBD) (Fox & Avramidis, 2003; Berger, 2006).

The essence of inclusivity lies at the heart of special education. Consequently, the realm of environmental education within this sphere necessitates adaptable and inclusive programs tailored to accommodate the diverse needs of students with disabilities. Achieving this demands a tapestry of innovative teaching methodologies, easily accessible learning materials, and robust support systems, ensuring that each student, irrespective of their abilities, can actively engage in and derive benefits from environmental education initiatives.

At its core, environmental education surpasses conventional classroom teachings; it is a beacon guiding individuals toward a deeper understanding of their roles within the larger ecosystem. Beyond the academic purview, it encompasses values, attitudes, and behaviors that nurture a profound respect for nature and instigate sustainable actions. This holistic approach is equally pertinent within the landscape of special education, warranting an integration that not only imparts knowledge but also cultivates a sense of stewardship toward the environment among students of all abilities.

The infusion of environmental topics into special education curricula catalyzes multifaceted growth. It not only augments the academic repertoire but also fosters skills essential for navigating the complexities of the modern world. Students develop critical thinking abilities, problem-solving skills, and an enhanced capacity to comprehend the interconnectedness of environmental issues with broader societal challenges. Moreover, it nurtures empathy, resilience, and a profound sense of responsibility toward the environment, nurturing well-rounded individuals poised to contribute meaningfully to a sustainable future.

Inclusivity in Environmental Education

Inclusive environmental education demands a paradigm shift in instructional strategies. It necessitates an amalgamation of varied teaching methodologies that accommodate diverse learning styles and cognitive abilities. Implementing multimodal approaches, such as incorporating visual aids, interactive sessions, tactile experiences, and adaptive technologies, ensures that every student can actively engage with and internalize environmental concepts. Additionally, fostering a supportive and inclusive classroom environment, where students feel empowered to express themselves without constraints, forms a cornerstone of successful environmental education in special needs settings.

The availability of accessible learning materials holds paramount significance in enabling inclusive education. These materials should cater to diverse needs, ensuring that content is comprehensible and engaging for all students. Braille resources, audiovisual materials with descriptive narration, simplified texts, or adaptive technologies facilitating interactive learning experiences serve as indispensable tools to ensure accessibility and equal participation for students with disabilities.

Furthermore, collaboration among educators, specialists, caregivers, and communities plays a pivotal role in nurturing an inclusive environment for environmental education. Establishing partnerships that harness collective expertise and resources fosters a supportive ecosystem, empowering students with disabilities to actively engage in environmental education initiatives. Moreover, parental involvement and community engagement serve as pillars fortifying the inclusivity and effectiveness of these programs, fostering a shared commitment toward environmental stewardship.

THEORETICAL FRAMEWORK

Incorporating, the Theory of Multiple Intelligence, Universal Design of Learning and Ecological Systems Theory into the chapter on environmental education for students with disabilities would guide educators in creating inclusive and holistic learning experiences. By designing instruction that addresses diverse needs and acknowledges the complex interplay between individuals and their environments, educators can foster meaningful engagement and promote environmental stewardship among all students.

Multiple Intelligences by Howard Gardner

The integral approach of Outdoor Education can be seen in the aspect, that learning outdoors "elevates the physical, emotional, cognitive, social and spiritual level of the individual." (Gilbertson, 2006, p. 5). This opens new possibilities for learning, as traditional education focuses on the cognitive level. Here a potential benefit for children with special needs can be realized. For children who have problems learning on a cognitive level, such as children with learning disabilities, for instance, new levels of learning emerge with Outdoor Education. Every person has an individual style of how they learn most effectively. This can be explained by the theory of multiple intelligences by Howard Gardner, who describes several forms of intelligence within each person. This means some persons have a strong verbal-linguistic core area whereas others have a logical-mathematical or a naturalist core area (as cited in Gilbertson, 2006, p. 38). Experiential and place-based learning activates various learning styles and is thus favorable for children with special needs, who may have problems in adapting content via a specific sensory channel, for instance, deaf or blind people.

Universal Design for Learning (UDL) by David H. Rose and Anne Meyer

Universal Design for Learning (UDL) is a framework aimed at creating instructional environments that accommodate the diverse needs of all learners. In environmental education, applying UDL entails incorporating multiple means of representation, engagement, and expression. This means presenting information in various formats such as visual, auditory, and tactile to cater to different learning preferences. For instance, providing diagrams, audio recordings, and hands-on activities ensures that students with diverse abilities can access and process the content effectively. Moreover, offering a range of activities and materials promotes engagement and participation among students with varying strengths and challenges. By embracing UDL principles, educators can design inclusive environmental education experiences that empower all learners to succeed and actively engage with the subject matter. There is a gap in the research and practice literature around the use of UDL in outdoor learning contexts, indeed, more generally, research into UDL as a practice is just beginning (University of California Berkeley, 2022). This is a potentially rich theme that needs further research evidence and stories from practice to determine its real potential as a tool for inclusion for all learners. Barriers to the implementation of UDL, as identified by Katz (2015), include the need for professional learning communities.

Ecological Systems Theory, Developed by Urie Bronfenbrenner

Ecological Systems Theory, developed by Urie Bronfenbrenner, underscores the dynamic interplay between individuals and their environment. In the realm of environmental education, this theory underscores the significance of acknowledging students' social and physical contexts. Environmental learning occurs within a framework of nested systems, including the microsystem (individual and immediate surroundings), mesosystem (interactions between microsystems), ecosystem (external settings influencing the microsystem), and macrosystem (cultural context). By recognizing these ecological systems, educators can tailor instructional approaches to facilitate meaningful learning experiences. For instance, understanding students' home environments and community resources allows for the integration of relevant and relatable environmental concepts into the curriculum. This approach fosters deeper engagement and comprehension by bridging the gap between abstract concepts and real-world contexts. Ultimately, leveraging Ecological Systems Theory in environmental education enables educators to create inclusive and impactful learning environments that empower students to connect with and understand the complexities of the natural world. Applying the ecological perspective to outdoor recreation can enhance our understanding of the broader context that shapes children's participation (Christiana et al., 2014; Manfredo et al., 2020).

HEIGHTENED NATURE AWARENESS

Elevated nature awareness surpasses mere appreciation of nature's aesthetics; it involves recognizing its intrinsic link with human life. Within the realm of special education, there exists a crucial role in nurturing this heightened awareness. It transcends superficial admiration, striving to instill a deep-seated understanding of nature's indispensable connection to human existence. Specially tailored programs within special education act as pathways for this profound comprehension, empowering students to comprehend the intricate relationship between the natural world and human life. Through these initiatives, a profound respect for nature's pivotal role in our lives is fostered, enabling students to embrace the significance of nature beyond its surface beauty, recognizing its intrinsic value in our existence.

This elevated awareness transcends the confines of traditional nature appreciation. It encompasses a comprehensive understanding of how nature intertwines with human life on multiple levels. Special education initiatives provide a platform for students to delve deeply into this relationship. They explore the ways in which ecosystems sustain life, how the natural world influences human health and well-being, and the interconnectedness between environmental factors and societal development. These programs serve as catalysts, nurturing a holistic understanding of how nature is not just an external entity but an integral part of our being.

By immersing students in specially curated programs that emphasize this holistic perspective, special education equips them with the tools to appreciate the intricate dance between humanity and the environment. Through experiential learning, students engage in activities that showcase nature's contributions to our lives – from providing essential resources to supporting mental and physical health. This hands-on approach deepens their connection with nature, fostering an emotional bond that transcends mere intellectual understanding.

Moreover, by nurturing this heightened nature awareness within special education, students develop a sense of responsibility toward environmental conservation. They understand that their actions impact the natural world and, in turn, influence human well-being. This realization instills a desire to become proactive stewards of the environment, advocating for its protection and preservation.

Sensory Experiences in Nature

Nature serves as an invaluable source of multisensory experiences, offering a diverse array of stimuli—textures, fragrances, and sounds—that collectively weave a vibrant sensory tapestry. For students with disabilities, these sensory encounters hold exceptional value. Within the realm of special education, tailoring initiatives to incorporate outdoor activities enables students to immerse themselves in nature's sensory richness on a profound level. This intentional engagement not only fosters sensory exploration but also nurtures a deeper connection between students and the natural world. Through this approach, special education facilitates a comprehensive understanding, empowering students with disabilities to actively engage with and appreciate nature's diverse sensory wonders.

The natural world presents a spectrum of sensory stimuli, each element contributing to a sensory symphony that transcends conventional classroom experiences. Textures—ranging from the rough bark of a tree to the softness of moss—offer tactile engagement, inviting students to explore and comprehend nature through touch. Fragrances emanating from flowers, earth, and foliage envelop students, igniting olfactory senses and providing a sensorial journey. Simultaneously, the melodic sounds of rustling leaves, chirping birds, and flowing water form an auditory tapestry, enriching the sensory mosaic of the natural environment. Through student engagement with the natural world, nature-based learning seeks to stimulate senses beyond sight and hearing, as is frequently done in the classroom (Gautheron, 2014). Acquiring knowledge, skills, beliefs, attitudes, and behaviors in areas such as academic success, personal development, and environmental concern are all included. It encompasses learning about the natural environment as well as engaging in any activity, interest, or skill while in a natural setting (Jordan & Chawla, 2019).

In special education, the intentional incorporation of outdoor pursuits harnesses these diverse sensory encounters, offering students a profound and holistic sensory experience. By orchestrating these engagements, educators provide a canvas for students to explore nature's sensory marvels firsthand. This immersive interaction transcends textbooks and lectures, enabling students to engage with nature through direct experiences that cater to their unique sensory needs.

Moreover, these sensory engagements foster a deeper bond between students and the natural realm. Through tactile exploration, students establish a tactile connection with nature, fostering a deeper understanding and appreciation for its textures and surfaces. The rich array of fragrances not only stimulates their olfactory senses but also evokes emotional connections and memories, grounding them in the natural environment. The symphony of sounds offers a captivating auditory experience, inviting students to attune their senses to the rhythm and melodies of nature.

This intentional immersion in nature's sensory offerings empowers students with disabilities to actively participate and embrace the diverse sensory palette that nature offers. It transcends mere observation; it's an experiential journey that fosters a profound connection, nurturing an appreciation for the intricate sensory marvels of the natural world. In essence, by customizing outdoor experiences within special education, educators provide a gateway for students to not just observe but fully engage with nature, fostering a deeper understanding and affinity for its sensory richness. Learning can be made more tailored and personalized using a nature-based learning method, which is precisely what special education is all about. Though it can benefit all kids in developing citizenship, diversity, and responsibility, students with EBD and/or ADHD may benefit most from nature-based learning. Studies reveal that children's symptoms of ADHD can be lessened by the peaceful surroundings that nature offers (Sailakumar & Naachimuthu, 2017). Research on the therapeutic benefits of outdoor education programs for disturbed kids, particularly those with mental health diagnoses (Hattie et al., 1997), has been conducted. Nature-based learning may be an opportunity for students with EBD, ADHD, and other types of disabilities to be more successful in school. When comparing time spent indoors versus time spent outside, the behavior of students with disabilities changes. Studies indicate a decline in disruptive behavior episodes; however, improvements are seen in inattention and disinterest (Wilson, 2014). By removing special education kids from the classroom, which they may already associate with failure, sending them outside can improve their learning environment (Lappin, 1984). Students with disabilities may have the opportunity to identify and accomplish their personal goals through NBL (Nature-Based Learning).

But among other things, nature can support one's sense of security. A 2017 study by Anderson, Trinh, Caldarella, Hansen, and Richardson found that children with emotional and/or behavioral problems frequently engage in inappropriate social activities and struggle academically. Nevertheless, studies on outdoor education programs for kids with behavioral disorders also show improvements in academic success, social adjustment, self-concept, and group cohesion (Lappin, 1984). Additionally crucial for students with emotional or behavioral difficulties are coping and problem-solving abilities. After spending a day in the forest, EBD kids report feeling happier and more energized and experiencing less stress and rage, according to Chawla (2018). Students with emotional and behavioral difficulties might also benefit from field trips that prioritize environmental education and awareness, conservation of natural resources, local history, community services, and physical and health education. Activities that present a challenge are beneficial for students who struggle with behavior. Activities like camping, hiking, rock climbing, rappelling, canoeing, rafting, and backpacking are all beginner-friendly and don't call for very strong physical skills. Students are challenged by nature-based activities like these, but they also allow them to push themselves and discover strengths they might not have discovered otherwise (Lappin, 1984). Children, particularly those with disabilities, may have the chance to see that learning and development may occur in settings other than the conventional classroom. Additionally, it has been demonstrated that natural environments successfully lessen the symptoms of ADHD, and schools that implement environmental education programs see increases in arithmetic, reading, writing, and listening scores on standardized tests (Martin et al., 2018). According to Chawla (2018), reintroducing nature and unstructured play into the curriculum can help students achieve more while reducing impulsivity and inattention. Cooper (2015) notes that prior studies have connected outdoor play, such as recess, to improving primary students' ability to focus on the classroom and lowering symptoms of ADHD.

Hands-On Learning

Immersing oneself in hands-on experiences amid nature acts as a gateway to a profound grasp of environmental principles. Within the domain of special education, educators assume a pivotal role in orchestrating immersive activities such as nature walks, garden projects, and wildlife observations. These initiatives serve as conduits, offering students direct and tactile exposure to the natural world, thereby enhancing their theoretical understanding, and kindling a sense of curiosity and attachment to their surroundings. Nature improves self-confidence, self-esteem, empathy, test scores, and connection to others in all children, not only those with special needs (Louv, 2008).

Nature walks, a cornerstone of these experiential learnings, allow students to actively witness ecological interactions. As they traverse natural landscapes, students become firsthand observers of the delicate balance inherent within ecosystems. Witnessing flora and fauna interweaving in their habitats fosters a deeper appreciation for the intricate harmonies prevailing in nature. These walks serve as living classrooms, unveiling the interconnectedness of various elements in the environment, thus nurturing a profound understanding of ecological dynamics.

Garden projects within special education initiatives emerge as fertile grounds for students to comprehend the life cycles of plants and embrace sustainable cultivation practices. Engaging in planting, nurturing, and tending to the garden not only acquaints students with the growth processes of flora but also instills a sense of responsibility toward nurturing and preserving the environment. The tactile experience of tilling soil, planting seeds, and witnessing the transformation from seedling to maturity offers a tangible connection to nature's cycles. Students, through these projects, develop a sense of stewardship by actively contributing to the cultivation and sustenance of green spaces.

Further enriching these experiences are wildlife observations, providing students with an up-close view of biodiversity. Observing various species in their natural habitats elucidates the interdependence among different life forms. Witnessing the intricate dance of interactions between species illuminates the web of life, emphasizing the critical role each organism plays in maintaining the delicate balance of ecosystems. Such direct encounters with wildlife spark a sense of wonder and appreciation for the diverse array of life forms, fostering empathy and understanding toward the natural world.

The hands-on approach not only fortifies academic understanding but also nurtures an emotional connection to nature. By immersing themselves in these experiences, students forge a deep and personal relationship with the environment. This emotional bond transcends mere theoretical knowledge, fostering a genuine affinity for the natural world. Empowered with both knowledge and emotional connection, students emerge as conscientious stewards of the environment, poised to advocate for and actively participate in preserving the delicate balance of our planet.

INTEGRATING SUSTAINABLE PRACTICES

The integration of sustainable practices within special education extends far beyond theoretical teachings; it embodies a deliberate endeavor to empower students with the seamless integration of sustainability into their daily lives. This comprehensive approach doesn't merely aim to impart knowledge but strives to cultivate a profound sense of responsibility and environmental mindfulness among students.

Teaching sustainable practices transcends the classroom through practical implementation, showcasing how eco-friendly habits effortlessly merge with everyday routines. By weaving these practices into the curriculum and daily activities, students not only absorb information about sustainability but actively put it into action. Whether it is practicing waste reduction, conserving energy, or engaging in sustainable gardening projects, students gain firsthand experience in understanding how their actions impact the environment.

This conscientious approach nurtures a heightened awareness within special education, enabling students to evolve into proactive and conscientious contributors to society. They don't just acquire knowledge; rather, they internalize a fundamental understanding of their pivotal role in preserving the environment. Thus, equipped with this awareness, they stand ready to make informed and sustainable choices as they navigate the world beyond the confines of the classroom. This empowerment extends beyond the theoretical realm, fostering a practical understanding of sustainability that becomes ingrained in their daily lives and future decisions.

Waste Reduction and Recycling

Special education programs act as a vital platform for educating students on the crucial aspects of waste reduction and recycling. The introduction of thorough recycling programs within school settings initiates a hands-on and immersive path toward sustainability. These endeavors not only foster environmental consciousness but also embolden students to take charge of environmental stewardship, extending their responsibility to their homes and broader communities.

The integration of comprehensive recycling initiatives within school environments serves as a cornerstone in nurturing sustainable practices. It goes beyond theoretical knowledge, empowering students to actively engage in actions that diminish waste. Through the cultivation of recycling values and waste reduction principles, students delve into analyzing their consumption patterns, comprehending the life cycle of materials, and playing an active role in lessening the ecological impact of waste.

Furthermore, these programs provide practical experiences that transcend mere theoretical teachings. Students actively participate in sorting and recycling materials, witnessing firsthand the transformation of waste into reusable resources. This hands-on learning instills a sense of efficacy and accountability, shaping a generation that comprehends waste reduction's significance and actively contributes to environmentally friendly initiatives. According to the Board of Cooperative Educational Services (2012), Boltz, Impola, Nyquist, & Torgerson (2012), and O'Conner (2011), recycling is a frequent practice in many special education groups. Many special needs students are participating in employment training as they approach graduation or make the switch from public education to the workforce. Many organizations and programs are supporting the shift from education to the workforce worldwide; some examples include Easter Seals & Poly Recovery, USA; Goodwill, USA; Ecommunity, Kibbutz Yas'ur, Israel; and Easter Seals New Hampshire, 2012; Goodwill, 2011; New Tang Dynasty Television, 2010; Scrap Autism, 2012).

By integrating these practices into their daily routines within the school environment, students internalize habits that extend beyond the classroom walls. They become advocates for sustainability, carrying these values into their homes and wider communities. Thus, special education programs serve as incubators, nurturing environmentally conscious individuals equipped with the knowledge, skills, and values to effect positive change in their surroundings.

Energy Conservation

In the pursuit of sustainability, instructing students about energy conservation emerges as a foundational pillar. Within special education, classrooms serve as dynamic arenas for imparting and implementing straightforward yet impactful practices, like conscientiously switching off lights and appliances when not in use. By weaving these practices into their daily routines, students become active participants in learning and applying energy conservation techniques. This engagement not only contributes to environmental preservation but also fosters a deep comprehension of how their actions directly influence energy consumption and the environment.

The integration of such practices within the daily rhythm of special education classrooms goes beyond theoretical learning; it cultivates a culture of responsibility and action. Students witness firsthand the tangible outcomes of their efforts in reducing energy waste, thereby reinforcing the significance of their contributions to a sustainable future. Moreover, this hands-on experience instills a sense of empowerment, as students recognize their pivotal role in effecting positive change within their immediate environment.

The impact extends beyond environmental benefits, showcasing tangible outcomes in reduced operational costs for educational institutions. These energy-conscious behaviors result in immediate financial savings, demonstrating the direct and practical advantages of adopting sustainable practices. This dual benefit not only reinforces the importance of energy conservation but also highlights the real-world implications and rewards of environmentally conscious behavior.

Sustainable Gardening and Food Production

Creating gardens within special education environments extends beyond being a mere educational tool; it serves as a pivotal platform for championing sustainable food production. Students actively participate in learning organic farming techniques, delving into the importance of locally sourced and seasonal food, and grasping the crucial impact of agriculture on environmental preservation. These immersive experiences offer more than theoretical knowledge; they immerse students in hands-on learning, providing a tangible understanding of sustainable food production while fostering a profound connection with the food cycle and the natural world. Over the past ten years, there has been a push to incorporate school gardens into the curriculum and a rise in interest in using them for educational purposes. (Coffey, A.,2001; Bell, A.C., 2001).

Real-world experiences and practical gardening methods are used to teach academic subjects including science, arithmetic, nutrition, and environmental education (Eames-Sheavly, 1999; Lineberger and Zajicek, 2000; Skelly and Zajicek, 1998; Waliczek et al., 2003). Since gardens usually include collections of plants and animals for teaching and display, outdoor education—and gardens in particular—has long been linked to environmental protection and education. For this reason, gardens are tasked with teaching kids with normal and special needs about themes related to conservation and worldwide environmental change (Mintz & Rode, 1999; Willison, 1997). According to Plaka (2016), gardens have an educational impact that is specifically seen in how they enhance environmental knowledge and encourage positive behavior. Students have a lot of opportunities to play, engage in creative activities, and connect with nature in schoolyards.

The cultivation of gardens within these settings nurtures a holistic understanding of sustainable food practices. Students engage in nurturing plants, observing their growth cycles, and tending to the garden's needs. Through these experiences, they not only comprehend the essence of sustainable agricultural practices but also internalize the intricate interplay between agriculture, environmental conservation, and food systems.

Moreover, these practical encounters instill a deeper connection between students and their food sources. They witness firsthand the journey of food from seed to table, fostering an appreciation for the origins of their meals and the impact of food choices on the environment. This experiential learning sparks a sense of responsibility and mindfulness in food consumption practices, empowering students to make informed and conscientious dietary choices.

By engaging in garden-based learning, students develop a comprehensive perspective on agriculture's pivotal role in environmental sustainability. They grasp the significance of cultivating food in ways that are harmonious with nature, minimizing environmental impact and promoting biodiversity. Through these experiences, students not only gain practical knowledge but also cultivate a sense of stewardship toward the environment, equipped with the understanding of how their choices can positively influence sustainability.

EMPIRICAL EVIDENCE AND CASE STUDIES

- The case study by Berger (2006) in Israel, utilizing outdoor activities to enhance communication, self-confidence, and other skills among students with special needs. Joseph, a 12-year-old with communication and social challenges, found traditional therapy settings uncomfortable. Instead, he led his therapist on walks near his school, eventually settling at a riverbank under a willow tree. Their therapy sessions centered around activities like brewing tea and building a small barrier around their spot. Joseph's meticulous attention to these rituals gradually improved his communication skills and deepened his connection with the therapist. The completion of the barrier marked a turning point, leading Joseph to express himself more freely. Even during winter, they returned to the riverbank, now called "Home-in-Nature," to reconnect and reflect on their progress. This physical and symbolic space became vital for Joseph's growth and therapeutic alliance. The building-a-home-in-nature method (Berger, 2004) implements the concept of nature as a therapeutic setting in a concrete embodied way, addressing it as a key point of reference in an essentially nonverbal therapeutic process. What Berger (2006) found from his ten-month long study on nature therapy was that nature could be used "as a media for experiential and non-verbal work" (p. 145). Berger suggests teachers use the outdoor environment to go beyond typical classroom management and behavioral approaches and instead use methods that can be better adapted for highly individualized needs (2006).
- In 1994, Wilson started a short list of action items that might help struggling teachers initiate an interest in EE in their classrooms, all of which are still applicable today. They include starting a recycling program within a school or classroom; decorating the classroom with items from nature; sharing books that are "pro-nature" with students (as opposed to ones that depict animals as dangerous or frightening characters); introducing living plants or animals into the class while demonstrating appropriate care for them; and celebrating Earth Day in some special way (Wilson, 1994). These suggestions are simple and can be appropriate for a wide variety of learners.
- The Kualapu'u program entitled PRISM (Providing Resolutions with Integrity for a Sustainable Molokai) was developed explicitly to engage students academically and provide the skills and experiences necessary to resolve important community issues, to become future community lead-

ers (Cheak et al, 2002; Smith, 2007). The PRISM program developed on Molokai worked with two grades from a public school and about 100 students (Volk & Cheak, 2003). PRISM students also exhibited notable academic increases in critical thinking skills, and environmental literacy, as well as knowledge of ecology, environment, and environmental and social issues. Oral and written communication skills also showed improvement in PRISM students exemplified by the reading of more complex texts to public verbal presentations about their investigations (Cheak, et al. 2002). In addition, teachers, parents, and community members found that participating students exhibited increased personal confidence, poise, and self-esteem in school and their daily lives (Volk & Cheak, 2003).

- In the Outdoor and Environmental Recreation and Therapeutic Recreation at Central Michigan University in Mt. Pleasant, Michigan, Instructors Lynn Dominguez, and Mary Lou Schilling offered a five-week environmental education program to high school students in a transition-to-work, special education class, the facilitators were university students majoring in environmental education and therapeutic recreation. These students, from 16 to 26 years of age, had a variety of disabilities, including Down syndrome, learning disabilities, fetal alcohol syndrome, mild and moderate mental retardation, emotional impairments, and mild physical impairments. The goal of the program was to increase the student's knowledge of environmental issues, such as waste management and water quality, and to help them understand the impact of their environmental decisions on community life. Both students and their classroom teachers were highly motivated and enthusiastic participants in our program. Teachers reported continued discussions and homework projects centered on environmental concerns during the program. Most exciting was the long-term involvement of the students in environmental activities after the program ended. The students developed and implemented a one-hour environmental education session on the water cycle for a second-grade classroom. Additionally, one group continued doing river cleanups.

Practical Examples

Many people with disabilities have made a significant contribution to the field of disability, not only by initiating rehabilitation services but also by helping to raise awareness about disability issues and campaigning for disability rights.

- Caroline, a visually impaired woman from Ireland, traveled on an elephant through the states of Kerala, Karnataka and Tamil Nadu from January- May 2001. She did this to raise funds for the Sight Savers Project (to support operations that would restore sight to people), and for projects that work for the conservation of the Asian elephant. The journey was about saying, 'We all can make a difference', proving the positivity of alternative ability in the workforce and in society." (www.the aislingproject.org)
- The Spastics Society of Northern India (SSNI) students extended aid to those affected by the Narmada Project, having learned about dam-related issues in class. Visiting a protest site in Delhi, they interacted with displaced villagers, discussing submergence and loss of livelihoods. Despite witnessing hardships, the students found solace in their challenges being lesser. One student, Mithun, expressed gratitude for his home not being washed away, highlighting the bittersweet nature of their involvement. This engagement underscored the importance of empathy and support in the face of adversity, fostering understanding between the students and the affected community.

- A laudable initiative is through an eco-education program by the National Botanical Research Institute (NBRI) in Lucknow. In July 2001 the NBRI inaugurated a garden for the blind. This was the first of its kind in Asia and the sixth in the world. Spread over 0.1 hectares, the garden has descriptions of the flora in braille and helps persons with visual impairment move around the park with minimum help. Small pebbled portions on the path, which are noted with the help of the cane, indicate an information board on the right. Turnings are indicated by pebbled spaces in the shape of an inverted 'T' (Banerjee, 2002). The initiative by the National Botanical Research Institute (NBRI) in Lucknow, establishing a garden for the blind and incorporating braille descriptions of flora to facilitate accessibility for visually impaired individuals.

THE WAY FORWARD AND ACTIONABLE RECOMMENDATIONS

Environmental education has the power to unite disparate audiences and build a community of capable leaders committed to promoting good change in the natural world and responsible stewardship of our natural resources in a world where people are growing more and more separated. Here, "diversified" is crucial. Since people with special needs are valuable contributors to our community, it is important to take their needs into account and include them in environmental education. Nature-based experiences benefit people with special needs and look at two frameworks that are in line with both their learning styles and environmental education. Camps offer a setting where the basis of their environmental literacy can be built upon. Environmental education-based camps can be a successful experience for all types of students because they combine a positive experience in nature with education about the environment (Dumais, 2016). In terms of children with special needs, who may be subject to an array of therapy appointments held indoors, contact with the outdoor environment could further their development, especially regarding stress relief (Chang & Chang, 2010). This is an important factor for individuals with special needs because they often have comorbid conditions of anxiety and depression. Additionally, these camps often include a recreational component, and this outdoor play helps to improve motor function challenges that individuals with special needs may face. The natural facets of rocks, hills, and trees that are used during the play experience help to develop strength, balance, and coordination (Muñoz, 2009). Chang & Chang (2010) report that the benefits of outdoor activities for children with special needs span six additional areas including social interaction, behavior, emotion, cognition, and sensory sensitivity. The natural environment holds many novel items and is ever-changing. This attracts attention and sparks peer conversations for improved social interaction, increases the content of speech for better communication, and drives acceptance of changing circumstances which is necessary in other aspects of life. Being outside is also sometimes seen as a reward, so the outdoor environment promotes positive behavior, the associated active play reduces agitations in vestibular systems, and the landscapes and natural environment can also help to calm emotions. Regarding cognition, there are chances for new learning experiences, for improving observation abilities, and for focusing attention. Finally, after repeated guided exposure, outdoor activities can help to decrease responses to sensitive stimulations.

Actionable Recommendations

The tips provided for exploring the outdoors with students with disabilities offer valuable insights into creating inclusive and enriching environmental education experiences.

- Buddy Systems: Establishing buddy systems enhances safety and inclusion by pairing students with disabilities with peers without disabilities. This approach provides instructional support and fosters a sense of belonging.
- Focus on Process: Emphasizing the process over the outcome allows students to engage meaningfully with nature regardless of their abilities. Encouraging awareness of the natural habitat promotes experiential learning.
- Repeat and Review: Repetition and varied teaching approaches are essential for reinforcing complex concepts. Employing multiple methods, such as visual aids and role-playing, accommodates diverse learning styles and reduces boredom.
- Identify Strengths: Recognizing students' strengths enables them to contribute meaningfully to activities. Building on individual abilities promotes teamwork and ensures active participation.
- Encourage Participation: Active participation reinforces learning and facilitates social interaction. Inclusive activities foster communication skills and promote a sense of community among students.
- Consider Mental and Chronological Age: Tailoring activities to students' mental and chronological age ensures appropriateness and engagement. Adapting games like Candy Land to reflect environmentally responsible actions maintains interest and relevance.
- Use Sensory Learning: Sensory experiences enhance environmental awareness and engagement. Encouraging students to use their senses promotes a deeper connection with nature.
- Provide a Normal Learning Environment: Creating functional learning environments facilitates the transfer of knowledge from the classroom to outdoor settings. Practical activities, such as water quality testing at the riverbank, enhance learning effectiveness.
- Adapt Activities: Adapting activities to accommodate physical impairments promotes success and inclusion. Using tactile boundaries and providing visual aids cater to diverse needs and abilities.
- Teach at Optimal Times: Timing instruction to align with students' alertness ensures engagement and learning effectiveness. Introducing new concepts at the beginning of sessions maximizes attention and retention.
- Use Step-by-Step Instruction: Breaking down activities into sequential steps accommodates attention deficits and ensures comprehension. Clear, structured instructions facilitate learning and success.
- Emphasize Fun and Success: Making activities enjoyable fosters enthusiasm and motivation. Celebrating success promotes a positive learning environment and encourages continued participation.

CONCLUSION

In conclusion, the realm of special education assumes a pivotal role in shaping environmental education, fostering nature awareness, and championing the adoption of sustainable practices. Through inclusive environmental education initiatives, students with disabilities are empowered to become proactive contributors to a sustainable future. These programs serve as transformative avenues, equipping individuals with the tools and consciousness to actively engage in environmental stewardship.

By nurturing heightened nature awareness and integrating sustainable practices into special education curricula, a pathway is paved to ensure inclusivity in the pursuit of a greener world. This approach transcends limitations, ensuring that every individual, irrespective of ability, becomes an integral part of the journey toward sustainability.

Inclusive environmental education serves as a catalyst, instilling in students with disabilities a profound understanding of their interconnectedness with the natural world. It nurtures a sense of responsibility and care toward the environment, fostering an ethos of stewardship that transcends individual capabilities. Through these initiatives, students not only absorb knowledge but internalize values, emerging as empowered individuals poised to actively contribute to a sustainable global community. Learning to advocate for oneself and others is a frequent and common goal in special education classes that involve life skill instruction (Cho, Wehmeyer, & Kingston, 2011; Fiedler & Danneker, 2007; Merlone & Moran, 2008; Weimer & Cappotelli, 1994). Since many students with disabilities find it challenging to gain self-determination skills that lead to self-advocacy, children with special needs should get support in developing these skills as members of a historically disenfranchised population (Carter, Sisco, & Lane, 2011). People with special needs must participate in decision-making as we move toward a more sustainable vision for our planet.

The integration of heightened nature awareness and sustainable practices within special education goes beyond addressing disabilities; it is a testament to nurturing the potential inherent in every student. It signifies an investment in cultivating responsible and compassionate stewards of our planet. By fostering a deep-rooted appreciation for nature and empowering students with practical skills to incorporate sustainable habits, special education becomes a platform for fostering a generation of conscientious individuals committed to preserving our environment.

Moreover, these initiatives do not solely benefit students with disabilities; they enrich the entire educational landscape. They encourage a culture of inclusivity and understanding, fostering empathy and cooperation among all students. By embracing diversity in environmental education, we cultivate a more holistic and impactful approach toward sustainability, ensuring that no one is marginalized or overlooked in the collective endeavor to create a greener, more sustainable world.

REFERENCES

Anderson, D. H., Trinh, S. M., Caldarella, P., Hansen, B. D., & Richardson, M. J. (2017). Increasing positive playground interaction for kindergarten students at risk for emotional and behavioral disorders. *Early Childhood Education Journal*, *46*(5), 487–496. doi:10.1007/s10643-017-0878-2

Artiles, A. J., Trent, S. C., & Palmer, J. D. (2004). In J. A. Banks & C. A. M. Banks (Eds.), *Handbook of research on multicultural education* (pp. 716–735). Jossey-Boss.

•Ashish. (2019). Environmental education and people with disabilities. In S. Padmanabhan (Ed.), *National Biodiversity Strategy and Action Plan*. Kalpavriksh. India. https://policycommons.net/artifacts/1614744/environmental-education-and-people-with-disabilities-sujatha-padmanabhan-national-biodiversity-strategy-and-action-plan/2304671/

Bell, A. C. (2001) The Pedagogical Potential of School Grounds. In: Grant, T. and Littlejohn, G., Eds., Greening School Grounds: Creating Habitats for Learning. New Society Publishers, Gabriola Island.

Berger, R. (2006). Using contact with nature, creativity and rituals as a therapeutic medium with children with learning difficulties: A case study. *Emotional & Behavioural Difficulties, 11*(2), 135–146. doi:10.1080/13632750600619430

Boltz, T., Impola, C., Nyquist, L., & Torgenson, A. (2012). *Say "yes!" to recycling at DCHS.* DCHS. http://www.dc.k12.mn.us/hs/ images/documents/2001-12/Sped-Newsletter-Website.pdf

Carter, E. W., Sisco, L. G., & Lane, K. L. (2011). Paraprofessional perspectives on promoting self-determination among elementary and secondary students with severe disabilities. *Research and Practice for Persons with Severe Disabilities : the Journal of TASH, 36*(1-2), 1–10. doi:10.2511/rpsd.36.1-2.1

Chang, Y., & Chang, C. (2010). *The Benefits of Outdoor Activities for Children with Autism.* Research Gate. https://www.researchgate.net/profile/Chun_Yen_Chang2/publication/266369570_The_Benefits_of_Outdoor_Activities_for_Children_with_Autism/links/54dd6246 0cf28a3d93f903c1/The-Benefits-of-Outdoor-Activities-for-Children- withAutism.pdf

Chawla, L. (2018). Aatc keynote address: Nature-based learning for student achievement and ecological citizenship. *Curriculum & Teaching Dialogue, 20*(1/2).

Cho, H., Wehmeyer, M., & Kingston, N. (2011). Elementary teachers' knowledge and use of interventions and barriers to promoting student self-determination. *The Journal of Special Education, 45*(3), 149–156. doi:10.1177/0022466910362588

Christiana, R. W., Davis, M., Wilson, M. G., McCarty, F. A., & Green, G. T. (2014). Factors related to rural young adolescents' participation in outdoor, noncompetitive physical activity. *Research Quarterly for Exercise and Sport, 85*(4), 509–518. doi:10.1080/02701367.2014.961049 PMID:25412133

Coffey, A. (2001) Transforming School Grounds. In: Grant, T. and Littlejohn, G., Eds., Greening School Grounds: Creating Habitats for Learning. New Society Publishers, Gabriola Island, 2-5.

Coffey, A. (2001) Transforming School Grounds. In: Grant, T. and Littlejohn, G., Eds., Greening School Grounds: Creating Habitats for Learning. New Society Publishers, Gabriola Island, 2-5.

Dietz, T., Rosa, E. A., & York, R. (2009). Environmentally efficient well-being: Rethinking sustainability as the relationship between human well-being and environmental impacts. *Human Ecology Review, 16*, 114–123.

Dumais, A. M. (2016). *Best Practices for Including Students with Autism, Disabilities, and Special Needs in Environmental Education* [Undergraduate Honors Thesis]. https://scholar.colorado.edu/honr_theses/1204/

Eames, M. (1999). *Sheavly sowing the Seeds of Success: How to Start and Sustain a Kids, Gardening Project in Your Community, Natl.* Gardening Assn.

Easter Seals New Hampshire. (2012). *Poly recovery partners with Easter Seals NH to boost staffing for material disassembly.* Easter Seals New Hampshire. http://nh.easterseals.com/site/PageServer?pagename=NHDR_Poly_Recovery_Partners

Edgar, E., Patton, J. M., & Day-Vines, N. (2002). Democratic dispositions and cultural competency: Ingredients for school renewal. *Remedial and Special Education, 23*(4), 231–241. doi:10.1177/07419325020230040601

Fedulova, I., Ivanova, V., Atyukova, O., & Nosov, V. (2019). Inclusive education as a basis for sustainable development of society. *Journal of Social Studies Education Research, 10*(3), 118–135. www.jsser.org

Fiedler, C. R., & Danneker, J. E. (2007). Self-advocacy instruction: Bridging the research-topractice gap. *Focus on Exceptional Children, 39*(8), 1–21. doi:10.17161/foec.v39i8.6875

Ford, B. A. (2004). Preparing special educators for culturally responsive school-community partnerships. Teacher Education and Special Education. *The Journal of the Teacher Education Division of the Council for Exceptional Children, 27*(3), 224–230.

Fox, P., & Avramidis, E. (2003). An evaluation of an outdoor education programme for students with emotional and behavioural difficulties. *Emotional & Behavioural Difficulties, 8*(4), 267–283. doi:10.1080/13632750300507025

Gautheron, O. (2014). *Teachers' perceptions of the effects of nature-based learning on junior level students.* Semantic Scholar. https://www.semanticscholar.org/paper/ Teachers%27-Perceptions-of-the-Effects-of-Learning-on-Gautheron/6f9dadff2d7aa 3324c45eb30094360612b8cc6ee

Goleman, D., Barlow, Z., & Bennett, L. (2010). Forging new norms in New Orleans. *Teacher Education Quarterly, 37*(4), 87–98.

Goodwill. (2011). *E-waste.* Goodwill. http:// www.gire.org/ewaste.htm

Hattie, J., Marsh, H. W., Neill, J. T., & Richards, G. E. (1997). Adventure education and Outward Bound: Out-of-class experiences that make a lasting difference. *Review of Educational Research, 67*(1), 43–87. doi:10.3102/00346543067001043

Irvine, J. J. (2012). Complex relationships between multicultural education and special education: An African American perspective. *Journal of Teacher Education, 63*(4), 268–274. doi:10.1177/0022487112447113

Jordan, C., & Chawla, L. (2019). A coordinated research agenda for nature-based learning. *Frontiers in Psychology, 10*, 766. doi:10.3389/fpsyg.2019.00766 PMID:31065246

Ju, E. J., & Kim, J. G. (2011). Using soil seed banks for ecological education in primary school. *Journal of Biological Education, 45*(2), 93–101. doi:10.1080/00219266.2010.546010

Katz, J. (2015). Implementing the three-block model of Universal Design for Learning: Effects on teachers' self-efficacy, stress, and job satisfaction in inclusive classrooms K-12. *International Journal of Inclusive Education, 19*(1), 1–20. doi:10.1080/13603116.2014.881569

Ladson-Billings, G. (1995). Toward a theory of culturally relevant pedagogy. *American Educational Research Journal, 32*(3), 465–491. doi:10.3102/00028312032003465

Lamar-Dukes, P. (2010). Reaching the hard to reach: A review of an initiative aimed at increasing participation and support for people of color with disabilities and their families in disability organizations. *Research and Practice for Persons with Severe Disabilities : the Journal of TASH, 34*(3-4), 76–80. doi:10.2511/rpsd.34.3-4.76

Lappin, E. (1984, September). *Outdoor education for behavior-disordered students* (pp. 1–4). ERIC. https://eric.ed.gov/contentdelivery/servlet/ERICServlet? accno=ED261811

Lewis, E., Mansfield, C., & Baudains, C. (2008). Getting down and dirty: Values in education for sustainability. *Issues in Educational Research, 18*(2), 138–155.

Lineberger, S. E., & Zajicek, J. M. (2000). School gardens: Can a hands-on teaching tool affect students' attitudes and behaviors regarding fruit and vegetables. *HortTechnology, 10*(3), 593–597. doi:10.21273/HORTTECH.10.3.593

Louv, R. (2008). *Last child in the Woods: Saving our children from nature-deficit disorder*. Algonquin Books of Chapel Hill.

Maclean, R., Jagannathan, S., & Sarvi, J. (Eds.). (2013). *Skills Development for Inclusive and Sustainable Growth in Developing Asia-Pacific (Technical and Vocational Education and Training: Issues, Concerns and Prospects Series)*. Springer Dordrecht. doi:10.1007/978-94-007-5937-4

Marouli, C. (2002). Multicultural environmental education: Theory and practice. *Canadian Journal of Environmental Education, 7*(1), 26–42.

Martin, H., Farrell, A., Gray, J., & Clark, T. B. (2018). Perceptions of the effect of recess on kindergartners. *Physical Educator, 75*(2), 245–254. doi:10.18666/TPE-2018-V75-I2-7740

Merlone, L., & Moran, D. (2008). Transition works: Self-awareness and self-advocacy skills for students in the elementary learning center. *Teaching Exceptional Children Plus, 4*(4), 1–10.

Mintz, S., & Rode, S. (1999). More than a walk in the park: Demonstration carts personalize interpretation'. *Roots, 18*, 24–26.

Muñoz, S. (2009). *Children in the Outdoors: A Literature Review*. Sustainable Development Research Centre.

New Tang Dynasty Television. (Producer). (2010). Electronic recycling by communities with special needs [Video]. YouTube. http:// www.youtube.com/watch?v=LnhGST1lryQ

Plaka, V., & Skanavis, C. (2016). The feasibility of school gardens as an educational approach in Greece: A survey of Greek schools. *International Journal of Innovation and Sustainable Development, 10*(2), 141. doi:10.1504/IJISD.2016.075546

Rambla, X., & Langthaler, M. (2016). The SDGs and inclusive education for all: From special education to addressing social inequalities (ÖFSE Briefing Paper, No. 14). *Austrian Foundation for Development Research (ÖFSE)*. http://www.oefse.at/publikationen/briefing-papers/detail-briefing- paper/publication/show/Publication/The-SDGs-and-inclusive-education-for-all-From-special-education-to-addressing-social-inequalities/

Riordan, M., & Klein, E. J. (2010). Environmental education in action: How expeditionary learning schools support classroom teachers in tackling issues of sustainability. *Teacher Education Quarterly, 37*(4), 119–137.

Sailakumar, S., & Naachimuthu, K. P. (2017). A phenomenological approach to understand the nature-based experiences and its influence on holistic development. *Indian Journal of Positive Psychology, 8*(2), 186–195.

Scrap Autism. (2012). *Home.* Scrap autism. http://scrapautism.org/

Skelly, S. M., & Zajicek, J. M. (1998). The effect of Project GREEN, an interdisciplinary garden program, on the environmental attitudes of elementary school students. *HortTechnology*, *8*(4), 579–583. doi:10.21273/HORTTECH.8.4.579

Spence, C. (2020, September 18). Senses of place: Architectural design for the multisensory mind. *Cognitive Research: Principles and Implications*, *5*(46), 46. doi:10.1186/s41235-020-00243-4 PMID:32945978

Stephenson, J., & Carter, M. (2011, August). The Use of Multisensory Environments in Schools for Students with Severe Disabilities: Perceptions from Teachers. *Journal of Developmental and Physical Disabilities*, *23*(4), 339–357. doi:10.1007/s10882-011-9232-6

Stone, M. (2007). Rethinking school lunch: Education for sustainability in practice. *Canadian Journal of Environmental Education*, *12*(1), 19–32.

UNESCO. (2018). Leicht, A., Heiss, J., & Byun, W. J. (Eds.). Issues and trends in education for sustainable development. United Nations Educational, Scientific and Cultural Organization. doi:10.54675/YELO2332

University of California Berkeley (2022). *Berkeley Universal Design for Learning*. University of California Berkeley.

Waliczek, T. M., Logan, P., & Zajicek, J. M. (2003). Exploring the impact of outdoor environmental activities on children using a qualitative text data analysis system. *HortTechnology*, *13*(4), 684–688. doi:10.21273/HORTTECH.13.4.0684

Weimer, B. B., Cappotelli, M., & DiCamillo, J. (1994). Self-advocacy: A working proposal for adolescents with special needs. *Intervention in School and Clinic*, *30*(1), 47–52. doi:10.1177/105345129403000108

Williams, D. (2008). Sustainability education's gift: Learning patterns and relationships. *Journal of Education for Sustainable Development*, *2*(1), 41–49. doi:10.1177/097340820800200110

Wilson, R. (2014). *Take it outside. Education Digest, 80*(2), 45–48. https://search.ebscohost.com/login.aspx?direct=true&AuthType=ip,url,geo,cpid,ui d&custid=kansas&geocustid=kansas&db=aph&AN=98510555&site=ehost-live

KEY TERMS AND DEFINITIONS

Disability: Disability refers to a diverse set of physical, cognitive, sensory, mental, or developmental impairments that may limit an individual's ability to engage in certain activities or interact with their environment in typical ways. Disabilities can vary widely in their nature, impact, and severity, and they can affect different aspects of a person's life.

Environmental Responsibility: Environmental responsibility refers to the ethical and moral obligation of individuals, organizations, and societies to act in ways that promote the conservation, protection, and sustainable use of the natural environment. It encompasses a sense of accountability for the impact of human actions on the environment, aiming to minimize negative effects and preserve natural resources for present and future generations.

Hands-On Learning: Hands-on learning, also known as experiential learning, is an educational approach that emphasizes active engagement and direct physical interaction with materials, objects, or phenomena to enhance understanding and retention of concepts.

Holistic Education: Holistic education is an approach that focuses on nurturing individuals' intellectual, emotional, social, physical, creative, and spiritual dimensions to foster their overall development. It emphasizes the interconnectedness of different aspects of human growth and aims to educate the whole person, not just academically but also in terms of personal fulfillment, well-being, and character development.

Inclusive Learning: Inclusive learning refers to an educational approach that fosters an environment where all students, regardless of their backgrounds, abilities, disabilities, or differences, feel valued, respected, and supported. It aims to create an inclusive culture within educational settings that acknowledges and accommodates diverse learning needs, styles, and preferences, ensuring equitable access to learning opportunities for every learner.

Nature Awareness: Nature awareness refers to an individual's conscious recognition, understanding, and appreciation of the natural world and its elements, encompassing both the physical environment and its ecosystems. It involves being attuned to the various aspects of nature, including flora, fauna, landscapes, climate, and ecological processes.

Sensory Experiences: Sensory experiences refer to the perceptions and responses generated through sensory stimuli that an individual encounters from their environment. These stimuli engage the senses, including sight, hearing, touch, taste, and smell, allowing individuals to interact with and interpret the world around them.

Special Education: Special education refers to a personalized and tailored educational approach designed to meet the specific learning needs of students who have exceptionalities or disabilities. It encompasses a range of educational interventions, strategies, and services aimed at supporting individuals with diverse learning requirements to achieve their academic, social, emotional, and behavioral potential.

Student Engagement: Student engagement refers to the extent to which students are actively involved, motivated, and invested in their learning experiences, both academically and socially. It encompasses the level of interest, participation, and commitment demonstrated by students in the learning process.

Compilation of References

Adhikari, A., & Saha, B. (2021). Contextualizing Social Theatre and its History: An Indian Perspective. [IJARESM]. *International Journal of All Research Education and Scientific Methods*, *9*(7), 497–502.

Adimihardja, K. (1999). Cosmology and Biodiversity of the Kasepuhan Community in the Mount Halimun Area of West Java, Indonesia. In D. A. Posey (Ed.), *Cultural and Spiritual Values of Biodiversity: A Complementary Contribution to the Global Biodiversity Assessment*. Intermediate Technology Publications Ltd.

Adorno, Th. (2003). *Negative Dialectics*. Continuum. doi:10.4324/9780203479605

Aghamelu, H. U. (2023). Community theatre and environmental education: the ukwu orji bus park project. *Ama: Journal Of Theatre And Cultural Studies, 16*(1). https://www.journals.ezenwaohaetorc.org/index.php/AMA/article/view/2270

Agrawal, A. (1995). Dismantling the Divide Between Indigenous and Scientific Knowledge. *Development and Change*, *26*(3), 413–439. doi:10.1111/j.1467-7660.1995.tb00560.x

Agung, A. A. G. (2005). *Bali Endangered Paradise? Tri Hita Karana and the Conservation of the Island's Biocultural Diversity*, [PhD Dissertation, *Leiden Ethnosystems And Development Programme (LEAD) Studies*.

Aiglsperger, J. (2014). *Yiatrosofia yia ton Anthropo*: Indigenous Knowledge and Utilisation of MAC Plants in Pirgos and Pretoria, Rural Crete: A Community Perspective on the Plural Medical System in Greece. [Thesis, PhD, *Leiden Ethnosystems And Development Programme (LEAD) Studies* Leiden, Leiden University.

Aikman, S. (1999). *Intercultural Education and Literacy: An Ethnographic Study of Indigenous Knowledge and Learning in the Peruvian Amazon*. John Benjamins Publishing Co. doi:10.1075/swll.7

Akhmetova, A. I., & Kolomiyets, O. M. A.I., Arpentieva, M. R., & Golubchikova, M. G.(2020). Environmental Education and Management: Exclusive, Inclusive, and Indigenous Doctrines, IOP Conference Series: Earth and Environmental Science. *Sci. 459 052086*. DOI 10.1088/1755

Alber, R. (2016). *New Teachers: Inspire Your Students to Write, Write, Write*. Edutopia. https://www.edutopia.org/blog/new-teachers-inspiring-your-students-write-write-write-rebecca-alber

Alcock, S., & Ritchie, J. (2018). Early childhood education in the outdoors in Aotearoa New Zealand. *Journal of Outdoor and Environmental Education*, *21*(1), 77–88. doi:10.1007/s42322-017-0009-y

Alexiades, A. V., Haeffner, M. A., Reano, D., Janis, M., Black, J., Sonoda, K., Howard, M., Fiander, C., & Buck, M. (2021). Traditional Ecological Knowledge and Inclusive Pedagogy Increase Retention and Success Outcomes of STEM Students. *Bulletin of the Ecological Society of America*, *102*(4), e01924. doi:10.1002/bes2.1924

Alfonso, S. (2024, April 11). *Deconstructing Settler Colonialism in Early Childhood: Nature-Based Preschool Teachers' Nature-Connectedness Journey & Perceptions of Nature. Disruption and Deconstruction with Early Childhood Teachers and Teacher Educators.* 2024 AERA Annual Meeting: Dismantling Racial Injustice and Constructing Educational Possibilities: A Call to Action, Philadelpia, PA.

Allsop, J., & Jones, K. (2008). Withering the citizen, managing the consumer: Complaints in healthcare settings. *Social Policy and Society, 7*(2), 233–243. doi:10.1017/S1474746407004186

Ambaretnani, P. (2012). *Paraji and Bidan in Rancaekek: Integrated Medicine for Advanced Partnerships among Traditional Birth Attendants and Community Midwives in the Sunda Area of West Java, Indonesia.* PhD Dissertation. *Leiden Ethnosystems And Development Programme (LEAD) Studies,* No. 7. Leiden: Leiden University.

Amel, E. L., Manning, C. M., & Scott, B. A. (2009). Mindfulness and Sustainable Behavior: Pondering Attention and Awareness as Means for Increasing Green Behavior. *Ecopsychology, 1*(1), 14–25. doi:10.1089/eco.2008.0005

Anderson, D. H., Trinh, S. M., Caldarella, P., Hansen, B. D., & Richardson, M. J. (2017). Increasing positive playground interaction for kindergarten students at risk for emotional and behavioral disorders. *Early Childhood Education Journal, 46*(5), 487–496. doi:10.1007/s10643-017-0878-2

Annas Little Farm. (n.d.). Seattle Nature Schools. https://www.annaslittlefarm.com

Anspach, N. M., & Carlson, T. N. (2020). What to believe? Social media commentary and belief in misinformation. *Political Behavior, 42*(3), 697–718. doi:10.1007/s11109-018-9515-z

Apffel-Marglin, F. (2011). *Subversive spiritualities: How rituals enact the world.* Oxford University Press.

Apffel-Marglin, F. (2011). *Subversive spritualities: how rituals enact the world.* Oxford University Press.

Aquinas, T. (2010). Humans as Moral End. In D. R. Keller (Ed.), *Environmental Ethics: the big questions* (pp. 63–65). Wiley Blackwell. doi:10.5422/fordham/9781531502041.001.0001

Arias Fuentes, H. J., & Zuleta Cotes, M. K. (2020). *Propuesta de lineamientos para la enseñanza de la filosofia arhuaca en las Instituciones Educativas De Nabusimake Sierra Nevada De Santa Marta* [Doctoral dissertation, Universidad Santo Tomás].

Ärlemalm-Hagsér, E., & Elliott, S. (2017). Special Issue: Contemporary Research on Early Childhood Education for Sustainability. *International Journal of Early Childhood, 49*(3), 267–272. doi:10.1007/s13158-017-0207-3

Armstrong, A. K., Krasny, M. E., & Schuldt, J. P. (2018). *Communicating climate change: A guide for educators.* Cornell University Press.

Artiles, A. J., Trent, S. C., & Palmer, J. D. (2004). In J. A. Banks & C. A. M. Banks (Eds.), *Handbook of research on multicultural education* (pp. 716–735). Jossey-Boss.

Asamblea Nacional Constituyente. (1991). *Constitución Política de Colombia. Bogotá, D.C.: Recuperado.* ANC. https://www.constitucioncolombia.com/indice.php

Asante, E. (2021). Participatory theatre in environmental health education: the woarabeba experience. *Social Science, 4*(12), 121-128.

Assagioli, R. (1976). *Psychosynthesis. A Collection of Basic Writings.* Penguin books.

Attenborough, D. F. (2021). *A Life on Our Planet: My Witness Statement and a Vision for the Future.* BBC Documentary.

Augustine, S. J. (1997). Traditional Aboriginal knowledge and science versus occidental science. *Paper prepared for the Biodiversity Convention Office of Environment Canada*. Research Gate.

Avery, M. (2012). *Our Vanishing Flora*. HarperCollins Publishers.

Bacon, F. (2010). The Mastery of Nature. In D. R. Keller (Ed.), *Environmental Ethics: the big questions* (pp. 65–69). Wiley-Blackwell.

Badiner, A. H. (Ed.). (1990). *Dharma Gaia: A Harvest of Essays in Buddhism and Ecology*. Parallax Press.

Bailie, P. (2015). Nature Preschools: The Cross Fertilization of Early Childhood and Environmental Education. In D. Sobel, P. Bailie, K. Finch, E. Kenny, & A. Stires (Eds.), Nature Preschools and Forest Kindergartens: The Handbook for Outdoor Learning (Illustrated edition, pp. 45–58). Redleaf Press.

Bailie, P., Larimore, R., & Pikus, A. (2023). *Evaluating Natureness: Measuring the Quality of Nature-based Classrooms in Pre-k Through 3rd Grade* (1st ed.). Gryphon House.

Baines, K., & Zarger, R. K. (2017). "It's Good to Learn about the Plants": Promoting social justice and community health through the development of a Maya environmental and cultural heritage curriculum in southern Belize. *Journal of Environmental Studies and Sciences*, *7*(3), 416–424. doi:10.1007/s13412-016-0416-3

Balakrishna, P. (2020). *Strategy on Education for Conservation (EfC): A Contribution from IUCN CEC and FLEDGE in Support of Realizing the Future Conservation Goals*. IUCN CEC and FLEDGE.

Balée, W. (1994). *Footprints of the Forest: Ka'apor Ethnobotany—The Historical Ecology of Plant Utilization by an Amazonian People*. Columbia University Press.

Bandura, A. (1982). Self-efficacy mechanism in human agency. *The American Psychologist*, *37*(2), 122–147. doi:10.1037/0003-066X.37.2.122

Bang, M., & Marin, A. (2015). Nature–culture constructs in science learning: Human/non-human agency and intentionality. *Journal of Research in Science Teaching*, *52*(4), 530–544. doi:10.1002/tea.21204

Banham, M. (Ed.). (2004). *A history of theatre in Africa*. Cambridge University Press.

Barker, D., Oguntoyinbo, J. S., & Richards, P. (1992). The Utility of the Nigerian Peasant Farmer's Knowledge in the Monitoring of Agricultural Resources. *General Report Series* No. 4, London, UK: Berkes, F. (1993). Traditional Ecological Knowledge in Perspective. In J. T. Inglis (Ed.), *Traditional Ecological Knowledge: Concepts and Cases*. Canadian Museum of Nature and International Development Research Centre.

Barnes, H. (2001). *Waldorf Education: An Introduction*. AWSNA.

Barnett, R. (Ed.). (1994). *Resistance and Reform in Tibet*. University of Indiana Press.

Barrable, A. (2019). Refocusing Environmental Education in the Early Years: A Brief Introduction to a Pedagogy for Connection. *Education Sciences*, *9*(1), 61–61. doi:10.3390/educsci9010061

Barrable, A., & Booth, D. (2020). Nature Connection in Early Childhood: A Quantitative Cross-Sectional Study. *Sustainability (Basel)*, *12*(1), 1. doi:10.3390/su12010375

Barrow, E. (2019). *Our Future in Nature: Trees, Spirituality, and Ecology*. Balboa Press.

Barstow, G. (Ed.). (2019). *The Faults of Meat: Tibetan Buddhist Writings on Vegetarianism*. Wisdom Publications.

Basil, C., & White, C. (2005). Environmental Literacy: Providing an Interdisciplinary Context for Young Children. In Readings in Environmental Education an Urban Model (1st ed., pp. 33–40).

Baskerville, D., & Anderson, D. M. (2021). Responding to climate change: Developing primary children's capability to engage with science through drama. *Science and drama: Contemporary and creative approaches to teaching and learning*, 93-105. . doi:10.1007/978-3-030-84401-1_6

Batchelor, M., & Brown, K. (Eds.). (1992). *Buddhism and Ecology*. Cassell.

Bateman, J. (2023). July 2023 brought record-high temperatures, devastating floods across the U.S. *NOAA*. https://www.noaa.gov/stories/july-2023-brought-record-high-temperatures-devastating-floods-across-us

Bateman, J. (2023). Record shattering: Earth had its hottest July in 174 years. *NOAA*. https://www.noaa.gov/news/record-shattering-earth-had-its-hottest-july-in-174-years

Battiste, M. (2002). *Indigenous knowledge and pedagogy in First Nations education: A literature review with recommendations*. Indian and Northern Affairs Canada.

Battle, C. P. (2019). *Climate change will displace millions. Here's how we prepare*. TEDTalk. https://www.ted.com/talks/colette_pichon_battle_climate_change_will_displace_millions_here_s_how_we_prepare

Baudon, P., & Jachens, L. (2021). A scoping review of interventions for the treatment of eco-anxiety. *Environmental Research and Public Health, 18*, 18. PMID:34574564

Baumeister, R. (1991). *Meanings of Life*. The Guilford Press.

Baxter, B. (1999). Environmental Ethics-Values or Obligation? A Reply to O'Neill. *Environmental Values, 8*(1), 107–112. doi:10.3197/096327199129341743

Beach, R., Share, J., & Webb, A. (2017). Teaching climate change to adolescents: Reading, writing, and making a difference. Taylor & Francis. doi:10.4324/9781315276304

Beckerman, W., & Pasek, J. (2010). In Defence of Anthropocentrism. In David R. Keller (eds.) Environmental Ethics: the big questions.(pp. 83-89). Wiley-Blackwell.

Beery, T. H., & Lekies, K. S. (2019). Childhood collecting in nature: Quality experience in important places. *Children's Geographies, 17*(1), 118–131. doi:10.1080/14733285.2018.1463431

Beery, T. H., & Wolf-Watz, D. (2014). Nature to place: Rethinking the environmental connectedness perspective. *Journal of Environmental Psychology, 40*, 198–205. doi:10.1016/j.jenvp.2014.06.006

Beery, T., Chawla, L., & Levin, P. (2020). Being and Becoming in Nature: Defining and Measuring Connection to Nature in Young Children. *International Journal of Early Childhood Environmental Education, 7*(3), 3–22.

Bell, A. C. (2001) The Pedagogical Potential of School Grounds. In: Grant, T. and Littlejohn, G., Eds., Greening School Grounds: Creating Habitats for Learning. New Society Publishers, Gabriola Island.

Bellamy, K., Garcia, V., Ahmed, S., & Archila, A. (2023). Ethnic Difference at the Center of Land Struggles in the Americas: A Complex History of Marginalization and Multidimensional Challenges among the Garifuna in the Northern Honduras. *Journal of Poverty, 27*(6-7), 474-494. https://www.tandfonline.com/doi/full/10.1080/10875549.2023.2173702

Bell, M., & Ashwood, L. (2016). *An Invitation to Environmental Sociology. Fifth Addition.* Sage Publications, Inc.

Benoit, L., Thomas, I., & Martin, A. (2022). Ecological awareness, anxiety, and actions among youth and their parents: A qualitative study of newspaper narratives. *Child and Adolescent Mental Health, 27*(1), 47–58. doi:10.1111/camh.12514 PMID:34687125

Berger, R. (2006). Using contact with nature, creativity and rituals as a therapeutic medium with children with learning difficulties: A case study. *Emotional & Behavioural Difficulties*, *11*(2), 135–146. doi:10.1080/13632750600619430

Berkes, F. (2012). *Sacred ecology*. Routledge Press. doi:10.4324/9780203123843

Berkes, F. (2018). *Sacred Ecology*. Routledge.

Bernal, I. (1994). Formación y desarrollo de Mesoamérica. In D. Cosío Villegas (Ed.), Historia general de México: volumen I (4th, reimpresión ed., pp. 125–164). El Colegio de Mexico. doi:10.2307/j.ctv47w8sq.6

Bernstorff, D., & von Welck, H. (Eds.). (2004). *Exile as Challenge: The Tibetan Diaspora*. Orient Longman. https://books.google.co.in/books?id=_lLiIvbf9S4C&printsec=frontcover&redir_esc=y#v=onepage&q&f=false

Berry, T. (1993). The Viable Human. Michael E. Zimmerman, J Baird Callicott, George Sessions G., Warren K. J. et al., (eds.) (1997). Environmental Philosophy From Animal Rights to Radical Ecology (pp.171-182) Prentice Hall

Betal, C. (2008). Conservation of ecology and environment through yogic lifestyle. *Journal of Environmental Research and Development*, *2*(4), 905–912. https://scholar.google.com/citations?view_op=view_citation&hl=en&user=SvszxRkAAAAJ&citation_for_view=SvszxRkAAAAJ:u5HHmVD_uO8C

BhatianiG.. (2022). Sustainable Urban Networks for Dynamic and Resilient (Sundar) India, Delhi: RTI International. DOI: doi:10.13140/RG.2.2.24783.76964

Bhatnagar, D. (2023). *Environmental Consciousness,* Medium. https://medium.com/@devanshi.art.bhatnagar/environmental-consciousness-63070dfd4d14

Bhattacharya, S., Bera, A., Dutta, A., & Ghosh, U. C. (2014). Effects of idol immersion on the water quality parameters of Indian water bodies: Environmental health perspectives. International Letters of Chemistry. *Physics and Astronomy*, *20*(2), 234–263.

Biba, J., & ... (2016). Traditional Knowledge systems in India for Biodiversity Conservation. *Indian Journal of Traditional Knowledge*, *15*(2), 304–312.

Bice, M. R., Ball, J. W., Bickford, N., Bickford, S., Hollman, A., Coughlin, A., Dinkel, D., Meyer, R., & Ranglack, D. H. (2018). Community Gardens: Interactions between Communities, Schools, and Impact on Students. *Health Educator : Journal of Eta Sigma Gamma*, *50*(1), 2–10. https://files.eric.ed.gov/fulltext/EJ1196092.pdf

Bilton, H., & Waters, J. (2017). Why Take Young Children Outside? A Critical Consideration of the Professed Aims for Outdoor Learning in the Early Years by Teachers from England and Wales. *Social Sciences (Basel, Switzerland)*, *6*(1), 1. Advance online publication. doi:10.3390/socsci6010001

Bladow, K., & Ladino, J. (Eds.). (2018). *Affective Ecocriticism: Emotion, Embodiment, Environment*. University of Nebraska Press. doi:10.2307/j.ctv75d0g8

Block, K., Gibbs, L., Staiger, P. K., Gold, L., Johnson, B., Macfarlane, S., Long, C., & Townsend, M. (2011). Growing community. *Health Education & Behavior*, *39*(4), 419–432. doi:10.1177/1090198111422937 PMID:22167317

Boas, F. (1928). *Anthropology and Modern Life*. Routledge.

Bodian, S. (1982). Simple in Means, Rich in Ends: A Conversation with Arne Naess. *OpenAirPhilosophy.org,* 3

Boileau, E. Y. S. (2013). Young Voices: The Challenges and Opportunities that Arise in Early Childhood Environmental Education Research. [CJEE]. *Canadian Journal of Environmental Education*, *18*, 142–154.

Boland, N. (2017). *Waldorf: An education of its time?* International Forum of Waldorf/Steiner Education.

Boltz, T., Impola, C., Nyquist, L., & Torgenson, A. (2012). *Say "yes!" to recycling at DCHS.* DCHS. http://www.dc.k12.mn.us/hs/ images/documents/2001-12/Sped-Newsletter-Website.pdf

Born, P. (2022). "It's All over the Map": A Report on Nature-Based Early Childhood Teacher Preparation Programs in the United States. *International Journal of Early Childhood Environmental Education, 9*(2), 69–82.

Boström, M., & Klintman, M. (2008). *Eco-standards, product labeling and green consumerism.* Palgrave Macmillan. doi:10.1057/9780230584006

Botero, E. U. (2005). *Natural Resource Conservation and Management in the Sierra Nevada of Santa Marta: Case Study. Universidad de los Andes, Facultad de Economía.* CEDE.

Boyd, D. (2019). Utilising place-based learning through local contexts to develop agents of change in Early Childhood Education for Sustainability. *Education, 47*(8), 983–997. doi:10.1080/03004279.2018.1551413

Brady, W. (1997). Indigenous Australian Education and Globalisation. *International Review of Education, 43*(5), 413–422. https://www.jstor.org/stable/3445057. doi:10.1023/A:1003077718880

Braun, S. (2023, Aug. 11). *Why is climate denial flourishing online?* Deutsche Welle. https://www.dw.com/en/climate-change-denial-still-thriving-online-but-why/a-66159857

Brayboy, B. M. J., & Maughan, E. (2009). Indigenous knowledge and the story of the bean. *Harvard Educational Review, 79*(1), 1–21. doi:10.17763/haer.79.1.l0u6435086352229

Breathing Space. (2023). *Yoga and Environmental Stewardship: Fostering a Connection with Nature in Children.* Breathing Space. https://breathingspacedc.com/yoga-and-environmental-stewardship-fostering-a-connection-with-nature-in-children/

Breidlid, A. (2009). Culture indigenous knowledge systems and sustainable development: Acritical view of education in an African context. *International Journal of Educational Development, 29*(2), 140–148. doi:10.1016/j.ijedudev.2008.09.009

Broda, H. (2002). Learning In and For the Outdoors. *Middle School Journal, 33*(3), 34–38. www.jstor.org/stable/23047005. doi:10.1080/00940771.2002.11494672

Bronowski, J. (1967). *The Common Sense of Science.* Harvard University Press.

Brown, J. C. (2017). A meta-synthesis of the complementarity of culturally responsive and inquiry-based science education in K-12 settings: Implications for advancing equitable science teaching and learning. *Journal of Research in Science Teaching, 54*(9), 11431173. doi:10.1002/tea.21401

Brown, K., Eernstman, N., Huke, A. R., & Reding, N. (2017). The drama of resilience: Learning, doing, and sharing for sustainability. *Ecology and Society, 22*(2), art8. doi:10.5751/ES-09145-220208

Bruchac, M. (2014). Indigenous knowledge and traditional knowledge. Encyclopedia of global archaeology, 10, 3814-24. doi:10.1007/978-1-4419-0465-2_10

Brundtland, G. H. (1987). Our Common Future, World Commission on Environment and Development (WECD). Oxford UK: Oxford University Press.

Brydon-Miller, M. (2018). Critical learning, community, and engagement: Elements for creating positive learning environments and opportunities for positive change. *Educational Action Research, 26*(3), 351–353. doi:10.1080/09650792.2018.1469651

Bula, G. (2008, May). Spinoza y el pensamiento ecológico. In *Ética y filosofía política, filosofía de la religión e historia de la filosofía: I Congreso Colombiano de Filosofía.* (p. 395). Universidad Jorge Tadeo Lozano. 10.2307/j.ctv23dxch6.23

Burchardt, J. (2021). How we can curb climate change by spending two percent more on everything. *TEDTalk*. https://www.ted.com/talks/jens_burchardt_how_we_can_curb_climate_change_by_spending_two_percent_more_on_everything

Burke, K. (1965). Terministic screens. *Philosophy and the Arts*, *39*, 87–102.

Burke, K. (1966). *Language as symbolic action: Essays on life, literature and method*. University of California Press. doi:10.1525/9780520340664

Burnaford, G. E., Fischer, J., & Hobson, D. (Eds.). (2001). *Teachers doing research: The power of action through inquiry*. Routledge.

Burns, H., Diamond-Vaught, H., & Bauman, C. (2015). Leadership for Sustainability: Theoretical Foundations and Pedagogical Practices that Foster Change. *International Journal of Leadership Studies*. https://www.regent.edu/acad/global/publications/ijls/new/vol9iss1/6-IJLS.pdf

Burt, K. G., Koch, P., & Contento, I. R. (2017). Development of the GREEN (Garden Resources, Education, and Environment Nexus) Tool: An Evidence-Based Model for School Garden Integration. *Journal of the Academy of Nutrition and Dietetics*, *117*(10), 1517–1527.e4. doi:10.1016/j.jand.2017.02.008 PMID:28389106

Burton, L. M. (1981). *A critical analysis and review of the research on Outward Bound and related programs*. [Unpublished doctoral dissertation, Rutgers University, New Brunswick, NJ].

Busch, K. C., Henderson, J. C., & Stevenson, K. T. (2019). Broadening epistemologies and methodologies in climate change education research. *Environmental Education Research*, *25*(6), 955–971. doi:10.1080/13504622.2018.1514588

Butterwick, S., & Selman, J. (2003). Deep listening in a feminist popular theatre project: Upsetting the position of audience in participatory education. *Adult Education Quarterly*, *54*(1), 7–22. doi:10.1177/0741713603257094

Cairns, R. (2020). *Dharmic environmentalism: Hindu traditions and ecological care* [Doctoral dissertation, Memorial University of Newfoundland].

Cajete, G. A. (2020). Indigenous science, climate change, and indigenous community building: A framework of foundational perspectives for indigenous community resilience and revitalization. *Sustainability (Basel)*, *12*(22), 9569. doi:10.3390/su12229569

Calamidas, E. G., Crowell, T. L., Engelmann, L., & Watkins-Jones, H. (2020). AtlantiCare healthy school edible garden startup grants: A content analysis of post-grant follow-up reports. *Health Education Journal*, *79*(6), 671–685. doi:10.1177/0017896920905622

Calis, D., & Yildirim, H. I. (2020). The Effect of Prediction, Observation, Explanation Supported Project-Based Environmental Education on the Levels of Attitude and Behavior Toward the Environment. *Educational Policy Analysis and Strategic Research*, *15*(1), 22–43. doi:10.29329/epasr.2020.236.2

Callicott, J. B. (2017). Notes on Self-Realization. *Worldviews*, *21*(3), 235-250.

Callicott, J. B. (1989). *In Defense of the Land Ethic*. SUNY Press.

Cambray, J. (2009). *Synchronicity: Nature and psyche in an interconnected universe* (Vol. 15). Texas A&M University Press.

Cameron-Faulkner, T., Melville, J., & Gattis, M. (2018). Responding to nature: Natural environments improve parent-child communication. *Journal of Environmental Psychology*, *59*, 9–15. doi:10.1016/j.jenvp.2018.08.008

Canella, G. S., & Lincoln, Y. S. (2009). Deploying critical methods for critical social purposes. In N. K. Denzin & M. D. Giardina (Eds.), *Qualitative inquiry and social justice: Toward a politics of hope* (pp. 53–72). Left Coast Press.

Capra, Fritjof (2013). Deep Ecology: Educational Possibilities for the Twenty-First Century. *The NAMTA Journal, 38*(1).

Cardinale, B. J., Duffy, J. E., Gonzalez, A., Hooper, D. U., Perrings, C., Venail, P., & Naeem, S. (2012). Biodiversity Loss and its Impact on Humanity. *Nature, 486*(7401), 59–67. https://doai.org/10.1038/nature11148. doi:10.1038/nature11148 PMID:22678280

Cargas, S., & Walsh Dilley, M. (2020). *Basic Needs Insecurity at UNM: 2020 Research Report.* https://basicneeds.unm.edu/basic-needs-at-unm-2020-research-report.pdf

Carr-Hill, R. (2016). The rural-urban gap in education in Honduras. *International Journal of Educational Development, 46*, 52–61.

Carr, V., Brown, R. D., Schlembach, S., & Kochanowski, L. (2017). Nature by Design: Playscape Affordances Support the Use of Executive Function in Preschoolers. *Children, Youth and Environments, 27*(2), 25–46. doi:10.1353/cye.2017.0002

Carter, E. W., Sisco, L. G., & Lane, K. L. (2011). Paraprofessional perspectives on promoting self-determination among elementary and secondary students with severe disabilities. *Research and Practice for Persons with Severe Disabilities : the Journal of TASH, 36*(1-2), 1–10. doi:10.2511/rpsd.36.1-2.1

Casimirri, G. (2003). Problems with Integrating Traditional Ecological Knowledge into Contemporary Resource Management, *Congress Proceedings of the III World Congress.* Quebec City, Canada: CCTL (2023). *Mission Statement of the Chicago Center for Teaching and Learning* (CCTL). Chicago: University of Chicago, Chicago Center for Teaching and Learning.

Castro, C. M. (2001). Pedagogía indígena: una visi6n diferente de hacer educaci6n en Colombia. *Nodos y Nudos, 2*(10).

CCE. (2019). *Finland Education Model.* Council for Creative Education. https://www.ccefinland.org/finland-education-model-c1yvm

Cernea, M. M. (1984). The role of sociological knowledge in planned rural development. *Sociologia Ruralis, 24*(3-4), 185–201. doi:10.1111/j.1467-9523.1984.tb00643.x

Chakrabarty, S. P., & Kaur, R. (2021). A Primer to Traditional Knowledge Protection in India: The Road Ahead. *The Liverpool Law Review, 42*(3), 401–427. doi:10.1007/s10991-021-09281-4 PMID:34099944

Chakravorty, S. (2021). Green Yoga and the Art of Living Harmoniously with Nature. *Livemint.* https://www.livemint.com/opinion/online-views/green-yoga-and-the-art-of-living-harmoniously-with-nature-11624164695671.html

Chambers, R., Pacey, A., & Thrupp, L. A. (1989). *Farmer First: Farmer Innovation and Agricultural Research.* Intermediate Technology Publications. doi:10.3362/9781780440149

Chandra, D. V. (2014). Re-examining the Importance of Indigenous Perspectives in the Western Environmental Education for Sustainability: From Tribal to Mainstream Education. *Journal of Teacher Education for Sustainability, 16*(1), 117–127. doi:10.2478/jtes-2014-0007

Chang, Y., & Chang, C. (2010). *The Benefits of Outdoor Activities for Children with Autism.* Research Gate. https://www.researchgate.net/profile/Chun_Yen_Chang2/publication/266369570_ The_Benefits_of_Outdoor_Activities_for_Children_with_Autism/links/54dd62460cf28a3d93f903c1/The-Benefits-of-Outdoor-Activities-for-Children- withAutism.pdf

Chantiluke, R., Kwoba, B., & Nkopo, A. (Eds.). (2018). *Rhodes must fall: The struggle to decolonise the racist heart of empire.* Zed. doi:10.5040/9781350222380

Chapple, C. K. (1993). *Nonviolence to Animals, Earth, and Self in Asian Traditions.* State University of New York Press.

Charles, C. (2009, October 2). The Ecology of Hope: Natural Guides to Building a Children and Nature Movement. *Journal of Science Education and Technology*. https://link.springer.com/article/10.1007/s10956-009-9193-z?error=cookies_not_supported&code=e548e227-9255-4ed1-979f-25a6849262ea

Chauhan S. & Chauhan S. V. S. (2019). *Worship and trees in India*. Sibirskij Lesnoj Zurnal (Sib. J. For. Sci.). . doi:10.15372/SJFS20190404

Chawla, L. (2018). Aatc keynote address: Nature-based learning for student achievement and ecological citizenship. *Curriculum & Teaching Dialogue*, *20*(1/2).

Chawla, S. (2018, September 25). From Illiteracy to PhD: How Exiled Tibetan Buddhist Nuns Are Ensuring Gender Parity. *The Wire*. https://thewire.in/religion/from-illiteracy-to-phd-how-exiled-tibetan-buddhist-nuns-are-ensuring-gender-parity

Chawla, S. (2019, March 15). The exiled daughters of Tibet seek equality. *Firstpost*. https://www.firstpost.com/world/the-exiled-daughters-of-tibet-seek-equality-6268131.html

Chawla, S., & Balasubramaniam, M. (2021, February 24). Writing more of India into Tibetology. *The Hindu*. https://www.thehindu.com/opinion/op-ed/writing-more-of-india-into-tibetology/article33917089.ece

Chawla, S. (2010). *"Like a Sudden Recovery of a Forgotten Memory": The Invention of Tradition Among Exiled Tibetans*. University of Delhi.

ChawlaS. (2015). The Female Renunciant in Exile: (Re)-Invention, Translation, Empowerment [University of Virginia]. doi:10.18130/V31H49

Cheeseman, A., & Wright, T. (2020). Perspectives from parents: An investigation of the longer-term benefits of an earth education summer camp program. *Applied Environmental Education and Communication*, *19*(2), 117–128. doi:10.1080/1533015X.2018.1530621

Cheng, L., Abraham, J., Trenberth, K. E., Boyer, T., Mann, M. E., Zhu, J., Wang, F., Yu, F., Locarnini, R., Fasullo, J., Zheng, F., Li, Y., Zhang, B., Wan, L., Chen, X., Wang, D., Feng, L., Song, X., & Liu, Y. (2024). New record ocean temperatures and related climate indicators in 2023. *Advances in Atmospheric Sciences*. https://link.springer.com/content/pdf/10.1007/s00376-024-3378-5.pdf?pdf=preview

Child and Nature Alliance, & Forest School Canada. (2020). *Principles of Forest/Nature School*.

Childs, G. (1991). *Steiner Education in Theory and Practice*. Floris Books.

Chinyowa, K. C. (2007). Helping them to help themselves: Case study of an integrated popular theatre approach in Africa. *Research in Drama Education*, *12*(2), 133–147. doi:10.1080/13569780701320957

Chirangi, M. M. (2013). '*Afya Jumuishi*': Towards Interprofessional Collaboration between Traditional and Modern Medical Practitioners in the Mara Region of Tanzania. PhD Dissertation. *Leiden Ethnosystems And Development Programme (LEAD) Studies* No. 8. Leiden: Leiden University.

Chivandikwa, N., & Makumbirofa, R. (2008). *Theatre*. Life Skills and Participatory Learning.

Choe, J.-H., Kim, C.-H., & Ri, G.-H. (2020). An investigation on the environmental knowledge and attitudes of senior middle school students in the Democratic People's Republic of Korea. *International Research in Geographical and Environmental Education*, *29*(2), 146–162. doi:10.1080/10382046.2019.1678276

Cho, H., Wehmeyer, M., & Kingston, N. (2011). Elementary teachers' knowledge and use of interventions and barriers to promoting student self-determination. *The Journal of Special Education*, *45*(3), 149–156. doi:10.1177/0022466910362588

Chowdhury, A., Iyer, A. S., Bhattacharyya, S., Khan, S. U., & Naz, A. (2022). Organic farming in the tropical, disaster-prone coasts: Interventions, major challenges and a success story from Sundarban Biosphere Reserve, India. []. IOP Publishing.]. *IOP Conference Series. Earth and Environmental Science*, *1084*(1), 012052. doi:10.1088/1755-1315/1084/1/012052

Chowdhury, A., Maiti, S. K., & Bhattacharyya, S. (2016). How to communicate climate change 'impact and solutions' to vulnerable population of Indian Sundarbans? From theory to practice. *SpringerPlus*, *5*(1), 1–17. doi:10.1186/s40064-016-2816-y PMID:27516957

Chowdhury, A., Nath, R., & Rosencranz, A. (2020). The sunderbans in crisis. *Economic and Political Weekly*, *55*(34), 5.

Chowdhury, A., Naz, A., Bhattacharyya, S., & Sanyal, P. (2018). Cost–benefit analysis of 'Blue Carbon' sequestration by plantation of few key mangrove species at Sundarban Biosphere Reserve, India. *Carbon Management*, *9*(6), 575–586. doi:10.1080/17583004.2018.1518105

Chowdhury, A., Prakash, R., Bhattacharyya, S., & Naz, A. (2023). Role of ponds as a local practice in Mitigating salinity intrusion threats at coastal aquifer: A case study from Sundarban Biosphere Reserve, India. In *Indigenous and Local Water Knowledge, Values and Practices* (pp. 287–306). Springer Nature Singapore. doi:10.1007/978-981-19-9406-7_17

Christiana, R. W., Davis, M., Wilson, M. G., McCarty, F. A., & Green, G. T. (2014). Factors related to rural young adolescents' participation in outdoor, noncompetitive physical activity. *Research Quarterly for Exercise and Sport*, *85*(4), 509–518. doi:10.1080/02701367.2014.961049 PMID:25412133

Chung, H. Q., Chen, V., & Olson, C. B. (2021). The impact of self-assessment, planning and goal setting, and reflection before and after revision on student self-efficacy and writing performance. *Reading and Writing*, *34*(7), 1885–1913. doi:10.1007/s11145-021-10186-x

Chu, T. (2014). *Human purpose and transhuman potential: a cosmic vision of our future evolution*. Red Wheel/Weiser.

CIKARD. LEAD & CIRAN(1990). Indigenous Knowledge and Development Monitor (IKDM), The Hague: CIRAN.

Clark, G. (1990). *Dialogue, dialectic and conversation: A social perspective on the function of writing*. siu press.

Clarkson, P. (1989). *Gestalt Counselling in Action*. SAGE Publications.

Clark, V. (2019). Gathering: An A/r/tographic Practice. In B. D. Hodgins (Ed.), *Feminist research for 21st-century childhoods: Common worlds methods* (pp. 121–130). Bloomsbury Publishing.

Clark, V., Pacini-Ketchabaw, V., & Hodgins, B. D. (2014). Thinking with paint: Troubling settler colonialisms through early childhood art pedagogies. *International Journal of Child, Youth and Family Studies, 5*(4.2). *Article*, *4*, 2. doi:10.18357/ijcyfs.clarkv.5422014

Claudia, M. (2012, April 9). *Notes from the Field: Environmental Education in Honduras*. Sustainable Development, Sustainable Livelihoods. https://treeswaterpeople.wordpress.com/2012/04/09/notes-from-the-field-environmental-education-in-honduras/

Clayton, S. (2020). Climate anxiety: Psychological responses to climate change. *Journal of Anxiety Disorders*, *74*, 102263. doi:10.1016/j.janxdis.2020.102263 PMID:32623280

Clugston, R., & Holt, S. (Eds.). (2012). *Exploring Synergies between Faith Values and Education for Sustainable Development*. Earth Charter International and Earth Charter Center for Education for Sustainable Development at University of Peace. https://www.iau-hesd.net/sites/default/files/documents/febr_2_2012.pdf

Coffey, A. (2001) Transforming School Grounds. In: Grant, T. and Littlejohn, G., Eds., Greening School Grounds: Creating Habitats for Learning. New Society Publishers, Gabriola Island, 2-5.

Cohen, M. (1999). *Section III - Philosophical Perspectives in Education.* Oregon State. https://oregonstate.edu/instruct/ed416/PP3.html

Cohen, M. (1997). *Reconnecting with Nature. Finding wellness through restoring your bond with the Earth.* Ecopress.

Cohen, M. J. (2010). *The Ecopsychology Interview.* Project NatureConnect Friday Harbor, Washington. *Ecopsychology, 2*(2), 53–57. doi:10.1089/eco.2010.0047

Collyer, R. (2023). How to supercharge renewables and energize the world. *TEDTalk.* https://www.ted.com/talks/rebecca_collyer_how_to_supercharge_renewables_and_energize_the_world

Comins Mingol, I. (2016). La Filosofía del Cuidado de la Tierra como Ecosofía. *Daimon, 67*, 133.

Comstock, A. B. (1911). *Handbook for Nature Study.* Comstock Publishing Company.

Confederacion Indigena Tairona CIT, O. d. (2015). *Plan de Salvaguarda del Pueblo Arhuaco* . Nabusimake.

Connolly, M., & Haughton, C. (2017). The perception, management and performance of risk amongst Forest School educators. *British Journal of Sociology of Education, 38*(2), 105–124. doi:10.1080/01425692.2015.1073098

Constantino, S. M., & Weber, E. U. (2021). Decision-making under the deep uncertainty of climate change: The psychological and political agency of narratives. *Current Opinion in Psychology, 42*, 151–159. doi:10.1016/j.copsyc.2021.11.001 PMID:34861621

Cook, J. (2019). Turning climate misinformation into an educational opportunity. In J. Fessmann (Ed.), *Strategic climate change communications: Effective approaches to fighting climate denial* (pp. 13–33). Vernon Press.

Cook, J. (2022). *Understanding and countering misinformation about climate change. Research anthology on environmental and societal impacts of climate change, 1633-1658.* IGI Global.

Cooper-Smithson, H. (2022). Arboromorphism. *Environmental Humanities, 14*(1), 233–236. doi:10.1215/22011919-9481539

Cotza, L. (2013, June 10). Improving education standards in Honduras: A long road ahead. *The Guardian.* https://www.theguardian.com/global-development-professionals-network/2013/jun/06/education-standards-honduras

Crain, W. (2005). *Theories of development: Concepts and applications* (5th ed.). Pearson Education, Inc.

Cramer, H., Sibbritt, D., Park, C. L., Adams, J., & Lauche, R. (2017). Is the practice of yoga or meditation associated with a healthy lifestyle? Results of a national cross-sectional survey of 28,695 Australian women. *Journal of Psychosomatic Research, 101*, 104–109. doi:10.1016/j.jpsychores.2017.07.013 PMID:28867414

Cree, J., & Robb, M. (2021). *The essential guide to Forest School and nature pedagogy.* Routledge. doi:10.4324/9780367853440

Cudworth, D. (2020). Promoting an emotional connection to nature and other animals via forest school: Disrupting the spaces of neoliberal performativity. *The International Journal of Sociology and Social Policy, 41*(3/4), 506–521. doi:10.1108/IJSSP-09-2019-0188

Cudworth, D., & Lumber, R. (2021). The importance of Forest School and the pathways to nature connection. *Journal of Outdoor and Environmental Education, 24*(1), 71–85. doi:10.1007/s42322-021-00074-x

Cultural Survival. (2013). Campaign Update: Honduras Gives Title to Lands to Miskito People. Cultural Survival. https://www.culturalsurvival.org/news/campaign-update-honduras-gives-title-lands-miskito-people

Dabaja, Z. F. (2022a). Reviewing two decades of research on the Forest School impact on children: The sequel. *Education, 50*(6), 737–750. doi:10.1080/03004279.2021.1905019

Dabaja, Z. F. (2022b). The Forest School impact on children: Reviewing two decades of research. *Education, 50*(5), 640–653. doi:10.1080/03004279.2021.1889013

Dabaja, Z. F. (2023). The strands of the Forest School implementation challenges: A literature review. *Prism.*

Dabrowski, K. (2015). *Personality Shaping Through Positive Disintegration*. Red Pill Press. (Original work published 1967)

Damico, J. S., Baildon, M., & Panos, A. (2020). Climate justice literacy: Stories-we-live-by, ecolinguistics, and classroom practice. *Journal of Adolescent & Adult Literacy, 63*(6), 683–691. doi:10.1002/jaal.1051

Daniel Boyd. (2021, December 18). *Lobo Gardens* [Video]. YouTube. https://www.youtube.com/watch?v=SXwTp0JqIrE

Das Gupta, S. (1997). Evolution of the Mother Worship in India. Swami Madhavananda & R. C. Majumdar, (eds.) The Great Women of India. Advaita Ashrama.

Das, L. (2022). Greenwash: What it is and how not to fall for it. *Greenpeace.* https://www.greenpeace.org.uk/news/what-is-greenwashing/

Das, A., Gujre, N., Devi, R. J., & Mitra, S. (2021). A review on traditional ecological knowledge and Its role in natural resources management: North east India, a cultural paradise. *Environmental Management*, 1–22. PMID:34791537

Dattatray, O. (2019). Yoga Education in India. *Northlines.* https://thenorthlines.com/yoga-education-in-india/

Davis, J. (2015). Retraction statement: LA sprouts randomized controlled nutrition, cooking and gardening program reduces obesity and metabolic risk in Latino youth. (2015). *Obesity (Silver Spring, Md.), 23*(12), 2522. doi:10.1002/oby.21390 PMID:26524103

De Bekker, J. C. M. (2020). *Transcultural Health Care Utilisation in Serengeti of Tanzania: Towards Applied Ethnoscience in Public Health Management*. PhD Dissertation. PhD Dissertation. *Leiden Ethnosystems And Development Programme (LEAD) Studies* No. 12. Leiden: Leiden University.

de la República de Colombia, C. (1994). Ley 115 del 8 de febrero. Recuperado de http://www.mineducacion.gov.co/1621/articles-85906_archivo_pdf.pdf

Decreto 804 de mayo 18 De 1995, Ministerio del interior. Recuperado de: Departamento Administrativo Nacional de Estadística (DANE). (2018). Censo Nacional de Población y Vivienda 2018.

Deleuze, G. (1994). *Difference and Repetition* (P. Patton, Trans.). Columbia University Press.

Deleuze, G. (1995). *Negotiations 1972–1990* (M. Joughin, Trans.). Columbia University Press.

DellaSala, D. A. (2021). Speaking truth to power for the Earth. In *Conservation Science and Advocacy for a Planet in Peril* (pp. 355–369). Elsevier. doi:10.1016/B978-0-12-812988-3.00010-7

Deringer, S. A., Hanley, A. W., Hodges, J. S., & Griffen, L. K. (2020). Improving Ecological Behavior in Outdoor Recreation Through Mindfulness Interventions: A Mixed Methods Inquiry. *Journal of Outdoor Recreation, Education, and Leadership, 12*(2), 2. doi:10.18666/JOREL-2020-V12-I2-9802

Derrida, J. (1992). Che cos'è la poesia? Points de suspension. Entretiens. Galilée.

Derr, V. (2020). Diverse perspectives on action for positive social and environmental change. *Environmental Education Research, 26*(2), 219–237. doi:10.1080/13504622.2020.1715925

Desai, S. & Smith, H. (2018). Kinship across species: Learning to care for non human others. *Feminist Review 118*(1), 41-60. DOI: 41305-018-0104-0. doi:10.1057/s

Descartes, R. (2010). Nonhumans as Machines. In D. R. Keller (Ed.), *Environmental Ethics: the big questions* (pp. 69–72). Wiley-Blackwell.

Deshmukh, V. D. (2019). Nava-Samkhya: New understanding, interpretation, and proposal for the classical Samkhya philosophy. *Journal of Alternative Medical Research, 11*(1), 53–59.

Devall, B., & Sessions, G. (1985). *Deep Ecology: Living As if Nature Mattered.* Gibbs M. Smith, Inc.

Devika, V. R., & Arulmani, G. (2014). Mahatma Gandhi's ideas for work, career, and life. In *Handbook of career development: International perspectives* (pp. 105–117). Springer New York. doi:10.1007/978-1-4614-9460-7_7

Dewall, B. (2021). *Living Deep Ecology: A Bioregional Journey* (S. C. Chew, Ed.). Lexington Books.

Dharma, A. P. (1981). *Indonesische Geneeskrachtige Planten.* De Driehoek.

Diehm, C. (2006). Arne Naess and the Task of Gestalt Ontology. *Environmental Ethics, 28*(1), 21–35. doi:10.5840/enviroethics200628137

Diener, E., & Suh, E. (2000). *Culture and Subjective Well-being.* The MIT Press. doi:10.7551/mitpress/2242.001.0001

Dietz, T., Rosa, E. A., & York, R. (2009). Environmentally efficient well-being: Rethinking sustainability as the relationship between human well-being and environmental impacts. *Human Ecology Review, 16*, 114–123.

Diffey, J., Wright, S., Uchendu, J. O., Masithi, S., Olude, A., Juma, D. O., Anya, L. H., Salami, T., Mogathala, P. R., Agarwal, H., Aboy, K. V., Cote, J., Saini, A., Mitchell, K., Kleczka, J., Lobner, N. G., Ialamov, L., Borbely, M., Hostetler, T., & Lawrance, E. (2022). "Not about us without us"–the feelings and hopes of climate-concerned young people around the world. *International Review of Psychiatry (Abingdon, England), 34*(5), 499–509. doi:10.1080/09540261.2022.2126297 PMID:36165749

Diya, V. S. (2023). *Environmental Consciousness and the Nine Schools of Indian Philosophy.* Cambridge Scholars Publishing.

Djen Amar, S. C. (2010). *Gunem Catur in the Sunda Region of West Java: Indigenous Communication on the MAC Plant Knowledge and Practice within the Arisan in Lembang.* PhD Dissertation. *Leiden Ethnosystems And Development Programme (LEAD) Studies* No. 6. Leiden: Leiden University.

Dkhar, M., & Tiwari, B. K. (2020). Traditional ecological knowledge of tribal communities of North East India. *Biodiversitas (Surakarta), 21*(7).

Doran, R. (2015). *The Theory of the Sublime from Longinus to Kant.* Cambridge University Press. doi:10.1017/CBO9781316182017

Dorn (2020). "A New Global Ethic»: A History of the United Nations International Environmental Education Program, 1975-1995". *Foro de Educación 18*(2), 83-108. DOI:. doi:10.14516/fde.808

Douglas, V., Purton, F., & Bascuñán, D. (2020). Possibility not difficulty: Difficult knowledge in K-12 classrooms as opportunities for renegotiating relationships with Indigenous perspectives and knowledge. *The Alberta Journal of Educational Research, 66*(3), 307–324. doi:10.55016/ojs/ajer.v66i3.56962

Drengson, A. (1995). The deep ecology movement. *The Trumpeter, 12*(3).

Drengson, A., Devall, B., & Schroll, M. A. (2011). The deep ecology movement: Origins, development, and future prospects (toward a transpersonal ecosophy). *International Journal of Transpersonal Studies*, *30*(1), 11.

Driver, B., Nash, R., & Haas, G. (1987). Wilderness benefits: a state of knowledge review. In *Proceedings: national wilderness research conference: issues, state-of-knowledge, future directions.* R. C. Lucas compiler. Gen. Tech. Rep. INT-220. Ogden, UT: U.S. Dept. of Agriculture, Forest Service, Intermountain Research Station.

du Prel, C. (1880). *Psychologie der Lyrik.* Günter Verlag.

Dumais, A. M. (2016). *Best Practices for Including Students with Autism, Disabilities, and Special Needs in Environmental Education* [Undergraduate Honors Thesis]. https://scholar.colorado.edu/honr_theses/1204/

Dumler-Winckler, E. (2015). "Romanticism as Modern Re-Enchantment: Burke, Kant, and Emerson on Religious Taste" Journal for the History of Modern Theology / Zeitschrift für Neuere Theologiegeschichte, 22(1). doi:10.1515/znth-2015-1001

Duncan, G. *The Psychological Benefits of Wilderness*. ePsychology. http://ecopsychology.athabascau.ca/Final/duncan.htm

Duram, L. A., & Klein, S. K. (2015). University food gardens: A unifying place for higher education sustainability. *International Journal of Innovation and Sustainable Development*, *9*(3/4), 282. doi:10.1504/IJISD.2015.071853

Dwivedi, O. P. (2001). *Classical India. A companion to environmental philosophy*, 37-51.

Dwivedi, O. P. (2004). Satyagraha for Conservation: Awakening the Spirit of Hinduism. In R. S. Gottlieb (Ed.), *This Sacred Earth: Religion, Nature, Environment* (2nd ed.). Routledge.

Dyczkowski, M. S. G. (2001). *The Cult of the Goddess Kubjika. A Preliminary Comparative Textual and Anthropological Survey of a Secret Newar Goddess.* Franz Steiner Verlag.

Dzambo, A. M., Mooney, M., Handlos, Z. J., Lindstrom, S., Hang, Y., & Ackerman, S. A. (2021). Improving Climate Literacy through an Interactive Online Undergraduate Course. *Bulletin of the American Meteorological Society*, *102*(6), 501–506. doi:10.1175/BAMS-D-19-0271.A

Eames, M. (1999). *Sheavly sowing the Seeds of Success: How to Start and Sustain a Kids, Gardening Project in Your Community, Natl.* Gardening Assn.

Easter Seals New Hampshire. (2012). *Poly recovery partners with Easter Seals NH to boost staffing for material disassembly.* Easter Seals New Hampshire. http://nh.easterseals.com/site/PageServer?pagename=NHDR_Poly_Recovery_Partners

Echeverría, J. (2011). Epistemopolis and knowledge communities. *Plenary Speech, International Conference on Technology, Knowledge and Society*, Bilbao, Spain.

Edgar, E., Patton, J. M., & Day-Vines, N. (2002). Democratic dispositions and cultural competency: Ingredients for school renewal. *Remedial and Special Education*, *23*(4), 231–241. doi:10.1177/07419325020230040601

Education & Training Foundation. (n.d.). *Education for sustainable development (ESD).* https://www.et-foundation.co.uk/resources/esd/

Elders, F. (2001). *The Sublime and the Beautiful: On Ontology and Creative Imagination.* VUB University Press.

Elliot, E., & Krusekopf, F. (2018). Growing a nature kindergarten that can flourish. *Australian Journal of Environmental Education.*, *34*(2), 115–126. doi:10.1017/aee.2018.27

Emma, P., & Sheila, D. (2016). Innovative pedagogical approaches in early childhood care and education (ECCE) in the Asia-Pacific region: a resource pack. *Asia-Pacific Regional Network for Early Childhood (Singapore)*. https://unesdoc.unesco.org/ark:/48223/pf0000246050

Enders, A. M., Uscinski, J. E., Seelig, M. I., Klofstad, C. A., Wuchty, S., Funchion, J. R., Murthi, M. N., Premaratne, K., & Stoler, J. (2023). The relationship between social media use and beliefs in conspiracy theories and misinformation. *Political Behavior*, *45*(2), 781–804. doi:10.1007/s11109-021-09734-6 PMID:34248238

Environmental and Energy Study Institute. (2023). *Public Polling on Climate Change*. EESI. https://www.eesi.org/briefings/view/020923camp

EPA. (2024). *Lesson plans, teacher guides and online environmental resources for educators*. EPA. https://www.epa.gov/students/lesson-plans-teacher-guides-and-online-environmental-resources-educators

Epelle, A., & Alafonye, U. O. (2010). Enhancing the Classroom Environment for Effective Teaching, Learning, *International Journal of Educational Development (IJOED)* (Jan 2010). https://www.researchgate.net/publication/362114552

Ernst, J., & Theimer, S. (2011). Evaluating the effects of environmental education programming on connectedness to nature. *Environmental Education Research*, *17*(5), 577–598. doi:10.1080/13504622.2011.565119

Ernst, J., Underwood, C., & Nayquonabe, T. (2023). Everyone Has a Piece of the Story: A Community of Practice Approach for Supporting Early Childhood Educators' Capacity for Fostering Empathy in Young Children through Nature-Based Early Learning. *International Journal of Early Childhood Environmental Education*, *11*(1), 34–62.

Erwina, W. (2019). *Iber Kasehatan* in *Tatar Sunda*, Indonesia: Utilisation of Traditional and Modern Health Information and Communication System in Sukamiskin, Bandung. PhD Dissertation. *Leiden Ethnosystem And Development Programme (LEAD) Studies* No. 10. Leiden: Leiden University.

European Union (2000). *European Biodiversity Programme Natura 2000* (2009/147/EG), Brussels: EU

Evans, B. (2014). Engaged Jain Traditions and Social Non-Violence: Ethnographic Case Studies of Lay Animal Activists and Service-Oriented Nuns. *Cross Currents*, *64*(2), 202–218.

Eyring, V., Gillett, N. P., Achuta Rao, K. M., Barimalala, R., Barreiro Parrillo, M., Bellouin, N., Cassou, C., Durack, P. J., Kosaka, Y., McGregor, S., Min, S., Morgenstern, O., & Sun, Y. (2021). Human influence on the climate system. In Masson-Delmotte, V., P. Zhai, A. Pirani, S.L. Connors, C. Péan, S. Berger, N. Caud, Y. Chen, L. Goldfarb, M.I. Gomis, M. Huang, K. Leitzell, E. Lonnoy, J.B.R. Matthews, T.K. Maycock, T. Waterfield, O. Yelekçi, R. Yu, and B. Zhou (Eds.), *Climate Change 2021: The Physical Science Basis. Contribution of Working Group I to the Sixth Assessment Report of the Intergovernmental Panel on Climate Change*. Cambridge University Press. https://www.ipcc.ch/report/ar6/wg1/chapter/chapter-3/

F. S. A. (n.d.). *What is the Forest School Association? | Forest School Association*. https://forestschoolassociation.org/the-forest-school-association/

F. S. C. (2012). *History of Forest and Nature School*. Forest School Canada. http://www.forestschoolcanada.ca/home/about-forest-school/history-of-forest-school

Fagan & Shephard. (1970). *Gestalt therapy now*. Harper & Row.

Fajaria, I. (2013, January). Outdoor Class Project: The Potential Benefits to Foster EFL Students' Language Proficiency. *Advances in Language and Literary Studies*, *4*(1), 140–144. Retrieved March 22, 2020, from https://eric.ed.gov/?id=EJ1129682. doi:10.7575/aiac.alls.v.4n.1p.140

Falkenberg, M., Galeazzi, A., Torricelli, M., Di Marco, N., Larosa, F., Sas, M., Mekacher, A., Pearce, W., Zollo, F., Quattrociocchi, W., & Baronchelli, A. (2022). Growing polarization around climate change on social media. *Nature Climate Change, 12*(12), 1114–1121. doi:10.1038/s41558-022-01527-x

Farm to School/New Mexico Grown. (2023). New Mexico Public Education Department. https://webnew.ped.state.nm.us/bureaus/student-success-wellness/nutrition/farm-to-%20school/#:~:text=The%20purpose%20of%20the%20New,the%20state%20for%20school%20meals

Febriyanti, M. (2021). *Ubar Kampung:* Indigenous Knowledge and Practice of Medicinal, Aromatic and Cosmetic (MAC) Plants Used for the Treatment of Diabetes Mellitus in the Tatar Sunda Region of West Java, Indonesia. PhD. Dissertation. *Leiden Ethnosystems And Development Programme (LEAD) Studies* No. 13. Leiden University.

Fedulova, I., Ivanova, V., Atyukova, O., & Nosov, V. (2019). Inclusive education as a basis for sustainable development of society. *Journal of Social Studies Education Research, 10*(3), 118–135. www.jsser.org

Ferris, E. (2008). *Natural disasters, human rights and the role of national human rights institutions.* The Brookings Institution. https://www.brookings.edu/articles/natural-disasters-human-rights-and-the-role-of-national-human-rights-institutions/

Fessmann, J. (2019). An unlevel playing field: A primer on the problems of climate change communications. In J. Fessmann (Ed.), *Strategic climate change communications: Effective approaches to fighting climate denial* (pp. 1–11). Vernon Press.

Fiedler, C. R., & Danneker, J. E. (2007). Self-advocacy instruction: Bridging the research-topractice gap. *Focus on Exceptional Children, 39*(8), 1–21. doi:10.17161/foec.v39i8.6875

Finch, K. (2016). The Dollars and Sense of Business Planning and Budgeting in D. Sobel, P. Bailie, K. Finch, E. Kenny, & A. Stires (Eds.), Nature Preschools and Forest Kindergartens: The Handbook for Outdoor Learning (Illustrated edition, pp. 158-190). Redleaf Press.

Finch, K. (2016). The Risks and Benefits of Nature Play in D. Sobel, P. Bailie, K. Finch, E. Kenny, & A. Stires (Eds.), Nature Preschools and Forest Kindergartens: The Handbook for Outdoor Learning (Illustrated edition, pp. 137-157). Redleaf Press.

Finch, K., & Bailie, P. (2016). Nature Preschools: Putting Nature at the Heart of Early Childhood Education. *Occasional Paper Series, 2015*(33). doi:10.58295/2375-3668.1020

Fisher, A. (2002). *Radical Ecopsychology. Psychology in the service of Life.* State University of New York Press. doi:10.1353/book4515

Fletcher, Nick. (1991, Jan. 31). Indians' plea to big chief Hanson. *The Times.*

Flood, G. D. (1996). *An Introduction to Hinduism.* Cambridge University Press.

Flores, A., & Mena, P. (2018). Cultural and Intercultural Education: Experiences of Ethno-Educational Teachers in Colombia [on line]. *The Australian Journal of Teacher Education, 43*(7), 62–81. doi:10.14221/ajte.2018v43n7.4

Florian, L. (2015). Conceptualising Inclusive Pedagogy: The Inclusive Pedagogical Approach in Action: Inclusive Pedagogy Across the Curriculum. In *International Perspectives on Inclusive Education* (Vol. 7). Emerald Group Publishing Limited., doi:10.1108/S147936362015000007001

Ford, B. A. (2004). Preparing special educators for culturally responsive school-community partnerships. Teacher Education and Special Education. *The Journal of the Teacher Education Division of the Council for Exceptional Children, 27*(3), 224–230.

Forsyth, M. (2013). How Can Traditional Knowledge Best Be Regulated? Comparing a Proprietary Rights Approach with a Regulatory Toolbox Approach. *University of Hawai'i Press, 25*(1), 1-31. https://www.jstor.org/stable/23725729

Fowler-Smith, L. (2009). *Hindu Tree Veneration as of Mode of Environmental Encounter*. The MIT Press Stable. https://www.jstor.org/stable/20532588

Fox, P., & Avramidis, E. (2003). An evaluation of an outdoor education programme for students with emotional and behavioural difficulties. *Emotional & Behavioural Difficulties*, *8*(4), 267–283. doi:10.1080/13632750300507025

Francis, D. (2013). You know the homophobic stuff is not in me, like us, it's out there". Using Participatory Theatre to challenge heterosexism and heteronormativity in a South African school. *South African Journal of Education*, *33*(4), 1–14. doi:10.15700/201412171338

Frank, G. (1985, Summer). Becoming the Other [University of Hawai'i Press on behalf of Center for Biographical Research Stable.]. *Empathy and Biographical Interpretation, Biography*, *8*(3), 189–210.

Frankl, V. (2000). *Man's Search For Meaning, An Introduction to Logotherapy*. Touchstone / Simon & Schuster.

Frankl, V. (2006). *Man's Search for Meaning*. Beacon Press.

Fraser, C., & Restrepo-Estrada, S. (1998). *Communicating for development: human change for survival*. IB Tauris and Co Ltd.

Freebody, K., & Finneran, M. (Eds.). (2015). *Drama and social justice: Theory, research and practice in international contexts*. Routledge. doi:10.4324/9781315755977

Freshwater, D., & Stickley, T. (2004). The heart of the art: Emotional intelligence in nurse education. *Nursing Inquiry*, *11*(2), 91–98. doi:10.1111/j.1440-1800.2004.00198.x PMID:15154888

Friedmann, H. (1930). *Die Welt der Formen. System eines morphologischen Idealismus*. C.H. Beck'sche Verlagsbuchhandlung.

Fullan, M. G. (2001). *The new meaning of educational change* (3rd ed.). Teachers College Press. doi:10.4324/9780203986561

Gaard, G. (2017). *Critical ecofeminism*. Lexington Books.

Gagliano, M., Ryan, J. C., & Vieira, P. (Eds.). (2017). *The Language of Plants. Science, Philosophy, Literature*. University of Minnesota Press.

Galeana, F. & Pantoja, E (2013). *Strengthening Indigenous Peoples' Land Tenure Governance: the Case of the Miskito People in Honduras*. Academia. https://www.academia.edu/4314700/Strengthening_Indigenous_Peoples_Land_Tenure_Governance_the_Case_of_the_Miskito_People_in_Honduras

Ganatsios, H. P., Filippou, A. F., Mpekiri, S., & Danahy, K. E. (2021). Education for cultivation of awareness as an important factor of raising environmental consciousness. *International Journal of Higher Education and Sustainability*, *3*(3), 231–265. https://www.inderscienceonline.com/doi/pdf/10.1504/IJHES.2021.117900. doi:10.1504/IJHES.2021.117900

Gandhi, G. (2014, October 14). Honoured Guest: The gem that India will not call by its name. *The Telegraph Online*. https://www.telegraphindia.com/opinion/honoured-guest-the-gem-that-india-will-not-call-by-its-name/cid/1312859

Gandini, L. (2003). Values and Principles of the Reggio Emilia Approach. In Insights and inspirations from Reggio Emilia: Stories of teachers and children from North America.

Gandini, L. (1993). Fundamentals of the Reggio Emilia Approach to Early Childhood Education. *Young Children*, *49*(1), 4–8.

Gautheron, O. (2014). *Teachers' perceptions of the effects of nature-based learning on junior level students.* Semantic Scholar. https://www.semanticscholar.org/paper/ Teachers%27-Perceptions-of-the-Effects-of-Learning-on-Gautheron/6f9dadff2d7aa 3324c45eb30094360612b8cc6ee

Gavin, M. (2017). *Accountability: Education in Honduras.* UNESCO. https://unesdoc.unesco.org/ark:/48223/pf0000259546

Gazizova, A. S. (2020). Protection of Traditional Knowledge: The Work and the Role of International Organisations and Conferences. *International Journal of Higher Education, 9*(8), 95–99. doi:10.5430/ijhe.v9n8p95

Geary, D. (2008). Destination Enlightenment: Branding Buddhism and Spiritual Tourism in Bodhgaya, Bihar. *Anthropology Today, 24*(3), 11–14. doi:10.1111/j.1467-8322.2008.00584.x

Gere, A. R. (2009). *Writing Outside of School.* NCTE. http://www.ncte.org/library/NCTEFiles/Resources/Journals/CC/0191-sep09/Chron0191BriefWriting.pdf

Gheneti, Y. (2007). *Microcredit Management in Ghana: Development of Co-operative Credit Unions among the Dagaaba,* PhD dissertation, *Leiden Ethnosystems And Development Programme LEAD Studies* No. 2. Leiden: Leiden University.

Ghose, L. (2007). Karma and the possibility of purification: An ethical and psychological analysis of the doctrine of karma in Buddhism. *The Journal of Religious Ethics, 35*(2), 259–290. doi:10.1111/j.1467-9795.2007.00306.x

Gilbert, L. (2005). What Helps Beginning Teachers? *Educational Leadership, 62*(8), 36–39.

Gillen, J. (2021). A New Approach to Environmental Education: Yoga Calm Earth Warriors. *Yoga Calm.* https://www.yogacalm.org/a-new-approach-to-environmental-education-yoga-calm-earth-warriors/

Giraldo Jaramillo, N. (2010). Camino en espiral. Territorio sagrado y autoridades tradicionales en la comunidad indígena iku de la Sierra Nevada de Santa Marta, Colombia. *Revista Pueblos y fronteras digital, 5*(9), 180-222.

Glackin, M. (2018). 'Control must be maintained': Exploring teachers' pedagogical practice outside the classroom. *British Journal of Sociology of Education, 39*(1), 61–76. doi:10.1080/01425692.2017.1304204

Glavin, C. (2017). *Prussian Education System.* K12 Academics. https://www.k12academics.com/Education Worldwide/Education in Germany/History/prussian-education-system

Gleason, T. (2021). *A Critical Ethnography of an Outdoor School: Reimagining the Relationship Between Science Education and Climate Change Politics.* Routledge. doi:10.4324/9780429331626

Godrej, F. (2010, September). *Gandhi's Body: Asceticism, Pain and Suffering in Environmental Political Discourse.* 2010 Conference of the American Political Science Association, Washington, DC.

Goldstein, T. R., & Lerner, M. D. (2018). Dramatic pretend play games uniquely improve emotional control in young children. *Developmental Science, 21*(4), e12603. doi:10.1111/desc.12603 PMID:28913920

Goleman, D., Barlow, Z., & Bennett, L. (2010). Forging new norms in New Orleans. *Teacher Education Quarterly, 37*(4), 87–98.

Gómez, F. F. G. (2013). Learning and adaptation as conservation practices in resilient traditional socio-ecological systems: The Elder Brothers of Sierra Nevada de Santa Marta. *Revista de Tecnología, 12*(1), 99–109.

Gonzales, T., Chambi, N., & Machaca, M. (1999). Agriculture and Cosmovision in the Contemporary Andes: The Nurturing of the Seeds. In D. A. Posey (Ed.), *Cultural and Spiritual Values of Biodiversity: A Complementary Contribution to the Global Biodiversity Assessment* (pp. 167–260). Intermediate Technology Publications Ltd.

Goodall, J. (1992). Roots and Shoots Programme, Washington D.C., USA: Jane Goodall Institute (JGI).

Goodall, J. (1999). *In the Shadow of Man.* Mariner Books.

Goodenough, W. H. (1957). "Cultural Anthropology and Linguistics"., in Garvin, P.C. (Ed.), *Report of the Seventh Annual Roundtable Meeting on Linguistics and Language Study*, 167-173. Washington, D.C: Georgetown University.

Goodwill. (2011). *E-waste.* Goodwill. http:// www.gire.org/ewaste.htm

Gordon, S. (2010). Major Asian Rivers of the Plateau of Tibet: The Basics. *Education About Asia, 15*(3), 15–18.

Gorman, S., & Gorman, J. M. (2019, Jan. 12). Climate change denial: Facing a reality too big to believe. *Psychology Today.* https://www.psychologytoday.com/us/blog/denying-the-grave/201901/climate-change-denial

Gough, A. (2006). A Long, Winding (and Rocky) Road to Environmental Education for Sustainability in 2006. *Australian Journal of Environmental Education, 22*(1), 71–76. doi:10.1017/S0814062600001671

Gough, A., & Scott, W. (2016). Environmental education policy research: A systematic review. *Environmental Education Research, 22*(6), 800–813.

Griffith, R. (Ed.). (1896) *The Rig Veda.* https://sacred-texts.com/hin/rigveda/index.htm

Grosvenor, I., & Myers, K. (2020). "Dirt and the child": A textual and visual exploration of children's physical engagement with the urban and the natural world. *History of Education, 49*(4), 517–535. https://proxy.marybaldwin.edu:2238/10.1080/0046760X.2019.1701097. doi:10.1080/0046760X.2019.1701097

Guerrero, O. M. B. (2005). *El diálogo de saberes y la educación ambiental* (Vol. 7). Univ. Nacional de Colombia.

Guha, R. (1989). Radical Environmentalism and Wilderness Preservation: A Third World Critique. *Environmental Ethics, 11*(1), 71–83. doi:10.5840/enviroethics198911123

Gupta, A. (2013). Altruism in Indian religions: embracing the biosphere. In *Altruism in cross-cultural perspective* (pp. 101–112). Springer New York. doi:10.1007/978-1-4614-6952-0_8

Gupta, S. M. (1971). *Plant Myths and Traditions in India.* E.J Brill. doi:10.1163/9789004611542

Gyatso, J. (2010). Female Ordination in Buddhism: Looking into a Crystal Ball, Making a Future. In T. Mohr & J. Tsedroen (Eds.), *Dignity and Discipline: Reviving Full Ordination for Buddhist Nuns* (pp. 1–21). Wisdom Publications.

Gyatso, J., & Havnevik, H. (Eds.). (2005). *Women in Tibet.* Columbia University Press.

Haberman, D. L. (2010). *Faces in the Trees.* The Journal of Religion. *Nature and Culture, 4*(2), 173–190.

Haberman, D. L. (2013). *People Trees. Worship of Trees in Northern India.* Academic. doi:10.1093/acprof:oso/9780199929177.001.0001

Hackett, S., & Dissanayake, S. T. (2014). *Environmental and natural resources economics: Theory, policy, and the sustainable society.* Routledge. doi:10.4324/9781315704586

Hale, J., Conerty, J., & Lindgren, S. (2023, October 12). *Environmental and Sustainability Education Programs at AAUs: A Review of the Field* [Comboned Panel Paper Presentation]. North American Association of Environmental Education Research Symposium, PheedLoop Online Conference Platform.

Ham, H. S., & Catillo, L. (1990). Elementary Schools in Rural Honduras Problems in Exporting Environmental Education Models &om the United States. *The Journal of Environmental Education, 21*(4), 27–32. doi:10.1080/00958964.1990.9941935

Handler, D., & Epstein, A. S. (2010). Nature education in preschool. *Highscope Extensions, 25*(2), 1–7.

Hannah, M. (2012). Colliding Gender Imaginaries: Transnational Debates about Full Ordination for Tibetan Buddhist Nuns. *Asian Journal of Women's Studies*, *18*(4), 7–44. doi:10.1080/12259276.2012.11666134

Haraway, D. J. (1985). A manifesto for cyborgs: Science, technology, and socialist feminism in the 1980s. *Socialist Review*, *5*(2), 65–107.

Haraway, D. J. (2016). *Staying with the Trouble: Making Kin in the Chthulucene*. Duke University Press.

Hargrove, E. C. (1992). Weak Anthropocentric Intrinsic Value. *The Monist*, *75*(2), 183–207. doi:10.5840/monist19927529

Haring, N. (2021). *Growing a Foest School: From the roots up!* (1st ed.). Forest School Association.

Harisha, R. P., Padmavathy, S., & Nagaraja, B. C. (2016). Traditional ecological knowledge (TEK) and its importance in south India: Perspecive from local communities. *Applied Ecology and Environmental Research*, *14*(1), 311–326. doi:10.15666/aeer/1401_311326

Harisha, R.P & others. (2015). Traditional ecological knowledge (TEK) and its importance in South India. *Applied Ecology and Environmental Research.*

Haristha, J. (2019). Sacred groves of north Kerala: The last refuge for biodiversity amongst urbanisation. *Mongabay*. https://india.mongabay.com/2019/01/sacred-groves-of-north-kerala-the-last-refuge-for-biodiversity-amongst-urbanisation/

Härkönen, M. (2023). *Power and Agency in the Lives of Contemporary Tibetan Nuns: An Intersectional Study*. Equinox. doi:10.1558/isbn.9781800503021

Harper, A. B. (Ed.). (2009). *Sistah vegan: Black female vegans speak on food, identity, health, and society*. Lantern Books.

Harrington, B. (2013). Using Indigenous Educational Research to Transform Mainstream Education: A Guide for P–12 School Leaders. *American Journal of Education*, *119*(4), 487–511. doi:10.1086/670962

Harris, K. I. (2016). Let's play at the park! Family pathways promoting spiritual resources to inspire nature, pretend play, storytelling, intergenerational play and celebrations. *International Journal of Children's Spirituality*, *21*(2), 90–103.. do i:10.1080/1364436X.2016.1164669

Harvey, G. (2019). Animism and ecology: Participating in the world community. *The Ecological Citizen*, *3*, 79-84. https://www.mininterior.gov.co/la-institucion/normatividad/decreto-804-de-1995

Harwood, D., Boileau, E., Dabaja, Z., & Julien, K. (2020). Exploring the national scope of outdoor nature-based early learning programs in Canada: Findings from a large-scale survey study. *The International Journal of Holistic Early Learning and Development*, *6*, 1–24.

Hatala, A. R., Njeze, C., Morton, D., Pearl, T., & Bird-Naytowhow, K. (2020). Land and nature as sources of health and resilience among Indigenous youth in an urban Canadian context: A photovoice exploration. *BMC Public Health*, *20*(1), 1–14. doi:10.1186/s12889-020-08647-z PMID:32312240

Hatfield, S., Marino, E., & Whyte, K. P. (2018). Indian time: Time, seasonality, and culture in Traditional Ecological Knowledge of climate change. *Ecological Processes*, *25*(1), 7. doi:10.1186/s13717-018-0136-6

Hattie, J., Marsh, H. W., Neill, J. T., & Richards, G. E. (1997). Adventure education and Outward Bound: Out-of-class experiences that make a lasting difference. *Review of Educational Research*, *67*(1), 43–87. doi:10.3102/00346543067001043

Havnevik. (1994). The Role of Nuns in Contemporary Tibet. In R. Barnett (Ed.), *Resistance and Reform in Tibet* (pp. 259–266). University of Indiana Press.

Hay-Edie, T. (1999). Landscape perception and sensory emplacement in decision-making. In D. A. Posey (Ed.), *Cultural and spiritual values of biodiversity* (pp. 65–74). United Nations Environment Programme.

Hays, D., & Ledford, L. (2020). *Collaboration for Impact: Modeling an Integrated Partnership.* Virginia Association for Environmental Education Conference. Amherst, Virginia.

Hazzard, E. L., Moreno, E., Beall, D. L., & Zidenberg-Cherr, S. (2011). Best Practices Models for implementing, sustaining, and using instructional school gardens in California. *Journal of Nutrition Education and Behavior*, *43*(5), 409–413. doi:10.1016/j.jneb.2011.05.005 PMID:21906553

Healthline. (n.d.). Everything You Need to Know About Tantric Yoga. https://www.healthline.com/health/tantric-yoga

Healthline. (n.d.). What Is Bhakti Yoga? https://www.healthline.com/health/fitness/bhakti-yoga

Healthline. (n.d.). What Is Kundalini Yoga and What Are the Benefits? https://www.healthline.com/health/kundalini-yoga

Hebinck, P. (2018). Poaching: Between Conservation from Below, Livelihoods and Resistance. In Nature Conservation in Southern Africa (pp. 257-292). Brill.

Hegel, F. (1966). *Science of Logic*. George Allen and Unwin.

Helms, M. W. (1969). The Cultural Ecology of a Colonial Tribe. *Ethnology*, *8*(1), 76–84. doi:10.2307/3772938

Henderson, P. J. (2013). *A Multi-Case Study of Teens Who Write Outside of School for Their Own Purposes.* TRACE. trace.tennessee.edu/cgi/viewcontent.cgi?article=3634&context=utk_graddiss

Heras, M., & Tàbara, J. D. (2014). Let's play transformations! Performative methods for sustainability. *Sustainability Science*, *9*(3), 379–398. doi:10.1007/s11625-014-0245-9

Heras, M., & Tàbara, J. D. (2016). Conservation theatre: Mirroring experiences and performing stories in community management of natural resources. *Society & Natural Resources*, *29*(8), 948–964. doi:10.1080/08941920.2015.1095375

Hernández-Martínez, A., et al. (2021). Ethno-education: Health Education Based on Cultural Diversity, *City Diversity Research* (2022). Volume 3 Issue 2, .1–10.

Hesse, H. (2017). *Demian*. Penguin Classics.

Heywood, V. H., & Watson, R. T. (Eds.). (1995). Global Biodiversity Assessment, United Nations Environment Programme. Cambridge, UK: Cambridge University Press.

Hill, R., Adem, Ç., Alangui, W. V., Molnár, Z., Aumeeruddy-Thomas, Y., Bridgewater, P., Tengö, M., Thaman, R., Adou Yao, C. Y., Berkes, F., Carino, J., Carneiro da Cunha, M., Diaw, M. C., Díaz, S., Figueroa, V. E., Fisher, J., Hardison, P., Ichikawa, K., Kariuki, P., & Xue, D. (2020). Working with Indigenous, local and scientific knowledge in assessments of nature and nature's linkages with people. *Current Opinion in Environmental Sustainability*, *43*, 8–20. doi:10.1016/j.cosust.2019.12.006

Holifield, R. (2001). Defining Environmental Justice and Environmental Racism. *Urban Geography*, *22*(1), 78–90. https://www.researchgate.net/profile/Ryan_Holifield/publication/250171256_Defining_Environmental_Justice_and_Environmental_Racism/links/5694093108ae425c6896255b/Defining-Environmental-Justice-and-Environmental-Racism.pdf. doi:10.2747/0272-3638.22.1.78

Holzhauer, R. W., & Gellaerts, S. L. (2022). Ideas and the Law. In G. Enthoven, S. Rudnicki, & R. Sneller (Eds.), *Towards a Science of Ideas: About the Emergence, Evolution and Expansion of Ideas and Their Translation into Action.* Vernon Press.

Honwad, S. (2018). Learning in Schools about Traditional Knowledge Systems in the Kumaon Himalayas. *Journal of Folklore and Education, 5*, 180–193. https://jfepublications.org/wp-content/uploads/2018/09/honwad-JFE20185.2-11.pdf

Howard, V. R. (2023). The Nonviolence Conundrum: Political Peace and Personal Karma in Jain and Hindu Traditions. *Religions, 14*(2), 178. doi:10.3390/rel14020178

Huang, M. H., & Rust, R. T. (2011). Sustainability and consumption. *Journal of the Academy of Marketing Science, 39*(1), 40–54. doi:10.1007/s11747-010-0193-6

Huber, T. (2001). Shangri-la in Exile: Representations of Tibetan Identity and Transnational Culture. In T. Dodin & H. Rather (Eds.), *Imagining Tibet—Perceptions, Projections, and Fantasies*. Wisdom Publications.

Hume, J. (2022). Meeting in Forest: Education in Posthuman Era. *Cultural and Pedagogical Inquiry, 14*(1), 118-127. http://ejournals.library.ualberta.ca/index.php/cpi/index

Hurley, S. (2009). *The Benefits of Taking Students Outside to Inspire Writing*. Edutopia. https://www.edutopia.org/student-motivation-writing-outdoors

Husamah, H., Miharja, F. J., & Hidayati, D. A. (2020). *Environmental Literacy of Islamic Boarding School Students: Study in MA Bilingual-Sidoarjo*. Prisma.

Husserl, E. (1962/1936). Die Krisis der europäischen Wissenschaften und die transzendentale Phänomenologie. Husserliana Band VI2. Martinus Nijhoff. Trans. The Crisis of European Sciences and Transcendental Phenomenology: An Introduction to Phenomenological Philosophy. Northwestern University Press, 1989.

Ibui, A. K. (2007). *Indigenous Knowledge, Belief and Practice of Wild Plants among the Meru of Kenya*. PhD Dissertation. *Leiden Ethnosystems And Development Programme (LEAD) Studies*, No. 3. Leiden: Leiden University.

Id. (1950). *Sinnvolle Odyssee. Geschichte eines Lebens und einer Zeit 1873-1950*. Beck'sche Verlagsbuchhandlung.

Id. (1956). *Das Gemüt. Gedanken zu einer Thymologie*. C.H. Beck.

Id. (1981. 1929-1932). *Der Geist als Widersacher der Seele*. Bouvier Verlag.

Inter-American Development Bank. (n.d.). *Save the Planet Initiative: An Education Pilot Project on Climate Change for Indigenous and Afro-Honduran communities*. IADB. https://www.iadb.org/en/whats-our-impact/HO-T1245

International Environmental Partnership. (n.d.). *Global Environmental Education Partnership*. IEP. https://www.iep-global.org/programs/global-environmental-education-partnership-geep/

IPBS. (2020). *Global Assessment Report on Biodiversity and Ecosystem Services*, Intergovernmental Science-Policy Platform on Biodiversity and Ecosystem Services (IPBES), Bonn, Germany, IPBES: www.ipbes.net

IPCC (1923). *Sixth Assessment Synthesis Report* (AR6) *of the Intergovernmental Panel on Climate Change* (IPCC). Geneva, Switzerland: ipcc-media@wmo.int.

Irvine, J. J. (2012). Complex relationships between multicultural education and special education: An African American perspective. *Journal of Teacher Education, 63*(4), 268–274. doi:10.1177/0022487112447113

Ismail, S. M., & Cazden, C. B. (2005). Struggles for Indigenous Education and Self-Determination: Culture, Context, and Collaboration. *Anthropology & Education Quarterly, 36*(1), 88–92. doi:10.1525/aeq.2005.36.1.088

IUCN. (1980). *World Conservation Strategy: Living Resource Conservation for Sustainable Development*. IUCN.

IUCN. (2023). Commission on Education and Communications (IUCN-CEC). Gland, Switzerland: IUCN

IUCN/NEWP/WWF. (1991). Caring for the Earth: A strategy for Sustainable Living. North American Wetlands Conservation Council (NEWP), International Union for Conservation of Nature (IUCN) and World Wildlife Fund (WWF). Gland, Switzerland: IUCN.

Iyengar, B. K. S. (2012). *Light on the Yoga Sutras of Patanjali* [ebook edition]. HarperCollins UK. https://www.everand.com/book/234816673/Light-on-the-Yoga-Sutras-of-Patanjali

Izquierdo, D. A. N. (2017). El Pueblo Arhuaco, Defensor del Territorio. *Americanía: Revista de Estudios Latinoamericanos*, 86-102.

Jackson, A. (2000). Inter-acting with the past-the use of participatory theatre at museums and heritage sites. *Research in Drama Education*, *5*(2), 199–215. doi:10.1080/713692887

Jackson, T. (Ed.). (1993). *Learning through theatre: New perspectives on theatre in education*. Psychology Press.

Jacob, M., Sabzalian, L., Jansen, J., Tobin, T. J., Vincent, C. G., & LaChance, K. M. (2018). The Gift of Education: How Indigenous Knowledges Can Transform the Future of Public Education. *International Journal of Multicultural Education*, *20*(1), 157–185. doi:10.18251/ijme.v20i1.1534

Jacobson, T. L. (1993). A pragmatist account of participatory communication research for national development. *Communication Theory*, *3*(3), 214–230. doi:10.1111/j.1468-2885.1993.tb00071.x

Jain, R. (2022). How to Practice the Yamas – Applied Yoga Philosophy. *Arhanta Yoga*. Retrieved from https://www.arhantayoga.org/blog/how-to-practice-the-yamas/

Jain, R. (2023). *What Is Jnana Yoga? Meaning, Benefits & How to Practice the Yoga of Knowledge,* Arhanta Yoga. https://www.arhantayoga.org/blog/jnana-yoga-guide/

Jain, R. (2023). *What Is Karma Yoga? Key Principles, Benefits, & Tips for Daily Practice,* Arhanta Yoga. https://www.arhantayoga.org/blog/four-principles-of-karma-yoga/

Jakovich, K., & Murray, J. A. (2019). Yoga in China. *The Asia Dialogue*. Retrieved from https://theasiadialogue.com/2019/12/18/yoga-in-china/

Jakubczak, M. (2017) The Woman-and-Tree motif in the ancient and contemporary India. Retracing the Past, Historical Continuity in aesthetics from a global perspective. International Association for Aesthetics.

Jalais, A. (2008). Bonbibi: Bridging Worlds. *Indian Folklife*, *28*, 6–8.

Jalais, A. (2014). *Forest of tigers: people, politics and environment in the Sundarbans*. Routledge. doi:10.4324/9780203085516

Jaspers, K. (1919). *Psychologie der Weltanschauungen*. Springer. doi:10.1007/978-3-662-05505-2

Jelavic, H. M. E. (2014). *Education for sustainable development: assessing the development impact of environmental education (EE) on Roatan, Honduras*. Universitet i Agder / University of Agder. http://hdl.handle.net/11250/222386

Jenkins, W. (2012). Religion and Environmental Rights. In J. Witte & M. C. Green (Eds.), *Religion and Human Rights: An Introduction* (pp. 330–345). Oxford University Press.

Jha, R. (2019). Evolution of Theatre as a means of Social Development. SOCIAL SCIENCE GAZETTEER.

Johnsen, L. (1999). *The Living Goddess. Reclaiming the Tradition of the Mother of the Universe*. Yes International Publishers.

Johnstone, C. (2020). *Nature Rx*. Chesapeake Bay Program. https://www.chesapeakebay.net/news/blog/nature_rx

Johnstone, A., Martin, A., Cordovil, R., Fjørtoft, I., Iivonen, S., Jidovtseff, B., Lopes, F., Reilly, J. J., Thomson, H., Wells, V., & McCrorie, P. (2022). Nature-Based Early Childhood Education and Children's Social, Emotional and Cognitive Development: A Mixed-Methods Systematic Review. *International Journal of Environmental Research and Public Health, 19*(10), 10. doi:10.3390/ijerph19105967 PMID:35627504

Jordan, C., & Chawla, L. (2019). A Coordinated Research Agenda for Nature-Based Learning. *Frontiers in Psychology, 10*, 766. doi:10.3389/fpsyg.2019.00766 PMID:31065246

Jorgenson, S. N., Stephens, J. C., & White, B. (2019). Environmental education in transition: A critical review of recent research on climate change and energy education. *The Journal of Environmental Education, 50*(3), 160–171. doi:10.1080/00958964.2019.1604478

Joshi, A., Azuma, A. M., & Feenstra, G. (2008). Do Farm-to-School programs make a difference? Findings and future research needs. *Journal of Hunger & Environmental Nutrition, 3*(2–3), 229–246. doi:10.1080/19320240802244025

Ju, E. J., & Kim, J. G. (2011). Using soil seed banks for ecological education in primary school. *Journal of Biological Education, 45*(2), 93–101. doi:10.1080/00219266.2010.546010

Kaiser, W. K., & Meyer, J. H. (2016). *International Organizations and Environmental Protection: Conservation and Globalization in the Twentieth Century,* New York, USA. *Berghahn., 2017.* Advance online publication. doi:10.3167/9781785333620

Kala, C. P. (2018). Cultural significance and current conservation practices of the Ganga's ecosystem and environment. *Applied Ecology and Environmental Sciences, 6*(4), 128–136.

Kanchibhotla, D., Nair, A., & Sangle, P. Gupte & P., Kulkarni, S. (2021). The Correlation between Yoga Practice, Connectivity with Nature and Climate Change Awareness in Indian Population. *International Journal of Innovative Research in Science, Engineering and Technology, 2*(11), 001-005. https://www.iomcworld.org/articles/the-correlation-between-yoga-practice-connectivity-with-nature-and-climate-change-awareness-in-indian-population.pdf

Kanojia, S. (2024). Guardians of Atolls: Examining the Paris agreement's role in climate change mitigation and resilience in the Maldives. In Bridging Health, Environment, and Legalities: A Holistic Approach (pp. 271-291). IGI Global.

Kant, I. (2010). Indirect Duties to Nonhumans. In D. R. Keller (Ed.), *Environmental Ethics: the big questions* (pp. 82–83). Wiley-Blackwell.

Kaplan, R., & Kaplan, S. (1989). *The experience of nature: a psychological perspective*. Cambridge University Press.

Kapoor, S. K. (2016). Hinduism: the faith eternal. Advaita Ashrama (A publication branch of Ramakrishna Math, Belur Math).

Kapstein, A., & Goldstein, T. R. (2019). Developing wonder: Teaching theatre for the very young through collaboration with developmental psychology. *Youth Theatre Journal, 33*(1), 52–69. doi:10.1080/08929092.2019.1580648

Kaptani, E., & Yuval-Davis, N. (2008). Participatory theatre as a research methodology: Identity, performance and social action among refugees. *Sociological Research Online, 13*(5), 1–12. doi:10.5153/sro.1789

Kara, A. (2008). *And the Thunder Said DA: Stories of Dharma in the Mythology, Philosophy, and Religion of India.* iUniverse.

Katz, J. (2015). Implementing the three-block model of Universal Design for Learning: Effects on teachers' self-efficacy, stress, and job satisfaction in inclusive classrooms K-12. *International Journal of Inclusive Education, 19*(1), 1–20. doi:10.1080/13603116.2014.881569

Kaza, S., & Gottlieb, R. S. (2006). The Greening of Buddhism. In *The Oxford Handbook of Religion and Ecology* (pp. 184–206). Oxford University Press., doi:10.1093/oxfordhb/9780195178722.003.0008

Keleş, Ö. (2015). Creative drama in the teaching of education for sustainable development: Sample activity about sustainability and sustainable production. *Journal of Environmental Sustainability*, *4*(4), 4.

Kellert, S. R., & Farnham, T. (Eds.). (2013). *The good in nature and humanity: connecting science, religion, and spirituality with the natural world.* Island Press.

Kemp, J. (2009). Exploring empowerment issues with student midwives using forum theatre. *British Journal of Midwifery*, *17*(7), 438–439. doi:10.12968/bjom.2009.17.7.43058

Kent, E. F. (2013). *Sacred groves and local gods: religion and environmentalism in South India.* Oxford university press. doi:10.1093/acprof:oso/9780199895465.001.0001

Kerr, R. B., Young, S. L., Young, C., Santoso, M. V., Magalasi, M., Entz, M., & Snapp, S. S. (2022). Farming for change: developing a participatory curriculum on agroecology, nutrition, climate change and social equity in Malawi and Tanzania. In *Critical Adult Education in Food Movements* (pp. 29–46). Springer Nature Switzerland. doi:10.1007/978-3-031-19400-9_3

Keskin, C., Akcay, H., & Kapici, H. O. (2020). The effects of environmental science projects on middle school students' behaviors and attitudes. [IJTES]. *International Journal of Technology in Education And Science*, *4*(2), 160–167. doi:10.46328/ijtes.v4i2.84

Khan, S. R., & Naqvi, A. (2000). Root Causes of Environmental Degradation. In *The Environment-Poverty Nexus: An Institutional Analysis* (pp. 1–7). Sustainable Development Policy Institute., https://www.jstor.org/stable/resrep00631.4

Kimmerer, R. (2013). *Braiding Sweetgrass: Indigenous Wisdom, Scientific Knowledge and the Teachings of Plants.* Milkweed Editions.

Kimmerer, R. W. (2002). Weaving traditional ecological knowledge into biological education: A call to action. *Bioscience*, *52*(5), 432–438. doi:10.1641/0006-3568(2002)052[0432:WTEKIB]2.0.CO;2

Kim, S., & Park, S. (2020). Garden-Based Integrated intervention for improving children's eating behavior for vegetables. *International Journal of Environmental Research and Public Health*, *17*(4), 1257. doi:10.3390/ijerph17041257 PMID:32075303

Klaniecki, K., Leventon, J., & Abson, D. J. (2018). Human–nature connectedness as a 'treatment' for pro-environmental behavior: Making the case for spatial considerations. *Sustainability Science*, *13*(5), 1375–1388. doi:10.1007/s11625-018-0578-x

Klein, N. (2019). On Fire: The (Burning) Case for a Green New Deal. New York: Simon & Schuster. Kaufman, C. (September 16, 2013). "Time to Write? Go Outside." New York Times. https://opinionator.blogs.nytimes.com/2013/09/16/time-to-write-go-outside/

Kleinman, A. (1987). Anthropology and psychiatry. The role of culture in cross-cultural research on illness. *The British Journal of Psychiatry*, *151*(4), 447–454. doi:10.1192/bjp.151.4.447 PMID:3447661

Klenk, N., & Meehan, K. (2015). Climate change and transdisciplinary science: Problematizing the integration imperative. *Environmental Science & Policy*, *54*, 160–167. doi:10.1016/j.envsci.2015.05.017

Knapp, B. L. (1997). The Popol Vuh: Primordial Mother Participates in the Creation. *Confluencia, 12*(2), 31–48. https://www.jstor.org/stable/27922440

Knapp, C. N., Fernandez-Gimenez, M., Kachergis, E., & Rudeen, A. (2011, March). Using Participatory Workshops to Integrate State-and-Transition Models Created with Local Knowledge and Ecological Data. *Rangeland Ecology and Management*, *64*(2), 158–170. doi:10.2111/REM-D-10-00047.1

Knight, S. (2013). Forest School and Outdoor Learning in the Early Years. *Forest School and Outdoor Learning in the Early Years*, 1–176.

Kochetkova, T., & Segalerba, G. (2024). Environmental Personhood: Philosophical Foundations in Tantric Ecology. Bridging Health, Environment, and Legalities: A Holistic Approach. 1-24.

Kochetkova, T. (2007). *Search for Authentic Spirituality in Modern Russian Philosophy*. Edwin Mellen Press.

Kochetkova, T. (2022). Environmental Management in Transition: Lessons from Tantra. In S. P. Sahni, T. Bhatnagar, & P. Gupta (Eds.), *Spirituality and Management. From Models to Applications* (pp. 77–97). Springer. doi:10.1007/978-981-19-1025-8_6

Kodirekkala, K. (2017). Internal and External Factors Affecting Loss of Traditional Knowledge: Evidence from a Horticultural Society in South India. *Journal of Anthropological Research*, *73*(1), 22–40. https://www.journals.uchicago.edu/doi/epdf/10.1086/690524. doi:10.1086/690524

Koller, D., & Farley, M. (2019). Examining elements of children's place attachment. *Children's Geographies*, *17*(4), 491–500. doi:10.1080/14733285.2019.1574336

Kopnina, H., Washington, H., Taylor, B., & Piccolo, J. (2021). Anthropocentrism: More than Just a Misunderstood Problem. *The International Journal of Ecopsychology*, *3*, 3–23. https://digitalcommons.humboldt.edu/ije/vol3/iss1/4

Korteweg, L., & Russell, C. (2012, May). Decolonizing + Indigenizing = Moving Environmental Education Towards Reconciliation. *Canadian Journal of Environmental Education*, •••, 17.

Kothari, A., & Das, P. (1999). "Local Community Knowledge and Practice"s. In D. A. Posey (Ed.), *Cultural and Spiritual Values of Biodiversity: A Complementary Contribution to the Global Biodiversity Assessment*. Intermediate Technology Publications Ltd.

Krishnamurti, J. (1953). *Education and the Significance of Life* (M. Lutyens, Ed.). Harper & Row.

Kubisch, S., Parth, S., Deisenrieder, V., Oberauer, K., Stötter, J., & Keller, L. (2020). From transdisciplinary research to transdisciplinary education—The role of schools in contributing to community well-being and sustainable development. *Sustainability (Basel)*, *13*(1), 306. doi:10.3390/su13010306

Kubota, R. (2020). Confronting epistemological racism, decolonizing scholarly knowledge: Race and gender in Applied Linguistics. *Applied Linguistics*, *41*(5), 712–732. doi:10.1093/applin/amz033

Kudryavtsev, A., Stedman, R. C., & Krasny, M. E. (2012). Sense of place in environmental education. *Environmental Education Research, 18*(2), 229 - 250. 10.1080/13504622.2011.609615

Kuhn, J. L. (2001). Toward an Ecological Humanistic Psychology. *Journal of Humanistic Psychology*, *41*(2), 9–24. doi:10.1177/0022167801412003

Kumrai, R. R., Chauhan, V., & Hoy, J. (2011). Boundary crossings: Using participatory theatre as a site for deepening learning. *Teaching in Higher Education*, *16*(5), 517–528. doi:10.1080/13562517.2011.570443

Kuthe, A., Körfgen, A., Stötter, J., & Keller, L. (2020). Strengthening their climate change literacy: A case study addressing the weaknesses in young people's climate change awareness. *Applied Environmental Education and Communication*, *19*(4), 375–388. doi:10.1080/1533015X.2019.1597661

Kyte, D. G. (2017). Toward a Sustainable Sense of Self in Teaching and Teacher Education: Sustainable Happiness and Well-Being Through Mindfulness. *McGill Journal of Education, 51*(3), 1143–1162. https://mje.mcgill.ca/article/view/9321. doi:10.7202/1039632ar

La Belle, T. J. (1982). Formal, Nonformal and Informal Education: A Holistic Perspective on Lifelong Learning. *International Review of Education / Internationale Zeitschrift Für Erziehungswissenschaft / Revue Internationale de l'Education, 28*(2), 159–175. https://www.jstor.org/stable/3443930

Ladson-Billings, G. (1995). Toward a theory of culturally relevant pedagogy. *American Educational Research Journal, 32*(3), 465–491. doi:10.3102/00028312032003465

Lamar-Dukes, P. (2010). Reaching the hard to reach: A review of an initiative aimed at increasing participation and support for people of color with disabilities and their families in disability organizations. *Research and Practice for Persons with Severe Disabilities : the Journal of TASH, 34*(3-4), 76–80. doi:10.2511/rpsd.34.3-4.76

LancasterOnline. [LNP]. (2024, February 2). *This week's good things: Environmental education takes root, helping hands in Honduras, and F&M College volunteerism.* Lancaster Online. https://lancasteronline.com/opinion/editorials/this-weeks-good-things-environmental-education-takes-root-helping-hands-in-honduras-and-f-m/article_d81ef58c-c15c-11ee-9da0-7be190c33228.html

Langton, M., & Rhea, Z. M. (2005). Traditional indigenous biodiversity-related knowledge. *Australian Academic and Research Libraries, 36*(2), 45–69. doi:10.1080/00048623.2005.10721248

Lappin, E. (1984, September). *Outdoor education for behavior-disordered students* (pp. 1–4). ERIC. https://eric.ed.gov/contentdelivery/servlet/ERICServlet? accno=ED261811

Larimore, R. (2016). Defining Nature-Based Preschools. *International Journal of Early Childhood Environmental Education, 4*(1), 32–36.

Larimore, R. A. (2018). Using Principles of Nature-Based Preschools to Transform Your Classroom. *YC Young Children, 73*(5), 34–41.

Larr and Neidell (2016). Pollution and Climate Change. *Children and Climate Change, 26.* doi:10.1353/foc.2016.0005

Lash, C. L. (2018). Making Americans: Schooling, Diversity, and Assimilation in the Twenty-First Century. *The Russell Sage Foundation Journal of the Social Sciences : RSF, 4*(5), 99–117. doi:10.7758/rsf.2018.4.5.05

Lawlor, L. (2008). Following the Rats: Becoming-Animal in Deleuze and Guattari. *SubStance, 37*(3).

Leakey, R. E., & Lewin, R. (1996). *The Sixth Extinction: Patterns of Life and the Future of Humankind.* Bantam Double Dell Publishing Group, Inc.

Leakey, R. E., & Slikkerveer, L. J. (1993). *Man-Ape Ape-Man: The Quest for Human's Place in Nature and Dubois' 'Missing Link.* Kenya Wildlife Service Foundation.

Leather, M. (2018). A critique of "Forest School" or something lost in translation. *Journal of Outdoor and Environmental Education, 21*(1), 5–18. doi:10.1007/s42322-017-0006-1

Lee, M. J., Oh, W., Jang, J. S., & Lee, J. Y. (2018). A pilot study: Horticulture-related activities significantly reduce stress levels and salivary cortisol concentration of maladjusted elementary school children. *Complementary Therapies in Medicine, 37,* 172–177. doi:10.1016/j.ctim.2018.01.004 PMID:29609930

Leeming, D. A., & Page, J. (1996). *God: myths of the male divine.* Oxford University Press.

Leiringer, R., & Cardellino, P. (2011). Schools for the twenty-first century: School design and educational transformation. *British Educational Research Journal, 37*(6), 915–934. www.jstor.org/stable/23077016. doi:10.1080/01411926.2010.508512

Leiserowitz, A., Maibach, E., Rosenthal, S., Kotcher, J., Goddard, E., Carman, J., Ballew, M., Verner, M., Marlon, J., Lee, S., Myers, T., Goldberg, M., Badullovich, N., & Thier, K. (2023). *Climate change in the American mind: Beliefs & attitudes. Yale Program on Climate Change Communication.* Yale University and George Mason University. https://climatecommunication.yale.edu/publications/climate-change-in-the-american-mind-beliefs-attitudes-fall-2023/

Leung, V. T. W., Yee Tiffany Tam, T., Pan, W.-C., Wu, C.-D., Candice Lung, S.-C., & Spengler, J. D. (2019). How is environmental greenness related to students' academic performance in English and Mathematics? *Landscape and Urban Planning, 181*, 118–124. doi:10.1016/j.landurbplan.2018.09.021

Leurs, L. N. (2010). *Medicinal, Aromatic and Cosmetic (MAC) Plants for Community Health and Bio-Cultural Diversity Conservation in Bali, Indonesia.* PhD Dissertation, *Leiden Ethnosystems And Development Programme (LEAD) Studies,* No. 5. Leiden: Leiden University.

Lewis, E., Mansfield, C., & Baudains, C. (2008). Getting down and dirty: Values in education for sustainability. *Issues in Educational Research, 18*(2), 138–155.

Liang, J. (2021, December 9). Buddhist nuns and female scholars are gaining new leadership roles, in a tradition that began with the ordination of Buddha's foster mother. *The Conversation.* http://theconversation.com/buddhist-nuns-and-female-scholars-are-gaining-new-leadership-roles-in-a-tradition-that-began-with-the-ordination-of-buddhas-foster-mother-172477

Liang, J., & Taylor, A. S. (2020). Tilling the Fields of Merit: The Institutionalization of Feminine Enlightenment in Tibet's First Khenmo Program. *Journal of Buddhist Ethics, 27*, 231–262.

Lidke, J. S. (2016) Towards a Theory of Tantra-Ecology. *Sutra Journal.* http://www.sutrajournal.com/towards-a-theory-of-tantra-ecology-by-jeffrey-s-lidke

Lidke, J. S. (2017). *The Goddess Within and Beyond the Three Cities, Sakta Tantra and the Paradox of Power in Nepala-Mandala.* DK Printworld Pvt. Ltd.

Lieberman, G. (2013). *Education and the Environment: Creating Standards-Based Programs in Schools and Districts.* Harvard Education Press.

Liebovich, B. (2018). The McMillan Sisters, The Roots of the Open-Nursery, and Breaking the Cycle of Poverty. *HSE–Social and Education History, 7*(1), 78–96. doi:10.17583/hse.2018.2925

Lineberger, S. E., & Zajicek, J. M. (2000). School gardens: Can a hands-on teaching tool affect students' attitudes and behaviors regarding fruit and vegetables. *HortTechnology, 10*(3), 593–597. doi:10.21273/HORTTECH.10.3.593

Lipps, T. (1883). *Grundtatsachen des Seelenlebens.* Cohen.

LitWorld. (2019, May 23). *Growing Environmental Awareness in Honduras.* LitWorld. https://www.litworld.org/blog/2019/5/23/un-mundo

Li, W. T., & Shein, P. P. (2023). Developing sense of place through a place-based Indigenous education for sustainable development curriculum. *Environmental Education Research, 29*(5), 692–714. doi:10.1080/13504622.2022.2098933

Locke, J. (2010). Nature as Economic Resource. In D. R. Keller (Ed.), *Environmental Ethics: the big questions* (pp. 77–82). Wiley-Blackwell.

Lomawaima, K. T., & McCarty, T. L. (2006). *"To remain an Indian": Lessons in democracy from a century of Native American education*. Teachers College Press.

Longinus. (2000). *Het Sublime (The sublime)*. Historische Uitgeverij.

Lopez, D. S., Jr. (2018). *Prisoners of Shangri-La: Tibetan Buddhism and the West*. University of Chicago Press. https://press.uchicago.edu/ucp/books/book/chicago/P/bo26798165.html

Loreman, T. (2017). *Pedagogy for Inclusive Education, Oxford Research Encyclopedia of Education*. Oxford University Press.

Louv, R. (2005). *Last Child in the Woods. Saving Our Children from Nature-Deficit Disorder*. Algonquin Books.

Louv, R. (2005). *Last Child in the Woods: Saving Our Children from Nature Deficit Disorder*. Algonquian Books of Chapel Hill.

Louv, R. (2008). *Last child in the Woods: Saving our children from nature-deficit disorder*. Algonquin Books of Chapel Hill.

Louv, R. (2016). Nature Smart. [World Edition]. *National Wildlife, 54*(2), 40–43. proxy.marybaldwin.edu:2060/login.aspx?direct=true&db=a9h&AN=112738502&site=ehost-live

Luomi, M. (2020). *Global Climate Change Governance: The search for effectiveness and universality*. International Institute for Sustainable Development (IISD). https://www.jstor.org/stable/resrep29269

Lynas, M., Houlton, B. Z., & Perry, S. (2021). Greater than 99% consensus on human caused climate change in the peer-reviewed scientific literature. *Environmental Research Letters, 16*(11), 114005. doi:10.1088/1748-9326/ac2966

MacEachren, Z. (2013). The Canadian Forest School Movement. *LEARNing Landscapes, 7*(1), 1. doi:10.36510/learnland.v7i1.639

Macedo, D. (Ed.). (2019). *Decolonizing foreign language education: The misteaching of English and other colonial languages*. Routledge.

Macekura, S. J. (2017). *Of Limits and Growth: The Rise of Global Sustainable Development in the Twentieth Century*. Cambridge University Press.

Machan, T. R. Do Animals have Rights. In J. Rachels & S. Rachels (Eds.), *The Right Thing to Do: Basic Readings in Moral Philosophy* (pp. 45–59). Mcgraw Hill.

Maclean, R., Jagannathan, S., & Sarvi, J. (Eds.). (2013). *Skills Development for Inclusive and Sustainable Growth in Developing Asia-Pacific (Technical and Vocational Education and Training: Issues, Concerns and Prospects Series)*. Springer Dordrecht. doi:10.1007/978-94-007-5937-4

MacNeill, T. (2020). Indigenous food sovereignty in a captured state: The Garifuna in Honduras. *Third World Quarterly, 41*(9), 1537–1555. doi:10.1080/01436597.2020.1768840

MacQuarrie, S., Nugent, C., & Warden, C. (2015). Learning with nature and learning from others: Nature as setting and resource for early childhood education. *Journal of Adventure Education and Outdoor Learning, 15*(1), 1–23. doi:10.1080/14729679.2013.841095

Maftuh, B., & Adriany, V. (2018). Bringing Social Justice into Research in Education. In E. Malihah, T. Aryanti, V. Adriany, H. Yulindrasari, & A. Izharuddin (Eds.), *Research for Social Justice*. Routledge Taylor and Francis Group.

Magni, G. (2017). Indigenous knowledge and implications for the sustainable development agenda. *European Journal of Education, 52*(4), 437–447. doi:10.1111/ejed.12238

Mahale, P., & Sorée, H. (1999). Cosmovisions and Agriculture in India. In D. A. Posey (Ed.), *Cultural and Spiritual Values of Biodiversity: A Complementary Contribution to the Global Biodiversity Assessment*. Intermediate Technology Publications Ltd.

Makley, C. E. (2005). The Body of a Nun: Nunhood and Gender in Contemporary Amdo. In J. Gyatso & H. Havnevik (Eds.), *Women in Tibet* (pp. 259–284). Columbia University Press.

Malhotra, A. K. (2001). *An Introduction to Yoga Philosophy: An annotated translation of theYoga Sutras*. Routledge.

Malihah, E., Aryanti, T., Adriany, V., Yulindrasari, H., & Izharuddin, A. (Eds.). (2018). *Research for Social Justice*. Routledge Taylor and Francis Group.

Mallik, B. (2022). Beauty, Wonder, and Sustainability: Reconnecting Humans to Nature for "Durable Harmony". In *Legends in Gandhian Social Activism: Mira Behn and Sarala Behn: Addressing Environmental Issues By Dissolving Gender And Colonial Barriers* (pp. 437–488). Springer International Publishing. doi:10.1007/978-3-030-95431-4_9

MAMO. (1997). *Universo arhuaco No. 10.* Antioquia, Medellín, junio.

Marley, S. A., Siani, A., & Sims, S. (2022). Real-life research projects improve student engagement and provide reliable data for academics. *Ecology and Evolution, 12*(12), e9593. doi:10.1002/ece3.9593 PMID:36514554

Marouli, C. (2002). Multicultural environmental education: Theory and practice. *Canadian Journal of Environmental Education, 7*(1), 26–42.

Martin, H., Farrell, A., Gray, J., & Clark, T. B. (2018). Perceptions of the effect of recess on kindergartners. *Physical Educator, 75*(2), 245–254. doi:10.18666/TPE-2018-V75-I2-7740

Maslow, A. (1987). *Motivation and Personality*. HarperCollins Publishers.

Mathai, W. (2023). The tree-growing movement restoring Africa's vital landscapes. *TEDTalk*. https://www.ted.com/talks/wanjira_mathai_the_tree_growing_movement_restoring_africa_s_vital_landscapes

Mathews, F. (2001). Deep Ecology. In D. Jamieson (Ed.), *A Companion to Environmental Philosophy* (pp. 218–232). Blackwell. doi:10.1002/9780470751664.ch15

Mathur, A. (2003). Who Owns Traditional Knowledge? *Economic and Political Weekly, 38*(42), 4471–4481. https://www.jstor.org/stable/4414163

May, A. (1958/2004). *Existence*. Rowman & Littelfield Publishers: Lanham.

Maynard, T. (2007). Forest Schools in Great Britain: An Initial Exploration. *Contemporary Issues in Early Childhood, 8*(4), 320–331. doi:10.2304/ciec.2007.8.4.320

May, S., & Aikman, S. (2003). Indigenous Education: Addressing Current Issues and Developments. *Comparative Education, 39*(2), 139–145. doi:10.1080/03050060302549

McArdle, R. (2021). Intersectional climate urbanism: Towards the inclusion of marginalised voices. *Geoforum, 126*, 302–305. doi:10.1016/j.geoforum.2021.08.005

McClimens, A., & Scott, R. (2007). Lights, camera, education! The potentials of forum theatre in a learning disability nursing program. *Nurse Education Today, 27*(3), 203–209. doi:10.1016/j.nedt.2006.04.009 PMID:16790301

McClintic, S., & Petty, K. (2015). Exploring Early Childhood Teachers' Beliefs and Practices About Preschool Outdoor Play: A Qualitative Study. *Journal of Early Childhood Teacher Education, 36*(1), 24–43. doi:10.1080/10901027.2014.997844

McCormick, J. (1991). *Reclaiming Paradise: The Global Environmental Movement.* Indiana University Press.

McCree, M., & Cree, J. (2017). Forest School: Core principles in changing times. *Children Learning Outside the Classroom from Birth to Eleven*, 222–232.

McDaniel, J. (2004). *Offering Flowers, Feeding Skulls. Popular Goddess Worship in West Bengal.* Oxford University Press. doi:10.1093/acprof:oso/9780195167900.001.0001

McDonagh, B., Brookes, E., Smith, K., Worthen, H., Coulthard, T. J., Hughes, G., Mottram, S., Skinner, A., & Chamberlain, J. (2023). Learning histories, participatory methods and creative engagement for climate resilience. *Journal of Historical Geography*, *82*, 91–97. doi:10.1016/j.jhg.2023.09.002

McDonald, B., Guldin, R., & Wetherhill, R. (1985). The spirit in the wilderness: The use and opportunity of wilderness experience for spiritual development. In R. Lucas, (Ed.) *Proceedings of the National Wilderness Research Conference: Issues, state of knowledge, future directions*, (pp. 294-319). IEEE.

Mcgovern, S. M. (2000). Reclaiming Education: Knowledge Practices and Indigenous Communities. *Comparative Education Review*, *44*(4), 523–529. doi:10.1086/447633

McGregor, D. (2004). Traditional ecological knowledge and sustainable development: Towards coexistence. In M. Blaser, H. Feit, & G. McRae (Eds.), *In the way of development, indigenous people, life projects and development* (pp. 72–91). Zed Books. doi:10.5040/9781350220720.ch-005

McMahan, D. L. (2008). *The Making of Buddhist Modernism.* Oxford University Press.

McNair, L. (2012). Offering Children first hand experiences through forest school: Relating to and learning about nature. In T. Bruce (Ed.), *Early Childhood Practice: Froebel Today* (pp. 86–99). SAGE Publications, Inc. doi:10.4135/9781446251287.n6

Mechler, H. M. (2016). Fostering Critical Thinking in Young Children. *Texas Child Care Quarterly, 40*(2). https://www.childcarequarterly.com/pdf/fall16_thinking.pdf

Mehrabian, A., & Russel, J. (1974). *An Approach to Environmental Psychology.* The MIT Press.

Melnick Dyer, A. (2022). *The Tibetan Nun Mingyur Peldrön: A Woman of Power and Privilege.* University of Washington Press.

Merlone, L., & Moran, D. (2008). Transition works: Self-awareness and self-advocacy skills for students in the elementary learning center. *Teaching Exceptional Children Plus*, *4*(4), 1–10.

Merrick, C. (Ed.). (2019). *Nature-Based Preschool Professional Practice Guidebook.* North American Association for Environmental Education.

Merrigan, K. A., & Bailey, M. R. (2008). The potential of Farm-to-College programs. *Nutrition Today*, *43*(4), 160–165. doi:10.1097/01.NT.0000303339.75605.9e

Mesoamerican Permaculture Institute. (n.d.). *Sowing the seeds helping it grow.* Mesoamerican Permaculture Institute. https://imapermaculture.org/

Metri, N. (2018). *Internet Society: How the Lenca Are Restoring the Past to Build Their Future.* Internet Society. https://www.internetsociety.org/blog/2018/04/lenca-people-restoring-past-build-future/

Meyer, J., Müller, U., & Macoun, S. (2017). Comparing Classroom Context and Physical Activity in Nature and Traditional Kindergartens. *Children, Youth and Environments*, *27*(3), 56–77. doi:10.1353/cye.2017.0020

Miao, R. E., & Cagle, N. L. (2020). The role of gender, race, and ethnicity in environmental identity development in undergraduate student narratives. *Environmental Education Research, 26*(2), 171–188. doi:10.1080/13504622.2020.1717449

Michaels, W. B. (1992). Race into culture: A critical genealogy of cultural identity. *Critical Inquiry, 18*(4), 655–685. doi:10.1086/448651

Middlewick, Y., Kettle, T. J., & Wilson, J. J. (2012). Curtains up! Using forum theatre to rehearse the art of communication in healthcare education. *Nurse Education in Practice, 12*(3), 139–142. doi:10.1016/j.nepr.2011.10.010 PMID:22094107

Mignolo, D. W., & Walsh, E. C. (2018). *On decoloniality: Concepts, analytics, praxis.* Duke University Press.

Millennium Ecosystem Assessment (MA). (2005). *Ecosystems and Human Well-being: A Framework for Assessment, MA* (Vol. 1). Island Press.

Miller, H. K. (2018). Developing a critical consciousness of race in place-based environmental education: Franco's story. *Environmental Education Research, 24*(6), 845–858. doi:10.1080/13504622.2017.1357802

Mill, J. S. (2010). The Amoral Status of Nature. In D. R. Keller (Ed.), *Environmental Ethics: the big questions* (pp. 73–77). Wiley-Blackwell.

Milstein, T., Sherry, C. S., Carr, J. E., & Siebert, M. (2023). "Got to get ourselves back to the garden": Sustainability transformations and the power of positive environmental communication. *Journal of Environmental Planning and Management*, 1–19. doi:10.1080/09640568.2023.2197140

Mindiola Romo, B. (2018). *El ser Arhuaco: transformaciones en las formas de enseñar-aprender los patrones culturales en la Comunidad de Nabusimake.*

Mintz, S., & Rode, S. (1999). More than a walk in the park: Demonstration carts personalize interpretation'. *Roots, 18*, 24–26.

Mironov, A. V. (2020). The Technique of Studying the World in Primary School. Moscow. *Pedagogy*. Advance online publication. doi:10.1088/1755-1315/459/5/052086

Mitchell, D. (2015). Inclusive Education is a Multi-Faceted Concept. *Center for Educational Policy Studies Journal, 5*(1), 9–28. doi:10.26529/cepsj.151

Miyazaki, Y. (2021). Walking in the Woods: Go back to nature with the Japanese way of shinrin-yoku, Hachette UK.

Mohi-ud-din, R., Farooq, S., Majeed, A., Khan, N. A., & Bhat, Z. A. (2019). Legal Framework on Protection of Traditional Knowledge: A Review. *International Journal of Advance Research in Science and Engineering, 8*(1), 100–108. https://www.researchgate.net/publication/331221999

Monferrer, L., Valentín, G. L., & Villanueva, M. S. (2022). Mathematical and Experimental Science Education from the School Garden: A Review of the Literature and Recommendations for Practice. *Education Sciences, 12*(1), 47. doi:10.3390/educsci12010047

Monroe, M. C., Andrews, E., & Biedenweg, K. (2008). A Framework for Environmental Education Strategies. *Applied Environmental Education and Communication, 6*(3–4), 205–216. doi:10.1080/15330150801944416

Monroe, M. C., Plate, R. R., Oxarart, A., Bowers, A., & Chaves, W. A. (2019). Identifying effective climate change education strategies: A systematic review of the research. *Environmental Education Research, 25*(6), 791–812. doi:10.1080/13504622.2017.1360842

Monroe, M. C., & Watkins, K. B. (2017). Culturally responsive environmental education: An indigenous perspective. *International Journal of Environmental and Science Education, 12*(3), 429–444.

Morello, D. (1997). *Honduras and Deforestation.* American. http://www1.american.edu/TED/honduras.html

Moreno-Fernández, O. (2021). Theatre as a Teaching Resource to Work on Socio-environmental Problems in Primary Education. *International Electronic Journal of Environmental Education, 11*(1), 79–90. doi:10.18497/iejeegreen.817608

More, T., & Valhe, V. (2020). Symbolic Significance of Sacred Nyagrodha Tree Worship in Indian Religion and Traditions. *Bulletin of the Deccan College Post-Graduate and Research Institute, 80*(October), 149–157.

Morgante, A. (1997). *Buddhist perspectives on the Earth Charter.* Boston Research Center for the 21st Century. https://search.worldcat.org/title/38283659

MovementS. https://www.sunrisemovement.org/

Mueller, M. P. (2009). Educational reflections on the "ecological crisis": Ecojustice, environmentalism, and sustainability. *Science & Education, 18*(8), 1031–1056. doi:10.1007/s11191-008-9179-x

Muñoz, S. (2009). *Children in the Outdoors: A Literature Review.* Sustainable Development Research Centre.

Muradova, L., Walker, H., & Colli, F. (2020). Climate change communication and public engagement in interpersonal deliberative settings: Evidence from the Irish citizens' assembly. *Climate Policy, 20*(10), 1322–1335. doi:10.1080/14693062.2020.1777928

Murdoch, J. (1991). Hindu and Muhammadan Festivals. Compiled from Wilson, Wilkins, Crooke, Sell, Hughes and Other Writers. Asian Educational Services.

Muthuswamy, V. (2018). Hinduism and Social Responsibility. *Religious Perspectives on Social Responsibility in Health: Towards a Dialogical Approach*, 91-105.

Myers, K. F., Doran, P. T., Cook, J., Kotcher, J. E., & Myers, T. A. (2021). Consensus revisited: Quantifying scientific agreement on climate change and climate expertise among Earth scientists 10 years later. *Environmental Research Letters, 16*(10), 104030. https://iopscience.iop.org/article/10.1088/1748-9326/ac2774/pdf. doi:10.1088/1748-9326/ac2774

NAAEE, North American Association for Environmental Education's Natural Start Alliance. (2019). *Nature-Based Preschool Professional Practice Guidebook. North American Association for Environmental Education., & Natural Start Alliance.*

NAAEE. (2023). *Natural Start.* Natural Start. https://naturalstart.org/

NAAEE. North American Association for Environmental Education, & Natural Start Alliance. (2019). Nature-Based Preschool Professional Practice Guidebook. North American Association for Environmental Education.

Naess, A. (1973). The shallow and the deep, long-range ecology movement. A summary. *Inquiry, 16*(1), 95 — 100. doi:10.1080/00201747308601682

Naess, A. (1986). *The Basic Principles of Deep Ecology.*

Naess, A. (1989). Ecosophy and Gestalt Ontology. *The Trumpeter 6*(4).

Naess, A. (1990). Ecology, community & lifestyle. London, UK: Cambridge.

Naess, A. (1993). Simpleis Means, Rich in Ends. Michael E. Zimmerman, J Baird Callicott, George Sessions, Karen J. Warren et al., (eds.) *Environmental Philosophy From Animal Rights to Radical Ecology*, (pp.182-193)

Naess, A. (1995). *Deep Ecology for the 21st Century, Readings on the Philosophy and Practice of the New Environmentalism.* George Sessions (Ed.) Shambhala.

Naess, A. (1989). *Ecosophy, Community, Lifestyle. Outline of an Ecosophy*. Cambridge University Press. doi:10.1017/CBO9780511525599

Naess, A. (1992). The encouraging richness and diversity of ultimate premises in environmental philosophy. The Trumpeter. *Journal of Ecosophy*, *9*(2), 53–60.

Napper, E. (2020, November 11). *The Education of Tibetan Buddhist Nuns* (Tibetan Nuns Project, Interviewer) [Interview]. https://tnp.org/the-education-of-tibetan-buddhist-nuns/?mc_cid=74edd6d899&mc_eid=987b3f085d

Narayanan, V. (2001). Water, wood, and wisdom: Ecological perspectives from the Hindu traditions. *Daedalus*, *130*(4), 179–206.

Natural Start Forum: Enrolling with Equity in Mind. (2021). Natural Start. https://naturalstart.org/bright-ideas/natural-start-forum-enrolling-equity-mind

Nature for the Blind. (2020). *What is a Braille Trail?* Nature for the Blind. <http://www.naturefortheblind.com/what-is-a-braille-trail>

Nazmi, A., Martinez, S. M., Byrd, A., Robinson, D., Bianco, S., Maguire, J., Crutchfield, R. M., Condron, K., & Ritchie, L. D. (2018). A systematic review of food insecurity among US students in higher education. *Journal of Hunger & Environmental Nutrition*, *14*(5), 725–740. doi:10.1080/19320248.2018.1484316

Neas, S., Ward, A., & Bowman, B. (2022). Young people's climate activism: A review of the literature. *Frontiers in Political Science*, *4*, 940876. doi:10.3389/fpos.2022.940876

Negi, V. S., & Maikhuri, R. K. (2013). Socio-ecological and religious perspective of agrobiodiversity conservation: Issues, concern and priority for sustainable agriculture, Central Himalaya. *Journal of Agricultural & Environmental Ethics*, *26*(2), 491–512. doi:10.1007/s10806-012-9386-y

Negi, V. S., Pathak, R., Thakur, S., Joshi, R. K., Bhatt, I. D., & Rawal, R. S. (2023). Scoping the need of Mainstreaming indigenous knowledge for sustainable use of bioresources in the Indian Himalayan region. *Environmental Management*, *72*(1), 135–146. doi:10.1007/s00267-021-01510-w PMID:34341866

Nelson, N., Pacini-Ketchabaw, V., & Nxumalo, F. (2018). Rethinking Nature-Based Approaches in Early Childhood Education: Common Worlding Practices. *Journal of Childhood Studies*, *43*(1), 4–14. doi:10.18357/jcs.v43i1.18261

Nepal, T. K. (2021). An Overview of Threats to Traditional Ecological Knowledge. *Journal of Research in Humanities and Social Science, 9*(11), 01-04. www.questjournals.org

Nesterova, I. A. (2020). Ethno-education as a Means of Enhancing the Health of the Population of Colombia. *Russian Journal of Humanities and Social Sciences*, *13*(4), 1602–1606.

New Tang Dynasty Television. (Producer). (2010). Electronic recycling by communities with special needs [Video]. YouTube. http:// www.youtube.com/watch?v=LnhGST1lryQ

Newburger, E. (2021, Nov 16). What the COP26 climate conference really accomplished. *CNBC*. https://www.cnbc.com/2021/11/16/un-cop26-climate-summit-what-was-accomplished.html

Newlyn, E. (n.d.). The 6 branches of yoga. *EkhartYoga*. Retrieved from https://www.ekhartyoga.com/articles/practice/the-6-branches-of-yoga

Newton, I. (2010). Mechanistic Metaphysics. In D. R. Keller (Ed.), *Environmental Ethics: the big questions* (pp. 72–73). Wiley-Blackwell.

Nhât Hanh, T. (2009). *Interbeing: Fourteen Guidelines for Engaged Buddhism* (Repr). Full Circle.

Nietzsche, F. (1993) The Birth of Tragedy Out of the Spirit of Music, ed. Michael Tanner, transl. Shaun Whiteside, Penguin Books Ltd.

Niño Izquierdo, R. C. (2011). *Organización y uso del territorio por la comunidad Indígena Arhuaca de Nabusímake Sierra Nevada de Santa Marta.*

Nitzke, S., & Braunbeck, H. G. (2021). Arboreal Imaginaries. An Introduction to the Shared Cultures of Trees and Humans. *Green Letters.*, *25*(4), 341–355. doi:10.1080/14688417.2021.2072633

Nkoana, E. M. (2020). Exploring the effects of an environmental education course on the awareness and perceptions of climate change risks among seventh and eighth grade learners in South Africa. *International Research in Geographical and Environmental Education*, *29*(1), 1, 7–22. doi:10.1080/10382046.2019.1661126

NOAA. (2024). *Toolbox for teaching climate and energy*. NOAA. https://www.climate.gov/teaching/toolbox

Norbu, D. (2001). Refugees from Tibet: Structural Causes of Successful Settlements. *The Tibet Journal*, *26*(2), 3–25.

Nuccitelli, D. (2020). How we know the Earth is warming and humans are responsible. *Bulletin of the Atomic Scientists*, *76*(3), 140–144. doi:10.1080/00963402.2020.1751969

Nxumalo, F., & Pacini-Ketchabaw, V. (2017). "Staying with the trouble" in child-insect-educator common worlds. *Environmental Education Research*, *23*(10), 1414–1426. doi:10.1080/13504622.2017.1325447

O'Brien, L. (2019). Learning outdoors: The Forest School approach. In S. Waite (Ed.), *Outdoor Learning Research.* Routledge.

O'Brien, L., & Murray, R. (2007). Forest School and its impacts on young children: Case studies in Britain. *Urban Forestry & Urban Greening*, *6*(4), 249–265. doi:10.1016/j.ufug.2007.03.006

O'Donnell, E. L. (2018). At the Intersection of the Sacred and the Legal: Rights for Nature in Uttarakhand, India. *Journal of Environmental Law*, *30*(1), 135–144. doi:10.1093/jel/eqx026

O'Neil, J. M., Newton, R. J., Bone, E. K., Birney, L. B., Green, A. E., Merrick, B., Goodwin-Segal, T., Moore, G., & Fraioli, A. (2020). Using urban harbors for experiential, environmental literacy: Case studies of New York and Chesapeake Bay. *Regional Studies in Marine Science*, *33*, 100886. doi:10.1016/j.rsma.2019.100886

O'Neill, J. (1992). The Varieties of Intrinsic Value. *The Monist*, *75*(2), 119–137. doi:10.5840/monist19927527

O'Neill, O. (1997). Environmental Values, Anthropocentrism and Speciesism. *Environmental Values*, *6*(2), 127–142. doi:10.3197/096327197776679121

OECD. (2021). Think green: Education and climate change. *Trends Shaping Education Spotlights*, *24*, 1–12.

Ohly, H., Gentry, S., Wigglesworth, R., Bethel, A., Lovell, R., & Garside, R. (2016). A systematic review of the health and well-being impacts of school gardening: Synthesis of quantitative and qualitative evidence. *BMC Public Health*, *16*(1), 286. doi:10.1186/s12889-016-2941-0 PMID:27716134

Ojala, M. (2012). Hope and climate change: The importance of hope for environmental engagement among young people. *Environmental Education Research*, *18*(5), 625–642. doi:10.1080/13504622.2011.637157

Ojala, M. (2018-19). Eco-anxiety. *RSA Journal*, *164*(4), 10–15.

Okafor, J. O., & Stella, O. (2018). *Hinduism and ecology: Its relevance and importance.*

Omprakash Team (2007-2021). *Ethno-Education in Barranquilla.* Colombia, Seattle, WA, USA: Omprakas

Orman, J. E. (2011). *Educational Psychology: Developing Learners* (7th ed.). Pearson Education, Inc.

Orr, S. H. (2015). Training the peer facilitator: Using participatory theatre to promote engagement in peer education. *Research in Drama Education, 20*(1), 110–116. doi:10.1080/13569783.2014.983466

Osho. (1991). *Vigyan Bhairav Tantra. The Book of Secrets.* Rebel Publishing House.

OSPI. (n.d.). *Since Time Immemorial: Tribal Sovereignty in Washington State.* Office of Native Education/OSPI. https://www.k12.wa.us/student-success/resources-subject-area/time-immemorial-tribal-sovereignty-washington-state

Ottuh, P. O. (2020). Religious Approach to Non-Anthropocentric Ethics in Environmental Philosophy. *Cogito-Multidisciplinary research Journal*, (1), 7-24.

Owls & Indigo. (n.d.). What Is Secular Spirituality? https://owlsandindigo.com/day-15-21daysofyule-what-is-secular-spirituality/

Ozer, E. J. (2006). The Effects of school gardens on students and Schools: Conceptualization and Considerations for Maximizing Healthy Development. *Health Education & Behavior, 34*(6), 846–863. doi:10.1177/1090198106289002 PMID:16861584

Pacini-Ketchabaw, V. (2013). Frictions in Forest Pedagogies: Common Worlds in Settler Colonial Spaces. *Global Studies of Childhood, 3*(4), 355–365. doi:10.2304/gsch.2013.3.4.355

Padma'tsho (Baimacuo), & Jacoby, S. (2020). Gender Equality in and on Tibetan Buddhist Nuns' Terms. *Religions, 11*(10). doi:10.3390/rel11100543

Palmo, D. (2019, November 1). Tibet's Rivers Will Determine Asia's Future. *The Diplomat.* https://thediplomat.com/2019/11/tibets-rivers-will-determine-asias-future/

Panda, R., & Maity, M. (2021). Global warming and climate change on earth: Duties and challenges of human beings. *International Journal of Research in Engineering. Science and Management, 4*(1), 122–125.

Panula, E., Jeronen, E., & Rodrigez-Aflecht, G. (2020). Nature Is Something We Can't Replace': Mexican Students' Views of the Landscape They Want to Conserve. *Education Sciences, 10*(13). https://files.eric.ed.gov/fulltext/EJ1241172.pdf>

Paquin, A. (2023). The Integration of Indigenous Knowledge in Education. M.Ed. Literature Reviews, 18.

Parry, S., McCarthy, S. R., & Clark, J. (2022). Young people's engagement with climate change issues through digital media–a content analysis. *Child and Adolescent Mental Health, 27*(1), 30–38. doi:10.1111/camh.12532 PMID:34904349

Pascal, B. (1995). *Pensees* (A. J. Krailsheimer, Trans.). Penguin Classics.

Patterson, M. E., Watson, A. E., Williams, D. R., & Roggenbuck, J. R. (1998). A hermeneutic approach to studying the nature of wilderness experiences. *Journal of Leisure Research, 30*(4), 423–452. doi:10.1080/00222216.1998.11949842

Paulson, S. (2019). Making Kin: An Interview with Donna Haraway. *Los Angeles Review of Books.* https://lareviewofbooks.org/article/making-kin-an-interview-with-donna-haraway/

Pérez Valverde, C. (2003). Theatre in education (TIE) in the context of educational drama. Oxford University Press.

Perls, F. S. (1972). *Gestalt Theory Verbatium.* Bantam Books.

Petitpierre, F. (1975). The Symbolic Landscape of the Muiscas. *Studies of Comparative Religion,* (Winter), 36–51.

Pierotti, R., & Wildcat, D. R. (1999). Traditional knowledge, culturally-based World-Views and Western Science. In D. A. Posey (Ed.), *Cultural and Spiritual Values of Biodiversity: A Complementary Contribution to the Global Biodiversity Assessment*. Intermediate Technology Publications Ltd.

Pierre, J. (2019). *Racism is a Public Health Crisis*. The Institute for Policy Studies. https://ips-dc.org/racism-is-a-public-health-crisis/

Pihkala, P. (2022). Towards a taxonomy of climate emotions. *Frontiers in Climate*, *3*, 1–22. doi:10.3389/fclim.2021.738154

Pincock, S. (2004). Poor communication lies at heart of NHS complaints, says ombudsman. *BMJ (Clinical Research Ed.)*, *328*(7430), 10. doi:10.1136/bmj.328.7430.10-d

Pintchman, T. (Ed.). (2007). *Women's Lives, Women's Rituals in the Hindu Tradition*. Oxford University Press. doi:10.1093/acprof:oso/9780195177060.001.0001

Plaka, V., & Skanavis, C. (2016). The feasibility of school gardens as an educational approach in Greece: A survey of Greek schools. *International Journal of Innovation and Sustainable Development*, *10*(2), 141. doi:10.1504/IJISD.2016.075546

Plato. (1966). Plato in Twelve Volumes, (trans. H. Fowler). Harvard University Press; London, William Heinemann Ltd.

Plumwood, V. (2002). *Environmental Culture: The ecological crisis of reason*. Routledge.

Policy 2100: Sustainability : University Policy. (2023, June 22). The University of New Mexico. https://policy.unm.edu/university-policies/2000/2100.html

Ponzetti, J. J. Jr, Selman, J., Munro, B., Esmail, S., & Adams, G. (2009). The effectiveness of participatory theatre with early adolescents in school-based sexuality education. *Sex Education*, *9*(1), 93–103. doi:10.1080/14681810802639905

Posey, D. A. (1998). The "Balance Sheet" and the "Sacred Balance": Valuing the Knowledge of Indigenous and Traditional Peoples. *Worldviews: Global Religions, Culture, and Ecology*, *2*(2), 91–106. doi:10.1163/156853598X00109

Posey, D. A. (1999). Cultural Landscapes, Chronological Ecotones and Kayapó Resource Management. In D. A. Posey (Ed.), *Cultural and Spiritual Values of Biodiversity: A Complementary Contribution to the Global Biodiversity Assessment*. Intermediate Technology Publications Ltd.

Posey, D. A. (Ed.). (1999). *Cultural and Spiritual Values of Biodiversity: A Complementary Contribution to the Global Biodiversity Assessment*. Intermediate Technology Publications Ltd.

Pound, L. (2015). Margaret and Rachel McMillan. *Nursery World*, *2015*(17), 34–35. doi:10.12968/nuwa.2015.17.34

Powell, R. B., Stern, M. J., Frensley, B. T., & Moore, D. (2019). Identifying and developing crosscutting environmental education outcomes for adolescents in the twenty-first century (EE21). *Environmental Education Research*, *25*(9), 1281–1299. doi:10.1080/13504622.2019.1607259

Price-Wallace, D. (2023). Buddhist Pro-Woman Attitudes Towards Full Ordination: Tibetan and Himalayan Monastics' Views. *Journal of Global Buddhism*, *24*(1), 1–24. doi:10.26034/lu.jgb.2023.3140

Pure Water for the World. (2016, May). *Honduras teacher Workshop*. Pure Water for the World. https://www.purewaterfortheworld.org/wpcontent/uploads/2015/10/PWW_Honduras_HygieneTeacherTraining-v2.pdf

Quality Assurance Agency. (n.d.). *Education for Sustainable Development*. QAA. https://www.qaa.ac.uk/the-quality-code/education-for-sustainable-development

Ragoff, B. (2003). *The Cultural Nature of Human Development* (1st ed.). Oxford University Press.

Rambla, X., & Langthaler, M. (2016). The SDGs and inclusive education for all: From special education to addressing social inequalities (ÖFSE Briefing Paper, No. 14). *Austrian Foundation for Development Research (ÖFSE).* http://www.oefse.at/publikationen/briefing-papers/detail-briefing- paper/publication/show/Publication/The-SDGs-and-inclusive-education-for-all-From-special-education-to-addressing-social-inequalities/

Ramsaroop, S., & Van Rooyen, H. (2013). Exploring Educator Competence in Teaching Environmental Education in Schools in Gauteng. *Africa Education Review, 10*(3), 595–613. doi:10.1080/18146627.2013.853555

Rao, M. (2022). Mental health impacts of the climate crisis: The urgent need for action. *International Review of Psychiatry (Abingdon, England), 34*(5), 439–440. doi:10.1080/09540261.2022.2128272 PMID:36165748

Rao, V. L. N., & Ramana, G. V. (2007). Indigenous knowledge, conservation and management of natural resources among primitive tribal groups of Andhra Pradesh. *Anthropol, 3*(Spec), 129–134.

Rawlinson, M. C. (2015). Food, Health, and Global Justice. *International Journal of Feminist Approaches to Bioethics, 8*(2), 1–9. doi:10.3138/ijfab.8.2.1

Rebellion, E. (2024). https://rebellion.global

Recio, M. E. (2015). *Honduras Resource Tenure and Sustainable Landscapes Assessment.* USAID Tenure and Global Climate Change Program. https://www.land-links.org/wpcontent/uploads/2016/09/USAID_Land_Tenure_TGCC_Honduras_Assessment_En.pdf

Recommendations for Outdoor Learning in New Mexico. (2021). NMLegis.gov. https://www.nmlegis.gov/handouts/ALESC%20111721%20Item%2010%20.1%20-%20Outdoor%20Classrooms%20Task%20Force%20Slides.pdf

Reichel-Dolmatoff, G. (1977). Cosmología como análisis ecológico: una perspectiva desde la selva pluvial. *Estudios antropológicos*, 355-375.

Reichel-Dolmatoff, G. (1982). Cultural Change and Environmental Awareness: A Case Study of the Sierra Nevada de Santa Marta, Colombia. *Mountain Research and Development, 2*(3), 289–298. doi:10.2307/3673093

Reid, J. L., Kagan, S. L., Hilton, M., & Potter, H. (2015). *A Better Start: Why Classroom Diversity Matters in Early Education.* PRRAC: Poverty Race and Research Council. https://www.prrac.org/pdf/A_Better_Start.pdf

Reid-Bowen, P. (2016). *Goddess as Nature: Towards a Philosophical Thealogy.* Routledge. doi:10.4324/9781315585338

Restall, B., & Conrad, E. (2015). A literature review of connectedness to nature and its potential for environmental management. *Journal of Environmental Management, 159*, 264–278. doi:10.1016/j.jenvman.2015.05.022 PMID:26087657

Resurrección, B. P. (2013, September). Persistent women and environment linkages in climate change and sustainable development agendas. *Women's Studies International Forum, 40*, 33–43. doi:10.1016/j.wsif.2013.03.011

Retallack, J. (2003). *The Poethical Wager.* University of California Press.

Retnaselvam, J., & Singh, P. (2019). *Sustainability and the sacred: a comparative study of Indian religious environmentalism with special reference to Christian and indigenous communities in the State of Kerala.* Research Gate.

Reyhner, J., & Jacobs, D. T. (2002). Preparing Teachers of American Indian and Alaska Native Students. *Action in Teacher Education, 24*(2), 85–93. doi:10.1080/01626620.2002.10734422

Richards, R. (2001). A New Aesthetic for Environmental Awareness Chaos Theory, The Beauty of Nature, and Our Broader Humanistic Identity. *Journal of Humanistic Psychology, 41*(2), 59–95. doi:10.1177/0022167801412006

Riebel, L. (2001). Consuming the Earth: Eating Disorders and Ecopsychology. *Journal of Humanistic Psychology, 41*(2), 38–58. doi:10.1177/0022167801412005

Riley, H. (2010, February 8). Students ask for sustainable garden. *Students Ask for Sustainable Garden - the Daily Lobo.* https://www.dailylobo.com/article/2010/02/students_ask_for_sustainable_garden

Riley, M. F., & Hendee, J. C. (1999). Wilderness vision quest. Clients, motivations, and reported benefits from an urban-based program 1988-1997. In A. E. Watson, G.H Aplet & J.C., Hendee, *Personal, societal and ecological values of wilderness: Sixth Word Wilderness Congress Proceedings on Research Management and Allocation*, Bangalore, India.

Riordan, K. M., MacCoon, D. G., Barrett, B., Rosenkranz, M. A., Chungyalpa, D., Lam, S. U., Davidson, R. J., & Goldberg, S. B. (2022). Does meditation training promote pro-environmental behavior? A cross-sectional comparison and a randomized controlled trial. *Journal of Environmental Psychology, 84*, 101900. Advance online publication. doi:10.1016/j.jenvp.2022.101900 PMID:36969767

Riordan, M., & Klein, E. J. (2010). Environmental education in action: How expeditionary learning schools support classroom teachers in tackling issues of sustainability. *Teacher Education Quarterly, 37*(4), 119–137.

Robbins, J. (2020). *How Emersion in Nature Benefits Your Health.* Yale School of Environmental Studies.

Robinson, D. (2022). Explainer: What is greenwashing and how to avoid it? *Earth.org.* https://earth.org/what-is-greenwashing/

Robinson-O'Brien, R., Story, M., & Heim, S. (2009). Impact of Garden-Based Youth Nutrition Intervention Programs: A review. *Journal of the American Dietetic Association, 109*(2), 273–280. doi:10.1016/j.jada.2008.10.051 PMID:19167954

Rodrigues, H. P. (2003). *Ritual Worship of the Great Goddess. The Liturgy of Durga Puja with Interpretations.* State University of New York Press. doi:10.1353/book4546

Rogayan, D. V. Jr. (2019). I heart nature: Perspectives of university students on environmental stewardship. *International Journal of Engineering Science and Technology, 1*(1), 10–16.

Roloff, C. (2020). *The Buddhist Nun's Ordination in the Tibetan Canon: Possibilities of the Revival of the Mūlasarvāstivāda Bhikṣuṇī Lineage.* Numata Center for Buddhist Studies.

Rolston, H. III. (1988). *Environmental Ethics.* Temple University Press.

Rome, D. (2009, August 5). The Green Buddha. *Tricycle: The Buddhist Review.* https://tricycle.org/trikedaily/the-green-buddha/

Rose, D., & Meyer, A. (2002). *Teaching Every Student in the Digital Age: Universal Design for Learning.* Association for Supervision and Curriculum Development.

Ross, A., Snodgrass, J. G., & Sherman, K. Delcore, Henry. D., & Sherman, R. (2011). Indigenous peoples and the collaborative stewardship of nature: Knowledge binds and institutional conflicts. Left Coast Press.

Ross, A., Friedmann, E., Bevans, M., & Thomas, S. (2013). National survey of yoga practitioners: Mental and physical health benefits. *Complementary Therapies in Medicine, 21*(4), 313–323. doi:10.1016/j.ctim.2013.04.001 PMID:23876562

Rossman, B., & Ulehla, Z. J. (1977). ' Psychological reward values associated with wilderness use: A bifunctional-reinforcement approach.'. *Environment and Behavior, 9*(1), 41–65. doi:10.1177/001391657791003

Roy, S. (2017). *How I Became a Tree.* Yale University Press.

Ruíz, R. (2010). Reorienting language-as-resource. In J. E. Petrovic (Ed.), *International perspectives on bilingual education: Policy, practice, and controversy* (pp. 155–172). Information Age Publishing.

Russ, A., & Krasny, M. (2017). *Urban Environmental Education Review*. Cornell University Press. www.jstor.org/stable/10.7591/j.ctt1qv5qhq

Russel, K. C., & Hendee, J. C. (1997). Testing Wilderness Discovery: A wilderness experience program for youth at risk in the Federal Job Corps. *Technical Report 24*, Moscow.

Ryff, C. D. (1989). Happiness is everything, or is it? Explorations on the meaning of psychological well-being. *Journal of Personality and Social Psychology*, *57*(6), 1069–1081. doi:10.1037/0022-3514.57.6.1069

Ryff, C. D. (2014). Psychological well-being revisited: Advances in the science and practice of eudaimonia. *Psychotherapy and Psychosomatics*, *83*(1), 10–28. doi:10.1159/000353263 PMID:24281296

Ryff, C. D., & Keyes, C. L. M. (1995). The structure of psychological well-being revisited. *Journal of Personality and Social Psychology*, *69*(4), 719–727. doi:10.1037/0022-3514.69.4.719 PMID:7473027

Sadruddin, M. M. (2023). Design and Pilot of an Inclusive Environmental Education Course for a Community Living Under the Unprecedented Environmental Emergencies in Pakistan- A Case Study. *Journal of Peace Education*. Advance online publication. doi:10.1080/17400201.2024.2314302

Saefullah, K. (2019). 'Gintingan in Subang': An Indigenous Institution for Sustainable Community-Based Development in the Sunda Region of West Java, Indonesia. PhD Dissertation. *Leiden Ethnosystems And Development Programme (LEAD) Studies* No. 11. Leiden: Leiden University.

Sagoff, M. (1974). On preserving the natural environment. *The Yale Law Journal*, *84*(2), 205–267. doi:10.2307/795401

Sailakumar, S., & Naachimuthu, K. P. (2017). A phenomenological approach to understand the nature-based experiences and its influence on holistic development. *Indian Journal of Positive Psychology*, *8*(2), 186–195.

Sajadi, N. (2021). Becoming conscious through the arts: A road to social sustainability. *Visions of sustainability for arts education: Value, challenge and potential*, 153-160. . doi:10.1007/978-981-16-6174-7_14

Salgado, N. S. (2013). Buddhist Nuns and Gendered Practice. In *Search of the Female Renunciant*. Oxford University Press. doi:10.1093/acprof:oso/9780199760022.001.0001

Salleh, A. (2009). *Eco-sufficiency & global justice: Women write political ecology*. Pluto Press London.

Salmón, E. (2000). Kincentric Ecology: Indigenous Perceptions of the Human–Nature Relationship. *Ecological Applications*, *10*(5), 1327–1332. doi:10.1890/1051-0761(2000)010[1327:KEIPOT]2.0.CO;2

Santamaría, C. N. (2020). Families, experiments, and nature: Learning science through project-based learning. *School Science and Mathematics*, *120*(8), 467–476. https://proxy.marybaldwin.edu:2238/10.1111/ssm.12438. doi:10.1111/ssm.12438

Sarkar, P. (1975). Jatra: The popular traditional theatre of Bengal. *Journal of South Asian Literature*, *10*(2/4), 87–107.

Sass, L. (2017, 1992). Madness and Modernism. Insanity in the Light of Modern Art, Literature, and Thought. Oxford University Press.

Savoie-Roskos, M. R., Wengreen, H. J., & Durward, C. (2017). Increasing Fruit and Vegetable Intake among Children and Youth through Gardening-Based Interventions: A Systematic Review. *Journal of the Academy of Nutrition and Dietetics*, *117*(2), 240–250. doi:10.1016/j.jand.2016.10.014 PMID:27964852

Schneider, Kirk J., Rediscovery of Awe: Splendour, Mystery and the Fluid Centre of Life. City, ST. (2004). Paragon House.

Schneider, N. (2022). A Revolution in Red Robes: Tibetan Nuns Obtaining the Doctoral Degree in Buddhist Studies (Geshema). *Religions, 13*(9), 838. doi:10.3390/rel13090838

Schultes, R. E., & Raffauf, R. F. (1990). *The Healing Forest: Medicinal and Toxic Plants of the Northwest Amazonia.* Dioscorides Press.

Schultz, P. W. (2002). Inclusion with Nature: The Psychology Of Human-Nature Relations. In P. Schmuck & W. P. Schultz (Eds.), *Psychology of Sustainable Development* (pp. 61–78). Springer US., doi:10.1007/978-1-4615-0995-0_4

Schuon, F. (2001). *Imágenes del espíritu: shinto, budismo, yoga*. Olañeta.

Schuyler, C. (2020). Nature Rx Presentation. *Foxcroft Environmental Education Conference*. Foxcroft School, Virginia.

Scott, J. T., Kilmer, R. P., Wang, C., Cook, J. R., & Haber, M. G. (2018). Natural Environments Near Schools: Potential Benefits for Socio-Emotional and Behavioral Development in Early Childhood. *American Journal of Community Psychology, 62*(3–4), 419–432. doi:10.1002/ajcp.12272 PMID:30222858

Scrap Autism. (2012). *Home.* Scrap autism. http://scrapautism.org/

Semetsky, I. (2011). Becoming-other: Developing the Ethics of Integration. *Policy Futures in Education, 9*(1), 138–144. doi:10.2304/pfie.2011.9.1.138

Sen, S., & Chakraborty, R. (2014). Traditional Knowledge Digital Library: a distinctive approach to protect and promote Indian indigenous medicinal treasure. *Current Science Association, 106*(10), 1340-1343. https://www.jstor.org/stable/24102476

Sengupta, M., Banerjee, P., & Wright, C. (2004). Worshipping the Tree. *Indian Literature, 48*(6).

Sessions, G. (1993). Introduction. Michael E. Zimmerman, J Baird Callicott, George Sessions, Karen J. Warren et al., (eds.) *Environmental Philosophy From Animal Rights to Radical Ecology*, (pp.161-170).

Sessions, G., & Devall, B. (1985). Deep ecology. Salt Lake City: Peregrine Smith Books.

Sevin, S. (2020). The Change of Secondary School Students' Environmental Consciousness, Attitude and Behaviors with Nature Education Project. *Higher Education Studies, 10*(2), 82-94.

Shafer-Landau, R. (2018). *Lliving Ethics: An Introduction with Readings*. Oxford University Press.

Shah, V. (2019). *Hindu Culture and Lifestyle: Living Indian traditions in the age of artificial intelligence*. Notion Press.

Shakya, T. (1994). The Development of Modern Tibetan Studies. In R. Barnett (Ed.), *Resistance and Reform in Tibet* (pp. 1–14). University of Indiana Press.

Sheldon, K. M., & Kasser, T. (2001). Goals, Congruence, and Positive Well-Being: New Empirical Support for Humanistic Theories. *Journal of Humanistic Psychology, 41*(1), 30–50. doi:10.1177/0022167801411004

Sherma, R. D. (1998). Sacred Immanence: Reflections of Ecofeminism in Hindu Tantra. In L. E. Nelson (Ed.), *Purifying the Earthly Body of God. Religion and Ecology in Hindu India* (pp. 89–132). SUNY Press.

Sherma, R. D. (1998). Sacred Immanence: Reflections of Ecofeminism in Hindu Tantra. In L. E. Nelson (Ed.), *Purifying the Earthly Body of God: Religion & Ecology in Hindu India* (pp. 89–132). State University of New York Press.

Sherry, C. S. (2022). Learning from the Dirt: Initiating university food gardens as a cross-disciplinary tertiary teaching tool. *Journal of Outdoor and Environmental Education, 25*(2), 199–217. doi:10.1007/s42322-022-00100-6

Shestov, L. (2011). *All Things Are Possible*. Girvin Press.

Shiva, V. (1988). *Staying Alive. Women, Ecology and Survival in India*. Kali for Women.

Shukla, S., & Gardner, J. S. (2006). *The Role of Traditional Ecological Knowledge in Education for Community-Based Resource Management*. Research Gate. https://www.researchgate.net/publication/42761744_The_Role_of_Traditional_Ecological_Knowledge_in_Education_for_Community-Based_Resource_Management

Shukla, S., & Gardner, J. (2006). Local knowledge in community-based approaches to medicinal plant conservation: Lessons from India. *Journal of Ethnobiology and Ethnomedicine*, *2*(20), 20. doi:10.1186/1746-4269-2-20 PMID:16603082

Shultes, R. E., Hofmann, A., & Ratsch, C. (2001). *Plants of the Gods: Their Sacred, Healing, and Hallucinogenic Powers*. Healing Arts Press.

Siegner, A., & Stapert, N. (2020). Climate change education in the humanities classroom: A case study of the Lowell school curriculum pilot. *Environmental Education Research*, *26*(4), 511–531. doi:10.1080/13504622.2019.1607258

Singer, P. (1989). Are all Animals Equal. In T. Regan & P. Singer (Eds.), *Animal Rights and Human Obligations* (pp. 215–226). Oxford University Press.

Singh, B. (2023). Integration of Traditional and Folk Knowledge with Current Education System: Indianizing Education System and Regaining the Status of Vishwa-guru. *AGPE The Royal Gondwana Research Journal*, *4*(3), 22-30. https://agpegondwanajournal.co.in/index.php/agpe/article/view/238

Singhal, A. (2001). *Facilitating community participation through communication*. UNICEF.

Singhal, A. (2003). Entertainment-education through participatory theater: Freirean strategies for empowering the oppressed. In *Entertainment-Education and Social Change* (pp. 399–420). Routledge. doi:10.4324/9781410609595-28

Singhal, A. (2004). Empowering the oppressed through participatory theater. *Investigación & Desarrollo*, *12*(1), 138–163.

Singh, N. K. (1997). *Vedic mythology*. APH Publishing.

Singh, R., Sharma, R. K., Babu, S., & Bhatnagar, Y. V. (2020). Indian Trans-Himalayas. *Pastoralism*, *15*(1), 15. doi:10.1186/s13570-020-00169-y

Singh, S., & Hageneder, F. (2007). *Tree Yoga. A Workbook, Strengthen Your Personal Yoga Practice Through the Living Wisdom of Trees*. Earthdancer.

Singleton, B., Gillette, M., & Green, C. (2021). Toward productive complicity: Applying 'traditional ecological knowledge' in environmental science. *The Anthropocene Review*, *10*(2), 393–414. Advance online publication. doi:10.1177/20530196211057026

Sinha, A., & Kumari, S. (2021). Integrating yoga with education in the modern schooling system. *Yoga Mimamsa*. 53(1), 46–58. https://journals.lww.com/yomi/fulltext/2021/53010/integrating_yoga_with_education_in_the_modern.9.aspx

Skelly, S. M., & Zajicek, J. M. (1998). The effect of Project GREEN, an interdisciplinary garden program, on the environmental attitudes of elementary school students. *HortTechnology*, *8*(4), 579–583. doi:10.21273/HORTTECH.8.4.579

Skinner, E. A., & Chi, U. (2012). Intrinsic Motivation and Engagement as "Active Ingredients" in Garden-Based Education: Examining models and measures derived from Self-Determination Theory. *The Journal of Environmental Education*, *43*(1), 16–36. doi:10.1080/00958964.2011.596856

Slikkerveer, L. J. (2016). LEAD: Leiden Ethnosystems And Development Programme. Verleden, Heden en Toekomst van het Leidse Speerpunt. Lokale Kennis Systemen en Ontwikkeling 1993-2016-2020. Internal Report, Leiden: LEAD Programme of Leiden University.

Slikkerveer, L. J. (2019). Towards a Model of Integrated Community-Managed Developmen Slikkerveer, L.J., Baourakis, G. & Saefullah, K. (Eds.) Integrated Community-Managed Development (ICMD): Strategizing Indigenous Knowledge and Institutions for Poverty Reduction and Sustainable Community Development in Indonesia - A Community-Based Contribution to the United Nations 2030 Agenda for Sustainable Development. Spinger Nature Switzerland AG. doi:10.1007/978-3-030-05423-6_4

Slikkerveer, L. J., & Dechering, W. H. J. C. (1995). LEAD: The Leiden Ethnosystems And Development Programme at Leiden University in The Netherlands. Warren, D. M., Slikkerveer, L. J. & Brokensha, D. (Eds.) (1995) The Cultural Dimension of Development: Indigenous Knowledge Systems. IT Studies on Indigenous Knowledge and Development, London, UK: Intermediate Technology Publications Ltd.

Slikkerveer, L. J., & Slikkerveer, M. K. L. (1995)."Taman Obat Keluarga (TOGA) IndigenusIndonesian Medicine fir Self-reliance. The Cultural Dimension of Development: Indigenous Knowledge Systems. IT Studies on Indigenous Knowledge and Development. London, UK: Intermediate Technology Publications Ltd.

Slikkerveer, L. J. (1989). Changing Values Attitudes of Social and Natural Scientist towards Indigenous Peoples and Their Knowledge Systems. In D. M. Warren, L. J. Slikkerveer, & S. O. Titilola (Eds.), *Indigenous Knowledge Systems: Implications for Agriculture and International Development. Studies in Technology and Social Change No. 11*. Iowa State University.

Slikkerveer, L. J. (1990). *Plural Medical Systems in The Horn of Africa: The Legacy of 'Sheikh' Hippocrates*. Kegan Paul International.

Slikkerveer, L. J. (1992). Indigenous Agricultural Knowledge Systems in Kenya: Towards Conservation of Bio-Cultural Diversity in East Africa. In W. M. Adams & L. J. Slikkerveer (Eds.), *Indigenous Knowledge an Change in African Agriculture. Studies in Technology and Social Change No. 9 Ames*. Iowa State University.

Slikkerveer, L. J. (1999). Ethnoscience, 'TEK', and its Application to Conservation. In D. A. Posey (Ed.), *Cultural and Spiritual Values of Biodiversity: A Complementary Contribution to the Global Biodiversity Assessment* (pp. 167–260). Intermediate Technology Publications Ltd. doi:10.3362/9781780445434.005

Slikkerveer, L. J. (2016). *Integrated Community-Managed Development: Reflections on the Ethnosystems Approach*. LEAD Programme of Leiden University.

Slikkerveer, L. J. (2018). Towards the Development of Ethno-Education: an Ethnoscience-Based Approach to Integrated Education Management in Indonesia. In E. Malihah, T. Aryanti, V. Adriany, H. Yulindrasari, & A. Izharuddin (Eds.), *Research for Social Justice*. Routledge Taylor & Francis Group.

Sloam, J., Pickard, S., & Henn, M. (2022). Young people and environmental activism: The transformation of democratic politics. *Journal of Youth Studies*, *25*(6), 683–691. doi:10.1080/13676261.2022.2056678

Sloman, A. (2012). Using participatory theatre in international community development. *Community Development Journal: An International Forum*, *47*(1), 42–57. doi:10.1093/cdj/bsq059

Smiley, K. T., Noy, I., Wehner, M. F., Frame, D., Sampson, C. C., & Wing, O. E. (2022). Social inequalities in climate change-attributed impacts of Hurricane Harvey. *Nature Communications*, *13*(1), 3418. doi:10.1038/s41467-022-31056-2 PMID:36008390

Smith, A. B. (2024, Jan 8). 2023: A historic year of U.S. billion-dollar weather and climate disasters. *NOAA and Climate.gov*. https://www.climate.gov/news-features/blogs/beyond-data/2023-historic-year-us-billion-dollar-weather-and-climate-disasters

Smith, M. A., Dunhill, A., & Scott, G. W. (2018). Fostering children's relationship with nature: Exploring the potential of Forest School. *Education 3-13, 46*(5), 525–534. doi:10.1080/03004279.2017.1298644

Smith, C., & Wobst, H. M. (2005). *Indigenous archaeologies: decolonizing theory and practice*. Routledge.

Smith, G. S., Anjum, E., Francis, C., Deanes, L., & Acey, C. (2022). Climate change, environmental disasters, and health inequities: The underlying role of structural inequalities. *Current Environmental Health Reports*, *9*(1), 80–89. doi:10.1007/s40572-022-00336-w PMID:35338470

Smith, L. (1999). *Decolonizing Methodologies: Research and Indigenous Peoples*. Zed Books Ltd.

Smith, L. T. (2012). *Decolonizing methodologies: research and indigenous peoples* (2nd ed.). Zed Books.

Sobel, D. (2016). Finding the right place to put down roots: Facilities and Infrastructure in D. Sobel, P. Bailie, K. Finch, E. Kenny, & A. Stires (Eds.), Nature Preschools and Forest Kindergartens: The Handbook for Outdoor Learning (Illustrated edition, pp. 191-210). Redleaf Press.

Sobel, D., Bailie, P., Finch, K., Kenny, E., & Stires, A. (2015). Nature Preschools and Forest Kindergartens: The Handbook for Outdoor Learning (Illustrated edition). Redleaf Press.

SomarathneE. A. S. K.GunathungaM. W.LokupitiyaE. (2023). How Does Meditation Relate to Quality of Life, Positive Lifestyle Habits and Carbon Footprint? SSRN. https://papers.ssrn.com/sol3/papers.cfm?abstract_id=4549372 doi:10.2139/ssrn.4549372

Soper, J. C., & Fetzer, J. S. (2007). Religious institutions, church–state history and Muslim mobilisation in Britain, France and Germany. *Journal of Ethnic and Migration Studies*, *33*(6), 933–944. doi:10.1080/13691830701432780

Spence, C. (2020, September 18). Senses of place: Architectural design for the multisensory mind. *Cognitive Research: Principles and Implications*, *5*(46), 46. doi:10.1186/s41235-020-00243-4 PMID:32945978

Sponberg, A. (1998). Green Buddhism and the Hierarchy of Compassion. In M. E. Tucker & D. R. Williams (Eds.), *Buddhism and Ecology: The Interconnection of Dharma and Deeds* (pp. 351–376). Harvard University Center for the Study of World Religions, Harvard University Press.

Sponsel, L. E. (2013). *Spiritual Ecology: A Quiet Revolution*. Praeger.

Spooner, S. L., Jensen, E. A., Tracey, L., & Marshall, A. R. (2019). Evaluating the impacts of theatre-based wildlife and conservation education at the zoo. *Environmental Education Research*, *25*(8), 1231–1249. doi:10.1080/13504622.2019.1569201

Spring, J. (2008). *The American School: From the Puritans to No Child Left Behind* (7th ed.). McGraw-Hill.

Stapleton, S. R. (2019). A case for climate justice education: American youth connecting to intragenerational climate injustice in Bangladesh. *Environmental Education Research*, *25*(5), 732–750. doi:10.1080/13504622.2018.1472220

Stapleton, S. R. (2020). Toward critical environmental education: A standpoint analysis of race in the American environmental context. *Environmental Education Research*, *26*(2), 155–170. doi:10.1080/13504622.2019.1648768

Staudt, L., & Howard, B. (2013). The added complications of climate change: Understanding and managing biodiversity and ecosystems. *Frontiers in Ecology and the Environment*, *11*(9), 494–501. doi:10.1890/120275

Stavenhagen, R. (2015). Indigenous Peoples' Rights to Education. *European Journal of Education*, *50*(3), 254–257. doi:10.1111/ejed.12136

Steele & Scott. (2016). Emotionality and Learning Stories: Documenting How We Learn How We Feel. *Canadian Journal of Environmental Education*, *21*(6), 1–19. https://cjee.lakeheadu.ca/article/view/1426/862

Steinbeck, A. (1836). *Der Dichter ein Seher, oder über die innige Verbindung der Poesie und der Sprache mit dem Hellsehen*. Verlag Georg Joachim Göschen.

Stephenson, J., & Carter, M. (2011, August). The Use of Multisensory Environments in Schools for Students with Severe Disabilities: Perceptions from Teachers. *Journal of Developmental and Physical Disabilities*, *23*(4), 339–357. doi:10.1007/s10882-011-9232-6

Stolnit, R. (2021, August 23). Big oil coined 'carbon footprints' to blame us for their greed. Keep them on the hook. *Guardian*. https://www.theguardian.com/commentisfree/2021/aug/23/big-oil-coined-carbon-footprints-to-blame-us-for-their-greed-keep-them-on-the-hook

Stone, M. (2007). Rethinking school lunch: Education for sustainability in practice. *Canadian Journal of Environmental Education*, *12*(1), 19–32.

Striniste, N. (2020). *"Nature Play at Home" presentation*. Virginia Association for Environmental Education Conference. Amherst, Virginia.

Striniste, N. (2019). *Nature Play at Home: Creating Outdoor Spaces That Connect Children with the Natural World*. Timber Press.

Student Sustainability Center Annual Impact Report. (2020). PDX.edu. https://www.pdx.edu/student-sustainability-center/sites/studentsustainabilitycenter.web.wdt.pdx.edu/files/2022-07/2020%20-%202021%20SSC%20Impact%20Report.pdf

Sturgeon, N. (1997). *Ecofeminist natures: Race, gender, feminist theory and political action*. Routledge.

Suara, I. D. M., Gelgel, I. P., & Yasa, I. W. S. (2018). Pitra yadnya implementation for Hindu heroes. *International Journal of Social Science and Humanity*, *2*(3), 136–148.

Sulistyani, H. D., & Rahrdjo, T. (2019). Ethno-Pedagogy Perspective on Ethnic Minority Discourse of Education, *Proceedings of the 2nd International Conference on Educational Sciences Series: Advances in Social Science, Education and Humanities Research*. 10.2991/ices-18.2019.53

Summer, J. (2018). Eating Our Way to Sustainability? Leisure, Food and Community Economic Development. *Sustainability*. https://www.mdpi.com/2071-1050/10/5/1422

Suryawati, E., Suzanti, F., Zulfarina, Z., Putriana, A. R., & Febrianti, L. (2020). The Implementation of Local Environmental Problem-Based Learning Student Worksheets to Strengthen Environmental Literacy. *Jurnal Pendidikan IPA Indonesia*, *9*(2), 169–178. doi:10.15294/jpii.v9i2.22892

Susanti, R. D. (2023). *'Kearifan Kesehatan Lokal'*: Indigenous Medical Knowledge Practice for Integrated Nursing of the Elderly with Cardiovascular Disease in Sumedang, West Java: Towards Transcultural Nursing in Indonesia. PhD Dissertation. *Leiden Ethnosystems And Development Programme (LEAD) Studies* No. 12. Leiden: Leiden University.

Suyanto. (2023). Digital Innovation in the Era of Society 5.0 Encourages Indonesia's Economic Recovery in Post-Pandemic, *International Journal of Integrative Sciences*, *2*(7), 1141-1148.

Swamy, R. N. (2014). Protection of Traditional Knowledge in the Present IPR Regime: A Mirage or Reality. *The Indian Journal of Public Administration*, *60*(1), 50. doi:10.1177/0019556120140103

Swiderska, K., Argumedo, A., Pant, R., Vedavathy, S., Munyi, P., Mutta, D., Heraclio, H., Song, Y., & Li, J. (2009). *Protecting traditional knowledge from the grassroots up*. International Institute for Environment and Development. http://www.jstor.com/stable/resrep01439

Tagare, G. V. (2003). *The Skanda-Purana*, Part 18. In G. P. Bhatt (Ed.), *Ancient Indian Tradition & Mythology* (Vol. 66). Motilal Banarasidass Publishers Pvt. Ltd.

Tang, R., & Gavin, M. C. (2016). A Classification of Threats to Traditional Ecological Knowledge and Conservation Responses. *Conservation & Society*, *14*(1), 57–70. https://www.jstor.org/stable/26393228. doi:10.4103/0972-4923.182799

Tauste, A. M. V. (1992). Funci6n metalingüística y uso del lenguaje. *Epos: Revista de filología*, (8), 123-123.

Taylor, P. W. (1984). Are Humans Superior to Animals and Plants. *Environmental Ethics 6*, 2. 149-160.

Taylor, A. (2013). *Reconfiguring the natures of childhood*. Routledge. doi:10.4324/9780203582046

Taylor, A., & Pacini-Ketchabaw, V. (2015). Learning with children, ants, and worms in the Anthropocene: Towards a common world pedagogy of multispecies vulnerability. *Pedagogy, Culture & Society*, *23*(4), 507–529. doi:10.1080/14681366.2015.1039050

The Fourteenth Dalai Lama of Tibet. (1987, September 21). *Five Point Peace Plan* [Address]. U.S. Congressional Human Rights Caucus, Washington, D.C. https://www.dalailama.com/messages/tibet/five-point-peace-plan

The Fourteenth Dalai Lama of Tibet. (1988, June 15). *Strasbourg Proposal* [Address]. European Parliament, Strasbourg, France. https://www.dalailama.com/messages/tibet/strasbourg-proposal-1988

The World Bank. (2023). *Betting on sustainability in the Moskitia, in Honduras, to create new livelihoods*. The World Bank. https://www.worldbank.org/en/news/feature/2023/04/24/betting-on-sustainability-in-the-moskitia-honduras-to-create-new-livelihoods

Thistlethwaite, J. E., & Ewart, B. R. (2003). Valuing diversity: Helping medical students explore their attitudes and beliefs. *Medical Teacher*, *25*(3), 277–281. doi:10.1080/0142159031000100346 PMID:12881050

Thompson, J. E. S. (1927). The Civilization of the Mayas. *Anthropology Leaflet, 25,* 1–96. https://www.jstor.org/stable/41444125

Thompson, C. (2005). Travel writing. In N. Roe (Ed.), *Romanticism: an Oxford guide*. Oxford University Press. doi:10.1093/oso/9780199258406.003.0035

Thornes (2016). Animals and Climate Change Author. *Journal of Animal Ethics*, *6*, 81-88. doi:10.5406/janimalethics.6.1.0081

Thunberg, G. (2019, May 4). It's 2019. Can we all now... [Twitter post]. https://twitter.com/gretathunberg/status/1124723891123961856

Tingley, B. (2024, Jan 12). 2023 was the hottest year on record, NASA and NOAA say. *Space Forums*. https://www.space.com/nasa-noaa-2023-hottest-year-on-record

Tinnell, J. (2011). Scripting just sustainability: Through green listing towards eco-blogging. *Environmental Communication*, *5*(2), 228–242. doi:10.1080/17524032.2011.562519

Tiny Trees Preschool. (n.d.). *Tiny Trees Preschool: Quality, Affordable, Outdoor Preschool in Seattle & Puget Sound*. Tiny Trees Preschool. https://tinytrees.org/

Toh, S.-H. (2012). Engaged Buddhism and its Contributions to Sustainable Development and ESD. In R. Clugston & S. Holt (Eds.), *Exploring Synergies between Faith Values and Education for Sustainable Development* (pp. 62–66). Earth Charter International and Earth Charter Center for Education for Sustainable Development at University of Peace. https://www.iau-hesd.net/sites/default/files/documents/febr_2_2012.pdf

Tomalin, E. (2016). *Biodivinity and biodiversity: The limits to religious environmentalism.* Routledge. doi:10.4324/9781315569352

Tomislav, K. (2018). The concept of sustainable development: From its beginning to the contemporary issues. *Zagreb International Review of Economics & Business, 21*(1), 67–94. doi:10.2478/zireb-2018-0005

Tomlinson, C. A. (2001). *How to Differentiate Instruction in Mixed-Ability Classrooms* (2nd ed.). Association for Supervision and Curriculum Development.

Toth, N., Little, L., Read, J. C., Fitton, D., & Horton, M. (2013). Understanding teen attitudes towards energy consumption. *Journal of Environmental Psychology, 34*, 36–44. https://www.sciencedirect.com/science/article/pii/S0272494412000722. doi:10.1016/j.jenvp.2012.12.001

Tov & Diener. (2013), *Subjective Well-Being*. Research Collection School of Social Sciences. https://ink.library.smu.edu.sg/soss_research/1395/ Archived 2020-12-02 at the Wayback Machine

Treen, K. M. D. I., Williams, H. T. P., & O'Neill, S. J. (2020). Online misinformation about climate change. *Wiley Interdisciplinary Reviews: Climate Change, 11*(5), e665. https://wires.onlinelibrary.wiley.com/doi/full/10.1002/wcc.665. doi:10.1002/wcc.665

Tsomo, K. L. (2015). Buddhist Nuns and Gendered Practice: In Search of the Female Renunciant [Review of *Buddhist Nuns and Gendered Practice: In Search of the Female Renunciant*, by N. S. Salgado]. *Journal of Global Buddhism, 16*, 1–7.

Tuck, E. (2009). Suspending damage: A letter to communities. *Harvard Educational Review, 79*(3), 409–428. doi:10.17763/haer.79.3.n0016675661t3n15

Tuck, E., McKenzie, M., & McCoy, K. (2014). Land Education: Indigenous, Post-Colonial, and Decolonizing Perspective on Place and Environmental Education Researc. *Environmental Education Research, 20*(1), 1–23. doi:10.1080/13504622.2013.877708

Tuck, E., & Yang, K. W. (2021). Decolonization is not a metaphor. *Tabula Rasa, 38*(38), 61–111. doi:10.25058/20112742.n38.04

Tucker, M. E., & Williams, D. R. (Eds.). (1998). *Buddhism and Ecology: The Interconnection of Dharma and Deeds.* Harvard University Center for the Study of World Religions, Harvard University Press.

Turner, L., Eliason, M., Sandoval, A., & Chaloupka, F. J. (2016). Increasing prevalence of US elementary school gardens, but disparities reduce opportunities for disadvantaged students. *The Journal of School Health, 86*(12), 906–912. doi:10.1111/josh.12460 PMID:27866388

Tylor, E. B. (1871). *Primitive culture: Researches into the development of mythology, philosophy, religion, art and custom* (Vol. 2). J. Murray.

Tynsong, H., Dkhar, M., & Tiwari, B. K. (2020). Review: Traditional ecological knowledge of tribal communities of North East India. *Biodiversitas (Surakarta), 21*(7), 3209–3224. doi:10.13057/biodiv/d210743

Tyutchev, F. (2000). *The Complete Poems of Tyutchev in An English Translation by F. Jude.* The University of Durham.

U.S. Global Change Research Program. (2009). *Climate Literacy: The Essential Principles of Climate Science*. US Global Change Research Program. https://www.globalchange.gov/reports/climate-literacy-essential-principles-climate-science

Uhrmacher, P. B. (1995). Uncommon Schooling: A Historical Look at Rudolf Steiner, Anthroposophy, and Waldorf Education. *Curriculum Inquiry*, *25*(4), 381–406. doi:10.1080/03626784.1995.11076190

Ullevig, S. L., Vasquez, L., Ratcliffe, L. G., Oswalt, S. B., Lee, N., & Lobitz, C. A. (2020). Establishing a campus garden and food pantry to address food insecurity: Lessons learned. *Journal of American College Health*, *69*(6), 684–688. doi:10.1080/07448481.2019.1705830 PMID:31916927

UN (1971). *Second United Nations Development Decade* (1971-1980). New York, USA: UN.

UN (1972). *Stockholm Declaration and Action Plan for the Human Environment*. New York, USA: UN.

UN (1992). *Earth Summit on Environment and Development* (Rio de Janeiro, Brasil). New York, USA: UN.

UN (2002). *World Summit on Sustainable Development* (Johannesburg). New York, USA: UN.

UN (2006). *Convention on the Rights of People with Disabilities*. New York, USA: UN.

UN (2014). *Decade of Education for Sustainable Development* (2005-2014). New York, USA: UN.

UN (2015). *2030 Agenda for Sustainable Development (SDG)*. New York. USA: UN.

UN (2016). *Global Gateway of IEEP: UN Office for Partnerships to Deliver the SDGs*. New York, USA: UN.

UN. (2019). Global Assessment Report on Biodiversity and Ecosystem Services. New York, USA: UN.

UN. (2020). G*lobal Biodiversity Outlook 5* (GBO-5), UN Convention on Biological Diversity. New York, USA: UN.

UNDP. (2020). *Environmental Education for Sustainable Development in Honduras: Challenges and Opportunities*. United Nations Development Programme.

UNDP. (n.d.). *The SDGs in Action*. UNDP. https://www.undp.org/sustainable-development-goals

UNEP. (2017). *Conservation through Cultural Survival: Indigenous Peoples and Protected Areas*. UNEP.

UNESCO. (1968). *Biosphere Conference*. Paris: UNESCO.

UNESCO. (1995). *World Conference on Special Needs Education*. Paris: UNESCO.

UNESCO. (2016). *Education for people and planet: creating sustainable futures for all. Global education monitoring report*. UNESCO.

UNESCO. (2018). Leicht, A., Heiss, J., & Byun, W. J. (Eds.). Issues and trends in education for sustainable development. United Nations Educational, Scientific and Cultural Organization. doi:10.54675/YELO2332

UNESCO. (2019). *Education for Sustainable Development in Latin America and the Caribbean: A Stocktaking Report on the Decade of Education for Sustainable Development (2005-2014)*. UNESCO Regional Bureau for Education in Latin America and the Caribbean.

UNESCO. (2021). *UNESCO urges making environmental education a core curriculum component in all countries by 2025*. UNESCO. https://www.unesco.org/en/articles/unesco-urges-making-environmental-education-core-curriculum-component-all-countries-2025

UNESCO. (2021). *World Conference on Education for Sustainable Development (Berlin)*. Paris: UNESCO.

UNESCO. (2023). *Greening every curriculum: Development of the greening curriculum guidance*. UNESCO. https://www.unesco.org/en/education-sustainable-development/greening-future/curriculum

UNESCO. (n.d.). *Sustainable Development Goals - Resources for educators*. UNESCO. https://en.unesco.org/themes/education/sdgs/material#:~:text=Its%20aims%20form%20one%20of,for%20present%20and%20future%20generations

UNESCO-UNEP. (1975). Education for Sustainable Development (ESD of IEEP). Paris/Nairobi, Kenya: UNEP.

UNESCO-UNEP. (1975). International Environmental Education Programme (IEEP) (1975-1983). Paris/Nairobi, Kenya: UNEP.

UNESCO-UNEP. (1975). *International Environmental Education Programme (IEEP)*. UNEP.

UNESCO-UNEP. (1977). *The* **Tbilisi Declaration,** *Tbilisi, Georgia (USSR)*. UNEP.

UNHCR. (2018). *World Directory of Minorities and Indigenous Peoples*. Minority Rights Group International. https://www.refworld.org/reference/countryrep/mrgi/2018/en/65136

UN-ICERD. (1965). *International Convention on the Elimination of all Forms of Racial Discrimination*. New York, USA: UN-ICERD.

Union of Concerned Scientists. (2018) *Heat Waves and Climate Change The Effects of Worsening Heat on People, Communities, and Infrastructure*. Union of Concerned Scientists. https://www.jstor.org/stable/resrep24150

United Nations. (2007). *United Nations Declaration of the Rights of Indigenous Peoples*. United Nations.

United Nations. (2015). Transforming our world: The 2030 agenda for sustainable development. *UN Department of Economic and Social Affairs: Sustainable Development*. UN. https://sdgs.un.org/2030agenda

United Nations. (2023). *Communities Organizing for Watersheds – Honduras*. UN. https://unfccc.int/climate-action/momentum-for-change/activity-database/communities-organizing-for-watersheds

United Nations. (2023). *Global Sustainable Development Report 2023: Times of crisis, times of change: Science for accelerating transformations to sustainable development*. UN. https://sdgs.un.org/gsdr/gsdr2023

United Nations. (2023). *The sustainable development goals report 2023, Special edition: Towards a rescue plan for people and planet*. United Nations. https://unstats.un.org/sdgs/report/2023/The-Sustainable-Development-Goals-Report-2023.pdf

United Nations. (2024). *Climate Action, Education is key to addressing climate change*. UN. https://www.un.org/en/climatechange/climate-solutions/education-key-addressing-climate-change

United Nations. (2024). *Department of Economic and Social Affairs, Population growth, environmental degradation and climate change*. UN. https://www.un.org/en/desa/population-growth-environmental-degradation-and-climate-change

United Nations. (2024). *The 17 Goals*. UN Department of Economic and Social Affairs: Sustainable Development. https://sdgs.un.org/goals

United States Environmental Protection Agency (EPA). (2018). *What is Environmental Education?* EPA. https://www.epa.gov/education/what-environmental-education>

University of California Berkeley (2022). *Berkeley Universal Design for Learning*. University of California Berkeley.

Vahlsing, C., & Liscow, Z. (2023, March 14). *The importance of measuring fiscal and economic costs of climate change*. The White House. https://www.whitehouse.gov/omb/briefing-room/2023/03/14/the-importance-of-measuring-the-fiscal-and-economic-costs-of-climate-change/

Valenzuela Pérez, L. (2020). *A Lenca Voice of Resistance Against a Culture of Extractivism: Donatila Girón Calix. Cultural Survival Quarterly*. https://www.culturalsurvival.org/publications/cultural-survival-quarterly/lenca-voice-resistance-against-culture-extractivism

Valera, L., & Vidal, G. (2022). Pantheism, Panentheism, and Ecosophy: Getting Back to Spinoza? *Zygon*, *57*(3), 545–563. doi:10.1111/zygo.12800

Van Rooyen, H. G. (2002). Education for a Sustainable Environment: A Potential Model. Johannesburg, SA:_ Rand Afrikaans University.

Vansina, J. (1985). *Oral tradition as history*. University of Wisconsin Press.

Vaske, J. J., & Kobrin, K. C. (2001). Place Attachment and Environmentally Responsible Behavior. *The Journal of Environmental Education*, *32*(4), 16–21. doi:10.1080/00958960109598658

Vavilov, N. I. (1951). *The Origin, Variation, Immunity and Breeding of Cultivated Plants*. Chronica Botanica Company. doi:10.1097/00010694-195112000-00018

Vedantatirtha, G. C. (Ed.). (1915). Kulachudamani Tantra. Sanskrit Text Depository & Luzac & Co.

Veenhoven, R. (2000). The Four Qualities of Life. *Journal of Happiness Studies*, *1*(1), 1–39. doi:10.1023/A:1010072010360

Verlie, B. (2019). Bearing worlds: Learning to live-with climate change. *Environmental Education Research*, *25*(5), 751–766. doi:10.1080/13504622.2019.1637823

Vijnananda, S. (1921-22) *The Srimad Devi Bhagavatam*. https://www.pdfdrive.com/srimad-devi-bhagavata-purana-english-by-sw-vijnanananda-d39957378.html (Last access: 3.02.2024)

Vizenor, G. (2008). Aesthetics of survivance. In G. Vizenor (Ed.), *Survivance: narratives of Native presence: 1-23*. University of Nebraska Press.

Vogler, J. (2012). *Self-Actualization and Peak Experiences in Outdoor Recreation*. TigerPrints. https://tigerprints.clemson.edu/all_theses/1485

WA DCYF. (2022). *Nature-Based Learning Competencies*. DCYF.

Wade, C. (2012). *The Harmony of the Abyss: Translating Tiutchev and Mandelstam*. Wesleyan University.

Wagler, R. (2011). *Human Extinction and the End of the World: An Analysis of the World's End Scenarios and Related Topics*. GRIN Verlag.

Waldau, P. (2003). The Specter of Speciesism: Buddhist and Christian Views of Animals. Oxford University Press. doi:10.1093/0195145712.001.0001

Walden University. (2019). *7 Ways to Teach Critical Thinking in Elementary Education*. Walden University. https://www.waldenu.edu/online-bachelors-programs/bs-in-elementary-education/resource/seven-ways-to-teach-critical-thinking-in-elementary-education

Waldron, J. W., Locock, A. J., & Botey, A. P. (2016). Building an outdoor classroom for field geology: The Geoscience Garden. *Journal of Geoscience Education*, *64*(3), 215–230. doi:10.5408/15-133.1

Waliczek, T. M., Logan, P., & Zajicek, J. M. (2003). Exploring the impact of outdoor environmental activities on children using a qualitative text data analysis system. *HortTechnology*, *13*(4), 684–688. doi:10.21273/HORTTECH.13.4.0684

Warden, C. H. (2015). Learning with nature: Embedding outdoor practice. *Learning with Nature*, 1–128.

Warner, J. (2015). The High School/College Writing Classroom Disconnect. *Inside Higher Ed.* https://www.insidehighered.com/blogs/just-visiting/high-schoolcollege-writing-classroom-disconnect

Warren, D. M., Slikkerveer, L. J., & Brokensha, D. (Eds.). (1995). *The Cultural Dimension of Development. Indigenous Knowledge Systems. IT Studies on Indigenous Knowledge and Development.* Intermediate Technology Publications Ltd. doi:10.3362/9781780444734

Watson, S. K. (2021, Dec. 4). 4 new myths about climate change – and how to debunk them. *Popular Science.* https://www.popsci.com/environment/climate-denier-myths/

Watson, P., & Doué, W. (2007). The Outdoor School Inspires Potentially Highly Qualified Middle Level Teachers. *Middle School Journal, 39*(2), 50–58. www.jstor.org/stable/23048339. doi:10.1080/00940771.2007.11461624

Watts, V. (2013). Indigenous Place-Thought and Agency Amongst Humans and Non Humans (First Woman and Sky Woman Go On a European World Tour!). *Decolonization, 2*(1), 1. https://jps.library.utoronto.ca/index.php/des/article/view/19145

Weber, T. (1999). Gandhi and Deep Ecology. *Journal of Peace Research, 36*(3), 3. doi:10.1177/0022343399036003007

Weeks, J. M., & Black, N. J. (1992). Notes on the Ethnopharmacology of the Lenca Indians of Western Honduras and Eastern El Salvador. *Mexico, 14*(4), 71–74. https://www.jstor.org/stable/23758612

Weimer, B. B., Cappotelli, M., & DiCamillo, J. (1994). Self-advocacy: A working proposal for adolescents with special needs. *Intervention in School and Clinic, 30*(1), 47–52. doi:10.1177/105345129403000108

Weinstein, N., Balmford, A., DeHaan, C. R., Gladwell, V., Bradbury, R. B., & Aman, T. (2015). Seeing the Community for the Trees: The Links Among Contact with Natural Environments, Community Cohesion, and Crime. *Bioscience, 65*(12), 1141–1153. doi:10.1093/biosci/biv151

Weiss, T., Kahn, P. H. Jr, & Lam, L.-W. (2023). Children's interactions with relatively wild nature associated with more relational behavior: A model of child-nature interaction in a forest preschool. *Journal of Environmental Psychology, 86*, 101941. doi:10.1016/j.jenvp.2022.101941

Weldemariam, K. (2017). Challenging and Expanding the Notion of Sustainability Within Early Childhood Education: Perspectives from Post-humanism and/or New Materialism. In O. Franck & C. Osbeck (Eds.), *Ethical Literacies and Education for Sustainable Development: Young People, Subjectivity and Democratic Participation* (pp. 105–126). Springer International Publishing. doi:10.1007/978-3-319-49010-6_7

Wells, N. M., & Evans, G. W. (2003). Nearby Nature: A Buffer of Life Stress among Rural Children. *Environment and Behavior, 35*(3), 311–330. doi:10.1177/0013916503035003001

Wells, N. M., & Lekies, K. S. (2006). Nature and the life course: Pathways from childhood nature experiences to adult environmentalism. *Children, Youth and Environments, 16*(1), 1–24. doi:10.1353/cye.2006.0031

Wheelright, E. G. (1974). *Vintage Medicinal Plants and Their History.* Dover Publications.

White, D., & Hendee, J. (2000) Primal hypotheses: The relationship between naturalness, solitude and the wilderness experience benefits of the development of self (DOS), development of community (DOC), and spiritual development (SD*).* In McCool, S., Cole, D., O'Loughlin, J. *Wilderness science in a time of change conference—Volume 3: Wilderness as a place for scientific inquiry.* Proceedings RMRS-P-15-VOL-3. Ogden, UT: U.S. Department of Agriculture, Forest Service, Rocky Mountain Research Station.

White, L. T. (1978). *Medieval religion and technology: collected essays* (Vol. 13). Univ of California Press.

White, L. Jr. (1967). The historical roots of our ecologic crisis. *Science*, *155*(3767), 1203–1207. doi:10.1126/science.155.3767.1203 PMID:17847526

White, S. A. (1999). *The art of facilitating participation: Releasing the power of grassroots communication*. Sage Pub.

Whyte, K. (2020). Too late for indigenous climate justice: Ecological and relational tipping points. *Wiley Interdisciplinary Reviews: Climate Change*, *11*(1), e603. doi:10.1002/wcc.603

Whyte, K. P. (2013). On the role of traditional ecological knowledge as a collaborative concept: A philosophical study. *Ecological Processes*, *7*(1), 2. doi:10.1186/2192-1709-2-7

Wicks, C., Barton, J., Orbell, S., & Andrews, L. (2022, August). Psychological benefits of outdoor physical activity in natural versus urban environments: A systematic review and meta-analysis of experimental studies. *Applied Psychology. Health and Well-Being*, *14*(3), 1037–1061. doi:10.1111/aphw.12353 PMID:35259287

Williams, D. (2008). Sustainability education's gift: Learning patterns and relationships. *Journal of Education for Sustainable Development*, *2*(1), 41–49. doi:10.1177/097340820800200110

Williams, D. R., & Dixon, P. (2013). Impact of Garden-Based Learning on academic outcomes in schools. *Review of Educational Research*, *83*(2), 211–235. doi:10.3102/0034654313475824

Williandri, R. (2024). 'Bakti and Sayan' Traditions among the Tenggerese People in East Java:The Role of Indigenous Institutions in Integrated Elderly Care Development in Indonesia. PhD Dissertation, *Leiden Ethnosystems And Development Programme (LEAD) Studies* No. 1. Leiden: Leiden University.

Wilson, R. (2014). *Take it outside. Education Digest, 80*(2), 45–48. https://search.ebscohost.com/login.aspx?direct=true&AuthType=ip,url,geo,cpid,ui d&custid=kansas&geocustid=kansas&db=aph&AN=98510555&site=ehost-live

Wilson, R. (2019). What Is Nature? *International Journal of Early Childhood Environmental Education*, *7*(1), 26–39.

Wilson, R. A. (1993). The Importance of Environmental Education at the Early Childhood Level. *International Journal of Environmental Education and Information*, *12*(1), 15–24.

Wlodkowski, R. J., & Galbraith, M. W. (2004). Creating motivating learning environments. *Adult learning methods: A guide for effective instruction, 3*, 141-164.

Wolfe, P. (2006). Settler colonialism and the elimination of the native. *Journal of Genocide Research*, *8*(4), 387–409. doi:10.1080/14623520601056240

Woodroffe, J. (2009). Sakti and Sakta. Essays and Addresses on the Sakta Tantrasastras. Celephaïs Press.

World Bank. (2019). *Education Sector Public Expenditure Review for Honduras*. World Bank Group.

World Intellectual Property Organisation. (2003). *Composite Study on the Protection of Traditional Knowledge* (5th session). Intergovernmental Committee on Intellectual Property and Genetic Resources, Traditional Knowledge and Folklore. https://www.wipo.int/edocs/mdocs/tk/en/wipo_grtkf_ic_5/wipo_grtkf_ic_5_8.pdf

World Land Trust. (n.d.). *Asociación ecológica de san marcos de ocotepeque (aesmo)*. World Land Trust. https://www.worldlandtrust.org/who-we-are-2/partners/aesmo/

Worrall, S. (2017). *We are Wired to be Outside*. National Geographic. https://news.nationalgeographic.com/2017/02/nature-fix-brain-happy-florence-williams/>

WRI/IUCN/UNEP (1992). *Global Biodiversity Strategy*. New York, USA: UN.

Wu, C. (2020). *Environmental education adapts to virtual learning*. Chesapeake Bay Program. chesapeakebay.net/news/blog/environmental_education_adapts_to_virtual_learning,

Yalcin, F., & Erden, F. (2021). A Cross-Cultural Study on Outdoor Play: Teachers' Beliefs and Practices. *Education in Science*, *46*(206), 131–159.

Young, C., & Jones, R. (2023, Nov. 1). *Denial is over: Climate change is happening, but why do we still act like it's not?* Phys.org. https://phys.org/news/2023-11-denial-climate.html

Young, R., & Crandall, R. (1984). Wilderness use and self-actualization. *Journal of Leisure Research*, *16*(2), 149–160. doi:10.1080/00222216.1984.11969582

Yunt, J. D. (2001). Jung's Contribution to Ecological Psychology. *Journal of Humanistic Psychology*, *41*(2), 96–121. doi:10.1177/0022167801412007

Zafeiroudi, A. (2018). Hatha Yoga: The importance of Yamas and their application from theory to practice. *Journal of Sport & Recreation Management*, *15*(1), 64–76. https://www.researchgate.net/publication/343452538_Hatha_Yoga_The_importance_of_Yamas_and_their_application_from_theory_to_practice

Zafeiroudi, A., & Hatzigeorgiadis, A. (2014). The effects of an outdoor pursuit's intervention program on adolescents' environmental beliefs. *International Journal on Advances in Education Research*, *1*(3), 106–118. https://www.researchgate.net/publication/343452414_The_effects_of_an_outdoor_pursuit's_intervention_program_on_adolescents'_environmental_beliefs

Zafeiroudi, A., Pipinia, M., Yfantidou, G., & Georgomanos, S. (2021). The Effects of Yoga Practice on Practitioners' Environmental Behaviours & Sustainability. *Environmental Management and Sustainable Development*, *10*(3), 76. Advance online publication. doi:10.5296/emsd.v10i3.18894

Zakopoulos, V., Makri, A., Ntanos, S., & Tampakis, S. (2023). Drama/Theatre Performance in Education through the Use of Digital Technologies for Enhancing Students' Sustainability Awareness: A Literature Review. *Sustainability (Basel)*, *15*(18), 13387. doi:10.3390/su151813387

Zambrano, María (2006/1939). *Filosofía y poesiá*, Fondo de cultura Económica Mexico. Trans. D. Ohmans (2008). *Philosophy and Poetry*.

Zamora, O. M. F. (1978). EL COLAPSO MAYA. *Anuario de Estudios Centroamericanos*, *4*, 489–507. https://www.jstor.org/stable/25661662

Zandalinas, S. I., Fritschi, F. B., & Mittler, R. (2021). Global warming, climate change, and environmental pollution: Recipe for a multifactorial stress combination disaster. *Trends in Plant Science*, *26*(6), 588–599. doi:10.1016/j.tplants.2021.02.011 PMID:33745784

Zapf, H. (Ed.). (2016). *Handbook of ecocriticism and cultural ecology* (Vol. 2). Walter de Gruyter. doi:10.1515/9783110314595

Zidny, R., Sjöström, J., & Eilks, I. (2020). A multi-perspective reflection on how indigenous knowledge and related ideas can improve science education for sustainability. *Science & Education*, *29*(1), 145–185. doi:10.1007/s11191-019-00100-x

Zimmerman, M. E. (1994). *Contesting Earth's future: Radical ecology and postmodernity*. University of California Press. doi:10.1525/9780520919228

About the Contributors

Tatjana Kochetkova is Dutch interdisciplinary researcher and writer. She has defended her Ph.D in philosophical anthropology at Radboud University of Nijmegen, Netherlands. She is publishing in Environmental Philosophy, Philosophical Anthropology, and Bioethics.She is the author of several books and a series of articles on environmental philosophy and the philosophy of man and technology, including books Search for Authentic Spirituality in Modern Russian Philosophy (2007) and The Plurality of the Arts of Living (2018).

***

Anasuya Agarwala was awarded her phd from the Centre of Philosophy, Jawaharlal Nehru University in 2022 in which she worked on the moral implications of indifference. Prior to that, she was awarded an Mphil degree in Philosophy where she worked on the critical potential of laughter. She completed her Bachelors and Masters degree from University of Delhi. Anasuya has previously worked as an Assistant Professor at Miranda House, Lady Shri Ram College for Women, Kirori Mal College and SRM University, Andhra Pradesh. She is currently Assistant Professor at Gargi College. Her research interests include existential philosophy, moral philosophy, feminism and environmental ethics.

Stacey M. Alfonso is an instructor of early childhood education and PhD candidate at the University of Washington in Seattle, WA, USA. She is a passionate advocate for nature-based learning and equitable access to the outdoors for all children. Her research focuses on early childhood environmental education, investigating how children form understanding and connections with their natural world. Stacey strives to enrich educators' and researchers' perspectives on child-nature relationships, shedding light on nature's role in development and learning.

Idris Bhat teaches at Department of Liberal Arts SRM University AP. Previously, he taught at Ashoka University.

Katrina Broughman (pronounced "Bruff-man") is a native of rural Rockbridge County, Virginia. A creative intellectual raised on arts, theater, dance, and music, Katrina actively uses her eclectic talents to advocate for hope with persistent compassion as a changemaker. Professionally trained as an educator, musician, and sociologist, her interests include youth leadership, community development, social justice, global issues, advocacy, the arts, environmental conservation, and empowering marginalized voices. Katrina taught English in Peace Corps Moldova (2015-2017) and wrote grants to develop successful youth-led programs. Katrina received a B.A. '13 is from Mary Baldwin College (MBU) in English and Sociology (minors in Education plus Peacemaking and Conflict Resolution) with emphasis on the environment. She received her Masters in Education (Environment-Based Learning) from MBU with additional M.A. Sociology coursework at Morehead State University (KY) in 2021. She has been certified in secondary English language arts since 2014 and as Central Blue Ridge VA Master Naturalist since 2018. Katrina aspires to begin a Ph.D program for sociology and nature-based learning in 2025 or 2026.

Abhiroop Chowdhury is Professor and Executive Dean at Jindal Global School of Environment and Sustainability. Abhiroop earned his, PhD degree from Indian Institute of Technology (Indian School of Mines) with MHRD fellowship, on the research question 'how development and climate change is impacting mangrove ecology at Indian Sundarbans?'. His post-doctoral research was on Kachchh mangrove ecology at Gujarat. He has published around 37 Scopus/ Web of Science indexed journal articles between 2016-22 with over 700 citations. He was awarded the Inder Mohan Thapar best researcher Award (2016) by IIT for his research contributions. He also received various awards, recognitions for research and academic achievements from 'Coursera', 'O.P. Jindal Global University', 'Elsevier (Publication house)' etc. He is on the editorial board of peer-reviewed journals such as BMC Public Health (Springer-Nature), BMC Environmental Sciences (Springer-Nature), Frontiers in Soil Science, and Toxin Reviews (Taylor & Francis). Abhiroop has been working with local communities in the Indian Sundarbans since 2007, to promote green livelihood initiatives, and mangrove conservation.

Sander Gellaerts studied law at Erasmus University in Rotterdam and worked as a Lecturer and Researcher in the field of Intellectual Property Rights. He received his PhD Degree from the Tilburg Institute for Law, Technology, and Society. He is an author of several books and articles about IP, Privacy, ICT, Education and Lego. In addition to his academic work, he has held several positions, such as Director of Legal Affairs, Consultant and Social Worker. He was also one of the Board Members who founded a Hospice inspired by Buddhism and is now a Committee Member of a Dutch Independent Accreditation Organization for Higher Education Programmes, a Member of the Board of the LEAD Foundation in Leiden and an Executive Coach.

Sibylle Gruber is a professor of Rhetoric, Writing, and Digital Media Studies at Northern Arizona University. She is the author of *Literacies, Experiences, and Technologies: Reflective Practices of an Alien Researcher*. She has edited several books and published numerous articles on ecofeminism, environmental rhetoric, the positionalities of international faculty, feminist rhetorics, technological literacies, composition theories and practices, and cultural studies. Her current work focuses on projects that explore the intersectionalities of climate change action, gender, sexuality, race, culture, and socio-economic status.

Shruti Jain is Associate Professor of German Language, Literature and Cultural Studies at the Centre for Foreign Languages, OEFL, O.P.Jindal Global University, Sonipat, India. In her PhD thesis she dealt with the Indian reception of Friedrich Nietzsche. Her areas of interest are Language teaching, Literary theory, Religion, Philosophy and Indo-German cultural dialogue.

Siddharth Kanojia is an Asst. Professor of Law at O.P Jindal Global University. He is also a visiting professor at Indian Institute of Management Amritsar and Sirmaur. He holds a Ph.D in Corporate Law & Governance, LL.M in International Business and Commercial Law, LL.B and BBS in Finance.

Before commencing the academic stints, he was associated with U.K Intellectual Property Management Company as Legal Consultant and Manchester based Start-up Company as Legal Advisor. Till date, he has various publication in the SCOPUS and WoS indexed journals. He is also an active member in Human Development and Capability Association, USA and Development Studies Association, U.K.

Aliya Naz is an Assistant Professor at 'Jindal School of Liberal Arts and Humanities', O.P. Jindal Global University, India. She has completed her Ph.D. from Indian Institute of Technology, (IIT) Dhanbad with MHRD fellowship after qualifying UGC-NET (with Junior Research Fellowship award) in Environmental Science. Her research topic focused on hexavalent Chromium pollution in water, soil and plants at Sukinda (the largest chromite mine of India) and its impact on human health in the mining belt. Along with Ph.D. research she was involved in enviro-social work project at Indian Sundarbans (the world's largest delta), working towards achieving UN Sustainable Development Goals (SDG's), primarily- SDG 1,2,3,6,7,8,10,13,15. She has extensively worked at Indian Sundarbans for different projects during 2017-2021 on the capacity of environmental consultant.

L. Jan Slikkerveer is Professor Emeritus of Anthropology and Applied Ethnoscience in the Faculty of Science of Leiden University in The Netherlands, and Adjunct Professor of Ethno-Economics in the Faculty of Economics and Business of Universitas Padjadjaran in Bandung, Indonesia. He received his PhD from Leiden University in 1983 and a Honorary PhD Degree from Universitas Padjadjaran in 2005. He is Founder and Chairman of the Leiden Ethnosystems And Development Programme (LEAD) of Leiden University (1987), transformed in (2022) into the Leiden Ethnoscience and Development Foundation (LEAD). He received substantial funding from the European Commission for his work on Indigenous Knowledge and Development in Sub-Saharan Africa, the Mediterranean Region and South-East Asia. In 2012, he co-established the International Master Course on Integrated Microfinance Management (IMM) at Universitas Padjadjaran in Bandung, in which he is still teaching and supervising post-graduate students.

Amara Szrom is an environmental educator and garden program manager with over a decade of classroom experience. During her master's coursework at the University of New Mexico, she managed the campus garden and outdoor classroom, teaching four semesters of the undergraduate course 'Lobo Gardens: Environment, Health, and Social Change.' Her research contributed to recommendations for how university campus gardens can strengthen their programs and establish themselves as vital student resource spaces.

Index

U

W

Y

www.ingramcontent.com/pod-product-compliance
Lightning Source LLC
LaVergne TN
LVHW060403300926
849162LV00006B/1332